Roland Kontermann   Stefan Dübel (Eds.)

Antibody Engineering

With 110 Figures, 2 in Color

 Springer

Dr. ROLAND KONTERMANN
Institut für Molekularbiologie und Tumorforschung
Universität Marburg
Emil-Mannkopff-Str. 2
35033 Marburg
Germany

Dr. STEFAN DÜBEL
Molekulare Genetik
Universität Heidelberg
Im Neuenheimer Feld 230
69120 Heidelberg
Germany

ISBN 3-540-41354-5 Springer-Verlag Berlin Heidelberg New York

Library of Congress Cataloging-in-Publication Data
Antibody engineering / Roland Kontermann, Stefan Dübel (eds.) p. cm. – (Springer lab manuals)
Includes bibliographical references and index.
ISBN 3540413545 (alk. paper)
1. Recombinant antibodies – Laboratory manuals. I. Kontermann, Roland, 1961-II. Dübel, Stefan. III. Springer lab manual.
QR186.87 .A56 2001
616.07'98–dc21

00-067931

Springer-Verlag Berlin Heidelberg New York
a member of BertelsmannSpringer Science + Business Media GmbH

http://www.springer.de

© Springer-Verlag Berlin Heidelberg 2001
Printed in Germany

Production: PRO EDIT GmbH, 69126 Heidelberg, Germany
Cover design: design & production GmbH, 69121 Heidelberg, Germany
Typesetting: Mitterweger + Partner, 68723 Plankstadt, Germany
Printed on acid free paper   SPIN 10709876   27/3130/SO   5 4 3 2 1 0

Preface

The major motivation for developing recombinant antibody technologies resulted from the possibility to generate human antibodies that are not accessible by conventional polyclonal or monoclonal approaches. However, despite the plethora of ideas and new concepts for diagnosis and therapies based on recombinant antibodies, it was not until late 1997 that the first recombinant antibody (Rituxan) was approved for clinical use in cancer treatment by the United States FDA, only followed by a few more until the end of the millennium (e.g. Zenapax, Synagis, Herceptin). These antibodies have not yet been generated using combinatorial approaches such as phage display technology, but are chimeric or humanised variants of mouse hybridoma antibodies. However, antibody therapeutics already represent the most abundant substance class (>60) in the FDA future approval pipeline. This growth will continue as about 400 antibody based drugs have entered the clinical evaluation phase. Keeping in mind the delay between research and even first clinical trials and the approval as a drug, it can be expected that recombinant antibody-based therapies will be a widespread and acknowledged tool in the hands of physicians by the year 2010. The rise in antibody-based therapeutics further illustrates the substantial change in the paradigms of pharmaceutical development, by utilising the body's own capabilities as a source for a drug rather than the reagent vessels of chemists.

In this book we have brought together experts in the field presenting state-of-the-art methods in antibody engineering. Although we are aware that this manual cannot describe every method available, we have tried to cover the essential technologies in antibody engineering. Besides the basic methods, various new protocols describing strategies established in the last couple of years have been included, as well as a few emerging technologies which we believe may become future standards. The present lab manual will help to stay up-to-date with the newest developments in this fast moving field. It is designed both to lead beginners in this technology in their first steps by supplying the most detailed and proven protocols and to supply professional antibody engineers with new ideas and approaches.

Marburg, Heidelberg, Spring 2001
ROLAND KONTERMANN
STEFAN DÜBEL

Contents

Expression and Purification of Antibody Fragments in Eukaryotic Cells

Determination of Affinities

Sequence and Structure Analysis and Modelling

Antibody Engineering to Improve Stability

Bivalent and Bispecific Antibody Fragments

Recombinant Antibody-Fusion Proteins

Intracellular Targeting of Antibody Fragments

Appendices

Introduction

Recombinant Antibodies

Stefan Dübel and Roland E. Kontermann

1 Introduction

Antibodies are our bodies modular defense system, used to identify and attack foreign intruders. To interact with as many foreign structures as possible, an immense number of different molecules, bearing different specificities, is required. This diversity is generated by somatic recombination and hypermutagenesis of a set of variant genes. The genetic information for this repertoire of different antibodies is stored in the B-cell pool of our lymphatic system.

The antigen-binding sites of antibodies are located at the upper tips of the Y- or T-shaped immunoglobulin molecules. Each tip includes 6 hypervariable loops, which constitute the surface of the antigen-binding site (Fig. 1). Their variability accounts for the large range of possible binding specificities and explains the prominent role of polyclonal and monoclonal antibodies in biochemical and cell biological research.

During the past decade, advances in molecular biology have greatly facilitated the genetic manipulation, recombinant production, identification, and conjugation of antibody fragments. The genetic manipulation of recombinant antibodies has improved our knowledge about the structure and functional organisation of immunoglobulins. Further, genetic fusion and recombinant expression has led to the development of a large variety of engineered antibody molecules for research, diagnosis, and therapy.

The most fascinating perspectives, however, have been opened up by the development of methods to screen for specific monoclonal antibodies

✉ Stefan Dübel, Universität Heidelberg, Institut für Molekulare Genetik, Im Neuenheimer Feld 230, 69120 Heidelberg, Germany (*e-mail* sd@uni-hd.de; *homepage* duebel.uni-hd.de)
Roland E. Kontermann, Universität Marburg, Institut für Molekularbiologie und Tumorforschung, Emil-Mannkopff-Straße 2, 35033 Marburg, Germany (*e-mail* rek@imt.uni-marburg.de)

in heterologous systems. This has, in particular, boosted the development of human antibodies, with all their benefits for therapy and *in vivo* diagnosis. It further allowed the generation of antibodies with specificities which were out of reach of conventional antibody technology. Once cloned, it is then possible to improve the affinity or specificity of antigen binding by mimicking the somatic hypermutation during an immune response.

Not surprisingly, the emergence of these methods opened a new chapter of employing antibodies in research, diagnosis, and therapy. This resulted in an exploding number of companies using this technology for novel immunotherapy approaches or in other prospering markets.

2 Recombinant expression of antibodies

Antibody molecules consist of light and heavy chains, each chain composed of one variable domain and between one and four constant domains, which assemble into molecules exhibiting two or more antigen-binding sites (Figure 1 and 2). The antigen-binding sites of immunoglobulins are embedded into the variable heavy and light chain domains (VH, VL) and are spacially separated from the effector function-mediating regions located in the Fc fragment (Fig. 1) (for review see: Burton, 1985; Burton & Woof, 1992). Before the development of recombinant technologies, antigen-binding fragments of immunoglobulins could only be generated by proteolytic cleavage. This produced F(ab')$_2$ or Fab fragments containing the variable and the first constant domains (Fig. 2). The Fv

Fig. 1. The Fv fragments contain the antigen-binding regions of antibodies. An IgG molecule (box upper left) consists of two identical Fv units held together by constant regions. The Cα-cartoon (lower box) visualises the antiparallel beta-sheet structure of the immunglobulin fold and the intramolecular disulphide bond in each of the variable regions, which are responsible for the stability of the antibody molecule. This "framework" structure supplies the scaffold for six loops, which define the antigen-binding surface. Sequence and length of the loops are variable, thereby constituting all of the remarkably different structures which are necessary to bind to millions of different antigens. The loops are labelled L1-3 and H1-3, according to the respective variable region of light or heavy chain. The terms "CDRs" (complimentary determining regions) and "hypervariable regions" are frequently used for these loops. It should be noted, however, that the term "CDR" is based on structural and functional analysis, and thus more appropriate to describe the antigen binding area than the term "hypervariable region", which is solely based on primary structure alignments. The Fv coordinates have been generated by A. Martin, as described in Liu et al. (1999). Structures were visualised with the programme RasMol.

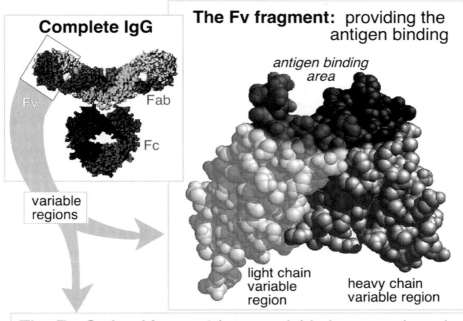

Complete IgG

Fv

Fab

Fc

variable
regions

The Fv fragment: providing the
antigen binding

*antigen binding
area*

light chain
variable
region

heavy chain
variable region

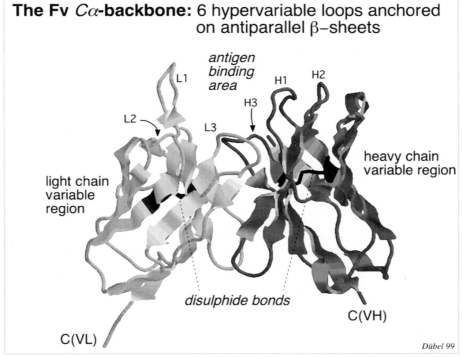

The Fv *Cα*-backbone: 6 hypervariable loops anchored
on antiparallel β–sheets

*antigen
binding
area*

L1

L2

L3

H3

H1

H2

light chain
variable
region

heavy chain
variable region

disulphide bonds

C(VH)

C(VL)

Dübel 99

fragment is the smallest fragment of an antibody molecule which is able to provide antigen specificity. However, this fragment is extremely difficult to produce by proteinase cleavage. Antigen-binding activity, although weaker than that of the entire binding site, has been described for a few individual VH or VL domains or even peptides derived from the complementarity determining regions (CDR) (for review see: Winter & Milstein, 1991). However, these examples are exceptions rather than the rule, as concluded from the growing database of three-dimensional antibody structures and analyses of antigen-antibody interactions.

The development of the hybridoma technology in 1975 allowed the generation and production of monoclonal antibodies with predefined specificities (Köhler & Milstein, 1975). Consequently, hybridoma cells provided the starting material for the first recombinant antibody molecules expressed as whole immunoglobulins, Fab, or Fv fragments in lymphoid or non-lymphoid cells (Neuberger et al., 1984 & 1985; Riechmann et al., 1988; Feys et al., 1988). However, initial experiments expressing antibodies in bacteria were hampered due to improper folding and aggregation of the polypeptides in the bacterial cytoplasm. This was mainly caused by the lack of disulphide bond formation in the reducing environment of the cytoplasm (Boss et al., 1984; Cabilly et al., 1984). These problems were solved by expressing only parts of the immunoglobulin molecule, i.e. Fv or Fab fragments. The breakthrough for efficient *E. coli* expression of antibody fragments, however, was brought by vectors providing the secretion into the periplasmic space with its oxidising milieu. This process allowed the correct formation of disulphide bonds within the immunoglobulin domains (Skerra & Plückthun, 1988; Better et al., 1988).

Further, it soon became evident that due to the non-covalent association of Fv fragments solubly expressed in *E. coli*, these fragments are quite unstable (Glockshuber et al., 1990). This drawback has been overcome by engineering single-chain Fv (scFv) fragments joining the VH and VL do-

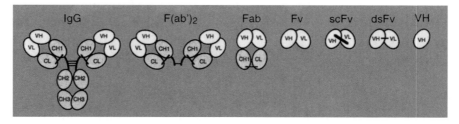

Fig. 2. Structure of IgG (human IgG1) and its antigen-binding fragments. Fv = variable fragment, scFv = single-chain Fv fragment, dsFv = disulphide-stabilised Fv fragment.

main with a flexible peptide linker of 15-20 amino acids. This increases the stability under physiological conditions (Huston et al., 1988; Bird et al., 1988). One of the most commonly used linkers is a stretch of glycine and serine residues of the format $(Gly_4Ser)_3$, but many other linker designs have been successfully employed. A different approach has been to covalently link the VH and VL domain by introducing additional cysteine residues at the interface, leading to disulphide-stabilised Fv (dsFv) fragments (Brinkmann et al., 1993) (Fig. 2).

Various other hosts have been employed for the expression of recombinant antibodies, including gram-positive bacteria (*Bacillus subtilis*), fungae (*Saccharmyces cerevisiae, Pichia pastoris, Trichoderma reesei*), plant cells (mainly *Nicotiana tabacum*), Bacculovirus-infected insect cells, and various mammalian cell lines (e.g. CHO, COS, HEK293). The choice of the best expression system depends on the antibody fragments expressed. While the scFv or Fab fragments are generally expressed in gram-negative bacteria such as *E. coli* without problems, larger fragments containing the CH3 or a Fc region of immunoglobulins or antibody-fusion proteins are often difficult to express in prokaryotic cells. Therefore, eukaryotic expression systems are methods of choice for such situations (for review, see Breitling & Dübel, 1999).

3 Accessing the diversity – generation of recombinant antibodies

Two routes are available for the generation of recombinant antibodies. Starting from hybridoma, genes encoding antigen-binding sites can be isolated from a cDNA pool by PCR and assembled into scFv or Fab fragments by genetic engineering. Alternatively, antibody fragments can be isolated by a combinatorial approach. To do this, one first has to construct an antibody gene library. This is usually achieved by PCR-amplification of the rearranged antibody genes from B-lymphocytes. The combinatorial libraries can be generated from the B cells of immunised animals (immune libraries) or from non-immunised donors (naive libraries). The use of human sources for repertoire cloning allows the generation of fully human antibody fragments. Alternatively, antibody genes can be assembled *in vitro* from overlapping randomised wobble-primers or by introducing randomised CDRs into germline V genes used as building blocks (for review see: Winter et al., 1994).

In order to isolate antibodies from libraries which contain millions of different clones, an extremely efficient selection system is required. This was achieved using a system analogous to the expression of the IgM antigen receptor on the surface of unactivated B-lymphocytes. Such a system displays antibody fragments on the surface of microorganisms containing antibody-encoding genes. Examples of these prokaryotic organisms are filamentous bacteriophage (phage display) (McCafferty et al., 1990; Breitling et al., 1991, Barbas, et al., 1991, Hoogenboom et al., 1991) or bacteria (Fuchs et al., 1991). The resulting surface display generates particles which physically combine the antigen-binding polypeptides with the genes encoding their function. Using selection methods based on an antibody-antigen interaction, particles recognising a specific antigen are isolated. Specific clones can then be amplified and used to produce the antibody fragments in *E.coli* or other suitable organisms. These powerful selection methods allow isolation of one out of more than 10^9 different expression clones. Since combinatorial antibody libraries, due to random light and heavy chain combination, may contain antibody fragments not found in nature, libraries can be used to isolate antibodies directed against antigens not suited for immunisation, e.g. recognising highly toxic substances or self-antigens.

The affinity of antibody fragments isolated from combinatorial libraries (especially when using naive libraries) strongly depends on the quality and diversity of the libraries. Often, these antibody fragments have only moderate affinity. In this case, the recombinant antibody technology further allows to improve the binding by using affinity maturation strategies, such as random or site-directed mutagenesis and chain shuffling followed by reselection of antibody fragments with improved binding properties under stringent conditions (Hawkins et al., 1992; Marks et al., 1992; Jackson et al., 1995).

To be used for therapy or for *in vivo* diagnosis, a human anti-mouse antibody (HAMA) response has to be avoided. To achieve this, well proven murine monoclonal antibodies can be humanised by various methods. These methods include the grafting of CDRs of the original monoclonal antibody onto a human framework (Jones et al., 1987) or a guided selection approach substituting the murine VL and VH domains by human equivalents from an antibody repertoire (Jespers et al., 1994). Furthermore, the stability of recombinant antibodies, another important aspect for *in vivo* applications, can be improved by genetic engineering.

4 New molecular architectures extend functions

4.1 Multivalent antibody fragments

The strength of antibody binding to its antigen is not only determined by the affinity but also by the valency of interaction, with multivalent interactions leading to an increase in functional affinity (avidity). Therefore, several strategies have been developed to generate di- or multivalent recombinant antibody fragments (for review see: Plückthun & Pack, 1997). These strategies can be divided into three general approaches: (i) di- or multivalent assembly of Fv domains, (ii) fusion of antibody fragments to immunoglobulin constant domains which are able to dimerise, and (iii) fusion of scFv fragments to di- or multimerising heterologous peptides or proteins (Fig. 3).

The first approach includes dimerisation of scFv fragments by C-terminal cysteine residues, resulting in disulphide-linked scFv fragments (Kipriyanov et al., 1994), the fusion of two scFv fragments in tandem linked by a flexible peptide sequence, (tandem scFv) (Mallander & Voss, 1994; Neri et al., 1995), or the formation of scFv dimers by crossover pairing of two VH-VL chains (diabodies) (Holliger et al., 1993). The formation of diabodies is forced by reducing the linker length joining the VH and VL domain from 15-20 amino acids to 5 or less amino acids. While fragments with a 5 amino acid linker generally result in the formation of dimeric molecules with two binding sites, further reduction often leads to the assembly of tri- or tetrameric molecules (triabodies, tetrabodies) (Iliades et al., 1997; Le Gall et al., 1999). The stability of these non-covalently linked dimeric diabody molecules can be improved by introducing disulphide bonds at the VH-VL interfaces (disulphide-stabilised diabodies; dsDb) (FitzGerald et al., 1997) or by converting the dimeric molecule into a single-gene encoded monomeric molecule (single-chain diabody, scDb) (Brüsselbach et al., 1999).

In the second approach, interaction of the immunoglobulin constant domains, i.e. CL homodimerisation (McGregor et al., 1994), CH1-CL heterodimerisation (Müller et al., 1998), and CH3 or Fc homodimerisation (Shu et al., 1993; Hu et al., 1996; Connelly et al., 1998; Alt et al., 1999), is applied to assemble scFv, tandem scFv, or single-chain diabody molecules into dimeric molecules with 2-4 binding sites. In this case, the antibody fragments are fused to the N-terminus of the constant domains. Alternatively, scFv fragments can be fused to the C-terminus of the CH1, CL, CH3, or the hinge region generating di- or tetravalent Fab-scFv, F(ab,)$_2$-scFv or IgG-scFv fusion proteins (Coloma & Morrison, 1997).

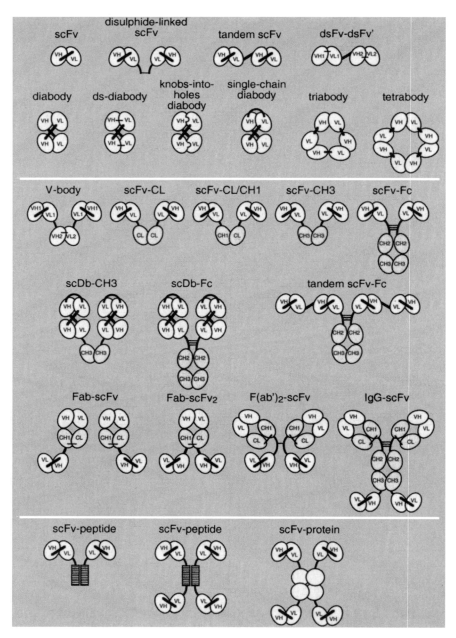

Fig. 3. Recombinant bivalent and bispecific antibody fragments. scFv = single-chain Fv fragment, dl-scFv = disulphide-linked scFv, ds-diabody = disulphide-stabilised diabody, scDb = single-chain diabody.

The third approach applies heterologous peptides or protein domains with the potential to di- or multimerise. Peptide sequences include leucine zippers and helix-loop-helix motifs forming four-helix bundle structures. Fusion of one or two scFv fragments to these peptides generates small di- or tetravalent molecules (miniantibodies) (Pack & Plückthun, 1992; Pack et al., 1995; Müller et al., 1998). Larger multivalent and multimeric scFv fusion proteins are formed using protein domains such as streptavidin, the tetramerisation domain of p53, or the octamerisation domain of the C4-binding protein (Dübel et al., 1995; Rheinecker et al., 1996; Libyh et al., 1997).

4.2 Bispecific and bifunctional antibody fragments

Some of the above described strategies have also been applied to generate bispecific molecules. Due to the dual specificity of bispecific antibody fragments, these molecules are able to mediate the recruitment of molecules or cells to specific targets (for review see: Fanger et al., 1992). The most commonly used bispecific antibody fragments are tandem scFv and diabody molecules. Bispecific tandem scFv are generated by linking two scFv of differing specificity. In contrast, bispecific diabodies are generated by expressing two fragments of the format VHA-VLB and VHB-VLA (directed against antigens A and B) in the same cell (Holliger et al., 1993). To avoid homodimerisation of these fragments, which would result in the assembly of inactive molecules, a knobs-into-holes structure can be engineered at the VH-VL interfaces which allows only formation of heterodimers (Zhu et al., 1997). Bispecific single-chain diabodies are generated by expressing a single polypeptide chain of the format VHA-VLB-linker-VHB-VLA (Brüsselbach et al., 1999). Furthermore, bispecific antibody fragments can be generated by fusing scFv fragments to heterodimerising domains such as the jun and fos leucine zipper sequences (Kostelny et al., 1992). Recently, two disulphide stabilised bispecific antibody formats, V-bodies (Dübel and Schmiedl, 2000) and dsFv-dsFv' (Schmiedl et al., 2000), have been generated, the latter utilising different positions in the two Fvs for the introduction of disulphide bonds at the VH and VL interface to prevent incorrect V pairing.

In contrast to bispecific antibody fragments, bifunctional molecules combine the antigen binding site of an antibody with a biological function encoded by a fusion partner. Among the proteins which have been fused to antibody fragments are enzymes, toxins, cytokines, growth factors, DNA-binding domains, and metal chelators. The genetic coupling of antibodies to heterologous proteins generates new possibilities for immunotargeting. Immunotargeting utilises the affinity of the antibody part of the fusion protein to increase the concentration/activity of the heterologous fusion partner at sites where antigen is present. Applications include immuno-toxins or immuno-radioagents, and, most recently, the tissue specific targeting of gene therapy vectors. The large variety of fusions that has been constructed demonstrates the potential of antibody engineering for generating new therapeutic and diagnostic agents.

4.3 Intrabodies and cell-surface-displayed antibody fragments

Potential therapeutic applications of recombinant antibodies not only include the administration of purified antibody fragments but also the intracellular expression of antibody fragments (scFv, Fab, scDb) in various subcellular compartments (intrabodies) (Marasco et al., 1993; Mhashilkar et al., 1995; Kontermann & Müller, 1999). Here, the antibody fragments can inhibit the activity of target molecules by directly blocking their function or indirectly by interfering with their subcellular trafficking (for review see: Marasco, 1995; Biocca & Cattaneo, 1995). Subcellular location is achieved by fusion of small localisation signals to the antibody fragments, such as an ER retention signal resulting in the accumulation of the antibody fragment in the endoplasmic reticulum or the fusion to a nuclear localisation signal directing the antibody fragment to the cell nucleus. However, due to the reductive redox milieu of the cytosol, antibody fragments directed to the cytoplasm or nucleus are often much less effective than fragments directed to the secretory pathway (Biocca et al, 1995, Kontermann & Müller, 1999). Thus, various attempts are currently made to engineer antibody fragments which fold correctly into stable molecules in a reducing environment (Martineau et al., 1998; Proba et al., 1998). The display of antibody fragments in the plasma membrane by fusion to a transmembrane domain generates cells expressing artifical receptors. Thus, cells can be produced exhibiting a defined novel specificity for soluble molecules or antigens present on other cells (e.g. cytotoxic T-lymphocytes recognising tumour cells). The presence of signal transducing regions in these chimeric constructs even allows the receptor-mediated activation of the cell as has been shown for antibody fragments fused to the γ and ζ subunits of the T cell receptor (Eshhar et al., 1993).

5 Summary

The rapid evolvement of antibody engineering in the last decade has led to powerful methods to isolate antibody fragments of desired specificity and provided efficient systems to produce recombinant antibodies in pro- and eukaryotic cells. Affinity, valency or size of antibodies can be engineered to improve the antibody-antigen interaction and the biochemical and pharmacokinetic properties. In addition, the antigen binding regions of immunoglobulin molecules have been fused to various polypeptides introducing novel biological functions, thus extending the antibodies abilities beyond nature.

6 Comments

A comprehensive review on antibody engineering is given by the book "Recombinant Antibodies" by F. Breitling & S. Dübel (1999). Additional and up-to-date information may be obtained from the internet web sites of the authors (http://aximt1.imt.uni-marburg.de/~REK/, http://www.duebel.uni-hd.de).

References

Alt M, Müller R, Kontermann RE (1999) Novel tetravalent and bispecific IgG-like antibody molecules combining single-chain diabodies with the immunoglobulin gamm1 Fc or CH3 region. FEBS Lett 454:90-94

Barbas III CF, Kang AS, Lerner RA, Benkovic SJ (1991) Assembly of combinatorial antibody libraries on phage surfaces: The gene III site. Proc Natl Acad Sci USA 88:7978-7982

Better M, Chang CP, Robinson RR, Horwitz AH (1988) *Escherichia coli* secretion of an active chimeric antibody fragment. Science 240:1041-1043

Bird RE, Hardman KD, Jacobson JW, Johnson S, Kaufman BM, Lee SM, Lee T, Pope SH, Riordan GS, Whitlow M (1988) Single-chain antigen-binding proteins. Science 242:423-426

Boss MA, Kenten JH, Wood CR, Emtage JS (1984) Assembly of functional antibodies from immunoglobulin heavy and light chains synthesised in *E. coli*. Nucl Acid Res 12:3791-3806

Breitling F, and Dübel S, (1999) Recombinant Antibodies. John Wiley and Sons, New York.

Breitling F, Dübel S, Seehaus T, Klewinghaus I, Little M (1991) A surface expression vector for antibody screening. Gene 104:147-153

Brinkmann U, Reiter Y, Jung SH, Lee B, Pastan I (1993) A recombinant immunotoxin containing a disulfide-stabilized Fv fragment. Proc Natl Acad Sci USA 90:7538-7542

Brüsselbach S, Korn T, Völkel T, Müller R, Kontermann RE (1999) Enzyme recruitment and tumor cell killing in vitro by a secreted bispecific single-chain diabody. Tumor Targeting 4:115-123

Burton DR (1985) Immunoglobulin G: Functional sites. Mol Immunol 22:161-206

Burton DR, Woof JM (1992) Human antibody effector function. Adv Immunol 51:1-84

Cabilly S, Riggs AD, Pande H, Shively JE, Holmes WE, Rey M, Perry LJ, Wetzel R, Heyneker HL (1984) Generation of antibody activity from immunoglobulin polypeptide chains produced in *Escherichia coli*. Proc Natl Acad Sci USA 81:3273-3277

Coloma MJ, Morrison SL (1997) Design and production of novel tetravalent bispecific antibodies. Nat Biotechnol 15:159-163

Connelly RJ, Hayden MS, Scholler JK, Tsu TT, Dupont B, Ledbetter KA, Kanner SB (1998) Mitogenic properties of a bispecific single-chain Fv-Ig fusion generated from CD2-specific mAb to distinct epitopes. Int Immunol 10:1863-1872

Dübel S, Breitling F, Kontermann R, Schmidt T, Skerra A, Little M (1995) Bifunctional and multimeric complexes of streptavidin fused to single chain antibodies (scFv). J Immunol Meth 178:201-209

Dübel S & Schmiedl A (2000) Antikörperkonstrukte mit variablen Regionen. German Patent App. 100 21 678.1

Eshhar Z, Waks T, Gross G, Schindler DG (1993) Specific activation and targeting of cytotoxic lymphocytes through chimeric singke-chains consisting of antibody-binding domains and the γ and ζ subunits of the immunoglobulin and T cell receptors. Proc Natl Acad Sci USA 90:720-724

Fanger MW, Morganelli PM, Guyre PM (1992) Bispecific antibodies. Crit Rev Immunol 12:101-124

Feys V, De-Waele P, van de Voorde A, Casneuf P, Fiers W (1988) Expression of functional mouse antibodies directed against the tumour marker human placental alkaline phosphatase in non-lymphoid cells. Int. J. Cancer 2:26-27

FitzGerald K, Holliger P, Winter G (1997) Improved tumour targeting by disulphide stabilized diabodies expressed in *Pichia pastoris*. Protein Eng 10:12221-1225

Fuchs P, Breitling F, Dübel S, Seehaus T, Little, M, (1991) Targeting recombinant antibodies to the surface of E. coli: Fusion to a peptidoglycan-associated lipoprotein. Bio/Technology 9:1369-1372

Glockshuber R, Malia M, Pfitzinger I, Plückthun A (1990) A comparison of strategies to stabilize immunoglobulin Fv-fragments. Biochemistry 29:1362-1367

Hawkins RE, Russell SJ, Winter G (1992) Selection of phage antibodies by binding affinity: Mimicking affinity maturation. J Mol Biol 226:889-896

Holliger P, Prospero TD, Winter G (1993) "Diabodies": small bivalent and bispecific antibody fragments. Proc Natl Acad Sci USA 90:6444-6448

Hoogenboom HR, Griffiths AD, Johnson KS, Chiswell DJ, Hudson P, Winter G (1991) Multi-subunit proteins on the surface of filamentous phage: methodologies for displaying antibody (Fab) heavy and light chains. Nucl Acids Res 19:4133-4137

Hu SZ, Shively L, Raubitschek A, Sherman M, Williams LE, Wong JYC, Shively JE, Wu AM (1996) Minibody: A novel engineered anti-carcinoembryonic antigen antibody fragment (single-chain Fv-CH3) which exhibits rapid, high-level targeting of xenografts. Cancer Res 56:3055-3061

Huston JS, Levnson D, Mudgett-Hunter M, Tai MS, Novotny J, Margolies MN, Ridge RJ, Bruccoleri RE, Haber E, Crea R, Oppermann H (1988) Protein engineering of antibody binding sites: recovery of specific activity in an anti-digoxin single-chain Fv analogue produced in Escherichia coli. Proc Natl Acad Sci USA 85:5879-5883

Iliades P, Kortt AA, Hudson PJ (1997) Triabodies: single chain Fv fragments without a linker form tirvalent trimers. FEBS Lett 409:437-441

Jackson JR, Sathe G, Rosenberg M, Sweet R (1995) In vitro antibody maturation: Improvement of a high affinity, neutralizing antibody against IL-1beta. J Immunol 154:3310-3319

Jespers LS, Roberts A, Mahler SM, Winter G, Hoogenboom HR (1994) Guiding the selection of human antibodies from phage display repertoires to a single epitope of an antigen. Bio/Technol 12:899-903

Köhler G, Milstein C (1975) Continuous cultures of fused cells secreting antibody of predefined specificity. Nature 256:495-497

Kontermann RE, Müller R (1999) Intracellular and cell surface displayed single-chain diabodies. J Immunol Meth 226:179-188

Kostelny SA, Cole MS, Tso JY (1992) Formation of a bispecific antibody by the use of leucine zippers. J Immunol 148:1547-1553

Le Gall F, Kipriyanov SM, Moldenhauer G, Little M (1999) Di-, tri and tetrameric single chain Fv antibody fragments against human CD19: effect of valency on cell binding. FEBS Lett 453:164-168

Libyh MT, Goossens D, Oudin S, Gupta N, Dervillez X, Juszczak G, Cornillet P, Bougy F, Reveil B, Philbert F, Tabary T, Klatzmann d, Rouger P, Cohen JHM (1997) A recombinant human scFv anti-Rh(D) antibody with multiple valences using a C-terminal fragment of C4-binding protein. Blood 90:3978-3983

Liu Z, Song D, Kramer A, Schneider-Mergener J, Martin ACR, Dandekar T, Bautz EKF, Dübel S (1999) Fine mapping of the antigen - antibody interaction of scFv215, a recombinant antibody inhibiting RNA polymerase II from Drosophila melanogaster. J Mol Recognit 12:103-111

Marks JD, Griffiths AD, Malqvist M, Clackson TP, Bye JM, Winter G (1992) By-passing immunization: building high affinity human antibodies by chain shuffling. Bio/Technol 10:779-783

Martineau P, Jones P, Winter G (1998) Expression of an antibody fragment at high levels in the bacterial cytoplasm. J Mol Biol 280:117-127

McCafferty J, Griffiths AD, Winter G, Chiswell DJ (1990) Phage antibodies: filamentous phage displaying antibody variable domains. Nature 348:552-554

McGregor DP, Molloy PE, Cunningham C, Harris WJ (1994) Spontaneous assembly of bivalent single chain antibody fragments in *Escherichia coli*. Mol Immunol 31:219-226

Müller KM, Arndt KM, Plückthun A (1998) A dimeric bispecific miniantibody combines two specificities with avidity. FEBS Lett 432:45-49

Müller KM, Arndt KM, Strittmatter W, Plückthun A (1998) The firstconstant domain (CH1 and CL) of an antibody used as heterodimerization domain of bispecific miniantibodies. FEBS Lett 422:259-264

Neubauer MS, Williams GT, Fox RO (1984) Recombinant antibodies possessing novel effector functions. Nature 312:604-608

Neubauer MS, Williams GT, Mitchell EB, Jouhal SS, flanagan JG, Rabbits TH (1985) Nature 314:268-270

Pack P, Müller K, Zahn R, Plückthun A (1995) Tetravalent miniantibodies with high avidity assembling in Escherichia coli. J Mol Biol 246:28-34

Pack P, Plückthun A (1992) Miniantibodies: Use of amphipathic helices to produce functional, flexibly linked dimeric Fv fragments with high avidity in Escherichia coli. Biochemistry 31:1579-1584

Plückthun A, Pack P (1997) New protein engineering approaches to multivalent and bispecific antibody fragments. Immunotechnol 3:83-105

Proba K, Wörn A, Honegger A, Plückthun A (1998) Antibody scFv fragments without disulfide bonds made by molecular evolution. J Mol Biol 275:245-253

Rheinnecker M, Hardt C, Ilag LL, Kufer P, Gruber R, Hoess A, Lupas A, Rottenberger C, Plückthun A, Pack P (1996) Multivalent antibody fragments with high functinal affinity for atumor-associated carbohydrate antigen. J Immunol 157:2989-2997

Riechmann L, Foote J, Winter G (1988) Expression of an antibody Fv fragment in myeloma cells. J. Mol. Biol. 203:825-828

Schmiedl A, Breitling F, Dübel S (2000) Expression of a bispecific dsFv-dsFv'antibody fragment in E. Coli. Protein Engineering 13

Shu L, Qi CF, Schloom J, Kashmiri SVS (1993) Secretion of a single-gene-encoded immunoglobulin from myeloma cells. Proc Natl Acad Sci USA 90:7995-7999

Skerra A, Plückthun A (1988) Assembly of functional immunoglobulin Fv fragments in *Escherichia coli*. Science 249:1038-1040

Winter G, Griffiths AD, Hawkins RE, Hoogenboom HR (1994) Making antibodies by phage display technology. Annu Rev Immunol 121:433-455

Winter G, Milstein C (1991) Man-made antibodies. Nature 349:293-299

Zhu Z, Presta LG, Zapata G, Carter P (1997) Remodeling domain interfaces to enhance heterodimer formation. Protein Sci 6:781-788

Generation of Recombinant
Antibodies from Hybridoma

Construction of scFv Fragments from Hybridoma or Spleen Cells by PCR Assembly

JÖRG BURMESTER and ANDREAS PLÜCKTHUN

Introduction

While today antibodies can be obtained from naive repertoires (Winter et al. 1994, Vaughan et al. 1996) or libraries of fully synthetic genes (Knappik et al. 2000), very many hybridomas have been generated, and are continuously being made, which produce monoclonal antibodies with very interesting properties. Even when naive and synthetic libraries may be a source of antibodies, often the immune response of an experimental animal may be of interest, requiring an analysis of the antibodies after immunization. Cloning retains and immortalizes these unique and extensively characterized specificities of mAbs, which can be crucial for the rescue of unstable hybridoma cell lines. Molecular cloning and sequencing of antibody variable domains forms the basis of antibody modelling (Rees et al., 1994), antibody engineering (Plückthun 1994, Dall'Acqua and Carter 1998) and experimental structure determination by NMR (Freund et al. 1994) or x-ray crystallography at high resolution (Braden et al. 1998). Moreover, once the variable region genes have been cloned, the antibody domains can be further engineered in a multitude of ways to produce antibody variants with lower immunogenicity (Thompson et al., 1998), higher affinity (Low et al. 1996, Schier et al. 1996, Hanes et al. 1998), altered antigenic specificity (Ohlin et al. 1996, Parsons et al. 1996), or enhanced stability (Martineau and Betton 1999, Wörn and Plückthun 1999). Furthermore, genetic fusions of scFv fragments to effector proteins and toxins are powerful tools for medicine and biotechnology (Hoogenboom et al. 1998).

Jörg Burmester, Universität Zürich, Biochemisches Institut, Winterthurerstr. 190, 8057 Zürich, Switzerland

✉ Andreas Plückthun, Universität Zürich, Biochemisches Institut, Winterthurerstr. 190, 8057 Zürich, Switzerland (*phone* +41-1-6355570; *fax* +41-1-6355712; *e-mail* PLUECKTHUN@BIOCFEBS.UNIZH.CH)

However, a prerequisite for the use of recombinant antibody technologies starting from hybridomas or immune repertoires is the reliable cloning of functional immunoglobulin genes. For this purpose, a standard phage display system was optimized for robustness, vector stability, tight control of the expression of the scFv-geneIII fusion, primer usage for PCR amplification of variable region genes, scFv assembly strategy and subsequent directional cloning using a single rare cutting restriction enzyme. Using this system, a number of variable antibody domains of hybridomas were accessible whose genes could not be cloned in previous experimental setups. This chapter essentially follows our earlier descriptions (Krebber et al. 1997, Plückthun et al. 1996), with some "evolution" in the conditions and primer sequences.

Note that, except for the initial preparation of the cells and the need for high ligation and transformation yields for library cloning, the construction of scFv fragment libraries from immunized mice and that of cloning one specific antibody from hybridomas is essentially the same, as described in this procedure. Why is an enrichment procedure even necessary, when the "monoclonal" antibody mRNA of a hybridoma cell is the starting material? It is not strictly required, but the present procedure provides an easy check that the right sequence has been cloned, and it is especially helpful if the hybridoma expresses more than one chain, as explained under the note of step 7.

Fig. 1. Scheme of the amplification and cloning procedure. The mRNA is derived from hybridoma or spleen cells and a random hexamer primer mixture ($pd(N)_6$) is used for cDNA synthesis. The cDNA is used as the PCR template for the amplification of VL and VH domains (the primers are listed in Fig. 2) which are assembled (SOE-PCR) into the scFv format by the outer primer pair *scback* and *scfor*. For antibody cloning into the phagemid the rare cutting enzyme *Sfi*I is the only enzyme used. Note that directional cloning of the *Sfi*I inserts is guaranteed because of the different *Sfi*I sites shown. In addition, self-ligation of insert or vector molecules is excluded by the asymmetry of the overhang. The transformed XL1-Blue cells are used for phage production by infection with helper phage. The enrichment of scFv antibody displaying phages by panning against the antigen will allow the detection and selection of functional antibody sequences in library settings or if the hybridoma cell line contained only a small fraction of mRNA specific for this antibody.

Outline

1. Isolation of mRNA (from hybridoma or spleen cells)

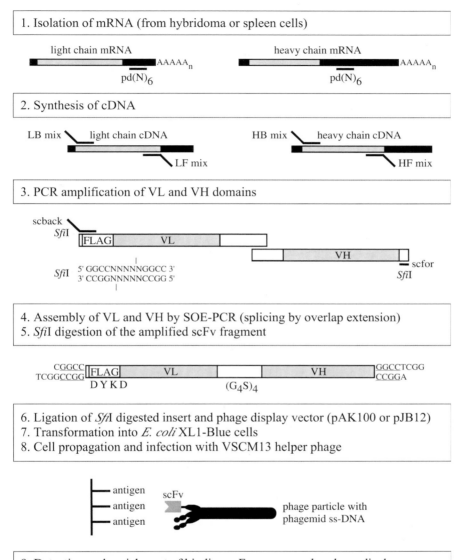

2. Synthesis of cDNA

3. PCR amplification of VL and VH domains

4. Assembly of VL and VH by SOE-PCR (splicing by overlap extension)
5. *Sfi*I digestion of the amplified scFv fragment

6. Ligation of *Sfi*I digested insert and phage display vector (pAK100 or pJB12)
7. Transformation into *E. coli* XL1-Blue cells
8. Cell propagation and infection with VSCM13 helper phage

9. Detection and enrichment of binding scFv sequences by phage display

Materials

- 1-5 x 10^6 cells from a growing or frozen hybridoma culture or spleen cells, respectively
- PCR primers (Fig. 2 and Fig. 3) and plasmids (Fig. 4 and Fig. 5)
- Helper phage (e.g. Stratagene VCSM13)
- F$^+$, *sup*E, *rec*A strain (e.g. Stratagene *E. coli* XL1-Blue)
- Anti M13-HRP conjugate (Pharmacia)
- PEG 6000

Fig. 2. This figure lists the primers used for assembling mouse scFv fragments in the orientation VL-(G$_4$S)$_4$-VH, which are compatible with the vectors presented in Fig 3. In this nomenclature, "back" refers to "toward the 3' end of the antibody gene" and "for" to "toward the 5' end of the antibody gene". The sequences are given using the IUPAC nomenclature of mixed bases (shown in underlined capital letters, R = A or G; Y = C or T; M = A or C; K = G or T; S = C or G; W = A or T; H = A or C or T; B = C or G or T; V = A or C or G; D = A or G or T), with a column listing the d-fold degeneration encoded in each primer. The "VL-back" primers *LB1-LB17* encode a stretch of 20 bases hybridizing to the mature mouse antibody κ sequences (in capital letters). Underlined is the preceding sequence which encodes the shortened FLAG sequence (Knappik and Plückthun 1994). Since the FLAG-tag uses the fixed N-terminal aspartate of the mature antibody (encoded by GAY), only three additional amino acids are necessary. The FLAG codons are in turn preceded by the codons specifying the end of the *pelB* signal sequence. The "VL-back" primer *LBλ* for mouse lambda chains is constructed in an analogous manner (the N-terminal glutamate of the mature mouse λ sequence is replaced by aspartate (encoded by GAT) to generate a FLAG tag). The "VL-for" primer sequences are complementary to the J-elements of kappa or lambda chains (capital letters) and also encode three repeats of the Gly$_4$Ser sequence, the terminal one (bold) of which has a different codon usage so that incorrect overlaps during the PCR assembly reaction are minimized. The "VH-back" primers encode the other part of the linker as well as a *Bam*HI recognition site (underlined), and the overlap with VL for in the sequence shown in bold. The 20 bases given in capital letters hybridize with the mature mouse VH sequences. The last 20 nt at the 3' end of the "VH-for" primers hybridize with the JH region. The first nt shown in capital letters will introduce a silent mutation at the end of VH in order to code for the first nt of the second *Sfi*I recognition site (underlined). The final assembly of the scFv gene by SOE-PCR is carried out with the *scback* and *scfor* primer set. The outer primer *scback* encodes the first *Sfi*I site (underlined).

```
                 5'           SfiI            FLAG      3'
scback  ttactcgcggcccagccggccatggcggactacaaaG                              d
                             5'          FLAG      VL  →              3'
LB1                gccatggcggactacaaaGAYATCCAGCTGACTCAGCC                   2
LB2                gccatggcggactacaaaGAYATTGTTCTCWCCCAGTC                   4
LB3                gccatggcggactacaaaGAYATTGTGMTMACTCAGTC                   8
LB4                gccatggcggactacaaaGAYATTGTGYTRACACAGTC                   8
LB5                gccatggcggactacaaaGAYATTGTRATGACMCAGTC                   8
LB6                gccatggcggactacaaaGAYATTMAGATRAMCCAGTC                  16
LB7                gccatggcggactacaaaGAYATTCAGATGAYDCAGTC                  12
LB8                gccatggcggactacaaaGAYATYCAGATGACACAGAC                   4
LB9                gccatggcggactacaaaGAYATTGTTCTCAWCCAGTC                   4
LB10               gccatggcggactacaaaGAYATTGWGCTSACCCAATC                   8
LB11               gccatggcggactacaaaGAYATTSTRATGACCCARTC                  16
LB12               gccatggcggactacaaaGAYRTTKTGATGACCCARAC                  16
LB13               gccatggcggactacaaaGAYATTGTGATGACBCAGKC                  12
LB14               gccatggcggactacaaaGAYATTGTGATAACYCAGGA                   4
LB15               gccatggcggactacaaaGAYATTGTGATGACCCAGWT                   4
LB16               gccatggcggactacaaaGAYATTGTGATGACACAACC                   2
LB17               gccatggcggactacaaaGAYATTTTGCTGACTCAGTC                   2
LBλ                gccatggcggactacaaaGATGCTGTTGTGACTCAGGAATC                1
```

Primer VL-for

```
                 5'      (Gly4Ser)3-linker          VL  →            3'
LF1'    ggagccgccgccgcc(agaaccaccaccacc)2ACGTTTKATTTCCAGCTTGG          2
LF4     ggagccgccgccgcc(agaaccaccaccacc)2ACGTTTTATTTCCAACTTTG          1
LF5     ggagccgccgccgcc(agaaccaccaccacc)2ACGTTTCAGCTCCAGCTTGG          1
LFλ     ggagccgccgccgcc(agaaccaccaccacc)2ACCTAGGACAGTCAGTTTGG          1
```

Primer VH-back

```
                 5'      (Gly4Ser)2-linker    BamHI  VH  →                3'
HB1     ggcggcggcggctccggtggtggtggatccGAKGTRMAGCTTCAGGAGTC                 8
HB2     ggcggcggcggctccggtggtggtggatccGAGGTBCAGCTBCAGCAGTC                 9
HB3'    ggcggcggcggctccggtggtggtggatccCAGGTGCAGCTGAAGSARTC                 4
HB4     ggcggcggcggctccggtggtggtggatccGAGGTCCARCTGCAACARTC                 4
HB5     ggcggcggcggctccggtggtggtggatccCAGGTYCAGCTBCAGCARTC                12
HB6'    ggcggcggcggctccggtggtggtggatccCAGGTYCARCTGCAGCARTC                 8
HB7'    ggcggcggcggctccggtggtggtggatccCAGGTCCACGTGAAGCARTC                 2
HB8'    ggcggcggcggctccggtggtggtggatccGAGGTGAASSTGGTGGARTC                 8
HB9'    ggcggcggcggctccggtggtggtggatccGAVGTGAWGSTGGTGGAGTC                12
HB10'   ggcggcggcggctccggtggtggtggatccGAGGTGCAGSTGGTGGARTC                 4
HB11'   ggcggcggcggctccggtggtggtggatccGAKGTGCAMCTGGTGGARTC                 8
HB12    ggcggcggcggctccggtggtggtggatccGAGGTGAAGCTGATGGARTC                 2
HB13'   ggcggcggcggctccggtggtggtggatccGAGGTGCARCTTGTTGARTC                 4
HB14'   ggcggcggcggctccggtggtggtggatccGARGTRAAGCTTCTCGARTC                 8
HB15'   ggcggcggcggctccggtggtggtggatccGAAGTGAARSTTGAGGARTC                 8
HB16'   ggcggcggcggctccggtggtggtggatccCAGGTTACTCTRAAASARTC                 8
HB17    ggcggcggcggctccggtggtggtggatccCAGGTCCAACTVCAGCARCC                 6
HB18'   ggcggcggcggctccggtggtggtggatccGATGTGAACTTGGAASARTC                 4
HB19'   ggcggcggcggctccggtggtggtggatccGAGGTGAAGGTCATCGARTC                 2
```

Primer VH-for

```
                 5' EcoRI            3'
scfor   ggaattcggccccgag
                 5' EcoRI   SfiI   VH  →                3'
HF1     ggaattcggccccgaggcCGAGGAAACGGTGACCGTGGT                          1
HF2     ggaattcggccccgaggcCGAGGAGACTGTGAGAGTGGT                          1
HF3     ggaattcggccccgaggcCGCAGAGACAGTGACCAGAGT                          1
HF4     ggaattcggccccgaggcCGAGGAGACGGTGACTGAGGT                          1
```

VL-κ	1	2	3	4	5	6	7
LB1	D	I	Q	L	T	Q	P
LB2	D	I	V	L	T, S	Q	S
LB3	D	I	V	I, L	T	Q	S
LB4	D	I	V	L	T	Q	S
LB5	D	I	V	M	T	Q	S
LB6	D	I	K, Q	I, M	N, T	Q	S
LB7	D	I	Q	M	I, T	Q	S
LB8	D	I	Q	M	T	Q	T
LB9	D	I	V	L	I, N	Q	S
LB10	D	I	E, V	L	T	Q	S
LB11	D	I	L, V	M	T	Q	S
LB12	D	I, V	L, V	M	T	Q	T
LB13	D	I	V	M	T	Q	A, S
LB14	D	I	V	I	T	Q	D, E
LB15	D	I	V	M	T	Q	I, M L, P
LB16	D	I	V	M	T	Q	P
LB17	D	I	L	L	T	Q	S

102	103	104	105	106	107	108	VL-κ
A, P S, T	K	L	E	I	K	R	LF1'
A, P S, T	K	L	E	I	K	R	LF4
A, P S, T	K	L	E	L	K	R	LF5

102	103	104	105	106	107	108	VL-λ
A, P S, T	K	L	T	V	L	G	LFλ

107	108	109	110	111	112	113	VH
T	T	V	T	V	S	S	HF1
T	T	L	T	V	S	S	HF2
T	L	V	T	V	S	A	HF3
T	S	V	T	V	S	S	HF4

VL-λ	1	2	3	4	5	6	7	8
LBλ	D	A	V	V	T	Q	E	S

VH	1	2	3	4	5	6	7
HB1	D, E	V	N, Q	L	Q	E	S
HB2	E	V	Q	L	Q	Q	S
HB3'	Q	V	Q	L	K	E, Q	S
HB4	E	V	Q	L	Q	Q	S
HB5	Q	V	Q	L	Q	Q	S
HB6'	Q	V	Q	L	Q	Q	S
HB7'	Q	V	H	V	K	Q	S
HB8'	E	V	K, N	L, V	V	E	S
HB9'	D, E	V	K, M	L, V	V	E	S
HB10'	E	V	Q	L, V	V	E	S
HB11'	D, E	V	H, Q	L	V	E	S
HB12	E	V	K	L	M	E	S
HB13'	E	V	Q	L	V	E	S
HB14'	E	V	K	L	L	E	S
HB15'	E	V	K	L, V	E	E	S
HB16'	Q	V	T	L	K	E	S
HB17	Q	V	Q	L	Q	Q	P
HB18'	D	V	N	L	E	E, Q	S
HB19'	E	V	K	V	I	E	S

- Standard molecular biology equipment and reagents to:
  - suspend spleen cells (Gibco 10 x EBSS Buffer)
  - determine the isotype of mAbs (Roche IsoStrip Mouse Monoclonal Antibody Isotyping Kit)
  - purify mRNA (Pharmacia QuickPrep mRNA Purification Kit)
  - perform a cDNA synthesis reaction (Pharmacia First Strand cDNA Synthesis Kit)
  - perform PCR reactions
  - cut and gel purify DNA
  - concentrate DNA (Amicon Microcon 30 for volumes less than 500 µl)
  - ligate and transform DNA
  - grow bacteria and phages
  - perform an ELISA
  - perform a Western blot

Fig. 3. Deduced amino acid sequence of the complementary part of primers from Fig. 2. The residues of VL and VH are numbered according to Kabat et al. (1991). The primers *LB1-17* and *HB1-19* determine only the first 2 nucleotides of residue 7, while the *LF* primers determine only the first 2 nucleotides of residue 102, and *LBλ* specifies only the first nucleotide of residue 8. Therefore, the original residues at those positions may not necessarily appear in the PCR amplified antibody gene.

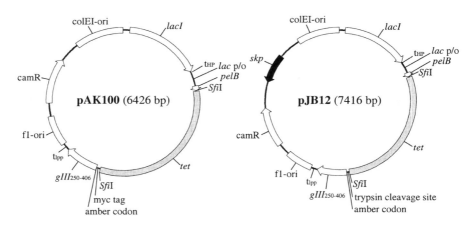

pAK / pJB vector series		phage display	Skp coexpression	enhanced expression	trypsin cleavage site	IMAC purification	C-terminal detection	direct detection	dimerization
pAK100	*Sfi*I *Sfi*I EcoRI HindIII — pelB — tet — myc * gIII 250-406 —	●							
pAK300	*Sfi*I *Sfi*I HindIII — pelB — tet — 6 his —					●	●		
pAK400	SDT7g10 *Sfi*I *Sfi*I HindIII — lac p/o — pelB — tet — 6 his —			●		●	●		
pAK500	*Sfi*I *Sfi*I EcoRI HindIII — pelB — tet — dHLX 5 his —					●	●		●
pAK600	*Sfi*I *Sfi*I EcoRI HindIII — pelB — tet — alkaline phosphatase —							●	●
pJB12	*Sfi*I *Sfi*I EcoRI HindIII — pelB — tet — KDIR * gIII 250-406 —	●	●		●				
pJB23	*Sfi*I *Sfi*I EcoRI HindIII — pelB — tet — 6 his —		●			●	●		
pJB33	SDT7g10 *Sfi*I *Sfi*I EcoRI HindIII — lac p/o — pelB — tet — 6 his —		●	●		●	●		

tet-cassette

Fig. 4. The pAK / pJB vector series (see also Fig.5) can be used either for phage display (pAK 100 and pJB12) by the strategy outlined in Fig. 1 or for the expression of the antibody in a variety of formats. All vectors contain a chloramphenicol resistance cassette (camR) and additionally a tetracyclin resistance "stuffer" cassette (*tetA* and *tetR*; 2101 bp), which will be replaced by the antibody gene (the *tet* cassette allows the monitoring of complete *Sfi*I digested vector by plating of transformed cells on tetracycline plates). At the bottom of the figure, the tet-cassette, which is simplified in the vector drawings, is shown in more detail. Furthermore, these vectors contain the *lacI* repressor gene, a strong upstream terminator (t_{HP}), the *lac* promoter/operator and the *pelB* (pectate lyase gene of *Erwinia carotovora*) leader sequence (modified to contain a *Sfi*I site) and a downstream terminator (t_{lpp}). The origins for phage replication and plasmid replication are as described in Ge et al. (1995). The antibody gene is alternatively fused in frame to geneIII$_{251-406}$ (for phage display), to a his tag for IMAC purification (Lindner et al. 1992) and C-terminal detection with another recombinant anti-his tag scFv-phosphatase fusion protein (Lindner et al. 1997), to dimerization helices (Pack et al. 1993, see also Chapter 43, Lindner and Plückthun) or to alkaline phosphatase for both, dimerization and direct detection (Lindner et al. 1997). In pAK100, the in-frame fusion contains a myc tag (Munro and Pelham 1986) to act as a detection handle, in addition to the short N-terminal 3-amino acid FLAG tag (Knappik et al. 1994) which is encoded by the primers (Fig. 2). The pJB12 contains a trypsin cleavage site (KDIR) and can therefore conveniently be used for selection of high affinity binders as described by Dziegiel et al. (1995) and Johansen et al. (1995). The asterisk in pAK100 and pJB12 represents an amber codon. The scFv expression level in pAK400 and pJB33 is enhanced due to the strong Shine Dalgarno sequence SDT7g10 (from T7 phage gene 10). In the pJB vector series the co-expressed periplasmic protein Skp (Bothmann and Plückthun 1998) or a similarly placed FkpA (Bothmann and Plückthun 2000) can increase the functional yield of antibody fragments expressed in the periplasm (for details, see also Chapter 23).

A

pAK100scFv, pAK300scFv, pAK500scFv, pAK600scFv, pJB12scFv, pJB23scFv

```
                                                      end lacI
...ATGCAGCTGGCACGACAGGTTTCCCGACTGGAAAGCGGGCAGTGAGC
...M  Q  L  A  R  Q  V  S  R  L  E  S  G  Q  *

              tHP terminator
GGTACCCGATAAAAGCGGCTTCCTGACAGAGGAGGCCGTTTTGTTTTGCAGC

                 CAP binding site
CCACCTCAACGCAATTAATGTGAGTTAGCTCACTCATTAGGCACCCCAGG

-35                -10            lac-operator
CTTTACACTTTATGCTTCCGGCTCGTATGTTGTGTGGAATTGTGAGCGGA
                                   | → mRNA

SD1          lacZ
TAACAATTTCACACAGGAAACAGCTATGACCATGATTACGAATTCTAGA
                          M  T  M  I  T  N  F  *

SD2        pelB signal sequence
TAACGAGGGCAAATCATGAAATACCTATTGCCTACGGCCAGCCGCTGATT
             M  K  Y  L  L  P  T  A  A  A  G  L

        SfiI              FLAG        VL
GTTATTACTCGCGGCCGGCCATGGCGCCGGCCATGGCGGACTACAAAGAY...
L  L  L  A  A  Q  P  A  M  A  D  Y  K  D ...
```

pAK400scFv, pJB33scFv

```
SD2        pelB signal sequence
...GAAGGAGATATACATATGAAATACCTATTGCCTACGGCCAGCC....
   T7g10    M  K  Y  L  L  P  T  A  A ...
```

B

pAK100scFv

```
VH    SfiI      EcoRI                         myc tag
...CGGCCTCGGGGGGCCGAATTCGAGCAGAAGCTGATCTCTGAGGAAGAC
... A  S  G  A  E  F  E  Q  K  L  I  S  E  E  D

        geneIII 250-406
CTGTAGGGTGGTGGCTCTGGTTCCGGTGATTTGATTATGAAAG....
L  *  G  G  G  S  G  S  G  D  F  D  Y  E  K ...
```

pJB12scFv

```
VH    SfiI      EcoRI          trypsin cleavage site
...CGGCCTCGGGGGGCCGAATTCGAGCAGAAGGATATCCGTGAGGAAGAC
... A  S  G  A  E  F  E  Q  K  D  I  R  E  E  D

        geneIII 250-406
CTGTAGGGTGGTGGCTCTGGTTCCGGTGATTTGATTATGAAAG....
L  *  G  G  G  S  G  S  G  D  F  D  Y  E  K ...
```

Fig. 5. A: upstream sequence of pAK100scFv, pAK300scFv, pAK400scFv, pAK500scFv, pAK600scFv, pJB12 and pJB23. The region from the end of the *lacI* repressor gene to the beginning of the antibody VL domain is shown. The *lacI* repressor gene, t$_{HP}$ terminator sequence, CAP binding site, *lac* operator region including the -35 and -10 sequence, Shine-Dalgarno (SD) sequence of *lacZ* (SD1), lacZ peptide, a second SD sequence (SD2), *pelB* signal sequence, N-terminal *SfiI* site, four amino acid FLAG tag and the start of the VL domain (bold) are indicated above the sequence. In pAK400 and pJB33, the 15 bp upstream from the *pelB* start codon are replaced by a sequence including the SD sequence of the phage T7 gene10. **B:** downstream sequence of pAK100scFv and pJB12scFv. The last two bases of VH (bold), *SfiI* and *EcoRI* restriction sites, myc tag or trypsin cleavage site and the start of geneIII$_{251-406}$ are indicated above the sequence.

C

pAK300scFv, pAK400scFv

```
       VH    SfiI        his tag                   HindIII
...CGGCCTCGGGGGCCGATCACCATCACCATCATTAGTAAGCTT..
 ...  A  S  G  A  D  H  H  H  H  H  H  *
```

pJB23scFv, pJB33scFv

```
       VH    SfiI   EcoRI   his tag
...CGGCCTCGGGGGCCGAATTCCACCACCATCACCACCATTAATGAAAG
 ...  A  S  G  .  A  E  F  H  H  H  H  H  H  *
HindIII
CTT....
```

D

pAK500scFv

```
       VH    SfiI   EcoRI                             dHLX
...CGGCCTCGGGGGCCGAATTCCCAAACTAGCACCCCCCTGGCA
 ...  A  S  G  A  E  F  P  K  P  S  T  P  P  G  S

GCAGTGGTGAACTGAAAGAGCTGCTTAAGCATCTTAAAGAACTTCTGAAG
  S  G  E  L  E  E  L  L  K  H  L  K  E  L  L  K

GGCCCCCGCAAAGGCGAACTCGAGGAACTGCTGAAACATCTGAAGGAGCT
  G  P  R  K  G  E  L  E  E  L  L  K  H  L  K  E  L
                                his tag
GCTTAAAGGTGGGAGCGGAGGCGCGCCGCCACCATCATCACCATTGACGTC
  L  K  G  G  S  G  G  A  P  H  H  H  H  H  H  *

HindIII
TAAGCTT...
```

pAK600scFv

```
       VH    SfiI   EcoRI    alkaline phosphatase (AP)
...CGGCCTCGGGGGCCGAATTCCGACACCAGAAATGCCTGTTCTG....
 ...  A  S  G  A  E  F  R  T  P  E  M  P  V  L  ...
                                   start AP

                                               HindIII
...CTCTTCTACACCATGAAAGCCGCTCTGGGGCTGAAATAAGCTT....
 ...  L  F  Y  T  M  K  A  A  L  G  L  K  *
                             end AP
```

C: downstream sequence of pAK300scFv, pAK400scFv, pJB23scFv and pJB33scFv. The last two bases of VH (bold), SfiI and EcoRI restriction sites and his$_6$ tags are indicated above the sequence.
D: sequences of EcoRI/HindIII fusion cassettes as used in pAK500 and pAK600. The dHLX dimerization motif was taken from Pack et al. (1993). The complete sequence of the mature *E. coli* alkaline phosphatase (AP) gene can be found in Shuttleworth et al. (1986). For the EcoRI/HindIII cloning cassette the two internal EcoRI sites of the AP gene have been removed by silent mutations.

▨ Procedure

Isolation of mRNA and cDNA synthesis

1. Take 1 to 5 million cells from a frozen or growing hybridoma (for iso-
    type determination use the Roche IsoStrip Mouse Monoclonal Anti-
    body Isotyping Kit) culture or spleen cells, respectively (see note),
    and perform a mRNA preparation as described in the QuickPrep
    mRNA Purification Kit (Pharmacia). According to the manufacturer,
    this kit can be used for up to 5×10^7 cells, but in order to get extremely
    pure mRNA take only 5×10^6 cells per oligo(dT)-cellulose column in-
    cluded in the kit, which will yield 1-10 µg of mRNA.

    Note: For mRNA preparation from spleen cells press the organ through
    a sieve (you may take a sterile strainer), suspend the cells in 10 ml EBSS
    buffer and separate off the connective tissue by centrifugation (15 s at
    200 g). Transfer the supernatant into a fresh tube, centrifuge (7 min at
    250 g) and resuspend the cells in 2 ml extraction buffer (Pharmacia
    QuickPrep mRNA Purification Kit). You typically obtain 5×10^7 B-cells
    (storage at -80°C in extraction buffer) from one mouse spleen.

    EBSS buffer (to suspend spleen cells)
    - 100 ml 10 x EBSS (Gibco)
    - 2.2 g $NaCO_3$
    - 25 ml 1 M HEPES (pH 7.3)
    - Adjust to pH 7.1 with HCl and fill up with water to 1 l

2. Separate the mRNA (step 1 yields 750 µl) in two aliquots and ethanol
    precipitate each. For an aliquot of 320 µl mRNA in elution buffer take
    800 µl ethanol (chilled to -20°C), 10 µl glycogen (10 mg/ml) and 32 µl
    potassium acetate (2.5 M, pH 5.0).

    Note: The mRNA can be stored in ethanol at -20°C for several months.

3. Collect the precipitated mRNA of one aliquot by centrifugation at
    16,000 g, 4°C for 30 min.

4. Wash with 1 ml ice-cold 90% ethanol.

5. Add 20 µl H_2O (diethylpyrocarbonate-treated). The mRNA solution is
    now ready for cDNA synthesis.

6. For reverse transcription take approximately 0.1-0.5 µg mRNA and 1 µl
    random hexamer primers (33 µl total reaction volume). The precise

procedure is described in the First Strand cDNA Synthesis Kit (Pharmacia).

PCR amplification

7. Take the primers (100 μM) which have been dissolved in TE buffer to prepare appropriate mixtures (LB mix, LF mix, HB mix and HF mix). Mix them according to the degree of degeneration (number of different unique sequences encoded by mixed bases) indicated in the column "d" in Fig 2. Take 1 μl from all non-degenerated primers (d = 1), 2 μl for d = 2-4, 3 μl for d = 6-9 and 4 μl for d = 12-16. To complete the LB mix (see note) add 2.8 μl of *LBλ* (= 5%) to the mixture of *LB1-17* (53 μl). For the LF mix take 19 μl of *LF1'*, 9.5 μl of *LF4*, 9.5 μl of *LF5* and 2 μl of *LFλ* (=5%). The total primer concentration of these mixtures is still 100 μM (ranging from 2.3 to 25 μM of each mixed synthesis primer).

Note: During the cloning of monoclonal antibodies problems can occur if the hybridoma transcribes more than one functional or even non-functional heavy or light chain variable region gene (Kütemeier et al. 1992). It was found that several kappa chain secreting hybridomas, where X63Ag8.653 myeloma cells were used as a fusion partner, are able to transcribe a functional lambda chain which competes with the kappa VL gene for in-frame scFv antibody assembly (Krebber et al. 1997). Therefore, it is highly recommended to leave out any lambda chain primer in the PCR reactions if the isotyping indicates that the hybridoma of interest secretes a kappa light chain.

8. For PCR amplification of VL and VH use the completed first-strand cDNA reaction and prepare the following mixes.

PCR mix for amplification of VL	PCR mix for amplification of VH
2 μl cDNA	2 μl cDNA
1 μl dNTP's (10 mM each)	1 μl dNTP's (10 mM each)
5 μl 10 x PCR Buffer	5 μl 10 x PCR Buffer
1 μl LB primer mix (100 μM)	1 μl HB primer mix (100 μM)
1 μl LF primer mix (100 μM)	1 μl HF primer mix (100 μM)
40 μl H₂O	40 μl H₂O
prepare 3 tubes of this mix	prepare 3 tubes of this mix

Note: The polymerase is added only after heating these mixes to 92°C. For Taq polymerase a MgCl₂ concentration of 2 mM is usually optimal. Although Taq polymerase works generally more robustly than other polymerases, the use of proof-reading polymerases is an alternative, but will require optimization steps such as Mg^{2+} titrations. This might be a problem, when the amount of cDNA template is limited.

9. Heat to 92°C for 3 min (use mineral oil if no lid heating is available), add 1 unit Taq polymerase (see note of step 8) and perform the following cycles (6 independent PCR reactions, 3 each for V_H and V_L with different annealing temperatures): 5 cycles of 1 min at 92°C, 1 min initial annealing at 45°C (alternatively, see note, 50°C and 55°C) and 1 min at 72°C, followed by 20 cycles of 1 min at 92°C, 1 min at 63°C and 1 min at 72°C.

   Note: For amplification of VL and/or VH, complete annealing of the 3'-ends of the primers with the template DNA is essential. Three different initial annealing temperatures (45°C, 50°C, 55°C) are recommended to be able to amplify the great majority of the antibody genes, as it is not clear a priori, which somatic mutations a given monoclonal antibody gene may carry in the primer regions. After 5 cycles the amplified PCR product will serve itself as template DNA. The annealing temperature of the last 20 cycles is therefore 63°C in all 3 cases.

10. Gel-purify the VL and VH genes and determine the DNA concentration of both chains.

    Note: Using the listed primer mixtures, the expected lengths of the PCR products of VL and VH are between 375-402 bp and 386-440 bp, respectively. The 3 different PCR reactions do not necessarily each yield product. However, these PCR products can be pooled if necessary.

11. Use approximately 10 ng of the PCR product of both domains for the assembly PCR (50 µl total volume). Take 200 µM dNTP's, 1 µM (each) *scback* and *scfor* primer. Use 1 unit DNA polymerase (see note of step 8) and perform 7 cycles of 1 min at 92°C, 30 s at 63°C, 50 s at 58°C and 1 min at 72°C, followed by 23 cycles of 1 min at 92°C, 30 s at 63°C and 1 min at 72°C.

    Note: Hot-start PCR and initial assembly of VL and VH in the absence of the primers is usually not necessary, but can be used.

Digestion and cloning of scFv genes

12. Perform a *Sfi*I digest of the amplified scFv for 3-4 hours at 50°C (overlay with mineral oil). In case of Taq polymerase the digestion can be conveniently carried out in the PCR buffer. To 40 μl PCR product add 11 μl 10 x buffer, 13 μl 10 x BSA, 64 μl H_2O and 2 μl (= 20 units) *Sfi*I.

    Note: This procedure might work with other PCR buffers as well but has not been tested yet.

13. Digest appropriate amounts of vector (pAK100 or pJB12, see Fig. 4) with *Sfi*I. Use 10 units *Sfi*I for 1 μg vector in 100 μl volume and incubate overnight at 50°C (overlay with mineral oil).

    Note: For pure preparations of a fully digested vector it is very important to not over-load the agarose gel. Furthermore, the gel electrophoresis has to be run long enough to separate small amounts of undigested vector from the digested vector band.

14. Purify and concentrate the digested scFv antibody genes and vector by preparative agarose gel (1%) electrophoresis in combination with the QIAEX II Gel Extraction Kit (Qiagen).

    Note: For large scale vector or insert preparation, electroelution is most efficient and convenient. For concentration you may use Microcon 30 columns (Amicon) as described in step 31.

15. Ligate 20 ng scFv gene fragment with the vector (ratio vector to insert 3:2) with 1 unit ligase in 10 μl volume. Incubate overnight at 16°C.

    Note: Compatible vector sets are available which allow an easy recloning of the scFv fragment into vectors for optimized soluble expression, generation of fusion proteins and other purposes (Fig. 4 and Fig. 5).

16. Transform 5-10 μl of the ligation into competent XL1-Blue (Stratagene) cells.

    Note: If many clones are required follow the instructions described in step 23-33.

17. Plate on 2 x YT, 1% glucose, chloramphenicol (30 μg/ml) agar plates and incubate overnight at 37°C.

    Note: You may check the ratio of desired ligation product to background by testing for tetracycline resistance. The portion of vector with unremoved or religated *tet* cassette (see note of step 13) is typically in the range of 0.01 to 0.1%.

Screening for binders

18. Pick 10 colonies and let them grow separately at 37°C in 2 ml 2 x YT, 1% glucose, chloramphenicol (30 μg/ml), until they reach an $OD_{550} = 0.5$. Add 2 ml 2 xYT, 1% glucose, chloramphenicol (30 μg/ml), 1 mM IPTG, 5×10^9 cfu helper phage (Stratagene) and shake overnight at 26°C or 37°C (for some scFv's, growth at 26°C after infection may be necessary). The phage titer after overnight incubation is in the range of 10^{11} cfu per ml supernatant.

    Note: The phage titer should be checked in order to rule out any problems during phage production. To determine the phage titer (in cfu) take a growing culture of XL1-Blue cells ($OD_{550} = 0.4$-0.8) and incubate aliquots of this culture with dilutions of your phage preparation. After 15 min incubation at 37°C, plate appropriate amounts (30-150 cfu/plate) on 2 x YT, 1% glucose, chloramphenicol (30 μg/ml) agar plates.

19. Centrifuge the culture. Take 1.6 ml supernatant and mix it with 0.4 ml 20% PEG 6000, 2.5 M NaCl in a 2 ml Eppendorf cap in order to precipitate the phages (Sambroook et al. 1989).

20. Incubate on ice for 15 min and spin at 16,000 g, 4°C for 20 min.

    Note: The size of the white pellet does not necessarily reflect a high or low phage titer.

21. Suspend the phage pellet in 400 μl PBS (2% milk) and use 100 μl phage solution per well in an ELISA assay (see step 22) to distinguish functional scFv antibody displaying phages from those which display a non-functional or non-productive antibody fragment.

    Note: Do not (!) centrifuge after suspending the phage pellet because you would spin down phages together with cell debris.

22. If soluble antigen is available include an ELISA control that shows that free antigen is able to compete with bound antigen for phage binding to distinguish non-specific "sticky" phage from specifically binding phage. In principle, the same ELISA protocol which was used for the hybridoma screening procedure can be used.

    Note: For weak binders it might be important to use more phages for ELISA analysis. In this case, the culture volume should be increased 10 times. If no functional clone shows up in ELISA of single clones, perform one round of phage panning in order to enrich the functional binders. The enrichment should be checked by comparison of eluted phages from a specific surface versus a surface without antigen.

Library cloning

23. For preparation of electrocompetent *E. coli* XL1-Blue (Stratagene) cells use 1 ml of a dense overnight pre-culture to inoculate 500 ml medium (2 x YT). Shake it at 37°C until an OD_{550} of 0.7 is reached. Then, chill the culture on ice as quickly as possible (cool the whole shake flask for 10 min in ice-cold water).

    Note: Sufficient agitation and aeration during growth seems to be very important for preparation of electrocompetent cells with reproducible efficiencies of 3-6 x 10^9 cfu/μg pUC19 DNA. Therefore, use 5 l baffled shake flasks with only 500 ml medium and make sure that the amplitude of the shaker is high enough to vigorously circulate the medium (if not, moderately increase the amount of medium).

24. Centrifuge 400 ml (8 x 50 ml in disposable tubes) for 9 min at 2500 g. Remove as much supernatant as possible (leave the tube upside down for 15-30 s on a clean tissue). Then, fill each tube with 50 ml ice-cold distilled water and remove the water immediately (the cell pellet is very solid after this first centrifugation step and will not be resuspended by the brief addition of distilled water).

    Note: It is recommended to carry out steps 24 to 28 in the coldroom.

25. Fill each tube with 25 ml distilled water (200 ml total) and resuspend each pellet carefully by pipetting the solution up and down with a 25 ml wide-gauge sterile plastic pipette (Falcon) which has to be placed very close to the pellet. Make sure that the cells are all taken up in a homogeneous suspension. Incubate for 10 min on ice.

26. Transfer the cells into 4 new 50 ml tubes and centrifuge at 2500 g for 12 min. Carefully remove the supernatant. Resuspend the four pellets in a total volume of 40 ml pre-chilled 10% (v/v) DMSO (Fluka). Incubate for 10 min on ice.

27. Centrifuge (2 x 20 ml) at 2500 g for 15 min and remove the supernatant (you might lose a small portion of cells - do not put the tubes upside down on tissue in this step!) and resuspend in 20 ml 10% DMSO. Incubate on ice for 10 min.

28. Centrifuge at 2500 g for 15 min. Carefully resuspend the cells in 2 ml 10% DMSO, freeze the cells (100 μl aliquots) by dipping the tubes in liquid nitrogen and store them at -80°C.

29. To check the transformation efficiency (see also step 32) of the cells add 1 µl of 10 pg/µl pUC19 DNA (in water). Fifty colonies per 1/1000 transformation correspond to an efficiency of 5 x 10^9 cfu/µg pUC19 DNA.

    Note: Using 2 mm electroporation cuvettes (2500 V) the time constant for the efficiency test with 10 pg pUC19 DNA should be 5.4-5.6 ms, reflecting properly washed cells.

30. For desalting prior to electroporation fill up the ligation mixture with water to 400 µl and heat it for 10 min at 65°C. Then load it on a Microcon 30 (Amicon) column and centrifuge at 11,000 g for 8 min. Discard the flow-through.

31. Fill up with 400 µl water, centrifuge at 11,000 g and discard the flow-through. Repeat this procedure 3 times. It might be necessary to prolong the last centrifugation step (the residual volume should be between 20 and 50 µl). Finally, turn the columns upside down in clean tubes and centrifuge at 960 g for 3 min.

32. For each transformation use desalted ligation mixture corresponding to 20-100 ng insert. Add the DNA to the barely thawed cells (on ice) and mix by flipping the tube shortly. Now, any further incubation is detrimental. Load it therefore directly into the chilled electroporation cuvette and trigger the pulse. Immediately add the medium (900 µl 2 x YT (room temperature) containing 10 mM $MgCl_2$, 2.5 mM KCl and 0.4% glucose).

    Note: For efficient transformation (= 10^8 clones per µg insert DNA) the time constant (using 2 mm cuvettes) should be $\geq$ 5.2 ms.

33. Plate on 2 x YT, 1% glucose, chloramphenicol (30 µg/ml) agar plates (do not exceed 10^8 clones per 0.06 m^2) and incubate overnight at 37°C. Scrape the colonies off the plates and subsequently store them at -80°C after addition of 30% glycerol.

    Note: Take care that your library is homogeneously mixed.

34. For screening proceed as described in step 18-22. For inoculation take at least 10 fold more viable cells than colonies obtained after transformation. Perform 3 rounds of phage panning (see note of step 22) before testing single clones.

    Note: The first panning round is the most crucial because you might lose any desired, but less abundant, antibody sequence by too extensive washing. Therefore, do not exceed 10 washing steps in the first panning round.

Troubleshooting

- In case of low transformation yields it should be analyzed if the problem is the transformation itself or rather the ligation. To check the ligation, an aliquot can be analyzed on an agarose gel which might indicate also any problems with nucleases. Furthermore, it might be instructive to compare the ligation efficiency of *Sfi*I digested PCR product with insert derived from plasmid digestion. In order to check both the ligation and the transformation efficiency, a defined amount of pUC19 DNA can be mixed into the ligation. Due to the chloramphenicol resistance of the cloning vector and the ampicillin resistance of pUC19 DNA it is possible to calculate the ligation efficiency by plating on ampicillin or chloramphenicol plates, respectively, and comparing the number of clones. The transformation efficiency (in presence of the ligation mixture) can be judged by comparison with pUC19 DNA alone.

- If no binders appear from this procedure the primers should be checked since the quality and presence of each of the oligonucleotide is most decisive for reliable cloning of various antibody genes or, in case of library cloning, for broad representation of the immune response. Any sequence absent from a complex mixture will obviously decrease the functional library size and single-base deletions present in any one of the oligonucleotides would be amplified into the final product. Take care, therefore, that the quality of the oligonucleotide primers is satisfactory. Because of the low profit margins, oligonucleotides are today frequently synthesized with a minimum of reagents excess and short reaction times, not reaching the coupling yields that would be possible. A low total yield of synthesized oligonucleotides indicates usually a low coupling yield and is a warning sign that the "full-length" oligonucleotide pool contains a significant portion of molecules with random single-base deletions. Gel purification can ameliorate but not solve the problem, which lies in poor DNA-synthesis quality. It is recommended, especially for library cloning, to sequence the genes of random clones or to check for full-length scFv by Western blot analysis.

Acknowledgements. We wish to thank Anke Krebber, Annemarie Honegger, Lutz Jermutus, Barbara Klinger and Cornelia Rinderknecht for helpful contributions to the reagents and procedures of this article.

References

Bothmann H, Plückthun A (1998) Selection for a periplasmic factor improving phage display and functional periplasmic expression. Nature Biotechnol 16:376-380

Bothmann H, Plückthun A (2000) The periplasmic E. Coli peptidylprolyl cis, trans-isomerase FkpA. Increased functional expression of antibody fragments with and without cis-prolines. J Biol Chem 275:17100-17105

Braden BC, Goldman ER, Mariuzza RA, Poljak RJ (1998) Anatomy of an antibody molecule: structure, kinetics, thermodynamics and mutational studies of the antilysozyme antibody D1.3. Immunol Reviews. 163:45-57

Dall'Acqua W, Carter P (1998) Antibody engineering. Curr Opin Struct Biol 8:443-450

Dziegiel M, Nielson LK, Anderson PS, Blancher A, Dickmeiss E, Engberg J (1995) Phage display used for gene cloning of human recombinant antibody against the erythrocyte surface antigen, rhesus D. J Immmunol Meth 182:7-19

Freund C, Ross A, Plückthun A, Holak TA (1994) Structural and dynamic properties of the Fv fragment and the single-chain Fv fragment of an antibody in solution investigated by heteronuclear three-dimensional NMR spectroscopy. Biochemistry 33:3296-3303

Ge L, Knappik A, Pack P, Freund C, Plückthun A (1995) Expressing antibodies in *Escherichia coli*. In: Borrebaeck CAK (ed) Antibody Engineering, 2nd edition. Oxford University Press, pp 229-236

Hanes J, Jermutus L, Weber-Bornhauser S, Bosshard HR, Plückthun A (1998) Ribosome display efficiently selects and evolves high-affinity antibodies *in vitro* from immune libraries. Proc Natl Acad Sci USA 95:14130-14135

Hoogenboom HR, de Bruine AP, Hufton SE, Hoet RM, Arends JW, Roovers RC (1998) Antibody phage display technology and its applications. Immunotechnology 4:1-20

Johansen LK, Albrechtsen B, Andersen HW, Engberg J (1995) pFab60: a new, efficient vector for expression of antibody Fab fragments displayed on phage. Prot Eng 10:1063-1067

Kabat EA, Wu TT, Reid-Miller M, Perry HM, Gottesman KS, Foeller C (1991) Sequences of Proteins of Immunological Interest. 5th edn. US Department of Health and Human Services, Public Service, NIH

Knappik A, Ge L, Honegger A, Pack P, Fischer M, Wellnhofer G, Hoess A, Wölle J, Plückthun A, Virnekäs B (2000) Fully Synthetic Human Combinatorial Antibody Libraries (HuCAL) based on modular consensus frameworks and CDRs randomized with trinucleotides. J Mol Biol, 296:57-86

Knappik A, Plückthun A (1994) An improved affinity tag based on the FLAG peptide for the detection and purification of recombinant antibody fragments. BioTechniques 17:754-761

Krebber A, Bornhauser S, Burmester J, Honegger A, Willuda J, Bosshard HR, Plückthun A (1997) Reliable cloning of functional antibody variable domains from hybridomas and spleen cell repertoires employing a reengineered phage display system. J Immunol Meth 201:35-55

Kütemeier G, Harloff C, Mocikat R (1992) Rapid isolation of immunoglobulin variable genes from cell lysates of rat hybridomas by polymerase chain reaction. Hybridoma 11:23-32

Lindner P, Guth B, Wülfing C, Krebber C, Steipe B, Müller F, Plückthun A (1992) Purification of native proteins from the cytoplasm and periplasm of *Escherichia coli* using IMAC and histidine tails: a comparison of proteins and protocols. Methods 4:41-56

Lindner P, Bauer K, Krebber A, Nieba L, Kremmer E, Krebber C, Honegger A, Klinger B, Mocikat R, Plückthun A (1997) Specific detection of his-tagged proteins with recombinant anti-his tag scFv-phosphatase or scFv-phage fusions. BioTechniques 22:140-149

Low NM, Holliger PH, Winter G (1996) Mimicking somatic hypermutation: affinity maturation of antibodies displayed on bacteriophage using a bacterial mutator strain. J Mol Biol 260:359-368

Martineau P, Betton JM (1999) In vitro folding and thermodynamic stability of an antibody fragment selected *in vivo* for high expression levels in *Escherichia coli* cytoplasm. J Mol Biol 292:921-929

Munro S, Pelham HR (1986) An Hsp70-like protein in the ER: Identity with the 78 kd glucose-regulated protein and immunoglobulin heavy chain binding protein. Cell 46:291-300

Ohlin M, Owman H, Mach M, Borrebaeck CAK (1996) Light chain shuffling of a high affinity antibody results in a drift in epitope recognition. Mol Immunol 33:47-56

Pack P, Kujau M, Schroeckh V, Knüpfer U, Wenderoth R, Riesenberg D, Plückthun A (1993) Improved bivalent miniantibodies, with identical avidity as whole antibodies, produced by high cell density fermentation of *Escherichia coli*. Biotechnology 11:1271-1277

Parsons HL, Earnshaw JC, Wilton J, Johnson KS, Schueler PA, Mahoney W, McCafferty J (1996) Directing phage selections towards specific epitopes. Prot Eng 9:1043-1049

Plückthun A (1994) Recombinant antibodies. In: Van Oss CJ, Van Regenmortel MHV (eds) Immunochemistry. Marcel Dekker, New York, pp 201-236

Plückthun A, Krebber A, Krebber C, Horn U, Knüpfer U, Wenderoth R, Nieba L, Proba K, Riesenberg D (1996) Producing antibodies in *Escherichia coli*: From PCR to fermentation. In: McCafferty J, Hoogenboom HR (eds) Antibody Engineering: A Practical approach. IRL press, Oxford, pp 203-252

Rees AR, Staunton D, Webster DM, Searle SJ, Henry AH, Pederson JT (1994) Antibody design: beyond the natural limits. Trends Biotechnol 12:199-206

Sambrook J, Fritsch E, Maniatis T (1989) Molecular Cloning. A Laboratory Manual, Cold Spring Harbor Laboratory Press, Cold Spring Harbor, NY

Schier R, Bye, J, Apell G, McCall A, Adams GP, Malmqvist M, Weiner LM, Marks JD (1996) Isolation of high-affinity monomeric human anti-c-erbB-2 single chain Fv using affinity-driven selection. J Mol Biol 255:28-43

Shuttleworth H, Taylor J, Minton N (1986) Sequence of the gene for alkaline phosphatase from *Escherichia coli* JM83. Nucleic Acids Res 14:8689

Thompson JE, Vaughan TJ, Williams AJ, Wilton J, Johnson KS, Bacon L, Green JA, Field R, Ruddock S; Martins S, Pope AR, Tempest PR, Jackson RH (1999) A fully human antibody neutralising biologically active human TGFbeta2 for use in therapy. J Immunol Meth 227:17-29

Vaughan TJ, Williams AJ, Pritchard K, Osbourn JK, Pope AR, Earnshaw JC, McCafferty J, Hodits RA, Wilton J, Johnson KS (1996) Human antibodies with sub-nanomolar affinities isolated from a large non-immunized phage display library. Nature Biotechnol 14:309-314

Winter G, Griffiths AD, Hawkins RE, Hoogenboom HR (1994) Making antibodies by phage display technology. Ann Rev Immunol 12:433-455

Wörn A, Plückthun A (1999) Different equilibrium stability behavior of scFv fragments: identification, classification, and improvement by protein engineering. Biochemistry 38:8739-8750

Abbreviations

BSA	bovine serum albumin
DMSO	dimethylsulfoxide
EBSS	Earle's buffered salt solution
HRP	horse radish peroxidase
IPTG	isopropylthiogalactoside
PEG	polyethylene glycol

Construction of scFv from Hybridoma by Two-Step Cloning

FRANK BREITLING, DIETER MOOSMAYER, BODO BROCKS, and STEFAN DÜBEL

Introduction

Despite the fact that a growing number of recombinant antibodies has been isolated from phage display libraries, still many antibody specificities are available from hybridoma cell lines. Here, a method is presented to obtain the genetic information for the antigen binding part of the antibody from hybridoma cells, and to assemble it into a functional bacterially expressed fusion protein (scFv fragment). To achieve this, vectors have been constructed which combine the two variable regions (Vh and Vl) with a peptide linker to yield an scFv fragment. The genetic information for Vh and Vl is amplified from hybridoma cells using the polymerase chain reaction (PCR) with antibody specific primers.

What are the reasons for cloning an scFv from a hybridoma? First, some hybridoma cell lines are very low producers, or antibody production is lost upon prolonged culture. In this case, a recombinant "hybridoma immortalization" can rescue a valuable antibody specificity. Second, the recombinant format can be required for the desired application. Ex-

Frank Breitling, Deutsches Krebsforschungszentrum, Im Neuenheimer Feld 280, 69120 Heidelberg, Germany

Dieter Moosmayer, Universität Stuttgart, Institut für Zellbiologie und Immunologie, Allmandring 31, 70569 Stuttgart, Germany (Current address: Schering AG, 13342 Berlin)

Bodo Brocks, Universität Stuttgart, Institut für Zellbiologie und Immunologie, Allmandring 31, 70569 Stuttgart, Germany (Current address: MorphoSys AG, Am Klopferspitz 19, 82152 Planegg, Germany)

✉ Stefan Dübel, Universität Heidelberg, Molekulare Genetik, Im Neuenheimer Feld 230, 69120 Heidelberg, Germany (*phone* +49-6221-545638; *fax* +49-6221-545678; *e-mail* sd@uni-hd.de; *homepage* duebel.uni-hd.de)

amples are complex fusion proteins, like immunotoxins and antibody-enzyme fusions, or bispecific antibodies, which cannot be prepared with defined stoichiometry and coupling points by conventional chemical modification. Another example is the application as an intrabody (for general review see: Breitling and Dübel, 1999). It has also been shown that the scFv format itself can be beneficial for the desired function. In one example, a monoclonal antibody against TNFα-Receptor (TNFR1), with limited TNF-agonistic activity was converted into a potent TNFα antagonist by producing it as monomeric scFv fragment, thus preventing both ligand binding and receptor crosslinking (Moosmayer et al., 1995). In another example, the scFv fragment but not the original antibody was able to act as an enzyme inhibitor, probably due to its smaller size avoiding sterical hindrance (Liu et al., 1999).

A problem frequently obstructing the functional cloning of V region DNA from hybridoma cell lines is their sequence heterogenicity. Point mutations and insertions as well as entirely different V regions may be found in the PCR products. Even if an antibody sequence has already been determined from the hybridoma, e.g. by PCR sequencing, it is not necessarily the sequence coding for the functional V regions, as shown e.g. for the Myc1-9E10 (anti c-myc) hybridoma cell line (Fuchs et al., 1997, Schiweck et al., 1997). Various explanations for this sequence heterogenicity can be considered. Mutations can accumulate in the hybridoma cell population upon prolonged culture, which are not evident from functional analysis of the supernatant as long as a sufficient fraction of the cells still produces the correct antibody chains. This type of mutation can be minimised by preparing the cDNA from a freshly produced hybridoma subclone. Entirely different V sequences might derive from traces of expressed mRNA from the myeloma fusion partner or even from the second allel of the B cell partner, since hybridoma cells are no longer controlled by the rigid regulation mechanisms of the immune system. Even antibody pseudogene transcripts with stop codons inside the V region have been found in PCR products. Further, point mutations at both ends of the sequence can result from base pair mismatches during PCR priming.

In conclusion, an assay for antigen binding function should be done as early as possible in the process of cloning. Creating a small phage display library from the PCR products and screening for function is recommended where possible, e.g. when sufficient amounts of soluble antigen are available to perform a panning (see Troubleshooting). This process is described in detail in Chapter 5. Here, we present a direct cloning approach for hybridoma antibodies recognising antigens which are not available in significant amounts, like cell surface antigens.

▨ Outline

The method comprises hybridoma subcloning, mRNA-Isolation, cDNA synthesis, PCR, stepwise cloning into a bacterial expression vector and initial characterisation steps for structure, production and function of the antibody. The key to successful cloning is the PCR primer set. Two different choices for PCR primers are given. The first is a minimal set which was empirically tested and evolved over more than 7 years (Dübel et al., 1994). To date, it has allowed successful amplification of V region DNA from over 40 hybridoma lines, including several rat hybridomas, and we have not observed a case so far where no PCR amplification was possible. However, this primer design strategy resulted in quite long oligonucleotides, thus introducing primer mutations at mismatch nucleotides which may interfere with antigen binding or folding. The second primer set has been designed based on more recent and extended knowledge of antibody sequences and also includes IgM and lambda primers. It has not been tested to a similar extent, but proved to amplify V region DNA from several hybridoma antibodies and it also has been successfully applied for cloning highly diverse repertoires from immunised mice (Brocks and Moosmayer, unpublished). In general, primers designed for the generation of murine V region libraries (see Chapter 5) might be used as well for cloning of V regions from hybridoma.

The entire procedure is outlined in Figure 1. Bacterial culture, DNA manipulations, transformation and gel electrophoresis methods are performed according to standard protocols (Sambrook et al., 1989).

▨ Materials

- ELISA reader   Equipment

- PCR thermocycler

- mRNA extraction Kit (Optiprep 1 & 2, Biometra)   Reagents

- M-MuLV Reverse Transcriptase (Roche) supplied with buffer concentrate

- Oligonucleotide primers, as described in Figure 2 or 3

- *vent*-DNA polymerase (New England Biolabs)

- PCR-reaction buffer (supplied with the enzyme)

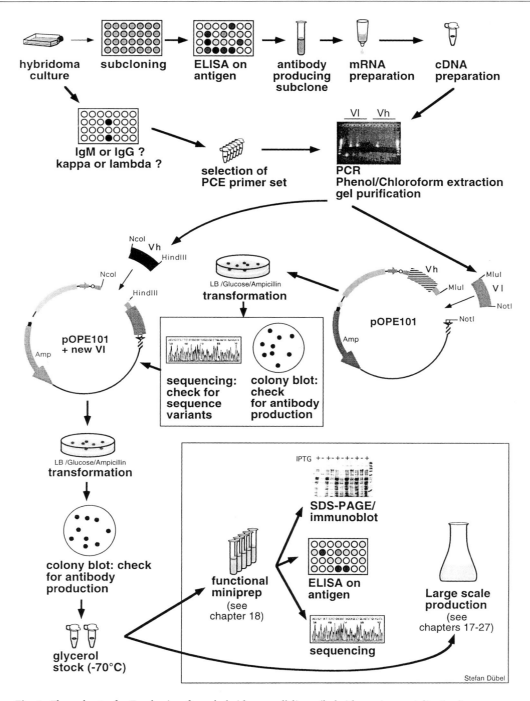

Fig. 1. Flow chart of scFv cloning from hybridoma cell lines (hybridoma immortalisation).

- Nucleotide stock solution containig 10 mM of each dNTP (Pharmacia)

- Bacterial culture, gel electrophoresis and agarose gel extraction equipment

- *Escherichia coli* K12, JM109, TG1 or XL1-blue competent cells

- Bacterial growth medium (LB) agar plates containing 100 mM glucose and 50 μg/mL ampicillin

- Bacterial growth medium (LB) agar plates containing 100 μM isopropyl-β-D-thiogalactopyranoside (IPTG). Media and agar plates are prepared according to standard protocols as previously described (Sambrook et al., 1989)

- Tris-NaCl-Tween (TNT): 20 mM Tris-HCl, 0.15 M NaCl, 0.05% Tween 20, pH 7.4

- Tris-buffered-saline (TBS): 20 mM Tris-HCl, 0.15 M NaCl, pH 7.4

- Nitrocellulose membrane filters fitting the petri dish used for plating of the transformed bacteria, e.g. BA85 0.45 μM 82mm, Schleicher und Schuell

- Indian Ink (Pelikan, Braunschweig)

- Chloroform

- Sodium azide powder

- 1% skimmed milk powder in TNT

- Antibody recognising a tag peptide coded by the expected fusion protein, here: mAb Myc1-9E10

- Enzyme-labelled antiserum recognizing the first antibody

- Precipitating substrate system for the enzyme labelled antiserum

- Autoclaved glycerol

Procedure

Isolation of antibody DNA

1. Prepare a freshly subcloned culture of the hybridoma cells and check the supernatant for antigen binding. Collect at least 10^6 hybridoma cells by centrifugation.

2. Isolate the mRNA using a kit.

   Note: Most Kits give reasonable results, we currently use the MPG Direct mRNA-Purification Kit (CPG, New Jersey, USA). In modification of the protocol given by the supplier, we change the final step to a precipitation with 70% (v/v) ethanol. The precipitate is collected by 30 min centrifugation at 15.000 xg and air dried.

3. Add 0.5 µl of a 1 mM solution of an Oligo-$(dT)_{15}$ primer to the complete mRNA dissolved in 10 µl water, heat to 65°C for 5 min and chill on ice.

4. Prepare the Reverse Transcriptase reaction:
   - 4 µl 5x Reverse Transcriptase buffer (supplied with enzyme)
   - 2 µl 10 mM dNTP (Pharmacia)
   - 2 µl DEPC water
   - 1 µl RNA-guard RNase inhibitor (Pharmacia)
   - 1 µl Reverse Transcriptase M-MuLV (Boehringer / Roche)
   - add Oligo-$(dT)_{15}$ primer / mRNA mix, incubate for 90 min at 37°C.

5. Heat Reverse Transcriptase reaction 5 min to 95°C, chill on ice, spin down drops.

   Note: A second strand synthesis is not necessary.

6. Mix the first strand cDNA with the PCR constituents on ice. Each 50 µl reaction contains 25pmol of each primer, polymerase buffer as described by the supplier and 250 µM dNTPs. Use 1 µl of cDNA for each 50 µl PCR reaction.

PCR 7. Preheat the thermocycler to 95°C. Add 0.5U per 50 µl reaction volume of *vent*-DNA polymerase on ice and mix. Avoid warming to room temperature, put the tubes quickly into the preheated thermocycler. Denature for max. 60 seconds.

   Note: *Taq*-DNA polymerase may also be used, but there is a higher risk of introducing mutations during amplification. Some enzyme products consist of a mixture of proofreading/non-proofreading enzymes, they may be used as well. Denaturation longer than 60sec at the start of the program is not necessary and may even result in loss of yield.

8. Perform 25-30 cycles of 30 sec denaturation at 95°C, 1 min hybridization at the appropriate primer hybridization temperature and 1 min polymerization at 72°C. Use the primers described in Figures 2 or 3

in individual reactions for each appropriate primer pair. After the end of the cycles, immediately cool down to 4°C.

Note: A hybridization temperature of 55°C should be tried initially. If no products are amplified, perform a set of 4 PCRs, differing only by their hybridization temperatures of 42, 45, 48 and 51°C. Buffer, nucleotides and primers may be stored as a premix at -20°C. Overcycling with *vent*-DNA polymerase may lead to product degradation.

9. Gel purify the PCR products. In case you have used the primer set of Figure 2, you can directly proceed to Step 10. In case you have used the primer set of Figure 3, you have to perform a second PCR to introduce the restriction sited necessary for cloning. To do this, use 1 μl of the gel purified first PCR reaction as a template in a reaction similar to step 7 and 8, but do only 9 cycles employing a hybridisation temperature of 57°C.

10. Collect 1/5 volume of the reaction for analysis on a 1.5% agarose gel containing ethidium bromide. Phenol extract the PCR product or freeze the PCR tubes immediately until you have time to extract it.

Note: The remaining activity of DNA polymerase needs to be removed after completion of the PCR reaction by phenol extraction to prevent digestion of 3' overhanging ends by the intrinsic 3'-5'-exonuclease activity of the Polymerases. This step is essential for efficient cloning after the subsequent restriction digest. Omission of this step may result in a drastically reduced ligation efficiency. Be aware that gel purification or spin columns do not remove this Polymease activity completely.

11. Double digest the purified PCR product with the appropriate restriction endonucleases.

Note: Calculate the amount of required enzyme carefully. Overdigestion may reduce the ligation efficiency.

12. Gel purify the digested PCR fragment. We recommend spin column kit systems. Alternatively, agarase digestion of low meting point agarose can be employed.

κ chain FR1 region: (EcoRV)

Bi6   5¥-GGT<u>GATATC</u>GTGAT(A/G)AC(C/A)CA(G/A)GATGAACTCTC

Bi7   5¥-GGT<u>GATATC</u>(A/T)TG(A/C)TGACCCAA(A/T)CTCCACTCTC

Bi8   5¥-GGT<u>GATATC</u>GT(G/T)CTCAC(C/T)CA(A/G)TCTCCAGCAAT

κ chain constant domain: (BamHI)

Bi5   5¥-GGGAAGAT<u>GGATCC</u>AGTTGGTGCAGCATCAGC

heavy chain FR1 region: (PstI, PvuII)

Bi3   5¥-GAGGTGAAG<u>CTGCAG</u>GAGTCAGGACCTAGCCTGGTG

Bi3b   5¥-AGGT(C/G)(A/C)AA<u>CTGCAG</u>(C/G)AGTC(A/T)GG

Bi3c   5¥-AGGT(C/G)(A/C)AG<u>CTGCAG</u>(C/G)AGTC(A/T)GG

Bi3d   5¥-AGGT(C/G)<u>CAGCTGCAG</u>(C/G)AGTC(A/T)GG

γ chain CH1 domain: (HindIII)

Bi4   5¥-CCAGGGGCCAGTGGATAGAC<u>AAGCTT</u>GGGTGTCGTTTT

<u>Reamplification primers for the introduction of other restriction sites</u>

heavy chain FR1 region: Bi3f   5¥- CAGCCGG<u>CCATGG</u>CGCAGGT(C/G)<u>CAGCTGCAG</u>(C/G)AG
                                                         NcoI   PvuII, PstI

κ chain constant domain: Bi5c   5¥- GAAGAT<u>GGATCC</u>AG<u>CGGCCGC</u>AGCATCAGC
                                                    BamHI   NotI

κ chain FR1 region: Bi8b

5¥- AATTTTCAGAAGC<u>ACGCGT</u>A<u>GATATC</u>(G/T)TG(A/C)T(G/C)ACCCAA(T/A)CTCCA
                      MluI   EcoRV

Cloning and colony screening

1. Ligate the appropriate dephosphorylated vector fragment with the di-   **Cloning**
   gested PCR product (see Figure 4).

2. Transform *E.coli* cells and plate on LB agar plates containing 100 mM
   glucose and 50 µg/mL ampicillin. Incubate overnight at 28-32°C to obtain small colonies.

   Note: The glucose should not be omitted since it is necessarry for the
   tight suppression of the synthetic promoter of pOPE vectors and thus
   for maintaining the stability of the insert.

3. When colonies with a diameter of about 0.5 mm have formed, put a   **Colony blot**
   nitrocellulose filter on the plate, wait a few seconds until it is entirely
   moistured.

4. Label the orientation of the filter on the agar plate by piercing a syringe
   needle dipped into indian ink through the filter into the agar at two
   different positions.

5. Use a scalpel or razor blade to cut out a section of about one fifth of the
   filter for the negative control.

6. Carefully remove both pieces of the filter with forceps, put them on
   new plates with the attached bacteria pointing upwards. Put the negative control onto selection medium with glucose, the major section
   onto selection medium containing 100 µM IPTG.

   Note: With pOPE-vectors in *E.coli* JM109, we achieved optimal protein
   secretion with 20 µM IPTG at 25°C. This optimal IPTG concentration
   can vary between different Fv-sequences by a factor of about two.
   Higher IPTG concentrations lead to higher amounts of total protein,
   but in this case most of the scFv fragments still carry the bacterial lea-

◄─────────────────────────────────────

Fig. 2. Minimal Oligonucleotide set for the amplification of mouse and rat immunoglobulin variable region DNA (according to Dübel et al., 1994). As a standard set, the combination
Bi3f + Bi4 should be used for VH and Bi8b + Bi5c for VL. If no product is found in this first
approach, other combinations can be tried. Preferentially, Bi7 should be tried instead of
Bi8b, and Bi3b/3c instead of 3f. In case internal restriction sites of enzymes essential
for cloning are present in the amplification products, they can be reamplified (not
more than 5-8 PCR cycles) with the primers containing alternative sites. Please note
that overcycling with *vent*-DNA polymerase may lead to a degradation of the correct product.

First PCR

heavy chain

gamma chain CH1 domain:(IgG)

Bi4   CCA GGG GCC AGT GGA TAG ACA AGC TTG GGT GTC GTT TT

mu chain CH1 domain:(IgM)

Bi4m   GGA GAC GAG GGG GAA AAG CTT TGG GAA GGA CTG ACT CTC

heavy chain FR1:

MHV.B1	GAT GTG AAG CTT CAG GAG TC
MHV.B2	CAG GTG CAG CTG AAG GAG TC
MHV.B3	CAG GTG CAG CTG AAG CAG TC
MHV.B4	CAG GTT ACT CTG AAA GAG TC
MHV.B5	GAG GTC CAG CTG CAA CAA TCT
MHV.B6	GAG GTC CAG CTG CAG C
MHV.B7	CAG GTC CAA CTG CAG CAG CCT
MHV.B8	GAG GTG AAG CTG GTG GAG TC
MHV.B9	GAG GTG AAG CTG GTG GAA TC
MHV.B10	GAT GTG AAC TTG GAA GTG TC
MHV.B11	GAG GTC CAG CTG CAA CAG TG
MHV.B12	GAG GTG CAG CTG GAG GAG TC

light chain

Kappa chains

kappa chain constant domain

MKC. F   GGA TAC AGT TGG TGC AGC ATC

kappa chain FR1

MKV.B1	GAT GTT TTG ATG ACC CAA ACT
MKV.B1	GAT GTT TTG ATG ACC CAA ACT
MKV.B2	GAT ATT GTG ATG ACG CAG GCT
MKV.B3	GAT ATT GTG ATA ACC CAG
MKV.B4	GAC ATT GTG CTG ACC CAA TCT
MKV.B5	GAC ATT GTG ATG ACC CAG TCT
MKV.B6	GAT ATT GTG CTA ACT CAG TCT
MKV.B7	GAT ATC CAG ATG ACA CAG ACT
MKV.B8	GAC ATC CAG CTG ACT CAG TCT
MKV.B9	CAA ATT GTT CTC ACC CAG TCT
MKV.B10	GAC ATT CTG ATG ACC CAG TCT

lambda chains

lambda chain constant domain

MLC.F GGT GAG TGT GGG AGT GGA CTT GGG CTG

lambda chain FR1 region

MLV.B CAG GCT GTT GTG ACT CAG GAA

second PCR

heavy chain

gamma chain CH1 domain:(IgG), Hind III site

Bi4   (identical to 1st PCR)

mu chain CH1 domain:(IgM), Hind III site

Bi4m   (identical to 1st PCR)

heavy chain FR1 with NcoI site:

MHV.B1.Nco   GAA TAG GCC ATG GCG GAT GTG AAG CTG CAG GAG TC
MHV.B2.Nco   GAA TAG GCC ATG GCG CAG GTG CAG CTG AAG GAG TC
MHV.B3.Nco   GAA TAG GCC ATG GCG CAG GTG CAG CTG AAG CAG TC
MHV.B4.Nco   GAA TAG GCC ATG GCG CAG GTT ACT CTG AAA GAG TC
MHV.B5.Nco   GAA TAG GCC ATG GCG GAG GTC CAG CTG CAA CAA TCT
MHV.B6.Nco   GAA TAG GCC ATG GCG GAG GTC CAG CTG CAG CAG TC
MHV.B7.Nco   GAA TAG GCC ATG GCG GAG GTC CAA CTG CAG CAG CCT
MHV.B8.Nco   GAA TAG GCC ATG GCG GAG GTG AAG CTG GTG GAG TC
MHV.B9.Nco   GAA TAG GCC ATG GCG GAG GTG AAG CTG GTG GAA TC
MHV.B10.Nco   GAA TAG GCC ATG GCG GAT GTG AAC TTG GAA GTG TC
MHV.B11.Nco   GAA TAG GCC ATG GCG GAG GTC CAG CTG CAA CAG TC
MHV.B12.Nco   GAA TAG GCC ATG GCG GAG GTG CAG CTG GAG GAG TC

light chain

Kappa chains

kappa chain constant domain with NotI site

MKC. F.Not   TGA CAA GCT TGC GGC CGC GGA TAC AGT TGG TGC AGC ATC

kappa chain FR1 with MluI site

MKV.B1.Mlu   TA CAG GAT CCA CGC GTA GAT GTT TTG ATG ACC CAA ACT
MKV.B2.Mlu   TA CAG GAT CCA CGC GTA GATATT GTG ATG ACG CAG GCT
MKV.B3.Mlu   TA CAG GAT CCA CGC GTA GAT ATT GTG ATA ACC CAG
MKV.B4.Mlu   TA CAG GAT CCA CGC GTA GAC ATT GTG CTG ACC CAA TCT
MKV.B5.Mlu   TA CAG GAT CCA CGC GTA GAC ATT GTG ATG ACC CAG TCT
MKV.B6.Mlu   TA CAG GAT CCA CGC GTA GAT ATT GTG CTA ACT CAG TCT
MKV.B7.Mlu   TA CAG GAT CCA CGC GTA GAT ATC CAG ATG ACA CA G ACT
MKV.B8.Mlu   TA CAG GAT CCA CGC GTA GAC ATC CAG CTG ACT CAG TCT
MKV.B9.Mlu   TA CAG GAT CCA CGC GTA CAA ATT GTT CTC ACC CAG TCT
MKV.B10.Mlu   TA CAG GAT CCA CGC GTA GAC ATT CTG ATG ACC CAG TCT

lambda chains

lambda chain constant domain with NotI site

MLC.F.Not   GA CAA GCT TGC GGC CGC CGGT GAG TGT GGG AGT GGA CTT
GGG CTG

lambda chain FR1 region with MluI site

MLV.B.Mlu   TA CAG GAT CCA CGC GTA CAG GCT GTT GTG ACT CAG GAA

Fig. 3. Extended oligonucleotide set for the two step amplification of mouse and rat immunoglobulin variable region DNA

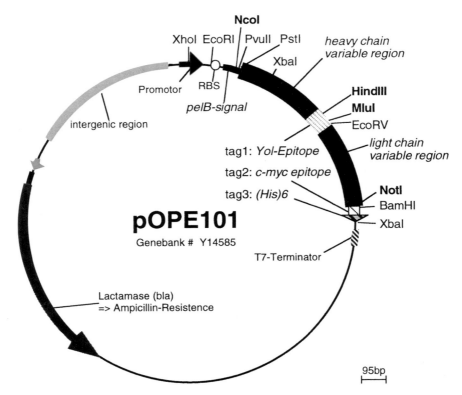

Fig. 4. The *E. coli* scFv expression vector pOPE101 (Genebank No. Y14585)

der sequence (Dübel et al., 1992) and form aggregates. However, for this analysis, it is not necessarry to discriminate between unprocessed and processed protein. Therefore, a higher IPTG concentration is used to increase the intensity of the staining.

7. Incubate for 3 h at 37°C.

8. Remove the filters from the petri dish and expose it for 15 min to chloroform vapour (in a glass chamber containing an open chloroform vessel).

Note: This step increases the staining intensity at the margin of a colony, thus improving the signal. It might be omitted in the case of strong reactions.

9. Wash the filters 2x in an excess of TNT (50ml for a filter with a diameter of 10cm) with the colonies pointing downwards. Remove the bacteria by gently rubbing the filters on the bottom of the washing vessel.

10. Wash the filters in TNT containing 0.01% NaN$_3$ (Caution: sodium azide is very toxic) with the colonies pointing downwards to kill residual bacteria.

    Note: NaN$_3$ should not be stored as a stock solution since it degrades rapidly in water.

11. Wash the filters 3x in an excess of TNT with the colonies pointing downwards.

    Note: Use a tray larger than the filter to allow it to slide a bit.

12. Block unspecific binding sites by shaking for 30 min in 1% milk powder in TNT.

    Note: Store 1% milk powder in TNT at 4°C for a maximum of 1 day, for longer storage freeze aliquots. For a filter of 10cm diameter, 4-5ml of blocking or antibody solution are usually enough to obtain an even staining. The best results are obtained with a platform shaker, which has a tilting motion in only one direction. The use of platform shakers with a tumbling movement requires larger incubation volumes to obtain an even staining. Only one filter should be used per incubation vessel.

13. Incubate in 1% milk powder /TNT containing the first antibody.

    Note: For scFv fragments cloned into pOPE51 or pOPE101, the monoclonal antibody 9E10 that binds to the internal epitope ..EQKLI-SEEDLN.. (Evan et al., 1985, commercially available from Cambridge Research Biochemicals) can be used. The recommended dilution is 1/10000. The most specific result is obtained after incubation overnight at 4°C. For most applications, however, a 1-2 h incubation at room temperature is sufficient. Alternatively, the His-tag can be utilised for a detection with Ni-NTA HRP conjugate (Qiagen, Hilden, Germany). In this case, a second antibody is not necessary; you can proceed directly with Step 17 after incubation.

14. Wash 3x for 5 min in TNT.

15. Incubate in 1% milk powder /TNT containing the second antibody.

Note: Commercially available anti-mouse IgG antisera labelled with horseradish peroxidase (HRP) usually require a dilution of 1:1000 to 1:5000 in a 1-2 h incubation.

16. Wash 3x for 5 min in TNT, and briefly in TBS to remove detergent.

17. Place the filters in substrate solution until the desired noise/signal ratio between the induced and the non-induced piece of the filter is reached.

Note: We use a diaminobenzidine / H_2O_2 - substrate system (Caution: diaminobenzidine is a cancerogene): Dilute 200 μl of diaminobenzidine stock solution (25mg/ml in water, stock solution should be stored at -20°C in aliqots for max. 6 months) in 10ml TBS containing 0.02% (w/v) $CoCl_2$ and add 1 μl of 30% (v/v) H_2O_2, use immediately. The addition of $CoCl_2$ enhances sensitivity about 30 fold. Commercially available premixed TMB substrate solutions may be used instead.

18. After sufficient substrate reaction (usually not more than 10 min), wash 3x for 5 min in tap water. Air dry for documentation.

Note: Scan or photograph within 1 day since bleaching may occur.

19. Pick a few positive colonies. Inoculate 1ml of LB_{GA} (LB medium containing 100 mM glucose and 50 μg/mL ampicillin). Grow overnight at 37°C. Add 250 μl Glycerol, mix and freeze at -30°C.

Note: The master plates should not be stored for longer than 1 day at 4°C. Cultures freshly inoculated directly from the frozen glycerol stock should be used for all subsequent experiments. Never thaw the glycerol stock; it is sufficielt to scrape a bit of ice from the surface for inoculation.

Troubleshooting

Frequently, more than one primer pair amplifies a PCR product of the correct size. The reasons are discussed in the introduction. In this case, it is recommended to pick, clone and sequence 5-10 clones of each product. In case different sequences are found, all have to be tested for function. A set of clones has to be generated containing all possible combinations of Vh and Vl regions. Alternatively, a phage display screening (panning) can be employed after cloning the PCR products directly into a phagemid surface expression vector (e.g. pSEX81, see Chapter 7) and screen for functional antibody fragments. The latter procedure is always recommended if soluble and purified antigen is available in amounts above a few microgram.

References

Breitling, F. und Dübel, S. (1999) Recombinant Antibodies. John Wiley and Sons, New York

Dübel, S., Breitling, F., Fuchs, P., Zewe, M., Gotter, S., Moldenhauer, G und Little, M. (1994) Isolation of IgG antibody Fv-DNA from various mouse and rat bybridoma cell lines using the polymerase chain reaction with a simple set of primers. J. Imm. Methods 175, 89-95

Dübel, S., Breitling, F., Klewinghaus, I., and Little, M. (1992) Regulated secretion and purification of recombinant antibodies in *E.coli*. Cell Biophys. 21, 69-79

Evan, G.I., Lewis, G.K., Ramsay, G. and Bishop, M. (1985) Isolation of monoclonal antibodies specific for human c-*myc* proto-oncogene product. Molec. and Cellular Biol. 5, 3610-3616

Fuchs, P., Breitling, F. , Little, M. und Dübel, S. (1997) Primary structure and functional scFv antibody expression of an antibody against the human protooncogen c-myc. Hybridoma 16, 227-233

Liu, Z., Schneider-Mergener, J., Martin, A., Dandekar, T., Bautz, E.K.F. und Dübel, S. (1999) Fine mapping of the antigen - antibody interaction of scFv215, a recombinant antibody inhibiting RNA polymerase II from Drosophila melanogaster. J. Mol. Recognit. 12, in press

Moosmayer, D., Dübel, S., Brocke, B., Watzka, H., Hampp, C., Scheurich, P., Little, M. und Pfizenmaier, K. (1995) A single chain TNF receptor antagonist is an effective inhibitor of TNF mediated cytotoxicity. Therap. Immunol. 2, 31-40

Sambrook, J., Fritsch, E. F. & Maniatis, T. (1989) Molecular Cloning. A Laboratory Manual, 2nd. Editon, Cold Spring Harbor Laboratory Press, Cold Spring Harbor, NY, USA

Schiweck W, Buxbaum B, Schatzlein C, Neiss HG, Skerra A (1997) Sequence analysis and bacterial production of the anti-c-myc antibody 9E10: the V(H) domain has an extended CDR-H3 and exhibits unusual solubility. FEBS Lett414:33-8

Cloning Hybridoma cDNA by RACE

ANDREW BRADBURY

Introduction

V region primer PCR is usually successful in the amplification of hybridoma V genes, especially if using diverse primer sets (Krebber et al. 1997). However, there are a number of potential pitfalls in using V region PCR. Mutations within the 5' or 3' ends of the V genes may inhibit primer annealing and so prevent amplification. Another problem is the presence of other V genes within the hybridoma which are preferentially amplified. These arise for two reasons. The first is non-productive rearrangments, which, not being mutated, are very good PCR templates, while the second, is probably caused by the fusion of more than one spleen cell to the myeloma cell line, resulting in multiple functional (as well as non-functional) V genes. In this situation, an alternative to V gene PCR is to use either traditional cDNA cloning or rapid amplification of cDNA ends (RACE) (Frohman et al. 1988). This technique relies on knowledge of a small part of gene sequence to amplify from that gene sequence to either end of the cDNA. For both cases, an oligo-dT primer containing a specific tag is used to amplify the cDNA end. In the case of the 3' end, the sequence to which it anneals is the naturally occurring poly-A tail, while in the case of the 5' end (which is that used when RACE is used to clone hybridoma V genes) a poly-A tail is added using terminal transferase. PCR specificity can be subsequently improved by using the specific tag primer and a nested sequence specific primer (Pescatore et al. 1995).

When applied to immunoglobulins, the isotype of the monoclonal to be cloned, provides the sequence knowledge, which can be used for the internal primer. This is used to create the cDNA. A poly-A tail is added to the 5' end of the cDNA using terminal transferase, and the complete V gene, including 5' untranslated region, leader sequence and a small

Andrew Bradbury, Los Alamos National Laboratory, Biosciences Division, MS-M888, Los Alamos, NM, 87545, USA (*phone* +1-505-665-0281/0287; *fax* +1-505-665-3024 /667-2891; *e-mail* amb@telomere.lanl.gov), and SISSA, Via Beirut 2-4, Trieste, 34014, Italy

part of the constant region, can then be amplified. Once amplified, the V genes can then be cloned into standard cloning vectors such as pUC, from which they may be sequenced or reamplified using V region primers for cloning into specific phage display (Hoogenboom et al. 1991) or eukaryotic antibody expression vectors (Persic et al. 1997; Persic et al. 1997). However, the annealing temperature required may be as low as 37°C if failure to amplify from total cDNA is due to mutations in the primer annealing sites (Ruberti et al. 1994). Mutations in the V genes can be avoided by using high amounts of cloned V gene (1 µg) and a DNA polymerase with proof reading activity (such as T. litoralis [Vent, New England Biolabs] or P. furiosus [Stratagene]) to reduce errors introduced by PCR.

Table 1 gives the sequences of hinge region primers of different heavy chain isotypes which can be used for cDNA synthesis. Additional restriction sites should be added to the 5' end with a nucleotide tail to allow efficient digestion. Hinge specific primers are used, as these tend to be the most isotype specific part of the constant regions genes. Nested primers which can be used are located at the 5' end of CH1 and these are also shown in Table 1. There is no equivalent to hinge region primers for the light chain. As a result, irrelevant light chains will also be amplified, although one can distinguish between mouse and rat light chains, and there are slight differences between the different isotypes. The light chain cDNA synthesis primer is found at the 3' end of the constant region and the PCR primer overlaps it with a 6 bp extension to preserve specificity.

Procedure

1. Prepare cytoplasmic mRNA from 5×10^6 hybridoma cells.

2. Synthesize cDNA[a] with the following protocol: denature 1 µg of poly(A) mRNA at 65°C for 5 min in DEPC treated water, put on ice and then add to a mixture containing 5 µl 5xRT buffer, 10 µl RNasin (Promega), 10 pmol cDNA synthesis specific primer (see Table 1), 250 µM of each of the four deoxynucleotide triphosphates (dNTPs) and 10U of Moloney murine leukemia virus reverse transcriptase in a total volume of 25 µl. The reaction mixture is incubated at 42°C for 60 min and then at 52°C for 30 min. After inactivation at 95°C for 5' the reverse transcription mixture is diluted with 2 ml of 0.1 TE (1 mM Tris pH7.0, 0.1 mM EDTA).

   Note: Although we use this protocol, which is based upon the original published method (Frohman et al. 1988), other protocols should also be effective.

Table 1. Primers for cDNA synthesis of mouse or rat immunoglobulin genes

cDNA primers specific for heavy chain isotypes (all priming in hinge)

RACEMOG1   TAT GCA AGG CTT ACA ACC ACA
(mouse IgG1)

RACEMOG2a   AGG ACA GGG CTT GAT TGT GGG
(mouse IgG2a)

RACEMOG2b   AGG ACA GGG GTT GAT TGT TGA
(mouse IgG2b)

RACEMOG3   GGG GGT ACT GGG CTT GGG TAT
(mouse IgG3)

RACERAG1   AGG CTT GCA ATC ACC TCC ACA
(rat IgG1)

RACERAG2a   ACA AGG ATT GCA TTC CCT TGG
(rat IgG2a)

RACERAG2b   GCA TTT GTG TCC AAT GCC GCC
(rat IgG2b)

RACERAG2c   TCT GGG CTT GGG TCT TCT GGG
(rat IgG2c)

Light chain primers (all prime at 3' end of CL)

CKFOR   CTC ATT CCT GTT GAA GCT CTT GAC
(mouse and rat K)

MOCKFOR   CTC ATT CCT GTT GAA GCT CTT GAC AAT
(mouse K)

RACKFOR   CTC ATT CCT GTT GAA GCT CTT GAC GAC
(rat K)

MOCKFOR and RACKFOR are identical to CKFOR except for the last three bases. If a V region from a rat mouse hybrid is to be cloned, and the mouse myeloma partner expresses a light chain V region mRNA, it can be excluded by the use of RACKFOR

CL1FOR   ACA CTC AGC ACG GGA CAA ACT CTT CTC
(mouse λ1 λ4; rat λ1)

CL2FOR   ACA CTC TGC AGG AGA CAG ACT CTT TTC
(mouse λ2, λ3; rat λ2)

These may be used individually if the lambda isotype isknown or as an equimolar mixture to prime all lambda chains.

Table 1. Continuous

RACE PCR primers

Heavy chain primers (all prime at 3' end of CH1)

MOCG12FOR   CTC AAT TTT CTT GTC CAC CTT GGT GC
(mouse IgG1, IgG2a; rat IgG1, IgG2a, IgG2b)

MOCG2bFOR   CTC AAG TTT TTT GTC CAC CGT GGT GC
(mouse IgG2b)

RACG2cFOR   CTC AAT TCT CTT GAT CAA GTT GCT TT
(rat IgG2c)

MOCG3FOR   CTC GAT TCT CTT GAT CAA CTC AGT CT
(mouse IgG3)

MOCMFOR   TGG AAT GGG CAC ATG CAG ATC TCT
(mouse IgM)

These may be used individually or as an equimolar mixture to prime all heavy chains.

Light chain primers (all prime at 3' end of CL)

CKRAsp   CTC ATT CCT GTT GAA GCT CTT GAC GAC GGG
(this is identical to CKFOR, except that at the 3' end it has 6 extra bases to increase its specificity for rat)

CKMOsp   CTC ATT CCT GTT GAA GCT CTT GAC AAT GGG
(this is identical to CKFOR, except that at the 3' end it has 6 extra bases to increase its specificity for mouse)

CL1FOR   ACA CTC AGC ACG GGA CAA ACT CTT CTC
(mouse λ1 λ4; rat λ1)

CL2FOR   ACA CTC TGC AGG AGA CAG ACT CTT TTC
(mouse λ2, λ3; rat λ2)

CL1FORsp   ACA CTC AGC ACG GGA CAA ACT CTT CTC CAC AGT
(mouse and rat λ1)

CL2FORsp   ACA CTC TGC AGG AGA CAG ACT CTT TTC CAC AGT
(mouse λ2, λ3; rat λ2)

CL4FORsp   ACA CTC AGC ACG GGA CAA ACT CTT CTC CAC ATG
(mouse λ4)

These may be used individually or in a pooled equimolar mixture. They are identical to the corresponding CLFOR primers, except that at the 3' end there are 6 extra bases to increase the specificity for each λ.

None of the constant region primers described above have restriction sites at the 5' end included for cloning. This should be inserted according to the vector to be subsequently used.

RACE PCR primer

XSCTnTag   GAC TCG AGT CGA CAT CGA TTT TTT TTT TTT TTT TT
Anneals to the poly A tail added by terminal transferase, and provides XhoI, SalI, ClaI sites at
the 5' end. This is the original primer described by Frohman et al., 1988. However, other
restriction sites can also be used.

3. Remove excess primer using a Centricon 100 spin filter (20 min at 1000 g, twice). The first retained liquid is collected and diluted to 2 ml before repeating the Centricon concentration. The second is concentrated to 10 μl and used in the following steps.

4. Synthesise a polyA tail at the 5' end of the cDNA by adding 4 μl 5X Tailing buffer (supplied by Promega with the enzyme), 4 μl dATP 1 mM and 10 U of Terminal deoxynucleotidyl transferase (Promega). The mix is incubated for 5 min at 37°C and then 5 min at 65°C. The volume of the cDNA/tailing reaction is adjusted to 500 μl.

5. Amplify 10 μl of reaction with Vent polymerase as follows: 1 precycle: 5 min 95°C, 5 min 60°C, 40 min 72°C; 40 cycles: 1 min 95°C, 1 min 60°C, 3 min 72°C. PCR is performed using the oligonucleotide XSCTnTag (Table 1), which hybridizes to the poly(A) tail added to the 5' end of cDNA and one PCR primer specific for the light or heavy chain (Table 1).

Results

After following this procedure, a single band should be obtained. This can be excised and purified from the agarose gel, digested with the appropriate enzymes and cloned into the vector of choice. Sometimes a smear may be obtained, this can be reduced by changing the PCR conditions, or can be ignored if not too strong.

Troubleshooting

We have found this method extremely successful in amplifying V regions which cannot be amplified by V region primers. As in any PCR reaction, varying the annealing temperature, the Mg concentration or the polymerase may improve the quality of the product.

In all cases, the specificity of amplified V genes should be confirmed by functional analysis of expressed antibodies or antibody fragments after expression in either bacteria or mammalian cells. If more than one V gene is obtained, both should be tested, in combination with the other V genes, to determine which is the correct combination.

References

Frohman M. A., Dush M. K. and Martin G. (1988) Rapid production of full length cDNAs from rare transcripts: Amplification using a single gene-specific oligonucleotide primer. Proc. Natl. Acad. Sci. U.S.A. 85: 8998-9002

Hoogenboom H. R., Griffiths A. D., Johnson K. S., Chiswell D. J., Hudson P. and Winter G. (1991) Multi-subunit proteins on the surface of filamentous phage: methodologies for displaying antibody (Fab) heavy and light chains. Nucl. Acids Res. 19: 4133-4137

Krebber A., Bornhauser S., Burmester J., Honeggar A., Willuda J., H.R. B. and Plückthun A. (1997) Reliable cloning of functional antibody variable domains from hybridomas and spleen cell repertoires employing a reengineered phage display system. J. Immunol. Methods. 201: 35-55

Persic L., Righi M., Roberts A., Hoogenboom H. R., Cattaneo A. and Bradbury A. (1997) Targeting vectors for intracellular immunisation. Gene 187: 1-8

Persic L., Roberts A., Wilton J., Cattaneo A., Bradbury A. and Hoogenboom H. (1997) An integrated vector system for the eukaryotic expression of antibodies or their fragments after selection from phage display libraries. Gene 187: 9-18

Pescatore M., Gargano N., Bradbury A. and Grassi A. (1995) The cloning of a cDNA encoding a protein (Latrodectin) which copurifies with the a-latrotoxin from the black widow spider Latrodectus tredecimguttatus (Theridiidae). Eur. J. Biochem. 230: 322-328.

Ruberti F., Cattaneo A. and Bradbury A. (1994) The use of the RACE method to clone hybridoma cDNA when V region primers fail. J. Imm. Methods 173: 33-39

Generation of Antibody Libraries

Phage - Display Libraries of Murine Antibody Fab Fragments

JAN ENGBERG, LISELOTTE BRIX JENSEN, ALI FAZIL YENIDUNYA, KATJA BRANDT, and ERIK RIISE

Introduction

In this chapter we describe efficient procedures for construction, expression and screening of comprehensive libraries of murine antibody Fab fragments displayed on the surface of filamentous phage. Phagemid vectors are used for placing randomly paired light (L) and heavy (H) chain coding regions under transcriptional control of P_{lac}. The L (or H) chain coding region is fused in-frame with the phage gene, $\Delta gIII$, coding for a truncated version of the phage surface protein pIII (ΔpIII). After superinfection with helper phage and induction of P_{lac}, Fd (composed of V_H and C_H1 domains) and κ L chains assemble into Fab fragments in the periplasm, and the Fab-ΔpIII protein complex is displayed at one end of the phage by displacing one (or more) of the wild-type pIII proteins. Enrichment of Fab phages with affinity for a specific antigen is then carried out by successive rounds of affinity purification using antigen-coated microtiter wells, immunotubes or plastic beads followed by reinfection of *E. coli* cells with the eluted bound phages (1-6). An outline of the method is illustrated in Figure 1.

✉ Jan Engberg, The Royal Danish School of Pharmacy, Department of Pharmacology, Universitetsparken 4, Copenhagen, 2100, Denmark (*phone* +45-3530-6384; *fax* +45-3530-6022; *e-mail* je@dfh.dk)
Liselotte Brix Jensen, The Royal Danish School of Pharmacy, Department of Pharmacology, Universitetsparken 4, Copenhagen, 2100, Denmark
Ali Fazil Yenidunya, The Royal Danish School of Pharmacy, Department of Pharmacology, Universitetsparken 4, Copenhagen, 2100, Denmark
Katja Brandt, The Royal Danish School of Pharmacy, Department of Pharmacology, Universitetsparken 4, Copenhagen, 2100, Denmark
Erik Riise, The Royal Danish School of Pharmacy, Department of Pharmacology, Universitetsparken 4, Copenhagen, 2100, Denmark

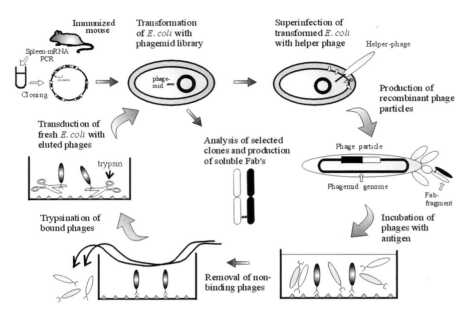

Fig. 1. Outline of the phage display system

We present protocols for making murine Fab libraries of the IgG iso-type and murine Fab libraries independently of IgH isotype. In the pro-tocol for making murine Fab libraries of the IgG isotype, Fd and κ coding regions are PCR amplified and assembled into a Fab gene cassette by the "jumping PCR" technique before being cloned (Subprotocol 1), or these coding regions are cloned separately (Subprotocol 4). In the protocol for making murine libraries independently of IgH isotype, V_H gene fragments are amplified using J region primers and joined to $C_H1_{\gamma1}$ gene fragments. In one variant of the protocol, V_H and $C_H1_{\gamma1}$ DNA is fused by PCR and linked to PCR-generated κ genes by jumping PCR before cloning (Subpro-tocol 2). In another variant, κ and V_H DNA is cloned in two steps into a specially developed vector that already contains $C_H1_{\gamma1}$ fused to Δg*III* (Sub-protocol 3). The present chapter represents an up-date of our previously published procedures (see references. 4, 7, 8 and 9).

■ Materials

Materials for preparation of total RNA, cDNA and cloning in the pFAB vectors

- Commercial kits for total RNA preparation and cDNA synthesis

- AmpliTaq polymerase, 10X Taq buffer I and II (Perkin-Elmer Cetus, Norwalk CT)

- Qiagen DNA purification columns (Qiagen GmbH, Hilden, Germany)

- Restriction enzymes *Asc*I, *Hind*III, *Kpn*I, *Nhe*I, *Not*I, *Pme*I, *Sfi*I, *Spe*I, *Srf*I and *Stu*I

- Endonuclease buffers: 10X NEB2, 3 and 4 (New England Biolabs, Beverly, MA), 10X Universal buffer (Stratagene). BSA: 10X BSA (0.1% from Amersham, Buckinghamshire, UK), 100X BSA (New England Biolabs)

- T4 DNA ligase (1 U/µl) and 5X T4 ligase buffer (Gibco/BRL)

- Glycogen (20 mg/mL from Boehringer Mannheim)

- Phagemid vectors pFab5c.His.T and pFab60 are depicted in Figures 2 and 3

- LB, 2xYT and SOC media and TE buffer are prepared as described in reference 10

- Antibiotics: ampicillin is used at a concentration of 50 µg/mL and tetracycline at 10µg/ml

- *E. coli* strains: TOP10F': *mcrA*, Δ(*mrr-hsdRMS-mcrBC*), θ80, *lacZ*ΔM15, Δ*lacX74*, *deoR*, *recA1*, *araD139*, Δ(*ara-leu*)7697, *galU*, *galK*, *rpsL*(StrR), *endA1*, *nupG*, F'{*lacI*^{q} Tn10 (TetR)} (Invitrogen). TG1: F'{*traD36*, *lacI*^{q},((*lacZ*)M15, *proA*$^+$*B*$^+$}/ ((*lac-proAB*), *thi-1*, *supE*, ((*hsdM-mcrB5*(r$_k^-$, m$_k^+$, *mcrB*$^-$)).

- Antibiotics: ampicillin is used at a concentration of 50 µg/mL and tetracycline at 10µg/mL

- PCR-primers for murine Fab libraries of the IgG isotype (see Table 1)

Table 1. PCR-primers for murine Fab libraries of the IgG isotype

Murine H-chain-variable region back primers

Var.	Name	Leader	Coding
4	MVH-1	G CCG GCC ATG GCC GAG GTR	**M**AG CTT CAG GAG TCA GGA C
4	MVH-2	G CCG GCC ATG GCC GAG GT**S**	CAG CT**K** CAG CAG TCA GGA C
4	MVH-3	G CCG GCC ATG GCC CAG GTG	CAG CTG AAG **SAS** TCA GG
2	MVH-4	G CCG GCC ATG GCC GAG GTG	CAG CTT CAG GAG TC**S** GGA C
4	MVH-5	G CCC GCC ATG GCC GA**R** GTC	CAG CTG CAA CAG TC**Y** GGA C
2	MVH-6	G CCC GCC ATG GCC GAG GTC	CAG CT**K** CAG CAA TCT GG
6	MVH-7	G CCG GCC ATG GCC CAG **STB**	CAG CTG CAG CAG TCT GG
4	MVH-8	G CCC GCC ATG GCC CAG GT**Y**	CAG CTG CAG CAG TCT GG**R** C
4	MVH-9	G CCG GCC ATG GCC GAG GT**Y**	CAG CT**Y** CAG CAG TCT GG
2	MVH-10	G CCG GCC ATG GCC GTC CA**R**	CTG CAA CAA TCT GGA CC
1	MVH-11	G CCG GCC ATG GCC CAG GTC	CAC GTG AAG CAG TCT GGG
4	MVH-12	G CCG GCC ATG GCC GAG GTG	AA**S STG** GTG GAA TCT G
6	MVH-13	G CCG GCC ATG GCC GA**V** GTG	AAG **Y**TG GTG GAG TCT G
4	MVH-14	G CCG GCC ATG GCC GAG GTG	CAG **SK**G GTG GAG TCT GGG G
4	MVH-15	G CCG GCC ATG GCC GA**K** GTG	CA**M** CTG GTG GAG TCT GGG
2	MVH-16	G CCG GCC ATG GCC GAG GTG	AAG CTG ATG GA**R** TCT GG
2	MVH-17	G CCG GCC ATG GCC GAG GTG	CA**R** CTT GTT GAG TCT GGT G
4	MVH-18	G CCG GCC ATG GCC GA**R** GT**R**	AAG CTT CTC GAG TCT GGA
4	MVH-19	G CCG GCC ATG GCC GAA GTG	AA**R STT** GAG GAG TCT GG
1	MVH-20	G CCG GCC ATG GCC GAA GTG	ATG CTG GTG GAG TCT GGG
8	MVH-21	G CCG GCC ATG GCC CAG GTT	ACT CT**R** AAA G**W**G T**S**T GGC C
6	MVH-22	G CCG GCC ATG GCC CAG GTC	CAA CT**V** CAG CA**R** CCT GG
4	MVH-23	G CCG GCC ATG GCC CAG GT**Y**	CA**R** CTG CAG CAG TCT G
1	MVH-24	G CCG GCC ATG GCC GAT GTG	AAC TTG GAA GTG TCT GG
1	MVH-25	G CCG GCC ATG GCC GAG GTG	AAG GTC ATC GAG TCT GG

| 88 | CONSENSUS: | G CCG GCC ATG GCC **SAN STN MHN** | **BKN VWN SWV YSN** G |
| | | **aa1**\|-----------------------| -----------------------> |

symbol	meaning	symbol	meaning
A	Adenine	W	A or T/U
C	Cytosine	S	C or G
G	Guanine	K	G or T/U
T	Thymine	D	A, G or T/U
U	Uracil	H	A, C or T/U
R	puRine (A or G)	V	A, C or G
Y	pYrimidine (C or T/U)	B	C, G or T/U
M	A or C	N	A, C, G or T/U

(The one-letter nucleotide symbols are used according to IUB nomenclature).

The MVH primers consist of 25 individually synthesized oligos representing 88 variants. The nucleotides starting with amino acid codon number 1 (aa1) correspond to the N-terminal part of the variable heavy (V_H) sequences. Degenerate nucleotide positions are boldfaced. Nucleotides 1-13 at the 5'-terminus correspond to the C-terminal part of the *pelB* leader. The concentration of each primer variant in the mixture used for PCR is 0.227 pmol/μL and the total concentration is 20 pmol/μL (20 μ*M*). (See Note 1.)

Table 1. Continuous

Murine C$_H$1 chain forward primer mix

MCH1-G1: CGACTAGTTTAGAATTCAAGCTGTCGAC <u>TCA</u> **ACA ATC CCT GGG CAC AAT TTT CTT GTC CACC**
MCH1-G2A: CGACTAGTTTAGAATTCAAGCTGTCGAC <u>TCA</u> **ACA GGG CTT GAT TGT GGG CCC TCT GGG**
MCH1-G2B: CGACTAGTTTAGAATTCAAGCTGTCGAC <u>TCA</u> **ACA GGG GTT GAT TGT TGA AAT GGG CCC G**
Nucleotides in bold are complementary to the sequence of the constant heavy (C$_H$1) and hinge junction regions of the γ1, γ2a and γ2b H-chain isotypes. The stop codon is underlined and the sequence positions 1 to 26 are complementary to the 5'-end of the Link-D (L-pelB) fragment, positions 1-26. The concentration of each primer in the MCH1-mix used for PCR is 6.67 μM and the total concentration is 20 μM.

Murine κ L-chain primers

Var:	Name	Coding
1	MVK-1	TTG GCT GCA CAA CCA GCA ATG GCA GAC ATT GTT CTC ACC CAG TCT CC
2	MVK-2	TTG GCT GCA CAA CCA GCA ATG GCA GAC ATT GTG CT**S** ACC CAG TCT CC
1	MVK-3	TTG GCT GCA CAA CCA GCA ATG GCA GAC ATT GTG ATG ACT CAG TCT CC
2	MVK-4	TTG GCT GCA CAA CCA GCA ATG GCA GAC ATT GTG CT**M** ACT CAG TCT CC
4	MVK-5	TTG GCT GCA CAA CCA GCA ATG GCA GAC ATT GTG **YTR** ACA CAG TCT CC
2	MVK-6	TTG GCT GCA CAA CCA GCA ATG GCA GAC ATT GT**R** ATG ACA CAG TCT CC
4	MVK-7	TTG GCT GCA CAA CCA GCA ATG GCA GAC ATT **M**AG AT**R** ACC CAG TCT CC
2	MVK-8	TTG GCT GCA CAA CCA GCA ATG GCA GAC ATT CAG ATG A**MC** CAG TCT CC
3	MVK-9	TTG GCT GCA CAA CCA GCA ATG GCA GAC ATT CAG ATG AC**D** CAG TCT CC
1	MVK-10	TTG GCT GCA CAA CCA GCA ATG GCA GAC ATT CAG ATG ACA CAG ACT AC
1	MVK-11	TTG GCT GCA CAA CCA GCA ATG GCA GAC ATT CAG ATG ATT CAG TCT CC
2	MVK-12	TTG GCT GCA CAA CCA GCA ATG GCA GAC ATT GTT CTC A**WC** CAG TCT CC
1	MVK-13	TTG GCT GCA CAA CCA GCA ATG GCA GAC ATT GTT CTC TCC CAG TCT CC
4	MVK-14	TTG GCT GCA CAA CCA GCA ATG GCA GAC ATT G**W**G CT**S** ACC CAA TCT CC
4	MVK-15	TTG GCT GCA CAA CCA GCA ATG GCA GAC ATT **S**TG ATG ACC CA**R** TCT C
4	MVK-16	TTG GCT GCA CAA CCA GCA ATG GCA GAC ATT **K**TG ATG ACC CA**R** ACT CC
1	MVK-17	TTG GCT GCA CAA CCA GCA ATG GCA GAC ATT GTG ATG ACT CAG GCT AC
3	MVK-18	TTG GCT GCA CAA CCA GCA ATG GCA GAC ATT GTG ATG AC**B** CAG GCT GC
2	MVK-19	TTG GCT GCA CAA CCA GCA ATG GCA GAC ATT GTG ACA AC**Y** CAG GAT G
1	MVK-20	TTG GCT GCA CAA CCA GCA ATG GCA GAC ATT GTG ATG ACC CAG TTT GC
1	MVK-21	TTG GCT GCA CAA CCA GCA ATG GCA GAC ATT GTG ATG ACA CAA CCT GC
1	MVK-22	TTG GCT GCA CAA CCA GCA ATG GCA GAC ATT GTG ATG ACC CAG ATT CC
1	MVK-23	TTG GCT GCA CAA CCA GCA ATG GCA GAC ATT TTG CTG ACT CAG TCT CC
1	MVK-24	TTG GCT GCA CAA CCA GCA ATG GCA GAC ATT GTA ATG ACC CAA TCT CC
1	MVK-25	TTG GCT GCA CAA CCA GCA ATG GCA GAC ATT GTG ATG ACC CAC ACT CC
50	CONS:	TTG GCT GCA CAA CCA GCA ATG GCA GAC ATT **NWD HTV WHN** CAV **NHT V**C

```
                                            |------------------------>
                                                       aa1
```

These primers consist of 25 individually synthesized oligos representing 50 variants. The concentration of each variant primer in the mixture used for PCR is 0.4 pmol/μL and the total concentration is 20 pmol/μL (20 μM). Nucleotides in bold correspond to the 5'-end of the κ chain sequences starting with aa1. Degenerate nucleotide positions are boldfaced. The N-terminal two amino acids of the V region are invariant. Sequence positions 1-24, corresponding to the 3'-end of the *pelB* leader, overlap the Link-D (L-pelB) fragment, positions 94-117.

Murine C$_K$ chain forward primer

MCK1: <u>TGC GGC CGC</u> **ACA CTC ATT CCT GTT GAA GCT CTT GAC**

Table 1. Continuous

The sequence in bold is complementary to the 3'-end of the Cκ gene. The *Not*I recognition site is underlined.

Extension (linker assembly) primers for Fd chains

Tag.Back1: CA GTC ACA GAT CCT CGC GAA TTG <u>GCC CA</u>**G CCG GCC** ATG GCC SAN G
Tag.Back2: CA GTC ACA GAT CCT CGC GAA TTG <u>GCC CA</u>**G CCG GCC** ATG GCC SAN C

Nucleotides in bold overlap with the MVH primers. The *Sfi*I recognition site is underlined.

The concentration of each primer variant in the PCR solution is 1.25 μM.

Link.For: GTC TGC CAT TGC TGG TTG TGC AGC CAA

This sequence is complementary to the 3'-end of the Link-D fragment. (See Note 2.)

Extension (linker assembly) primers for κ chains

Link.Back: CGA CAG CTT GAA TTC TAA ACT AGT CGA AGG CGC GCC AAG GAG CA GTC AT

This sequence overlaps the 5'-end of the Link-D fragment.

Tag.For: CAG TCA CAG ATC CTC GCG AAT TGG **TGC GGC CGC** ACA CTC ATT CCT G

The sequence in bold corresponds to that of MCK1, positions 1-22. The *Not*I site is underlined.

The Link-D (L-*pelB*) fragment

CGACAGCTTGAATTCTAAACTAGTCGAAGGCGCGCC<u>AAGGAG</u>ACAGTCATA <u>ATG</u> AAA TAC CTA TTG CCT ACG GCA GCC GCT GGA TTG TTA TTA **TTG GCT GCA CAA CCA GCA ATG GCA**

The bold-faced sequence at the 5'-end overlaps with the MCH1 primers, whereas the sequence in bold at the 3'-end is complementary to the MVK primers. The ribosome-binding site and the ATG triplet marking the start of the *pelB* leader sequence is underlined. The Link-D (L-*pelB*) fragment originates from the λc2 vector (referenced in 4) and was taken through several PCR cloning steps using different sets of tagged primers in order to introduce the sequence changes necessary for the present assembly system. The Link-D fragment was cloned in the Bluescript KS+ vector giving rise to pLink-D as described in reference (4) (See Note 3).

Assembly primer

Assembly: CA GTC ACA GAT CCT CGC GAA TTG G

This sequence is complementary to the 5'-end of Tag.For and Tag.Back. The assembly primer solution used for PCR is 5 μM.

Subprotocol 1
PCR-assembly Method for Constructing Murine Antibody Fab Libraries of the IgG isotype

For an overview of Subprotocol 1 see Figure 2.

■ ■ Procedure

Extraction of total RNA from spleen material of mice and synthesis of cDNA is done by using commercially available kits. The resulting cDNA is resuspended in sterile water at a concentration of 0.1 to 1 µg/µl.

The PCR-assembly method for making murine IgG antibody Fab libraries consists of five steps:

- Primary amplification of the Fd and κ L-chain genes.

- PCR assembly of each of the primary PCR fragments with Link-D.

- Final PCR assembly of pairs of Fd/Link-D and Link-D/L-chain fragments.

- Cloning of the final PCR product into the expression vector pFAB5-c.His.T.

- Electroporation, growth and storage of libraries (described in other chapters).

PCR reactions are hot-started and carried out in PCR tubes. All PCR reaction mixtures (100 µl) are covered with paraffin oil or run in a hot-lid apparatus.

1. Primary amplification of Fd gene fragments:

cDNA (0.1 to 1 µg/µl)	5 µl
10xTaq Buffer I	10 µl
10xdNTP (1.25 mM each)	8 µl
MVH1-25 (20 µM mix)	1 µl
MCH-mix (20 µM mix)	1 µl
Sterile water	75 µl

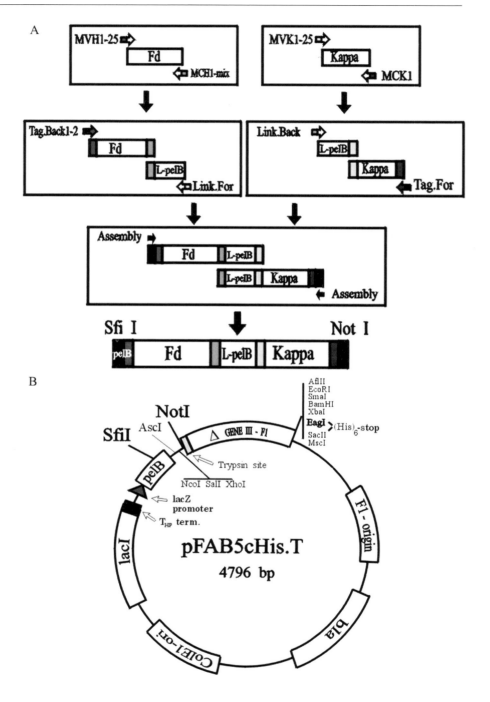

2. Primary amplification of L-chain genes:

cDNA (0.1 to 1 µg/µl)	5 µl
10xTaq Buffer I	10 µl
10xdNTP (1.25 mM each)	8 µl
MVK1-25 (20 µM mix)	1 µl
MCK1 (20 µM)	1 µl
Sterile water	75 µl

3. Amplification of the Link-D fragment:

plasmid pLink-D DNA (1 ng/µl)	1 µl
10xTaq Buffer I	10 µl
10xdNTP (1.25 mM each)	8 µl
Link.Back (20 µM)	1 µl
Link.For (20 µM)	1 µl
Sterile water	79 µl

◄────────────────────────────

Fig. 2. Overview of the primers and steps involved in PCR assembly of murine Fab gene fragments of the IgG isotype and a map of expression vector pFab5c.His.T. (**A**) Strategy for "Jumping-PCR" assembly. Diagram exemplifying the primary PCR amplification, the linker assembly and the final assembly; respectively. The boxes show different gene-segments included in the construction of the Fab expression cassette: Fd, includes the H chain from the N-terminal amino acid to the cysteine residue of the hinge region, which forms the disulphide bridge to the C-terminal cysteine of the kappa L chain; kappa corresponds to the entire variable and constant parts of the light chain; L-pelB = Link-D, a 117-bp fragment containing a translational stop codon for Fd translation, a ribosome binding site for L-chain expression and the coding region corresponding to the N-terminal part of the *pel*B leader. Primers depicted above boxes are back primers and identical to the DNA top strand, while those below are forward primers and complementary to the top strand. Shaded zones within boxes represent regions of overlapping complementarity. (**B**) pFab5c.His.T. Phagemid vector used for cloning and expression of murine antibody Fab gene fragments. The Fab antibody DNA cassette is introduced as a *Sfi*I-*Not*I fragment and is expressed from the inducible P_{lac} promoter. The pFAB5c.His.T vector contains several improvements to the pFAB5c vector described by us (4). In the polylinker region downstream of ΔgIII, an adaptor was introduced that contained an *Eag*I site followed by six histidine codons and a stop codon. This makes it possible to delete ΔgIII by digestion with *Eag*I, (*Eag*I recognizes a subset of the *Not*I recognition sequence present upstream of ΔgIII) and religation, which also regenerates a *Not*I site. Additionally, $(CAT)_6TGA$ is in reading frame with CK, and the $(His)_6$ tail facilitates easy purification of free Fab molecules on metal affinity columns (7). Finally, the transcription terminator T_{HP} (11) was introduced in front of the P_{lac} promoter.

4. PCR program: 94°C 5 min, suspend, add 0.3 µl AmpliTaq polymerase
   (5 U/µl); 94°C 1 min, 55°C 1 min, 72°C 1 min, 30 cycles, 72°C 10 min,
   refrigerate.

5. Gel-purify the primary amplification products as in steps 6-7 below
   (see Note 6). Fd gene fragments are about 720 bp, L-chain genes about
   650 bp and the Link-D fragment is 117 bp.

6. Both untreated and digested PCR products (and digested vector DNA)
   are purified on standard agarose gels and excised bands are purified
   using the Qiagen™ gel extraction kit according to the manufacturers
   recommendations.

7. Vector DNA (~5 kb) is purified on 1%, assembled Fd/L-chain DNA on
   1.5 %, Fd and L-chain DNA on 1.5% to 2%, and C_H1/Link-D, V_H and
   Link-D DNA on 2% TAE agarose gels.

8. Assembly of Fd and Link-D DNA (see Note 7):

1 to 50 ng purified Fd DNA	5.00 µl
1 to 20 ng purified Link-D DNA	2.00 µl
10xTaq Buffer I	10.00 µl
10xdNTP (1.25 mM each)	8.00 µl
Tag.Back1-2 (20 µM mix)	1.25 µl
Link.For (20 µM)	1.25 µl
Sterile water	72.50 µl

9. Assembly of L-chain and Link-D DNA:

about 5 ng purified L-chain DNA	5.00 µl
about 1 ng purified Link-D DNA	2.00 µl
10xTaq Buffer I	10.00 µl
10xdNTP (1.25 mM each)	8.00 µl
Link.Back (20 Mm)	1.25 µl
Tag.For (20 µM mix)	1.25 µl
Sterile water	72.50 µl

10. PCR program: 94°C 5 min, suspend, add 0.3 µl AmpliTaq polymerase
    (5 U/µl); 94°C 1 min, 65°C 1 min, 72°C 1 min, 25 cycles, 72°C 10 min,
    refrigerate.

11. Process the assembled Fd/Link-D and Link-D/L-chain fragments, which have increased about 100 bp in size, as described in steps 6-7 above.

12. Final assembly of Fd/Link-D and Link-D/L-chain DNA. It is important to use a relatively low concentration of assembly primer (10 pmol/100 μl) in this reaction:

1 ng purified Fd-Link-D DNA	5 μl
1 ng purified L-chain-Link-D DNA	5 μl
10xTaq Buffer I (Cetus)	10 μl
10xdNTP (1.25 mM each)	8 μl
Assembly (5 μM)	2 μl
Sterile water	70 μl

13. PCR program: 94°C 10 min, suspend, add 0.3 μl AmpliTaq polymerase (5 U/μl); 94°C 2 min, 72°C 2 min, 25 cycles, 72°C 10 min, refrigerate.

14. Treat the assembled 1.6 kb Fd/Link-D/L-chain fragments as in steps 6-7 above.

15. Digest purified PCR fragments with a 50-100 times surplus and vector DNA with about 10 times surplus of the appropriate enzymes for 2-4 h in buffers from New England Biolab at the recommended temperatures.

16. Mix in a microcentrifuge tube and incubate at 50°C for 3 h:

about 0.5 μg purified final-assembly DNA	37.5 μl
10X NEB2	5.0 μl
10X BSA	5.0 μl
*Sfi*I (50 units)	2.5 μl
Sterile water	7.5 μl

17. Add 3 μl *Not*I (25 units) and incubate at 37°C for 3 more h.

18. Gel-purify the DNA as in steps 6-7 above. The agarose gel electrophoresis combined with the Quiagen™ gel extraction step eliminates the small *Not*I and *Sfi*I end fragments. Dissolve the purified fragments in TE buffer at a final concentration of 50 ng/μl.

19. Make pFAB5c.His.T DNA from 150 mL cells grown overnight in rich medium, using a Qiagen maxiprep column according to the manufacturers guidelines. A portion is digested with *Sfi*I and *Not*I and processed as described in steps 15-18 above. Digest further with one of the enzymes, whose recognition sequence is present in the stuffer fragment to be excised (see Figure 2 and Note 8).

20. For ligation of insert to vector, mix in a microfuge tube and incubate overnight at 16°C:

insert DNA from step18 above (50 ng/μl)	6.0 μl
vector DNA from step 19 above (50 ng/μl)	6.0 μl
5xT$_4$ ligase buffer	10.0 μl
Sterile water	25.5 μl
T$_4$ DNA ligase (2.5 units)	2.5 μl

21. Add 100 μl TE buffer to the ligation mixture and extract once with phenol/chloroform/isoamyl alcohol (25:24:1). Add 0.1 vol 3 M NaOAc, pH 5.2 and 1 μl glycogen (see Note 9), mix, precipitate with 2-2.5 vol ethanol, and, depending on DNA concentration, centrifuge for 10-30 min at 15,000g in a microfuge. Rinse pellet with 70% alcohol 3-4 times to remove as much salt as possible, dry lightly in a SpeedVac centrifuge and dissolve the DNA in water at a concentration of 10-20 ng/μl.

22. Proceed with electroporations to produce the final IgG Fab library.

Comments

See Notes in Subprotocol 2.

Subprotocol 2
PCR-Assembly Method for Constructing Murine Antibody Fab Libraries Independent of IgH Isotype

An overview of the method is shown in Figure 3.

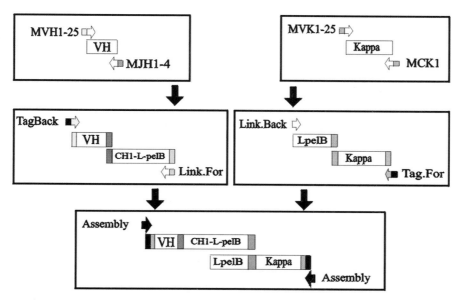

Fig. 3. Overview of primers and steps involved in the PCR assembly of murine Fab gene fragments that are independent of IgH isotype. CH1 = the first constant domain of the H-chain IgG1 isotype; CH1-L-pelB = CH1-LINK-D. (Compare to legend of Figure 2)

Materials

For PCR-primers see Table 2.

The remaining necessary primers are described in Subprotocol 1.

Table 2. PCR primers

PCR primers:

Murine J region forward primers:

> **MJH1** : *GGG GGA GTA GTC TTA* **GCT** ~~GAG~~ **GAG GAG ACG GTG ACC GTG G**
> **MJH2** : *GGG GGA GTA GTC TTA* **GCT GAG GAG ACT GTG AGA GTG G**
> **MJH3** : *GGG GGA GTA GTC TTA* **GCT GCA GAG ACA GTG ACC AGA G**
> **MJH4** : *GGG GGA GTA GTC TTA* **GCT GAG GAG ACG GTG ACT GAG G**

> Nucleotides in bold are complementary to the sequence of the J region of all known murine immunoglobulin genes. Italicized nucleotides are complementary to the 5'-end of the C_H1 gene fragment. The concentration of each primer in the mixture used for PCR is 5 µ*M*.

C_H1-Link-D (C_H1-L-pelB) back primers:
CH1-Link-D-Back: **GTC TCY KCA GCT** AAG ACT ACT CCC <u>CCG AG</u>C GTC TAT CCA CTG GCC CC

CH1-Link-D-Back is complementary to the 5'-end of the CH1-Link-D fragment and is degenerate at the underlined positions in order to match all MJH primers (see entry above). The concentration of the primer used for PCR is 20 µ*M*. Plasmid pCH1-Link-D contains the $C_H1_{\gamma1}$ region upstream of the Link-D fragment.

▪ ▪ Procedure

The PCR-assembly method (see Notes 4 and 5) for making murine antibody Fab libraries independent of the IgH isotype consists of five steps:

- Primary amplification of V_H and κ L-chain genes.
- PCR assembly of V_H and C_H1-Link-D fragments and of L-chain and Link-D fragments.
- Final PCR assembly of pairs of V_H/C_H1-Link-D and Link-D/L-chain fragments.
- Cloning of the final PCR products into pFab5c.His.T (Note 10 and reference 8).
- Electroporation, growth and storage of the library.

PCRs are hot-started and carried out in PCR tubes. All PCR mixtures (100 µl) are covered with paraffin oil. Plasmids pLink-D and pCH1-Link-D are available from the author.

1. Primary amplification of V_H gene fragments:

cDNA (0.1 to 1 μg/μl)	5.0 μl
10xTaq Buffer I	10.0 μl
10xdNTP (1.25 mM each)	8.0 μl
MVH1-25 (20 μM mix)	2.5 μl
MJH1-4 (20 μM mix)	2.5 μl
Sterile water	72.0 μl

2. Primary amplification of κ L-chain genes:

cDNA (0.1 to 1 μg/μl)	5.0 μl
10xTaq Buffer I	10.0 μl
10xdNTP (1.25 mM each)	8.0 μl
MVK1-25 (20 μM mix)	2.5 μl
MCK1 (20 μM)	2.5 μl
Sterile water	72.0 μl

3. Amplification of the LinkD fragment:

plasmid pLink-D DNA (1 ng/μl)	1 μl
10xTaq Buffer I	10 μl
10xdNTP (1.25 mM each)	8 μl
Link.Back (20 μM)	1 μl
Link.For (20 μM)	1 μl
Sterile water	79 μl

4. Amplification of the C_H1-Link-D fragment:

plasmid pCH1-Link-D DNA (1 ng/μl)	1 μl
10xTaq Buffer I	10 μl
10xdNTP (1.25 mM each)	8 μl
CH1-Link-D.Back (20 μM)	1 μl
Link.For (20 μM)	1 μl
Sterile water	79 μl

PCR program: 94°C 5 min, suspend, add 0.3 μl AmpliTaq polymerase (5 U/μl); 94°C 1 min, 55°C 1 min, 72°C 1 min, 30 cycles, 72°C 10 min, refrigerate. The C_H1-Link-D fragment is about 440 bp.

5. Gel-purify the primary amplification products as described in steps 6-7, Subprotocol 1.

6. Assembly of V_H gene fragments and the C_H1-Link-D fragment (see step 8 in Subprotocol 1 and Notes 7 and 11 below):

10 ng purified V_H DNA	5 µl
50 ng purified C_H1-Link-D DNA	5 µl
10xTaq Buffer I	10 µl
10xdNTP (1.25 mM each)	8 µl
Tag.Back1-2 (20 µM mix)	1 µl
Link.For (20 µM)	1 µl
Sterile water	70 µl

7. Assembly of L-chain gene fragments and the Link-D fragment:

10 ng purified L-chain DNA	5 µl
1 ng purified Link-D DNA	2 µl
10xTaq Buffer I	10 µl
10xdNTP (1.25 mM each)	8 µl
Link.Back (20 µM)	1 µl
Tag.For (20 µM)	1 µl
Sterile water	73 µl

8. PCR program: 94°C 5 min, suspend, add 0.3 µl AmpliTaq polymerase (5 U/µl); 94°C 1.0 min, 65°C 1.0 min, 72°C 1.0 min, 25 cycles, 72°C 10 min, refrigerate.

9. Gel-purify the assembled V_H/C_H1-Link-D and Link-D/L-chain fragments as in steps 6 and 7, Subprotocol 1.

10. Final assembly of V_H/C_H1-Link-D and Link-D/L-chain DNA. It is important to use a relatively low concentration of assembly primer (10 pmol/100 µl) in this reaction:

5 ng purified V_H/C_H1-Link-D DNA	5 µl
5 ng purified Link-D/L-chain DNA	5 µl
10xTaq Buffer I	10 µl
10xdNTP (1.25 mM each)	8 µl
Assembly (5 µM)	2 µl
Sterile water	70 µl

11. PCR program: 94°C 10 min, suspend, add 0.3 μl AmpliTaq polymerase (5 U/μl), 94°C 2 min, 72°C 2 min, 25 cycles, 72°C 10 min, refrigerate.

12. Gel-purify the assembled V_H/C_H1-Link-D/L-chain fragments (about 1.6 kb in size) as in steps 6 and 7, Subprotocol 1.

13. Continue with steps 15 to 22, Subprotocol 1 followed by electroporations to produce the IgH-independent Fab library.

Comments

Notes to Subprotocol 1 and 2

Note 1: Only high quality preparations that predominantly contain full-size product were used.

The rationale behind the design of our rather extensive series of PCR primers for the variable H and L chains has already been presented (4). In short, we believe that the primer sets used should match all available sequence data for these regions and we argued that highly degenerate primers for the variable regions are likely to generate biased libraries and to introduce amino acids not normally found in the variable regions. Ideally, separate PCR reactions should be performed with each individual VH primer and the MJH1-4 mix or MCH1-mix as well as with each individual VK primer and MCK1 followed by mixing of the individual reactions before proceeding to the subsequent PCR assembly reaction.

Note 2: To ensure efficient digestion with restriction enzymes, we generally equip final PCR products with long flanking regions (20 or more nucleotides between enzyme recognition sequence and ends of fragments).

Note 3: Third base substitutions starting at codon no 15 of the pelB leader sequence were introduced to reduce the chance of in vivo recombination between the upstream and the downsteam pelB sequences. Also in Link.D, an *Asc*I (including a *Bss*HII site) was introduced between the *Eco*I-site and the RBS-site as well as a AA-dinucleotide to accommodate non-templated A residues during the PCR assembly.

Note 4: The main advantage of the method outlined here and in the next subprotocol is that the entire cloning procedure requires the use of only two rare-cutting restriction endonucleases (*Sfi*I and *Not*I).

Note 5: Although the PCR assembly strategy used in this and the next subprotocol has been described as somewhat difficult (12), we find it attractive to have available complementary procedures for cloning of Fab gene fragments.

Note 6: If the yield of the primary PCR reactions is unsatisfactory, check the integrity of the RNA preparation by running a formaldehyde-agarose gel (10). If the bands representing rRNA look distinct and are of the correct molecular size, assume that the mRNAs for H and L chain are intact too. We recommend repeating the phenol/chloroform extractions involved in the preparation of RNA and cDNA to improve the success of the PCR reactions.

Note 7: This reaction sometimes needs optimization which is done by varying the relative concentrations of the two templates. We normally obtain the best results by using a 10 times molar excess of the Link-D fragment to the Fd gene fragment.

Note 8: As depicted in Figs. 2 and 4, the stuffer fragments between pairs of cloning sites in the murine vectors contain eight-bp recognition sequences for restriction enzymes that occur extremely rarely in murine Fab DNA. Self-ligation of the fraction of vector molecules only cut once can be reduced by subsequent digestion with the enzyme, whose recognition sequence is present in the stuffer fragment to be excised. Vectors prepared for ligation with inserts should always be checked for self-ligation as a measure of the extent of cleavage by the relevant restriction enzymes.

Although phage particles without Fab molecules displayed on their surface theoretically should be lost in the panning procedure, some unspecific binding of such phages to the antigen coat of the microtiter plate will occur. This may lead to an outgrowth of the specific binders because of the growth advantage of phages having a smaller genome size and not having to express large quantities of two foreign proteins. This can be prevented by gel-purifying the entire library as phagemid DNA before each round of panning. (unpublished results).

Note 9: Addition of (inert) glycogen to a standard ethanol precipitation, ensures quantitative recovery and that the pellet is always visible.

Note 10: The reasons for choosing the outlined approach rather than attempting to use Fd gene fragments made by MHV and C_H primers of all known isotypes or making single-chain F_V antibody fragments are detailed in reference 8.

Note 11: Successful assembly of VH and CH1-LINK-D PCR products requires a larger amount of template than does assembly of κ and LINK-D products (see step 8 in Subprotocol 1 and Note 4 in Subprotocol 2). The sluggishness of VH/CH1-LINK-D assembly may be caused by formation of strong secondary structures within the C_H1 gene fragment, which would also account for the relatively weaker amplification of Fd genes as compared to the V_H gene fragments observed previously (4).

Subprotocol 3
Direct Cloning Method for Constructing Murine Antibody Fab Libraries Independent of the IgH

For an overview see Figure 4.

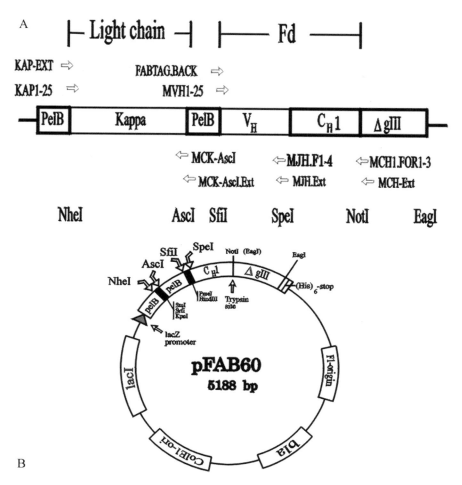

Fig. 4. Overview of the primers used to make murine V and Fd-region H-chains and κ L-chain DNA for separate cloning, and a map of expression vector pFab60, which is described in detail in reference (7). (**A**) Oligonucleotide primers for PCR production of Fab gene fragments. Primers above boxes are identical to the DNA top strand, while those below are complementary. Heavy boxes indicate genetic elements present in the vector; light boxes, elements to be cloned. The entire Fd gene fragment may be cloned between the *Sfi*I and *Not*I sites. Important restriction enzyme sites are indicated. (**B**) Phagemid, pFab60, used for cloning and expression of murine antibody Fab gene fragments. Relevant genetic elements are boxed. Black boxes indicate stuffer DNA between cloning sites: Not all unique restriction sites are indicated.

Materials

PCR-primers for direct cloning of murine antibody Fab libraries independent of IgH isotype are listed in Table 3.

Procedure

The direct cloning method (see Note 12) for making murine antibody Fab libraries independent of IgH isotype consists of five steps:

- Primary PCR amplifications of V_H gene fragments and κ genes.
- "Tagging" the primary amplification products with restriction enzyme recognition sequences (extension).
- Cloning of κ genes in pFab60 (equal to the pFab60/κ library).
- Cloning of V_H gene fragments using pFab60/κ DNA (see Note 13).
- Electroporation, growth and storage of the library.

Figure 4 shows an overview of the PCR primers, the steps of thePCR assembly procedure and the expression vector pFab60.

PCRs are hot-started and carried out in PCR tubes. All PCR mixtures (50 or 100 μl) are covered with paraffin oil.

1. Primary amplification of κ genes:

c-DNA (0.1 to 1 μg/μl)	5.0 μl
10 x Taq buffer II	10.0 μl
MgCl$_2$ (25 mM)	6.0 μl
dNTP (1.25 mM each)	8.0 μl
KAP1-25 (20μM)	1.0 μl
MCK-AscI (20 μM)	1.0 μl
H$_2$O	69.0 μl

Table 3. PCR-primers for direct cloning of murine antibody Fab libraries independent of IgH isotype

Murine κ L-chain-variable (V$_\kappa$) back primers

Name	partial pelB-leader	coding
KAP-1	CT GCA CAA CCA GCA ATG GCA GAC ATT	GTT CTC ACC CAG TCT CC
KAP-2	CT GCA CAA CCA GCA ATG GCA GAC ATT	GTG CTS ACC CAG TCT CC
KAP-3	CT GCA CAA CCA GCA ATG GCA GAC ATT	GTG ATG ACT CAG TCT CC
KAP-4	CT GCA CAA CCA GCA ATG GCA GAC ATT	GTG CTM ACT CAG TCT CC
KAP-5	CT GCA CAA CCA GCA ATG GCA GAC ATT	GTG YTR ACA CAG TCT CC
KAP-6	CT GCA CAA CCA GCA ATG GCA GAC ATT	GTR ATG ACA CAG TCT CC
KAP-7	CT GCA CAA CCA GCA ATG GCA GAC ATT	MAG ATR ACC CAG TCT CC
KAP-8	CT GCA CAA CCA GCA ATG GCA GAC ATT	CAG ATG AMC CAG TCT CC
KAP-9	CT GCA CAA CCA GCA ATG GCA GAC ATT	CAG ATG ACD CAG TCT CC
KAP-10	CT GCA CAA CCA GCA ATG GCA GAC ATT	CAG ATG ACA CAG ACT AC
KAP-11	CT GCA CAA CCA GCA ATG GCA GAC ATT	CAG ATG ATT CAG TCT CC
KAP-12	CT GCA CAA CCA GCA ATG GCA GAC ATT	GTT CTC AWC CAG TCT CC
KAP-13	CT GCA CAA CCA GCA ATG GCA GAC ATT	GTT CTC TCC CAG TCT CC
KAP-14	CT GCA CAA CCA GCA ATG GCA GAC ATT	GWG CTS ACC CAA TCT CC
KAP-15	CT GCA CAA CCA GCA ATG GCA GAC ATT	STG ATG ACC CAR TCT C
KAP-16	CT GCA CAA CCA GCA ATG GCA GAC ATT	KTG ATG ACC CAR ACT CC
KAP-17	CT GCA CAA CCA GCA ATG GCA GAC ATT	GTG ATG ACT CAG GCT AC
KAP-18	CT GCA CAA CCA GCA ATG GCA GAC ATT	GTG ATG ACB CAG GCT GC
KAP-19	CT GCA CAA CCA GCA ATG GCA GAC ATT	GTG ATA ACY CAG GAT G
KAP-20	CT GCA CAA CCA GCA ATG GCA GAC ATT	GTG ATG ACC CAG TTT GC
KAP-21	CT GCA CAA CCA GCA ATG GCA GAC ATT	GTG ATG ACA CAA CCT GC
KAP-22	CT GCA CAA CCA GCA ATG GCA GAC ATT	GTG ATG ACC CAG ATT CC
KAP-23	CT GCA CAA CCA GCA ATG GCA GAC ATT	TTG CTG ACT CAG TCT CC
KAP-24	CT GCA CAA CCA GCA ATG GCA GAC ATT	GTA ATG ACC CAA TCT CC
KAP-25	CT GCA CAA CCA GCA ATG GCA GAC ATT	GTG ATG ACC CAC ACT CC

```
                                         |--------------------------->
                                         aa1
```

KAP1-25, are identical to MVK1-25 described in Subprotocol 1, except that the KAP primers are four nucleotides shorter at their 5'-end.

L-chain-constant region (C$_\kappa$) forward primer
MCK-AscI: CTTCTCGAGG CGCGCCTCATCA **ACACTCAT TCCTGTTGAA GCTCTTG**
MCK-AscI adds on the recognition sequence for *AscI* (underlined) and tandem stop codons (small capitals) and has a 25 nucleotides (boldfaced) overlap with C$_\kappa$.

Murine L-chain extension primers
KAP.Ext: TACTACTACT ACTAGCTAGC TGCACAACCA GCAATGGCAG AC
KAP.Ext adds on the *NheI* recognition site (underlined) and has an overlap of 23 nt (italicized) with KAP1-25.
MCK-AscI.Ext: TACTACTACT CTTCTCGAGG CGCGCCTCATCAAC
MCK-AscI.Ext has a 24 nucleotides (italicized) overlap with MCK-AscI.

Murine H-chain-variable region back primers

Var.	Name	Leader	Coding
4	MVH-1	G CCG GCC ATG GCC GAG GTR MAG CTT CAG GAG TCA GGA C
4	MVH-2	G CCG GCC ATG GCC GAG GTS CAG CTK CAG CAG TCA GGA C
4	MVH-3	G CCG GCC ATG GCC CAG GTG CAG CTG AAG SAS TCA GG

Table 3. Continuous

```
2       MVH-4   G CCG GCC ATG GCC GAG GTG CAG CTT CAG GAG TCS GGA C
4       MVH-5   G CCC GCC ATG GCC GAR GTC CAG CTG CAA CAG TCY GGA C
2       MVH-6   G CCC GCC ATG GCC CAG GTC CAG CTK CAG CAA TCT GG
6       MVH-7   G CCG GCC ATG GCC CAG STB CAG CTG CAG CAG TCT GG
4       MVH-8   G CCC GCC ATG GCC CAG GTY CAG CTG CAG CAG TCT GGR C
4       MVH-9   G CCG GCC ATG GCC GAG GTY CAG CTY CAG CAG TCT GG
2       MVH-10  G CCG GCC ATG GCC GAG GTC CAR CTG CAA CAA TCT GGA CC
1       MVH-11  G CCG GCC ATG GCC CAG GTC CAC GTG AAG CAG TCT GGG
4       MVH-12  G CCG GCC ATG GCC GAG GTG AAS STG GTG GAA TCT G
6       MVH-13  G CCG GCC ATG GCC GAV GTG AAG YTG GTG GAG TCT G
4       MVH-14  G CCG GCC ATG GCC GAG GTG CAG SKG GTG GAG TCT GGG G
4       MVH-15  G CCG GCC ATG GCC GAK GTG CAM CTG GTG GAG TCT GGG
2       MVH-16  G CCG GCC ATG GCC GAG GTG AAG CTG ATG GAR TCT GG
2       MVH-17  G CCG GCC ATG GCC GAG GTG CAR CTT GTT GAG TCT GGT G
4       MVH-18  G CCG GCC ATG GCC GAR GTR AAG CTT CTC GAG TCT GGA
4       MVH-19  G CCG GCC ATG GCC GAA GTG AAR STT GAG GAG TCT GG
1       MVH-20  G CCG GCC ATG GCC GAA GTG ATG CTG GTG GAG TCT GGG
8       MVH-21  G CCG GCC ATG GCC CAG GTT ACT CTR AAA GWG TST GGC C
6       MVH-22  G CCG GCC ATG GCC CAG GTC CAA CTV CAG CAR CCT GG
4       MVH-23  G CCG GCC ATG GCC CAG GTY CAR CTG CAG CAG TCT G
1       MVH-24  G CCG GCC ATG GCC GAT GTG AAC TTG GAA GTG TCT GG
1       MVH-25  G CCG GCC ATG GCC GAG GTG AAG GTC ATC GAG TCT GG
```

```
88      CONSENSUS:  G CCG GCC ATG GCC SAN STN MHN BKN VWN SWV YSN G
                                       |------------------------------->
                                       aa1
```

MVH1-25, are described in Subprotocol 1.

Murine J forward primers

```
MJH.F1:  5´ ATCATCATCC ACTAGTTTTG GCTGAGGAGA CGGTGACCGT GG
MJH.F2:  5´ ATCATCATCC ACTAGTTTTG GCTGAGGAGA CTGTGAGAGT GG
MJH.F3:  5´ ATCATCATCC ACTAGTTTTG GCTGCAGAGA CAGTGACCAG AG
MJH.F4:  5´ ATCATCATCC ACTAGTTTTG GCTGAGGAGA CGGTGACTGA GG
```
26 nucleotides (boldfaced) of the MJH primers are complementary to the J and C$_H$1 spanning region. The *Spe*I recognition sequence is underlined. The concentration of each primer variant used for PCR is 5 µM.

Murine V$_H$ extension primers, Tag.Back1-2 and MJH.Ext

Tag.Back1: CA GTC ACA GAT CCT CGC GAA TTG GCC CAG CCG GCC ATG GCC SAN G
Tag.Back2: CA GTC ACA GAT CCT CGC GAA TTG GCC CAG CCG GCC ATG GCC SAN C

Nucleotides in bold overlap with the MVH primers. The *Sfi*I recognition site is underlined. The concentration of each primer variant in the PCR solution is 1.25 µM.

MJH.Ext: CAGGCGCCAG TGGATAGAATCATCATCCAC   TAGTTTTGGC T

MJH.Ext and the MJH.For primers overlap by 23-26 nt (italicized).

2. Primary amplification of V_H gene fragments:

c-DNA (0.1 to 1 μg/μl)	5.0 μl
10 x Taq buffer II	10.0 μl
$MgCl_2$ (25 mM)	6.0 μl
dNTP (1.25 mM each)	8.0 μl
MVH1-25 (20 μM)	1.0 μl
MJH.F1-4 (20 μM)	1.0 μl
H_2O	69.0 μl

3. PCR program: 94°C 10 min, suspend, add 0.3 μl AmpliTaq polymerase (5 U/μl); 94°C 1 min, 55°C 1 min, 72°C 1 min, 30 cycles, 72°C 10 min, refrigerate.

4. κ extension:

2 ng κ DNA	0.5 μl
10 x Taq buffer II	5.0 μl
$MgCl_2$ (25 mM)	5.0 μl
dNTP (1.25 mM each)	8.0 μl
Kap.Ext (20 μM)	0.5 μl
MCK-AscI.Ext (20 μM)	0.5 μl
H_2O	30.5 μl

5. PCR program: 94°C 10 min, suspend, add 0.3 μl AmpliTaq polymerase (Cetus 5 U/μl); 94°C 1 min, 60°C 1 min, 72°C 1 min, 10 cycles, 72°C 10 min, refrigerate. Extended κ products are about 710 bp.

6. V_H extension:

0.4 ng V_H DNA	0.5 μl
10 x Taq buffer II	5.0 μl
$MgCl_2$ (25 mM)	5.0 μl
dNTP (1.25 mM each)	8.0 μl
Tag.Back1-2 (20 μM)	0.5 μl
MJH.Ext (20 μM)	0.5 μl
H_2O	30.5 μl

7. PCR program: 94°C 10 min, suspend, add 0.3 μl AmpliTaq polymerase (5 U/μl); 94°C 1 min, 55°C 1 min, 72°C 1 min, 15 cycles, 72°C 10 min, refrigerate. Extended V_H products are about 450 bp.

8. Gel-purify the primary amplification products as in steps 6 to 7, Subprotocol 1.

9. Digest pFab60 and extended κ DNA with *Asc*I and *Nhe*I in a 1:1 mixture of NEB2 and NEB4 + BSA (see step 15, Subprotocol 1). Ethanol precipitate DNA and further digest with *Kpn*I in NEB1 + BSA or *Stu*I in NEB2. (see Note 14). Gel-purify DNA as in steps 6 and 7, Subprotocol 1.

10. Ligate the digested and purified pFab60 and κ DNA in a molar ratio of about 1:2 and extract and precipitate the ligation products as described in steps 20 and 21, Subprotocol 1.

11. Proceed with electroporations and make pFab60/κ library DNA. Use this DNA as "vector" to clone the V_H gene fragments.

12. Digest extended V_H DNA and pFab60/κ library DNA first with *Sfi*I at 50°C in NEB2 + 0.01% BSA, then with *Spe*I at 37°C in the same buffer. Digest the pFab60/κ DNA further with *Pme*I in NEB4 + BSA. Gel-purify as in steps 6 and 7, Subprotocol 1.

13. Ligate the digested pFab60/κ library and V_H DNA in a molar ratio of about 1:2 and extract and precipitate the ligation products as in steps 20 and 21, Subprotocol 1 (The vector harbors the $C_H1_{\gamma 1}$ gene fragment, see Figure 4.)

14. Proceed with electroporations to produce the IgH-independent Fab library (see Note 15).

Comments

See Notes in Subprotocol 4.

Subprotocol 4
Direct Cloning Method for Making Murine Antibody Fab Libraries of the IgG Isotype

▓ ▒ Materials

- The murine κ and extension primers are described in Subprotocol 1,   **PCR-primers**
  Table 1

- The murine V_H and extension back primers are described in Subprotocol 1, Table 1

- Primer mix (see Table 4)

- Forward extension primer (see Table 4)

- 10 x Taq buffer II (Perkin Elmer)

Table 4. Primer mix and forward extension primer

Murine C_H forward primer mix

MCHγ1.For:	CTCGCGAATT GGTGCGGCCG C**ACAATCCCT GGGCACAATT TTCTTGTCC**
MCHγ2a.For:	CTCGCGAATT GGTGCGGCCG C**GGGCTTGAT TGTGGGCCCT CTGGG**
MCHγ2b.For:	CTCGCGAATT GGTGCGGCCG C**GGGGTTGAT TGTTGAAATG GGCCCG**

The 24 to 28 distal nucleotides (boldfaced) of the MCHγ.For primers are identical to the corresponding $C_H1\gamma$ and hinge junction sequence. The *Not*I recognition sequence is underlined. The concentration of each primer in the MCHmix.For solution used for PCR is 6.67 μM giving a total concentration of μM.

Murine $C_H\gamma$ forward extension primer

MCH.Ext:	CAGTCACAGA TC*CTCGCGAA TTGGTGCGGC CGC*

MCH.Ext and the MCHγ.For primers overlap by 21 nucleotides (italicized). The *Not*I recognition sequence is underlined.

■■ Procedure

The direct cloning method for making murine antibody Fab libraries of the IgG isotype consists of five steps:

- Primary PCR amplifications of Fd gene fragments and κ genes.
- "Tagging" the primary amplification products with restriction enzyme recognition sequences (extension).
- Cloning of κ genes in pFab60 (equal to the pFab60/κ library).
- Cloning of Fd gene fragments using pFab60/κ DNA.
- Electroporation, growth and storage of the library.

Figure 4 shows an overview of the PCR primers and the expression vector pFab60.

PCRs are hot-started and carried out in PCR tubes. All PCR mixtures (50 or 100 µl) are covered with paraffin oil.

1. Extended κ gene fragments are produced as described in step 1 and steps 3 to 5, Subprotocol 3.

2. Primary amplification of Fd gene fragments:

cDNA (0.1 to 1 µg/µl)	5.0 µl
10 x Taq buffer II	10.0 µl
MgCl$_2$ (25 mM)	6.0 µl
dNTP (1.25 mM each)	8.0 µl
MVH1-25 (20 µM)	1.0 µl
MCHmix.For (20 µM)	1.0 µl
H$_2$O	69.0 µl

3. PCR program: 94°C 10 min, suspend, add 0.3 µl AmpliTaq polymerase (5 U/µl); 94°C 1 min, 55°C 1 min, 72°C 1 min, 30 cycles, 72°C 10 min, refrigerate.

4. Extension of Fd gene fragments:

1.6 ng Fd DNA	0.5 µl
10 x Taq buffer II	5.0 µl
MgCl$_2$ (25 mM)	5.0 µl
dNTP (1.25 mM each)	8.0 µl
Tag.Back 1-2 (20 µM)	0.5 µl
MCH.Ext (20 µM)	0.5 µl
H$_2$O	30.5 µl

5. PCR program: 94°C 10 min, suspend, add 0.3 µl AmpliTaq polymerase (5 U/µl); 94°C 1 min, 55°C 1 min, 72°C 1 min, 15 cycles, 72°C 10 min, refrigerate. Extended Fd products are about 750 bp.

6. Make the pFab60/(library as described in steps 8 to 11, Subprotocol 3.

7. Digest pFab60/κ and Fd DNA in NEB2 + BSA with *Sfi*I at 50°C, then with *Not*I and *Pme*I at 37°C. Gel-purify DNA as in steps 6 and 7, Subprotocol 1.

8. Ligate the digested pFab60/κ and Fd DNA in a molar ratio of about 1:2 and extract and precipitate the ligation products as described in steps 20 and 21, Subprotocol 1.

9. Proceed with electroporations to produce the IgG Fab library.

▪▪▪ Comments

Note 12: The main advantage of the direct cloning method is that the risk of cloning PCR artifacts is reduced, because considerably fewer rounds of PCR amplification are required (40-45 vs 75 rounds).

Note 13: Digestion of pFab60/κ library DNA with *Spe*I causes some vector background because 1% of known κ genes contain *Spe*I sites, which produces a subpopulation of linear phagemid DNA with two *Spe*I ends. (Similarly, digestion of pFab60/Fd library DNA with *Nhe*I, gives some background because 0.25% of known Fd gene fragments contain *Nhe*I sites.) The background problem can be avoided by cloning κ and V$_H$ DNA separately in pFab60 and by moving the resulting Fd library DNA as *Sfi*I-*Not*I fragments or using Subprotocol 4.

Note 14: Use the rare-cutting *Pme*I enzyme instead of *Hind*III, when pFab60/κ library DNA is used. Digest with the rare-cutting *Srf*I enzyme

Notes to Subprotocol 3 and 4

instead of *Kpn*I or *Stu*I, if pFab60/Fd library DNA is used instead of pFab60 DNA.

Note 15: To avoid reduction in diversity of the first-step library, it is necessary that the number of transformants in the second cloning step is several-fold higher than in the first.

References

1. Hoogenboom, R. H., Griffiths, A. D., Johnson, K. S., Chiswell, D. J., Hudson, P., and Winter, G. (1991) Multi-subunit proteins on the surface of filamentous phage: methodologies for displaying antibody (Fab) heavy and light chains. Nucleic Acids Res., 19, 4133-4137.
2. Kang, A. K., Barbas, C. F., Janda, K. D., Benkovic, S. J. and Lerner, R. A. (1991) Linkage of recognition and replication functions by assembling combinatorial antibody Fab libraries along phage surfaces. Proc. Natl. Acad. Sci. USA 88, 4363-4366.
3. Breitling, F., Dübel, S., Seehaus, T., Klewinghaus, I., and Little, M. (1991) A surface expression vector for antibody screening. Gene 104, 147-153.
4. Ørum, H., Andersen, P. S., Riise, E., Øster, A., Johansen, L. K., Bjørnvad, M., Svendsen, I., and Engberg, J. (1993) Efficient method for constructing comprehensive murine Fab antibody libraries displayed on phage. Nucleic Acids Res. 21(19), 4491-4498.
5. Hoogenboom, H. R., Marks, J. D., Griffiths, A. D., and Winter, G. (1992) Building antibodies from their genes. Immunol. Rev. 130, 41-68.
6. Winter, G., Griffiths, A. D., Hawkins, R. E. and Hoogenboom, H. R. (1994) Making antibodies by phage display technology. Annual Rev. Immunol. 12, 433-55.
7. Johansen, L. K., Albrechtsen, B., Andersen, H. W., and Engberg, J. (1995) pFab60: a new efficient vector for expression of antibody Fab fragments displayed on phage. Protein Engineering 8(10), 1063-1067.
8. Andersen, P. S., Ørum, H., and Engberg, J. (1996) One-step cloning of murine Fab gene fragments independent of IgH isotype for phage display libraries. BioTechniques 20(3), 340-342
9. Engberg, J., Andersen, P.S., Nielsen, L.K., Dziegiel, M., Johansen, L.K. and Albrechtsen, B. (1996) Phage-display libraries of murine and human antibody Fab fragments. Molecular Biotechnology, 6(3), 271-294.
10. Sambrook, J., Fritsch, E. F., and Maniatis, T. (1989) Molecular Cloning. A Laboratory Manual, 2nd. ed. Cold Spring Harbor Laboratory, Cold Spring Harbor, NY.
11. Krebber, A., Burgmester, J. and Plückthun, A.(1996) Inclusion of an upstream transcriptional terminator in phage display vectors abolishes background expression of toxic fusions with coat protein g3p. Gene 178, 71-74.
12. Engelhardt, O., Grabherr, R., Himmler, G., and Rüker, F. (1994) Two step cloning of antibody variable domains in a phage display vector. Biotechniques 17, 44-46.

Generation of Naive Human Antibody Libraries

CATHERINE HUTCHINGS, SARA CARMEN, and SIMON LENNARD

Introduction

The phage display of antibody fragments as Fabs or scFvs (McCafferty et al., 1990), has its origins in experiments demonstrating that both small peptides and folded proteins could be displayed on the surface of filamentous bacteriophage (Smith, 1985; Bass et al., 1990). Since the generation of the first human antibodies by phage display (Winter et al., 1994), the technology has developed to the point where large scFv repertoires have been created that yield antibodies with sub-nanomolar affinities (Vaughan et al., 1996; Xie et al., 1997). This is comparable with the best antibodies obtained using hybridoma technology. Phage display repertoires can also be used to isolate antibodies not easily obtained by hybridoma technology, such as those specific for toxic proteins and human anti-self antibodies (Griffiths et al., 1993; Vaughan et al., 1998). In addition, using a variety of selection and screening strategies, the same non-immunised "single pot" library can be used to simultaneously derive many high affinity antibodies with different specificities in only a matter of weeks. Furthermore, selection techniques have evolved to the point where strategies now exist that facilitate selections on complex antigens expressed on the cell surface (Chapter 12).

The advantages of a scFv library constructed from naturally rearranged V-genes in a phagemid vector include natural diversity in the length of the VHCDR3, a higher number of functional scFvs, and soluble antibody ex-

Principles and applications

Catherine Hutchings, Cambridge Antibody Technology, The Science Park, Melbourn, Cambridgeshire, SG8 6JJ, United Kingdom

✉ Sara Carmen, Cambridge Antibody Technology, The Science Park, Melbourn, Cambridgeshire, SG8 6JJ, United Kingdom (*phone* +44-1763-269284; *fax* +44-1763-263413; *e-mail* sara.carmen@camb-antibody.co.uk)

Simon Lennard, Cambridge Antibody Technology, The Science Park, Melbourn, Cambridgeshire, SG8 6JJ, United Kingdom

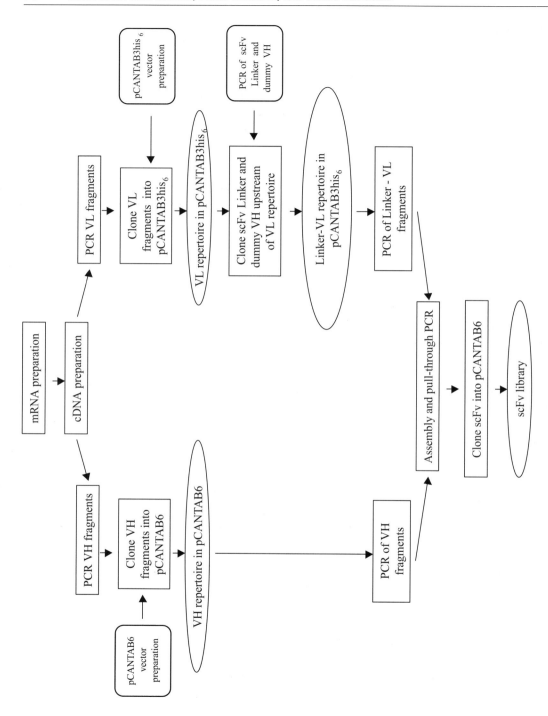

Fig. 1. Overview for construction of scFv library

pression without the need for sub-cloning (Vaughan et al., 1996). To ensure maximum diversity, V-genes are amplified from peripheral blood lymphocytes (PBL), or lymphoid tissue isolated from several non-immunised donors. The PCR primers employed are largely based on those published previously (Marks et al., 1991), with additional information on the most recent VH sequences obtained from the V-BASE directory (Tomlinson et al., MRC Centre for Protein Engineering), and correspond to all known VH, Vκ and Vλ gene sequences. Random hexamers are employed in the initial first-strand cDNA synthesis from PBL/lymphoid mRNA to ensure that all five antibody classes are likely to be represented. This also helps increase the overall size of the final library, which subsequently increases the chance of isolating high affinity antibodies. The library is constructed using an efficient two-fragment PCR assembly strategy followed by ligation and over one hundred electroporations into TG1 *E.coli* (Vaughan et al., 1996).

Antibodies selected from such large, non-immunised repertoires of scFv fragments have a multitude of applications by virtue of their specificity, affinity and the relative ease with which they can be derived. The range of selection strategies that can be employed allows for the rapid isolation of antibodies with specific properties, such as receptor agonists or antagonists, specific epitope recognition or high neutralisation potency (Vaughan et al., 1998). A technique known as guided selection can also be employed to derive human equivalents from these libraries of existing rodent mAbs (Jespers et al., 1994), and scFvs displayed on phage or as soluble fragments can be used as immunological reagents for ELISAs, immunocytochemistry, Western blotting or epitope mapping. The use of mAbs in the clinic has, until recently, been limited because they were originally derived by hybridoma technology and so were recognised by the human immune system as foreign. Attempts to humanise murine mAbs and the development of chimaeric antibodies has resulted in a number of these antibodies progressing through clinical trials (Johnson and Glover, 1998). However, both humanised and chimaeric mAbs are still a combination of human and mouse material and the presence of any mouse protein is a disadvantage that could potentially result in a human anti-mouse antibody (HAMA) reaction. Phage display technology, therefore, provides a means by which high affinity, fully human mAbs can be rapidly isolated and also provides a starting point from which a selected antibody can, if necessary, be affinity matured for improved neutralisation potency or binding kinetics. The use of naturally rearranged germline fragments in the construction of the library ensures that these fully human scFv fragments are unlikely to be immunogenic making them ideal therapeutic agents. The first fully human therapeutic mAbs isolated from a phage displayed library for the treatment of rheumatoid arthritis and ocular scarring are currently progressing through phase II clinical trials (Glover, 1999).

▩ Materials

Equipment
- Water baths
- Microcentrifuge
- DNA Thermocycler for PCR
- Agarose gel electrophoresis tank/gel trays
- Power pack for agarose gel electrophoresis
- DNA thermocycler for PCR
- BioRad Gene Pulser™
- Benchtop centrifuge
- 37°C and 30°C Incubators
- Sorvall RC5C plus centrifuge
- 2 l shake flasks

Reagents and solutions
- Restriction enzymes: *Sfi*I, *Not*I, *Xho*I, *Hind*III, *Apa*LI
- Buffer saturated Phenol *Ultra*Pure™ (Gibco BRL)
- Chloroform
- Abs. ethanol (-20°C)
- 70% (v/v) ethanol (-20°C)
- Water A.C.S. reagent (Aldrich Cat. No. 32, 007-2)
- *Taq* DNA polymerase (Roche Molecular Biochemicals)
- 10X *Taq* DNA polymerase buffer (Roche Molecular Biochemicals)
- 5 mM each of dNTPs (Pharmacia)
- PCR mineral oil
- Water A.C.S Reagent (Aldrich Cat. No 32, 007-2)

- Amersham Ligation Kit (Amersham Cat No. RPN 1507)

- 2TY (for recipe see Sambrook et al., 1990)

- Glucose 20% (w/v) sterile filtered

- 5X Ligation Buffer (500 mM Tris pH 7.4, 25 mM $MgCl_2$)

- Ampicillin 100 mg/ml stock, filter sterilised

- Kanamycin 50 mg/ml stock, filter sterilised

- 20% (w/v) PEG 8000, 2.5 M NaCl

- TE buffer (10 mM Tris pH 8.0, 0.1mM EDTA pH 8.0)

Procedure

Library generation overview

The VH and VL gene segments are amplified in separate PCR reactions and are initially cloned into two different vectors; pCANTAB6 and pCAN-TAB3his_6. The latter is used for cloning the VL repertoire as it has the appropriate polylinker cloning sites for the ApaLI/NotI digested VL fragments, while the VH repertoire is cloned into the SfiI/XhoI sites of pCAN-TAB6. A linker from an existing scFv is cloned (together with an irrelevant or "dummy" VH) into the VL repertoire upstream of the VL fragments. The VH and VL-linker repertoires are then amplified from their respective vectors and the scFv construct prepared using a simple two-fragment PCR assembly. This construct is cloned into the SfiI and NotI sites of pCAN-TAB6 to create the naive scFv library (Vaughan et al., 1996).

Preparation of cDNA template

For naive human repertoires, peripheral blood lymphocytes are isolated from whole blood as detailed by Marks et al. (1991), where 50 ml of blood should yield approximately 1 x 10^7 cells. Immediately isolate the mRNA from the cell pellet using an oligo (dT)-purification system, such as the 'Quickprep mRNA kit' (Pharmacia). A total of 1 x 10^7 cells provides in the region of 10 µg total RNA, of which 1-5% is mRNA. Synthesise first strand cDNA from the mRNA template using a 'First-strand cDNA synthesis kit' (Pharmacia), with random hexamers according to the manufacturer's instructions (Vaughan et al., 1996). At this stage, ensure that there is enough cDNA for all the VH and VL PCR reactions as 0.5 ng cDNA is used per reaction.

Construction of a naive VH repertoire

PCR primers are based on those published by Marks et al., (1991) and in the V-BASE directory (Tomlinson et al. MRC Centre for Protein Engineering). The 5' and 3' VH primers are tagged with *Sfi*I and *Xho*I restriction sites respectively to facilitate cloning (Table 1). All PCR reactions in this chapter are performed using primers at a concentration of 10 μM, dNTPs at 5mM and 5 units of *Taq* polymerase, unless otherwise stated. All PCR products should be checked on 1-2% (w/v) TAE agarose to ensure that a product of the expected size has been generated. Always include "no template" controls.

1. Prepare separate 50 μl PCR reactions, for each of the VH Back primers, in 0.4 ml Eppendorf tubes as follows:

5.0 μl	10x *Taq* Buffer
2.5 μl	5 mM dNTPs
2.5 μl	(10 μM) of an individual VH Back*Sfi* primer
2.5 μl	(10 μM) JH1-6For*Xho* primer mix
5.0 μl	first strand cDNA (typically 0.5 ng)
31.5 μl	PCR water
1.0 μl	*Taq* polymerase (5 units)

2. Add one drop mineral oil per tube (or use a PCR thermocycler that has a heated lid) and cycle samples according to the following reaction profile: 94°C 1 min, 55°C 1 min, 72°C 2 min for 30 cycles, final extension: 72°C 10 min. Hold at 4°C if necessary.

3. Check for a PCR product from each primer mix on a 1% (w/v) TAE agarose gel before pooling the PCR products from each reaction.

4. Concentrate the DNA by phenol/chloroform extraction followed by ethanol precipitation (Sambrook et al., 1990).

5. Resuspend the purified product in A.C.S water (Sigma).

6. Digest sequentially with *Sfi*I and *Xho*I according to the method described in Sambrook et al., (1990). Note that after digestion the VH segments will be approx 350bp.

Table 1. Oligonucleotide primers for amplification of human VH sequences

HUMAN VH Back*Sfi*I PRIMERS

VH1b/7a Back*Sfi*I

5'- GTC CTC GCA ACT GCG GCC CAG CCG GCC ATG GCC CAG (AG)TG CAG CTG GTG CA(AG) TCT GG - 3'

VH1c Back*Sfi*I

5'- GTC CTC GCA ACT GCG GCC CAG CCG GCC ATG GCC (GC)AG GTC CAG CTG GT(AG) CAG TCT GG - 3'

VH2b Back*Sfi*I

5' - GTC CTC GCA ACT GCG GCC CAG CCG GCC ATG GCC CAG (AG)TC ACC TTG AAG GAG TCT GG - 3'

VH 3b Back*Sfi*I

5' - GTC CTC GCA ACT GCG GCC CAG CCG GCC ATG GCC (GC)AG GTG CAG CTG GTG GAG TCT GG - 3'

VH3c Back*Sfi*I

5' - GTC CTC GCA ACT GCG GCC CAG CCG GCC ATG GCC GAG GTG CAG CTG GTG GAG(AT)C(TC) GG - 3'

VH4b Back*Sfi*I

5' - GTC CTC GCA ACT GCG GCC CAG CCG GCC ATG GCC CAG GTG CAG CTA CAG CAG TGG GG - 3'

VH4c Back*Sfi*I

5'-GTC CTC GCA ACT GCG GCC CAG CCG GCC ATG GCC CAG (GC)TG CAG CTG CAG GAG TC(GC) GG-3'

VH5b Back*Sfi*I

5'-GTC CTC GCA ACT GCG GCC CAG CCG GCC ATG GCC GA(AG) GTG CAG CTG GTG CAGTCT GG-3'

VH 6a Back*Sfi*I

5'-GTC CTC GCA ACT GCG GCC CAG CCG GCC ATG GCC CAG GTA CAG CTG CAG CAG TCA GG-3'

HUMAN VH For*Xho*I PRIMERS

HuJHFor1-2*Xho*I

5'-ACC GCC TCC ACC ACT CGA GAC GGT GAC CAG GGT GCC (TC), (TC), (GT) GCC CCAñ3'

HuJHFor3*Xho*I

5'-ACC GCC TCC ACC ACT CGA GAC GGT GAC CAT TGT CCC (TC), (TC), (GT) GCC CCAñ3'

HuJHFor4-5*Xho*I

5'-ACC GCC TCC ACC ACT CGA GAC GGT GAC CAG GGT TCC (TC), (TC), (TG) GCC CCAñ3'

HuJHFor6*Xho*I

5'-ACC GCC TCC ACC ACT CGA GAC GGT GAC CGT GGT CCC (TC), (TC), (TG) CCC CCAñ3'

7. Gel purify using 'GeneClean' (Bio101 Inc.) as per the manufacturers recommended protocol and estimate the concentration of the VH insert by comparing on a 1% (w/v) TAE agarose gel against a DNA marker (typically λ-*Hind*III) of known concentration. The VH gene segments are then ready to be ligated into the vector.

For preparation of digested vector (pCANTAB6), plasmid DNA is prepared either by the alkali lysis method and caesium-banded, as detailed in Sambrook et al., (1990), or by using the Qiagen vector preparation system (see manufacturer's protocol):

1. Digest vector with *Sfi* and *Xho*I restriction enzymes according to the protocol detailed in Sambrook et al., (1990).

2. Concentrate the cut vector by performing a phenol/chloroform extraction followed by ethanol precipitation. (Sambrook et al.,1990).

3. Estimate the concentration of recovered DNA by comparison with known DNA markers as described previously.

In general about 20 μg of caesium-banded vector gives about 5-10 μg of purified cut vector. Efficient digestion is crucial to prevent high background from non-recombinant clones.

The digested VH gene segments are ligated into the previously *Sfi*I/ *Xho*I digested pCANTAB6 phagemid vector using the Amersham Ligation System. A series of ligation reactions should be set up which cover a range of molar insert to vector ratios, such as 1:1, 2:1, 4:1, 1:2, etc. (see manufacturer's guidelines). This will help to ascertain the optimal ratio of insert to vector. Generally a repertoire of approximately 1×10^7 to 1×10^8 recombinants can be generated if 0.5 μg of digested VH segments are ligated with 1.5 μg of digested vector. Also include a vector only "ligation control" in order to assess background due to non-recombinants. Protocols for the preparation of electrocompetent *E.coli* TG1 cells and subsequent electroporations are as described in Sambrook et al. (1990):

1. Perform at least 40 electroporations, each at 2.5V, 200 Ohms, 25μF.

2. Allow cells to recover for 1 hour in 2TYG (2% w/v glucose) with shaking at 200 rpm, 37°C incubator.

3. Pool all electroporations prior to taking a small aliquot for a titration series and plate 10-fold serial dilutions onto 2TYAG plates (100 μg/ml ampicillin, 2% w/v glucose, 1.5% w/v agar), to determine the size of the library.

4. Centrifuge the rest of the cells at 3,500 rpm for 10 min in a benchtop centrifuge. Remove supernatant and resuspend in 1-2 ml 2TY. Spread the total library on 4 large 243 mm x 243 mm 2TYAG plates (100 μg/ml ampicillin, 2% w/v glucose, 1.5% w/v agar).

Grow overnight, inverted at 30°C. Estimate library size from the titre plates, which should be in the region of 1×10^7 to 1×10^8 individual recombinants. See later section for details on the preparation of glycerol stocks.

Construction of a naive VL repertoire

Human VL kappa (κ) and VL lamda (λ) gene fragments are amplified separately using each back primer in combination with the appropriate equimolar mixture of the Jκ or Jλ forward primers (Table 2). The same cycling parameters are employed for the PCR reactions as outlined for the VH fragments:

1. Perform 50 µl PCR reactions using 5 µl of first strand cDNA as template using the following primer combinations:
    VκBack*Apa*LI + Jκ 1-5For*Not*I primer mix
    and:
    VλBack*Apa*LI + Jλ1-5For*Not*I primer mix.

2. Concentrate the PCR products as described for the VH gene segments and digest with *Apa*LI followed by *Not*I (Sambrook et al., 1990). The size of the resulting fragments will be around 350bp.

3. Gel purify VL fragments and estimate the concentration of purified insert as described previously. The VL gene fragments are then ready for ligation.

Caesium-banded vector plasmid DNA (pCANTAB3his_6) is prepared and digested sequentially with *Apa*LI and *Not*I restriction enzymes which along with the subsequent ligations and electroporations, are all performed as described for the VH repertoire. This should generate Vκ and Vλ repertoires of between 1×10^5 and 1×10^6 individual recombinants.

The next stage is to PCR amplify the (Gly$_4$Ser)$_3$ scFv linker from an existing scFv together with an irrelevant (or "dummy") VH and introduce this fragment into the finished VL repertoire upstream of the VL gene segments as a *Hind*III-*Apa*LI digested fragment:

1. Prepare 50 µl PCR reactions using the pUC19rev and fdtetseq primers. For the template DNA; using a sterile pipette tip, pick a colony of an irrelevant clone that possesses the required scFv linker and PCR amplify using the cycling parameters described previously.

2. Digest the "dummy" VH and linker fragment with *Apa*LI followed by *Hind*III (Sambrook et al., 1990) to create a product of around 450 bp.

Table 2. Oligonucleotide primers for amplification of human VL sequences

HUMAN Vλ Back*ApaLI* PRIMERS:

Huλ1a Back*ApaLI*
5'-ACC GCC TCC ACC AGT GCA CAG TCT GTG CTG ACT CAG CCA CC-3'

Huλ1b Back*ApaLI*
5'-ACC GCC TCC ACC AGT GCA CAG TCT GTG (TC)TG ACG CAG CCG CC-3'

Huλ1c **Back*ApaLI***
5'-ACC GCC TCC ACC AGT GCA CAG TCT GTC GTG ACG CAG CCG CC-3'

Huλ2 Back*ApaLI*
5'-ACC GCC TCC ACC AGT GCA CA(AG) TCT GCC CTG ACT CAG CCT-3'

Huλ3a Back*ApaLI*
5'-ACC GCC TCC ACC AGT GCA CTT TCC TAT G(AT)G CTG ACT CAG CCA CC-3'

Huλ3b **Back*ApaLI***
5'-ACC GCC TCC ACC AGT GCA CTT TCT TCT GAG CTG ACT CAG GAC CC-3'

Huλ4 Back*ApaLI*
5'-ACC GCC TCC ACC AGT GCA CAC GTT ATA CTG ACT CAA CCG CC-3'

Huλ5 Back*ApaLI*
5'-ACC GCC TCC ACC AGT GCA CAG GCT GTG CTG ACT CAG CCG TC-3'

Huλ6 Back*ApaLI*
5'-ACC GCC TCC ACC AGT GCA CTT AAT TTT ATG CTG ACT CAG CCC CA-3'

Huλ7/8 Back*ApaLI*
5'-ACC GCC TCC ACC AGT GCA CAG (AG)CT GTG GTG AC(TC) CAG GAG CC-3'

Huλ9 Back*ApaLI*
5'-ACC GCC TCC ACC AGT GCA C(AT)G CCT GTG CTG ACT CAG CC(AC) CC-3'

HUMAN Vκ Back*ApaLI* PRIMERS:

Huκ1b Back*ApaLI*
5'-ACC GCC TCC ACC AGT GCA CTT GAC ATC CAG (AT)TG ACC CAG TCT CC-3'

Huκ2 Back*ApaLI*
5'-ACC GCC TCC ACC AGT GCA CTT GAT GTT GTG ATG ACT CAG TCT CC-3'

Huκ3b Back*ApaLI*
5'-ACC GCC TCC ACC AGT GCA CTT GAA ATT GTG (AT)TG AC(AG) CAG TCT CC-3'

Huκ4b Back*ApaLI*
5'-ACC GCC TCC ACC AGT GCA CTT GAT ATT GTG ATG ACC CAC ACT CC-3'

Huκ5 Back*ApaLI*
5'-ACC GCC TCC ACC AGT GCA CTT GAA ACG ACA CTC ACG CAG TCT CC-3'

Huκ6 Back*ApaLI*
5'-ACC GCC TCC ACC AGT GCA CTT GAA ATT GTG CTG ACT CAG TCT CC-3'

Table 2. Continuous

HUMAN Vλ For*NotI* PRIMERS:

HuJλ1For*NotI*

5'-GAG TCA TTC TCG ACT TGC GGC CGC ACC TAG GAC GGT GAC CTT GGT CCC-3'

HuJλ2-3 For*NotI*

5'-GAG TCA TTC TCG ACT TGC GGC CGC ACC TAG GAC GGT CAG CTT GGT CCC-3'

HuJλ4-5 For*NotI*

5'-GAG TCA TTC TCG ACT TGC GGC CGC ACT TAA AAC GGT GAG CTG GGT CCC-3'

HUMAN Vκ For*NotI* PRIMERS:

HuJκ1For*NotI*

5'-GAG TCA TTC TCG ACT TGC GGC CGC ACG TTT GAT TTC CAC CTT GGT CCC-3'

HuJκ2 For*NotI*

5'-GAG TCA TTC TCG ACT TGC GGC CGC ACG TTT GAT CTC CAG CTT GGT CCC-3'

HuJκ3 For*NotI*

5'-GAG TCA TTC TCG ACT TGC GGC CGC ACG TTT GAT ATC CAC TTT GGT CCC-3'

HuJκ4 For*NotI*

5'-GAG TCA TTC TCG ACT TGC GGC CGC ACG TTT GAT CTC CAC CTT GGT CCC-3'

HuJκ5 For*NotI*

5'-GAG TCA TTC TCG ACT TGC GGC CGC ACG TTT AAT CTC CAG TCG TGT CCC-3'

3. Gel purify using 'Geneclean' (Bio101 Inc.) and estimate DNA concentration by comparison with known DNA markers.

4. Prepare a plasmid DNA mini-prep of the pCANTAB3his_6 vector containing the VL repertoire (Wizard mini-prep kit, Promega).

5. Sequentially digest 10 µg of plasmid DNA with *Apa*LI and *Hind*III restriction enzymes as described and concentrate the cut vector by phenol/chloroform extraction followed by ethanol precipitation (Sambrook et al., 1990).

6. Estimate the concentration of recovered DNA by comparison with known DNA markers.

7. Ligate approximately 0.4 µg of *Hind*III/*Apa*LI digested dummy VH-linker DNA fragment with 1 µg of pooled Vκ and Vλ repertoire DNA using the Amersham Ligation System as per manufacturer's protocol.

8. Electroporate into *E.coli* TG1 cells and plate out. This should generate a VL repertoire with an upstream scFv linker of between $1x10^6$ and $1x10^7$ recombinants.

Library construction

The VH and linker-VL DNA fragments are amplified separately from each of the cloned repertoires (VH in pCANTAB6 and linker-VL in pCAN-TAB3his_6), assembled on the JH region and amplified by pull-through PCR (refer to Table 3 for pull-through PCR primers). The resulting assembled scFv constructs (VH-linker-VL) are then sequentially digested with *Sfi*I and *Not*I and ligated into *Sfi*I/*Not*I digested pCANTAB6.

1. Prepare 50 µl PCR reactions and, using the cycling parameters described previously, amplify the VH repertoire with pUC19rev and JHFor primers; and the VL repertoire with Reverse JH and fdtetseq primers. Gel purify using 'Geneclean' (Bio101 Inc.) and estimate DNA concentration by comparison with known DNA markers.

2. Mix equal amounts of the VH and VL PCR products (5-20 ng of each) and make up the volume to 100 µl with sdH$_2$O.

3. Ethanol precipitate and resuspend in 25 µl sdH$_2$O.

4. For the assembly reaction add the following reagents to the pooled VH and VL fragments and perform 25 cycles of 94°C for 1 min, followed by 65°C for 4 min:
   - 3.0 µl 10x *Taq* Buffer (Roche Molecular Biochemicals)
   - 1.5 µl 5 mM dNTP mix (Pharmacia)
   - 0.5 µl *Taq* polymerase (2.5 units)

5. Once the assembly has been performed, prepare at least two 50 µl pull-through PCR reactions per VH Back*Sfi* primer, combined with either the Jκ1-5For*Not* primer mix or the Jλ1-5For*Not* primer mix. Use 5.0 µl assembly DNA per reaction and amplify using the PCR cycling parameters described previously. The correct size of the assembled construct is around 700 bp.

6. Pool the PCR products and after phenol/chloroform extraction followed by ethanol precipitation, sequentially digest with *Sfi*I and *Not*I restriction endonucleases (Sambrook et al., 1990)

Table 3. Oligonucleotide primers for pull-through PCR

JH Forward primers

HuJH1-2For

5'-TGA GGA GAC GGT GAC CAG GGT GCC-3'

HuJH3For

5'-TGA AGA GAC GGT GAC CAT TGT CCC-3'

HuJH4-5For

5'-TGA GGA GAC GGT GAC CAG GGT TCC-3'

HuJH6For

5'-TGA GGA GAC GGT GAC CGT GGT CCC-3'

RevJH PRIMERS

RHuJH1-2

5'-GCA CCC TGG TCA CCG TCT CCT CAG GTG G-3'

RHuJH3

5'-GGA CAA TGG TCA CCG TCT CTT CAG GTG G-3'

RHuJH4-5

5'-GAA CCC TGG TCA CCG TCT CCT CAG GTG G-3'

RHuJH6

5'-GGA CCA CGG TCA CCG TCT CCT CAG GTG C-3'

7. Gel purify the digested scFv assembly construct and ligate with *Sfi*I/ *Not*I digested pCANTAB6, having already determined the optimum insert to vector ratio as described previously. To ensure maximum transformation efficiency, it is necessary to perform at least one hundred electroporations divided into batches, with each batch ultimately spread on large 243 mm x 243 mm 2TYAG plates (100 μg/ml ampicillin, 2% w/v glucose, 1.5% w/v agar). To calculate the total size of the library, take aliquots from each batch and plate out serial dilutions on titre plates. The size of the final library should be in the region of 1x10^8 to 1x10^9 individual recombinants.

8. Once the library size has been determined, scrape the large plates using 5 ml 2TY per plate and pool the cells in a 50 ml Falcon tube prior to the addition of 0.5 vol of 50% (v/v) glycerol. Ensure homogeneous resuspension of the cells by rotating the tube on a mixing wheel for 30 min before determining cell density by OD measurement at 600 nm. Store aliquoted library glycerols at –70°C.

Library QC analysis

Both the percentage of recombinant clones present in the final library and library diversity are confirmed by *Bst*NI fingerprinting and sequence analysis. PCR amplify the scFv inserts from 50 randomly picked clones per repertoire with the vector primers pUC19 reverse and fdtetseq, as described previously. For *Bst*NI fingerprint analysis, digest 20 µl PCR product per clone with *Bst*NI restriction enzyme, as described Sambrook et al. (1990), and electrophorese 10 µl on a 1% (w/v) agarose gel to visualise the restriction profiles. Purify the PCR products using the 'Wizard' PCR prep system (Promega) before sequence analysis with fluorescent dideoxy chain terminators (Applied Biosystems) using PCR H-link and mycseq10 as heavy and light chain primers. Sequence analysis is clearly the best method and should give 50 different sequences, while *Bst*NI analysis will give an indication of germline diversity but is limited by the size resolution of the agarose gel.

Preparation of library phage stocks

To generate the phage displayed library for use in selections, the phagemid particles are rescued with the helper phage M13K07 (Pharmacia). The resultant library phage are PEG precipitated and caesium-banded and so can be stored at 4°C for 2 years:

Day 1 1. Inoculate 500 ml 2TYAG (100 µg/ml ampicillin, 2% w/v glucose) with 10^{10} cells from the library glycerol stock and incubate at 37°C with shaking at 250 rpm until OD at 600 nm = 0.5 - 1.0.

2. Add M13KO7 helper phage to a final concentration of 5 x 10^9 pfu/ml.

3. Incubate cells for 30 min at 37°C without shaking, then for 30 min with gentle shaking (200 rpm) to allow infection.

4. Centrifuge cultures at 4,500 rpm for 15 min using a GSA rotor in a Sorvall centrifuge. Resuspend pellet in the same volume of 2TYAK (100 µg/ml ampicillin, 50 µg/ml kanamycin). Incubate overnight in 2 l flasks with rapid shaking (300 rpm) at 30°C.

Day 2 5. Transfer the 500 ml culture into a large centrifuge pot and centrifuge at 8,000 rpm for 15 min at 4°C. Collect the supernatant in prechilled 1 l bottles.

6. Add 0.3 volume of PEG/NaCl (20% w/v polyethylene glycol 8000, 2.5 M NaCl). Swirl to mix. Allow to precipitate for at least 1 hour on ice.

7. Concentrate the phage pellet by centrifuging at 8,000 rpm at 4°C for another 15 min, centrifuging twice in the same bottle (collect all phage).

8. Decant as much supernatant as possible. Resuspend each pellet in 7.5 ml TE buffer (pH 8.0).

9. Re-centrifuge the phage stock in smaller tubes (SS34 rotor) at 10,000 rpm for 10 min. Remove the supernatant taking care to ensure that bacterial pellet is not disrupted.

10. Add 3.6 g of caesium chloride to the phage and make up to 9 ml total volume.

11. Ultracentrifuge samples at 40,000 rpm, 23°C for at least 24 hours.

12. Remove phage band using a 19G syringe.

13. Dialyse samples against 2 x 1 l TE (pH 8.0) at 4°C for 24 hours.

14. Titre phage stocks and store aliquoted at 4°C.

Results

By using a two-fragment PCR assembly strategy combined with over one hundred electroporations of ligated material, a naive scFv library of between 1×10^8 and 1×10^9 individual recombinants can be constructed. Sequence analysis should confirm that a full-length scFv insert is present in >95% of the clones constructed by this approach. The affinities of antibodies directly isolated from scFv repertoires constructed in this manner, without antibody engineering, can be in the sub-nanomolar range and tend to have slower off-rates than those derived from rodent immune responses, smaller scFv repertoires or large synthetic Fab libraries (Vaughan et al., 1996).

Once the library has been constructed, large, highly stable stocks of the phage displayed scFv repertoire can be prepared and stored in aliquots at 4°C. This "single pot" library can then be used as required, to isolate high affinity antibodies to many, many different antigens in a matter of weeks. These antibodies may then be used for a wide range of applications, including human antibody therapeutics.

References

Bass S, Greene R and Wells JA (1990) Hormone phage: an enrichment method for variant proteins with altered binding properties. Proteins 8:309-314

Glover DR, (1999) Fully human antibodies come to fruition. SCRIP Magazine (May): 16-19

Griffiths AD, Malmqvist M, Marks JD, Bye JM, Embleton MJ, McCafferty J et al (1993) Human anti-self antibodies with high specificity from phage display libraries. EMBO J. 12:725-734

Jespers LS, Roberts A, Mahler SM, Winter G and Hoogenboom HR (1994) Guiding the selection of human antibodies from phage display repertoires to a single epitope of an antigen. BioTechnology 12: 899-903

Johnson KS and Glover DR (1998) Antibodies come in from the cold. Innovations in Pharmaceutical Technology p82-85

Marks JD, Hoogenboom HR, Bonnert TP, McCafferty J, Griffiths AD and Winter G (1991) By-passing immunization: Human antibodies from V-gene libraries displayed on phage. J Mol Biol 222:581-597

McCafferty J, Griffiths AD, Winter G and Chiswell D (1990) Phage antibodies: filamentous phage displaying antibody variable domains. Nature 348:552-554

Sambrook J, Fritsch EF and Maniatis T (1990) Molecular cloning - a laboratory manual. Cold Spring Harbor Laboratory. New York.

Smith GP (1985) Filamentous phage: novel expression vectors that display cloned antigens on the virion surface. Science 228(4705): 1315-1317

Vaughan TJ, Williams AJ, Pritchard K, Osbourn JK, Pope AR, Earnshaw JC, McCafferty J, Hodits RA, Wilton J and Johnson KS (1996) Human antibodies with sub-nanomolar affinities isolated from a large non-immunized phage display library. Nature Biotechnology 14: 309-314

Vaughan TJ, Osbourn JK and Tempest PR (1998) Human antibodies by design. Nature Biotechnology 16: 535-539

Xie M-H, Yuan J, Adams C and Gurney A (1997) Direct demonstration of MuSK involvement in acetylcholine receptor clustering through identification of agonist scFv. Nature Biotechnology 15:768-771

Generation of Antibody Gene Libraries from Seropositive Human Donors

JOACHIM KOCH and STEFAN DÜBEL

Introduction

Being a major part of the immune system, antibodies represent a powerful weapon system defending our body against non-self agents. In some situations, however, antibodies to threatening antigens cannot be produced, as in the case of most cancer cells, or they appear too late to control the antigen, as e. g. in the case of intoxication. In this situation, therapeutic antibodies can be applied to help our own immune system. Further, therapeutic antibodies against self-antigens can be employed to modulate targets not related to an immune challenge, e. g. to interfere with endogenous regulation processes.

The origins of current therapeutic antibodies are mainly equine, bovine or mouse. Non-human passive vaccines are well established for the treatment of patients with acute infections or intoxication. However, these passive vaccines are recognized as non-self by the body and repeated application finally results in an immune response towards the therapeutic agent. The use of human antibodies is the optimal solution to evade this problem. During the past decade, phage display technology has supplied new straightforward methods to obtain human antibodies to almost every antigen, by using a heterologous system for the clonal selection and the production of antibodies (Barbas et al., 1991, Breitling et al., 1991, Marks et al., 1991). Highly complex universal ("naive") or synthetic antibody gene libraries allow to generate human antibodies to toxic or highly pathogenic antigens. On the other hand, antibodies with useful specificities may already exist in a patient's blood, where they e. g. helped to defeat

✉ Joachim Koch, Universität Heidelberg, Molekulare Genetik, Im Neuenheimer Feld 230, 69120 Heidelberg, Germany (*phone* +49-6221-545649; *fax* +49-6221-545678; *e-mail* j-koch@sirius.mgen.uni-heidelberg.de)
Stefan Dübel, Universität Heidelberg, Institut für Molekulare Genetik, Im Neuenheimer Feld 230, 69120 Heidelberg, Germany

the pathogen. The genes encoding for these antibodies can be obtained from the patient's leukocytes. In addition to the generation of passive vaccines from that source, human antibody gene libraries allow the analysis of autoimmune reactions, by providing a new method for the identification and analysis of antibodies related to the autoimmune process. These "patient libraries" or "immunized donor libraries" can be less complex than universal libraries by order of magnitude, thus dramatically reducing the effort for generating and maintaining the library. This protocol therefore describes the generation of an scFv antibody fragment phage display library from human donor blood.

A key step in generating the library is the PCR amplification of the variable regions. The primers employed were designed to amplify as much as possible antibody V regions, since the sequences providing the desired antigen binding specificities are not known. To minimize a bias against particular sequences, the PCR is done in two steps. The first provides the unbiased amplification, in the second the restriction endonuclease sites necessary for cloning was introduced. The described set of primers (Welschof et al., 1995) was designed from the amino acid sequences of immunoglobulins. The respective nucleotide sequences were then deduced from the codon usage at every individual amino acid position. To allow sequence variability, wobble nucleotides were included to represent the percentage of occurrence of the respective amino acids in this position.

After amplification, the DNA-fragments (expected size: κLC: 360 bp, HC and λLC: 400 bp) are cloned into the phagemid *pSEX81* (Welschof et al., 1997) to allow affinity screening ("panning"). The restriction endonuclease cleavage sites introduced for this purpose are NcoI (5') and HindIII (3') for human variable heavy chains and MluI (5') and NotI (3') for human variable light chains. They have been selected for having a low probability to cut within human variable light and variable heavy chain sequences. Further, they produce overhangs of 4 nucleotides or more to get optimal cloning efficiency, they are not dependent on methylation and their efficiency in the recommended double digests is greater than 90 %. The described primer set and the pSEX81 phage display vector have been successfully employed for the isolation of various human scFv fragments from a "seropositive" donor as well as a "universal" library (Welschof et al., 1997, Dörsam et al., 1997).

Outline

The method describes the construction of a human scFv antibody fragment library for the display on phage (Fig. 1). First, Leukocytes are prepared for mRNA extraction and cDNA synthesis. A set of specially designed primers is used to amplify the DNA fragments coding for the VL and VH chains of immunoglobulins. Subsequently a second set of primers is used to add restriction sites at the end of the VL and VH gene fragments to allow cloning into the expression vector pSEX81.

Subprotocol 1
PCR-Amplification of the Rearranged Variable Light (κ, λ) and Variable Heavy (μ, γ) Antibody Chains

Materials

Preparation of leukocyte cDNA

- 5 x buffer for reverse transcriptase (Roche Diagnostics, Mannheim, Germany)   **Reagents**

- 75% ethanol in DEPC-treated water

- HEPES (N-2-Hydroxyethylpiperazine-N'-2-Ethane Sulfonic Acid), (Gibco BRL, Karlsruhe, Germany)

- Chloroform (MERCK, Darmstadt, Germany)

- dNTPs (2.5 mM each), (Fermentas, St. Leon-Rot, Germany)

- EDTA (Ethylenediamintetraacetic acid Disodium Salt), (AppliChem, Darmstadt, Germany)

- Fetal Calf Serum (FCS), (Gibco BRL, Karlsruhe, Germany)

- Ficoll (Seromed, Berlin, Germany)

- Isopropyl alcohol (MERCK, Darmstadt, Germany)

- Oligo-dT primer (17 nucleotides in length), (Roche Diagnostics, Mannheim, Germany)

- Penicillin-Streptomycin Solution (Gibco BRL, Karlsruhe, Germany)

- M-MuLV Reverse transcriptase (Roche Diagnostics, Mannheim, Germany)

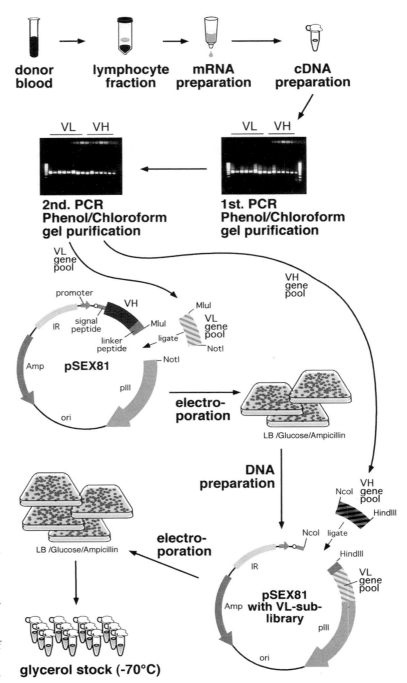

Fig. 1. Construction of a human scFv antibody library from donor blood. Abbreviations: Amp, Ampicillin resistance gene; pIII, Gene 3 of M13 coding for a minor surface protein; IR, Intergenic Region of f1 filamentous phage; VL, gene for the variable region of the light chain; VH, gene for the variable region of the heavy chain.

- RNase inhibitor ("RNA guard"), (Roche Diagnostics, Mannheim, Germany)

- RNase-free water (SIGMA, Steinheim, Germany)

- RPMI 1640 medium (Gibco BRL, Karlsruhe, Germany)

- TRIZOL-Reagent (monophasic solutions of phenol and guanidine isothiocyanate for isolating RNA), (Gibco BRL, Karlsruhe, Germany)

- Trypan blue (SIGMA, Steinheim, Germany)

- Cell centrifuge   Equipment

▪ ▪ ▪ Procedure

1. All steps are done at room temperature. Take 15 ml of donor blood and mix with an equal volume of RPMI medium supplemented with 25 mM HEPES, EDTA 5 % (w/v), 100 U/ml penicillin, 100 µg/ml streptomycin and 10 % (v/v) FCS.

2. Divide the diluted blood into two 50 ml tubes by carefully adding it on top of a cushion of 15 ml 1.077 g/ml Ficoll in isotonic NaCl. Centrifuge for 30 min at 1300 x g. Let the centrifuge run out without braking.

3. The erythrocytes sediment through the Ficoll cushion, whereas the leukocytes are concentrated at the Ficoll/buffer-interphase. Collect five milliliters of the turbid grayish interphase containing the leukocyte fraction. Wash the cells two times with 30 ml of supplemented RPMI medium by centrifugation for 7 min at 1000 x g followed by a washing step with 30 ml of supplemented RPMI medium without EDTA for 5 min at 500 x g.

4. Resuspend the leukocyte-containing pellet to a final density of 10^6 cells/ml in supplemented RPMI medium without EDTA. Determine the cell number in the presence of 0.5% (w/v) trypan blue to stain dead cells.

Note: The viability of the cells should be greater than 95%.

5. Isolate the RNA out of 10^6 cells with TRIZOL-Reagent according to the manufacturers protocol. The resulting RNA (resuspended in 9.5 µl RNase-free water) is completely used up for cDNA synthesis.

Note: Many other RNA purification kits work as well.

6. Add 0.5 µl of 1 mM oligo-dT primer to 9.5 µl of RNA (see above) and incubate for 5 min at 65°C and immediately place on ice afterwards.

7. Add 4 µl 5 x buffer for the reverse transcriptase, 2 µl of 10 mM dNTPs, 2 µl RNase-free water, 1 µl of 37,000 U/µl RNA guard and 1 µl of 20 U/µl reverse transcriptase and incubate at 37°C for 90 min. Place immediately on ice after an incubation at 95°C for 5 min.

8. The resulting first strand cDNA can be used directly in PCR reactions or stored at -20°C.

Subprotocol 2
PCR-Amplification of the Rearranged Variable Amtobody Chain Genes ("1st PCR")

▩▩   Materials

Reagents
- 5 x DNA loading buffer (according to Sambrook et al., 1989)

- dNTPs (2.5 mM each), (New England Biolabs, Inc., Beverly, USA)

- 100 bp "ladder" DNA molecular weight marker (Gibco BRL, Karlsruhe, Germany)

- 10 x PCR-buffer (containing 15 mM $MgCl_2$) (Roche Diagnostics, Mannheim, Germany)

- Agarose (e.g. LE-agarose, Biozym, Oldendorf, Germany)

- Bovine Serum Albumin (non-acetylated), (SIGMA, Steinheim, Germany)

- Chloroform (p.a., Merck, Darmstadt, Germany)

- Ethanol (p.a., Merck, Darmstadt, Germany)

- High fidelity polymerase (Roche Diagnostics, Mannheim, Germany)

- Isoamyl alcohol (p.a., Merck, Darmstadt, Germany)

- Isopropanol (p.a., Merck, Darmstadt, Germany)

- Liquid nitrogen

- TAE buffer (0.04 M Tris-acetate, 0.001 M EDTA)

- TE-saturated Phenol (Roth, Karlsruhe, Germany)
- PCR-Thermocycler   Equipment

▪ ▪ Procedure

1. Set up 15 separate PCR reactions for the amplification of VH and VL
   DNA using the primers given in Table 1. Each one of the variable heavy
   chain primers (1VH1-1VH3) is employed in combination with the corresponding 1IgM or 1IgG primer, and each κ (1κ1/4-1κ3) and λ (1λ1-
   1λ6) variable light chain primer with the corresponding constant region primer (1κC or 1λC). For a 50 µl PCR reaction, use 0.5 µl of leukocyte cDNA, 0.5 µg BSA, 5 µl PCR reaction buffer, 0.3 mM $MgCl_2$,
   0.75 U high fidelity polymerase and 25 pmol of each of the two primers.

2. Run 30 PCR cycles, each consisting of 95°C for 0.5 min (denaturation),
   55°C for 1 min (annealing), and 75°C for 1 min (extension). At the
   beginning of the first cycle incubate at 95°C for 3 min and at the
   end of the last cycle incubate at 75°C for 5 min.

   Note: Low stringency conditions are used in order to allow mismatches
   to amplify a maximum number of different immunoglobulin genes.

3. Analyze the PCR reaction by electrophoresis on a 1.5 % agarose gel (use
   5 µl for analysis, 45 µl for a preparative gel) and stain with ethidium
   bromide (expected size for the HC and lLC is 400bp, for the kLC
   360bp).

4. Excise the DNA fragments corresponding to the HC and LC under low
   wavelength UV-light (around 360 nm) and do two cycles of freezing in
   liquid nitrogen and thawing at 65°C.

5. Add an equal volume of TE-saturated phenol to the pieces of agarose,
   vortex and freeze again in liquid nitrogen.

6. Centrifuge the frozen samples at 11000 x g for 15 min at room temperature. Transfer the upper liquid phase into fresh tubes and add an
   equal volume of chloroform/isoamyl alcohol (24:1), vortex.

7. Centrifuge at 11000 x g for 5 min at room temperature. Transfer the
   upper liquid phase into fresh tubes and add 0.1 volume of sodium acetate (3 M, pH 5,2) and 0.8 volumes of isopropanol, mix by inverting and
   centrifuge in a microcentrifuge for 1 h at 4°C. Wash once with 500 µl of
   ethanol. Discard the supernatant, air-dry the DNA pellet and resuspend each pellet in 10 µl of ddH_2O.

Subprotocol 3
Introduction of Restriction Sites into the Purified Constructs of the First PCR Reaction ("2nd PCR")

▓▓ Procedure

1. The second PCR is done using the gel purified antibody gene fragments obtained from the first PCR as templates. Each individual primer contains the sequence of its corresponding primer used in the first PCR (see table 1), extended by the restriction site sequences necessary for the subsequent cloning into pSEX81. Perform 15 separate PCR reactions. Each variable heavy chain primer (2VH1-2VH3) is combined with the corresponding 2IgM or 2IgG primer and each κ (2κ1/4-2κ3) and λ (2λ1-2λ6) variable light chain primer with the corresponding constant region primers (2κC or 2λC) (see table 2). For each 50 μl PCR reaction use 2 μl of the purified product of the first PCR samples (approximately 10-20 ng), 0.5 μg BSA, 5 μl PCR reaction-buffer, 300 mM dNTP, 0.75 U high fidelity polymerase and 25 pmol of each of the two primers (see table 2).

2. Run 30 PCR cycles on a thermal cycler. A thermal cycle consists of 95°C for 0.5 min (denaturation), 57°C for 1 min (annealing), and 75°C for 1 min (extension). At the beginning of the first cycle, incubate at 95°C for 3' and at the end of the last cycle, incubate at 75°C for 5'.

   Note: For some templates, 10-15 PCR cycles might be enough. Over-cycling can lead to exhaustion of the primer pool, resulting in single strand DNA fragments which can not be cloned.

3. The amplified DNA fragments are analyzed by electrophoresis on a 1.5% agarose gel (5 ml for analysis, 45 ml for a preparative gel) and stained with ethidium bromide (expected size for the HC and λLC is 400bp, for the κLC 360bp).

   Note: One clear band should be obtained. In any other case different cycle numbers or more careful band excision after the first PCR is recommended.

4. Excise the DNA fragments corresponding to the HC and LC and follow the purification protocol mentioned above (see points 5-9, 1st PCR protocol). Always do a phenol extraction between the PCR reactions and the following digest (see next protocol), even if you choose a dif-

ferent gel extraction protocol. Omission of the phenolization step frequently results in reduced library size, probably due to residual polymerase activity during the restriction digest which affects the overhanging ends needed for ligation.

Subprotocol 4
Cloning of the VH and VL Chains into the Expression Vector pSEX81

▨▨ Materials

- 1 M EDTA (pH 8.0)   Reagents

- 5 x DNA loading buffer (according to Sambrook et al., 1989)

- Calf Intestinal Phosphatase (CIP) (Pharmacia, Freiburg, Germany)

- Electrocompetent *E.coli* XL1-blue cells (Stratagene, Amsterdam, Netherlands)

- Electroporation cuvettes (2 mm gap, 400 µl volume) (Eppendorf, Hamburg, Germany)

- Glycerol (AppliChem, Darmstadt, Germany)

- Glycogen (Roche Diagnostics, Mannheim, Germany)

- LE-Agarose (Biozym, Oldendorf, Germany)

- Midiprep columns (e.g. CONCERT™ High Purity Plasmid Midiprep System, Gibco BRL, Karlruhe, Germany)

- T4 DNA Ligase (Gibco BRL, Karlsruhe, Germany)

- T4 DNA Ligase buffer (Gibco BRL, Karlsruhe, Germany)

- NucleoSpin Extract kit for the extraction of nucleic acids from agarose gels (Macherey & Nagel, Düren, Germany)

- Plastic boxes for agar plates (400 cm^2) (Karl Bock GmbH, Kunstofftechnik, Lauterbach, Germany)

- Reaction buffers for restriction enzymes (BioLabs, Schwalbach, Germany)

- Restriction enzymes: NotI, MluI, HindIII, NcoI (BioLabs, Schwalbach, Germany)

Table 1. "1st" PCR primers for the amplification of human immunoglobulin variable (A) heavy chain and (B) light chain subgroups.

n = number of sequences that the primer was based on. Bases in parentheses represent substitutions at a given position and the number indicates the percentage of the substituted nucleotides.

(A) Heavy Chain

Variable (FR1; amino acid positions: 1-7)

VH1 (n=124): 5'C(G^{15})AG GTG CAG CTG GTG CAG TCT 3'

VH2 (n=106): 5'CAG GTG(A^{40}) CAG CTG CAG G(C^{40})AG TC 3'

VH3 (n=259): 5'GAG GTG CAG CTG G(T^{30})TG GAG TCT 3'

Constant (CH1: N-terminus; amino acid positions: 121-115)

IgM (n=9):   5'AAG GGT TGG GGC GGA TGC ACT 3'

IgG (n=42):   5'GAC C(G^{50})GA TGG GCC CTT GGT GGA 3'

(A) Light Chain

Variable (FR1; amino acid positions: 1-10)

Kappa

κ1/4 (n=171): 5'GAC ATC C(G^{50})A(T^{50})G ATG ACC CAG TCT CC 3'

κ2 (n=51):   5'GAT ATT GTG ATG ACT(C^{50}) CAG T(A^{30})CT CCA CTC T 3'

κ3 (n=171):   5'GAA ATT GTG(A^{25}) T(A^{20})TG ACG(A^{25}) CAG TCT CCA 3'

Lambda

λ1 (n=43):   5'CAG TCT GTG T(C^{25})TG ACG(T^{25}) CAG CCG CCC TCA 3'

λ2/5 (n=47): 5'CAG TCT GCG CTG ACT CAA(G^{50}) CCG G(C^{50})CC TCT 3'

λ3 (n=63):   5'TCC TAT GAA CTG ACT CAG CCA CCC(T^{33}) T 3'

λ4a (n=9):   5'TCT GAA CTG ACT CAG CCG(A^{33}T^{33}) C(G^{33})CC TC 3'

λ4b (n=9):   5'TCT GAA CTG ACT CAG GAC CCT GC(T^{33})T 3'

λ6 (n=15):   5'A(G^{33})AT TTT ATG CTG ACT CAG CCC CAC TCT 3'

Constant (CL: N-terminus; amino acid positions: 109-122)

Kappa (n=12):   5'GAA GAC AGA TGG TGC AGC CAC AGT 3'

Lambda (n=36):   5'AGA GGA G(C^{25})GG C(T^{30})GG GAA CAG AGT GAC 3'

Table 2. "2nd" PCR primers with NcoI (5') and HindIII (3') for human variable heavy chains(A), MluI (5') and NotI (3') for human variable light chains (B).
n = number of sequences that the primer was based on. Bases in parentheses represent substitutions at a given position and the number indicates the percentage of the substituted nucleotides.

(A) Heavy Chain

Variable (NcoI)

VH1: 5'GAA TAG G**CC ATG G**CG C(G^{15})AG GTG CAG CTG GTG CAG TCT 3'

VH2: 5'GAA TAG G**CC ATG G**CG CAG GTG(A^{40}) CAG CTG CAG G(C^{40})AG TC 3'

VH3: 5'GAA TAG G**CC ATG G**CG GAG GTG CAG CTG G(T^{30})TG GAG TCT 3'

Constant (HindIII)

IgM: 5'CA GTC **AAG CTT** TGG GGC GGA TGC ACT 3'

IgG: 5'CA GTC **AAG CTT** TGG GCC CTT GGT GGA 3'

(B) Light Chain

Variable (MluI)

Kappa

κ1/4: 5'TA CAG GAT CC**A CGC GT**A GAC ATC C(G^{50})A(T^{50})G ATG ACC CAG TCT CC 3'

κ2: 5'TA CAG GAT CC**A CGC GT**A GAT ATT GTG ATG ACT(C^{50}) CAG T(A^{30})CT CCA 3'

κ3: 5'TA CAG GAT CC**A CGC GT**A GAA ATT GTG(A^{25}) T(A^{20})TG ACG(A^{25}) CAG TCT CCA 3'

Lambda

λ1: 5'TA CAG GAT CC**A CGC GT**A CAG TCT GTG T(C^{25})TG ACG(T^{25}) CAG CCG CCC TCA 3'

λ2/5: 5'TA CAG GAT CC**A CGC GT**A CAG TCT GCG CTG ACT CAA(G^{50}) CCG G(C^{50})CC TCT 3'

λ3: 5'TA CAG GAT CC**A CGC GT**A TCC TAT GAA CTG ACT CAG CCA CCC(T^{33}) T 3'

λ4a: 5'TA CAG GAT CC**A CGC GT**A TCT GAA CTG ACT CAG CCG(A^{33}T^{33}) C(G^{33})CC TC 3'

λ4b: 5'TA CAG GAT CC**A CGC GT**A TCT GAA CTG ACT CAG GAC CCT GC(T^{33})T 3'

λ6: 5'TA CAG GAT CC**A CGC GT**A A(G^{33})AT TTT ATG CTG ACT CAG CCC CAC TCT 3'

Constant (NotI)

Kappa: 5'TGA CAA GCT T**GC GGC CGC** GAA GAC AGA TGG TGC AGC CAC AGT 3'

Lambda: 5'TGA CAA GCT T**GC GGC CGC** AGA GGA G(C^{25})GG C(T^{30})GG GAA CAG AGT GAC 3'

– RNaseA (BioLabs, Schwalbach, Germany)

– SOB plates (according to Sambrook et al., 1989) containing 100 µg/ml of Ampicillin and 100 mM of Glucose ("SOB-GA plates")

– SOC-Medium (according to Sambrook et al., 1989)

Equipment   – Electroporator (e.g. Gene Pulser II, BioRad, München)

Procedure

Preparation of the pSEX81 vector for the insertion of the light chain sublibrary

1. Digest 10 µg of pSEX81 DNA with 50 units of MluI and 50 units of NotI for 4 h to 16 h. The reaction mix should contain 0.5 units of alkaline phosphatase (CIP), restriction buffer according to the manufacturers instructions, and 10 µg RNaseA. After the end of the incubation time add another 0.5 units of CIP to the restriction mix and incubate for additional 2 h at 37°C.

   Note: Concentrations of glycerol higher than 5 % (v/v) may cause star activity of the restriction enzymes. Therefore, the total volume of restriction enzyme stock added to the reaction should not exceed 10 % (v/v).

2. Inactivate the CIP by adding EDTA (pH 8.0) to a final concentration of 5 mM and incubation for 20 min at 65°C.

3. Add DNA loading buffer to the sample and separate 2 µg of the digest per lane in a 1 % agarose gel.

4. Excise the larger DNA fragments (corresponding to the linearized pSEX81 vector) and follow the purification protocol mentioned above (see points 5-9, 1st PCR protocol).

Cloning of the light chain PCR products into pSEX81

1. Digest 1 µg of the purified and phenolized κ- and λ-light chain PCR products with 19 units of MluI and 14 units of NotI for 4 h to 16 h at 37°C. Heat inactivate the enzymes for 20 min at 65°C (see point 1 "Preparation of the pSEX81 Vector for the Insertion of the Light Chain Sublibrary").

2. Purify the digested PCR products with spin columns (NucleoSpin Extract kit) for DNA extraction from agarose gels according to the manufacturers protocol.

3. Ligate the vector to the insert using a molecular ratio of 1:1 to 1:3. The total amount of vector plus insert DNA should be approximately 100 ng of DNA. Add 1 unit of T4 ligase, ligation buffer according to the manufacturer and H_2O to a final volume of 10-15 µl. The reaction is carried out for 16 h at 16°C.

4. Precipitate the DNA by adding H_2O to a final volume of 50 µl and 500 µl of water-saturated butanol. Mix by inverting the tube and centrifuge at 11000 x g for 15 min at 4°C. Dry the DNA-pellet at room temperature or in a speed vac for 5 min and resuspend the DNA in 1-5 µl of H_2O.

   Note: i. For visualization and easy recovery of pellets containing only small amounts of DNA, 20 µg of Glycogen can be added to the ligation reaction prior to precipitation without any adverse effect on following reactions. ii. Alternatively to the butanol precipitation, the ligated DNA can be precipitated with ethanol followed by twice washing with 70 % ethanol. However, precipitation with butanol is more efficient.

5. Pre-warm the SOC-Medium to 37°C and pre-cool the DNA solution, cells and the electroporation cuvettes on ice. Transform 50 µl of electrocompetent *E.coli* XL1-blue bacteria with the ligations, by setting the electroporator to 2,5 kV, 200 Ω and 25 µF. Add 1 ml pre-warmed SOC medium to the 50 µl electroporation sample and incubate for 45 min at 37°C.

6. Plate the bacteria of each transformation by distributing them onto 3 SOB-GA plates and incubate for 16 h at 37°C.

   Note: i. TG1 bacteria may be used instead. ii. Cultivation of transformed bacteria on plates instead of liquid culture reduces the risk of loosing slowly growing clones. iii. To increase the transformation efficiency, the precipitated ligation products can be splitt up and used in 3-5 different electrotransformations with the appropriate amount of bacteria. iv. One ligation reaction should yield 10^6-$5x10^6$ single colonies. The number of transformants indicates the maximum number of different antibody gene fragments, thus indicating the complexity of the light chain sublibrary.

7. Resuspend the bacteria by adding 10 ml of SOB-GA medium or dYT medium to each plate and mix together. To prepare glycerol stocks,

take 1/10 of the suspension, add glycerol to a final concentration of 20 % (v/v), mix, and freeze at -80°C.

8. The remaining 9/10 of the bacteria are used for the purification of the light chain sublibrary DNA via midiprep columns according to the manufacturers protocol. Use one column for 5×10^6-2.5×10^7 colonies.

Note: Further amplification after this step should be avoided to minimize loss of complexity of the library by different growth rates of the clones.

Construction of the VH+VL-library: preparation of the heavy chain PCR products and cloning into pSEX81 containing the light chain sublibrary

1. Prepare the DNA of the light chain sublibrary in pSEX81 as described above (see sub-protocol "Cloning of the Light Chain PCR Products into pSEX81").

2. Digest 1µg of the purified and phenolized IgM and IgG heavy chain PCR products with 32 units of NcoI and 19 units of HindIII. The reaction is proceeded for 4 h to 16 h at 37°C. The restriction enzymes are heat inactivated for 20 min at 65°C (see point 1 "Cloning of the Light Chain PCR Products into pSEX81").

3. Purify the digested PCR products with the NucleoSpin Extract purification kit according to the manufacturers protocol.

4. Ligate the heavy chain fragments into the pSEX81 vector containing the light chain sublibrary following the procedure described (see points 3-4 "Cloning of the Light Chain PCR Products into pSEX81").

5. The transformation of electrocompetent *E.coli* XL1-blue bacteria and their collection and storage is done as described in "Cloning of the Light Chain PCR Products into pSEX81" (see points 5-7). The resulting library can be used directly for packaging (see Chapter 6) into M13 phages or stored in aliquots at -70°C.

References

Barbas, C. F., Kang, A. K., Lerner, R. A. and Benkovic, S. J. (1991) Assembly of combinatorial antibody libraries on phage surfaces: the gene III site. Proc. Natl. Acad. Sci., USA, 88, 7978

Breitling, F., Dübel S., SeehausT., Klewinghaus, I., and Little, M. (1991) A surface expression vector antibody screening. Gene 104, 147.

Dörsam, H., Rohrbach, P., Kurschner, T., Kipriyanov, S., Renner, S., Braunagel, M., Welschof, M. and Little, M. (1997) Antibodies to steroids from a small human naive IgM library. FEBS Lett. 414(1), 7.

Marks, J. D., Hoogenboom, H. R., Bonnert, T. P., McGafferty, J., Griffiths, A. D. and Winter, G. (1991) By-passing immunization. Human antibodies from V-gene libraries displayed on phage. J. Mol. Biol., 222, 581.

Sambrook, J., Frisch, E. F. and Maniatis, T., (1989) Molecular Cloning. Second Edition. Cold Spring Harbor Laboratory Press.

Welschof, M., Terness P., Kipriyanov S. M., Stanescu D., Breitling F., Dörsam H., Dübel S., Little M. and Opelz G. (1997) The antigen-binding domain of a human IgG-anti-F(ab')2 autoantibody, Proc. Natl. Acad. Sci., U S A, 94(5), 1902

Welschof, M., Terness P., Kolbinger F., Zewe M., Dübel S., Dörsam H., Hain C., Finger M., Jung M., Moldenhauer G., et al. (1995) Amino acid sequence based PCR primers for amplification of rearranged human heavy and light chain immunoglobulin variable region genes. Immunol. Methods, 179(2), 203

Generation of Rabbit Immune Libraries

RÜDIGER RIDDER and HERMANN GRAM

Introduction

The combinatorial cloning of antibody variable domains and the display of antibody fragments on the surface of filamentous phage by fusion to phage coat proteins provides an alternative strategy for the rapid generation of monoclonal antigen-binding proteins. Various antibody-displaying phage libraries have been described, which are based on the B cell repertoire of rearranged immunoglobulin genes from spleen and bone marrow of previously immunized mice (see Chapter 2, Chapter 3). However, the combinatorial library technique may also be applied for the generation of monoclonal antibody fragments from species, which are not easily amenable to conventional methodology based on eukaryotic cell fusion techniques, e.g. humans, rabbits, and chicken. The numerous reports about the isolation of functional antibody fragments from combinatorial libraries prepared from peripheral blood lymphocytes from immunized or non-immunized human donors exemplify the general applicability of this approach (see Chapter 6, Chapter 7).

For a long time, rabbits have been used for the generation of polyclonal antibodies of excellent quality. However, the routine production of monoclonal antibodies from these animals has been hampered by the lack of efficient strategies for the stable immortalization of antibody-expressing B lymphocytes, although some progress has recently been made in the generation of suitable fusion partners (Raman et al., 1994; Spieker-Polet et al., 1995). Therefore, we and others have successfully used the combinatorial library and phage display technologies to produce monoclonal

✉ Rüdiger Ridder, MTM Laboratories AG, Im Neuenheimer Feld 519,
69120 Heidelberg, Germany (*phone* +49-6221-6496614; *fax* +49-6221-6496610;
e-mail Ridder@MTM-Laboratories.com)
Hermann Gram, Novartis Pharma AG, Arthritis and Bone Metabolism, Bldg. 386-927,
4002 Basel, Switzerland

antibody fragments from rabbits (Ridder et al., 1995; Lang et al., 1996; Foti et al., 1998).

Rabbits show exceptional features with respect to the mechanisms, which are utilized for the generation of antibody diversity. Studies of the VH gene rearrangements in rabbit B cells revealed that although there are more than one hundred germline genes, preferentially a single VH gene (VH1) is rearranged in 70-90% of all VHa allotype immunoglobulin expressing B-lymphocytes (Knight and Becker, 1990; Raman et al., 1994). Furthermore, the similarity of the individual VH genes is generally more than 80%, indicating the affiliation to a single VH gene family (reviewed in Knight, 1992; Knight and Crane, 1994). The diversity of the VH antibody repertoire was shown to be mainly generated by somatic gene conversion and hypermutation (reviewed in Mage, 1998; Becker and Knight, 1990; Knight and Becker, 1990; Raman et al., 1994). Similarily, rabbit light chain generation is characterized by the preferential expression of a single C gene (C1), which represents more than 90% of all light chains (Heidmann and Rougeon, 1983; Knight and Crane, 1994). Additionally, there is only a small number of functional J-segments present in rearranged VL genes (one functional *b4* and two functional *b9* allotype J-segments, Emorine et al., 1983; Akimenko et al., 1986).

Therefore, the limited number of germline immunoglobulin genes used for antibody generation significantly facilitates the molecular cloning of the rabbit antibody repertoire. In contrast to the high number of degenerative oligonucleotide primers necessary for the PCR amplification of both murine and human immunoglobulin genes (see Chapters 2-7), only very few primers are required for the representative cloning of the rearranged immunoglobulin gene repertoire from rabbits (Ridder et al., 1995). For this reason, immunized rabbits represent rather ideal candidate animals for the generation of monoclonal antibody fragments by the combinatorial library and phage display approach.

Outline

In this chapter, we focus on the description of a general strategy for the PCR amplification and cloning of cDNAs coding for the variable heavy and light chain domains of rabbits. The combinatorial cloning of these VH- and VL-specific PCR fragments allows the subsequent display of rabbit scFv antibody fragments on the surface of filamentous bacteriophage and thus the affinity selection of phage antibodies with the desired antigen specificity. The general outline for the generation and phage display selection of scFv antibody fragments from rabbits is rather comparable to the protocol described for the production of murine scFv antibodies (Chapter 5). The essential difference solely exists in the selection of specific primers necessary for the PCR amplification of cDNAs coding for rabbit VH and VL antibody domains (listed in Table 1).

The combinatorial cloning approach we used for the generation of rabbit scFv antibody fragments is outlined in Fig. 1. In general, any of the various cloning strategies described in the literature might be employed, but may require modifications of the oligonucleotide primers (e.g. by introducing different restriction endonuclease recognition sites) dependent on the used vector and scFv linker sequences.

Table 1. Oligonucleotide primers for the amplification of rabbit VL and VH antibody domains. Wobble positions are indicated according to the IUPAC code: M, A/C; K, G/T; R, A/G; Y, C/T; S, C/G; W, A/T. The *Sac*I restriction endonuclease recognition site used for the cloning of the PCR fragments comprising the VL domain is underlined.

VL	
VL--1	5′-AGC ACC GAG CTC GTG MTG ACC CAG ACT CCA-3′
VL--2	5′-AGC ACC GAG CTC GAT MTG ACC CAG ACT CCA-3′
VL--b4J2	5′-TTT GAC GAC CAC CTC GGT CCC-3′
VL--b9J2	5′-TAG GAT CTC CAG CTC GGT CCC-3′
CL-	5′-CTG CGG TGT KTT ACT GTT CTC G-3′
VH	
VH1	5′-CAG TCG GTG GAG GAG TCC RGG-3′
VH2	5′-CAG TCG GTG AAG GAG TCC GAG-3′
VH3	5′-CAG TCG YTG GAG GAG TCC GGG-3′
VH4	5′-CAG SAG CAG CTG GWG GAG TCC GG-3′
CH1-	5′-GAC TGA YGG AGC CTT AGG TTG C-3′

Additionally, we include primer sequences which we found suitable for the amplification of the VL and CL domains as well as the VH and CH1 domains for the cloning of rabbit Fab antibodies.

Materials

– Tissue homogenizer for the extraction of RNA from frozen spleen samples (e.g. Ultra Turrax T8, IKA)   **Equipment**

– Refrigerated microcentrifuge

– PCR thermocycler (e.g. Perkin-Elmer 9600)

– Rabbit generally used for the production of polyclonal antibodies (e.g.   **Reagents** New Zealand white rabbit)

– Adjuvant (e.g. incomplete Freund's adjuvant)

– Liquid nitrogen (if available) or ethanol/dry ice bath

– *Taq* DNA polymerase and PCR reaction buffer (supplied with the enzyme)

– RNA extraction kit (e.g. Trizol Reagent, Life Technologies Gibco BRL)

– Gel extraction kit (e.g. GENECLEAN II Kit, BIO 101 Inc.)

– PCR product purification kit (e.g. High Pure PCR Product Purification Kit, Roche Diagnostics)

– Oligonucleotide primers for the initial PCR amplification of rabbit VL and VH domains (listed in Table 1) as well as for subsequent modifying of PCR reactions for the incorporation of both linker and restriction endonuclease recognition sequences into the PCR fragments (listed in Table 2).

Table 2. Rabbit VL and VH domain modifying oligonucleotide primers. Wobble positions are indicated according to the IUPAC code: S, C/G; W, A/T. The *Xho*I restriction endonuclease recognition site introduced into the linker sequence as well as the *Nhe*I site added to the 5'-end of the CH1- specific reverse PCR primer are underlined. Linker sequences are indicated by small letters.

b4J2-link	5´- acc aga agt aga acc TTT GAC GAC CAC CTC G -3´
b9J2-link	5´- acc aga agt aga acc TAG GAT CTC CAG CTC G -3´
J2-link-Xho	5`- atc cct cga gct tcc cag aac cag aag tag aac cT- 3´
link-VH1-3-Xho	5´- gga agc tcg agg aag gaa aag gcC AGT CGG TGG AGG AG -3´
link-VH4-Xho	5´- gga agc tcg agg aag gaa aag gcC AGS AGC AGC TGG WG -3´
CH1--Nhe	5´- GAG GGC TAG CGA CTG AYG GAG CCT AGG TT G -3´

Procedure

Immunization of rabbits

1. Immunize the rabbit according to standard immunization protocols (typically by subcutaneous injection of 200 µg antigen dissolved in adjuvant, followed by subsequent injections of 100 µg antigen after 4 and 6 weeks).

2. Control for specific antibody titer of the immunized rabbit by performing ELISA assays with serial dilutions (1/1.000 - 1/20.000) of pre-immune control serum and immune serum.

Preparation of B cell fractions

For the preparation of antibody producing B cells, either spleen (Ridder et al., 1995), bone marrow (Lang et al., 1996), or peripheral blood lymphocytes (Foti et al., 1998) may be used. If sufficient serum has been collected, the extraction of RNA from spleen and/or bone marrow from the sacrificed rabbit is recommended.

1. After taking the spleen from the sacrificed rabbit with steril forceps, put the spleen into a petridish plate and cut it into 10 to 12 pieces of about equal size. Transfer each of the pieces into an individual polypropylen tube, and freeze the tubes immediately by placing them into

liquid nitrogen or into an ethanol/dry ice bath. Generally, we use 2 ml cryo-tubes to avoid leaking during freezing and storage.

2. Store the tubes at either -80°C or in liquid nitrogen until use.

RNA extraction and first strand cDNA generation

For total RNA isolation, we generally utilized the acid guanidinium thiocyanate-phenol-chloroform based extraction protocol as described by Chomczynski and Sacchi (1987). However, any commercially available RNA isolation system may be used. Furthermore, mRNA instead of total RNA may be directly isolated from the antibody producing B cells, however, the purification of mRNA is not necessary to obtain material of sufficient quality for the subsequent RT-PCR amplification of the immunoglobulin gene repertoire. In the following, we give some advice concerning critical steps in the preparation of total RNA of adequate quality.

1. Add the guanidinium thiocyanate (GTC) containing RNA extraction buffer directly to a single sample of frozen spleen, thus eliminating any RNAse activity during thawing of the sample in the absence of GTC.

    Note: Wear gloves and protection glasses during this and the following steps of the RNA extraction protocol.

2. Homogenize the rabbit spleen sample for about 30 seconds using a tissue homogenizer. Alternatively, vortex the tube vigourously until the spleen sample is completely disintegrated.

3. Use sufficient amounts of the denaturing RNA extraction buffer. Insufficient quantities of the extraction buffer will result in low RNA yield as well as strong contaminations of the final RNA solution with genomic DNA. Generally, we obtained satisfying results by using about 2 ml of extraction buffer per spleen sample representing one tenth of total rabbit spleen.

4. Control for integrity of RNA by agarose gel electrophoresis. Employing a denaturing sample loading buffer containing formamid, extracted RNA may be separated in 0.8% agarose minigels using standard trisborate-EDTA buffer (1x TBE).

5. Prepare first strand cDNA by reverse transcription of 3 µg of total RNA using either AMV or M-MLV reverse transcriptase according to the

instructions of the manufacturer. For priming of the reverse transcription reaction, either oligo (dT)18, random hexanucleotides pd(N)6, or rabbit immunoglobulin constant domain specific oligonucleotides CL- (5'-CTG CGG TGT TKT ACT GTT CTC G -3') and CH1- (5'-GAC TGA YGG AGC CTT AGG TTG C -3') may be used. We generally obtained good results using 0.2 µg oligo (dT)18 for priming of the reverse transcription reaction of about 3 µg of total RNA.

PCR amplification of rabbit VL and VH antibody domains

1. For the amplification of rabbit immunoglobulin cDNA, use standard PCR reaction mixtures (total volume: 50 µl) as follows: 20 mM Tris-HCl (pH 8.4), 50 mM KCl, 1.5 mM MgCl2, 0.2mM dNTP, 25 pmole of each primer, 2.5 U *Taq* DNA polymerase, and about 1/20 of the first strand cDNA reaction as template.

   Note: Hot start conditions (e.g. by using TaqStart Antibody, Clontech Laboratories, Palo Alto) significantly increased the yield of PCR products specific for VL and VH antibody domains in our hands.

2. Oligonucleotide primers listed in Table 1 are used in individual PCR reactions in the following combinations: VL–1/VL–b4J2, VL–1/VL–b9J2, VL–2/VL–b4J2, and VL–2/VL–b9J2 for the amplification of rabbit VL specific cDNAs; VH1/CH1-, VH2/CH1-, VH3/CH1-, and VH4/CH1- for the amplification of rabbit VH specific cDNAs.

3. After an initial denaturation step for 3 min at 94°C, perform 25 – 30 cycles of PCR amplification as follows: 94°C for 30 s, 60°C for 30 s, and 72°C for 45 s, followed by a final elongation step at 72°C for 7 min.

4. Check 20 µl of the PCR products on 1.2% TAE-buffered agarose gels. PCR products specific for rabbit immunoglobulin variable chain cDNA will result in DNA fragments of 330 to 340 bp (VL) and of about 400 bp (VH), respectively.

5. Extract the VL and VH specific PCR fragments from the agarose gel by using any commercial gel extraction kit (e.g. Geneclean II, BIO101 Inc.; or QIAquick Gel Extraction kit, Qiagen, Hilden, Germany).

Modification of the VL and VH cDNA fragments by subsequent PCR reactions

For cloning of the antibody variable domain encoding PCR fragments, we generally use a two-step cloning strategy. In contrast to the experimental approach described in Chapter 2, we usually generate the DNA sequence coding for the scFv linker peptide by 1-2 subsequent rounds of PCR using modifying oligonucleotide primers (listed in Table 2). Furthermore, we use a linker initially described by Colcher et al. (1990), which codes for a peptide consisting of 14 amino acids (NH2-gly-ser-thr-ser-gly-ser-gly-lys-ser-ser-glu-gly-lys-gly-COOH). By replacing both serine residues in codon positions 9 and 10 (underlined) with leu-glu residues, a *Xho*I restriction endonuclease recognition site was introduced into the linker sequence. This site was subsequently utilized for joining of the VL and VH encoding cDNA domains.

For the modification of the VL specific cDNA fragments, two subsequent rounds of PCR reactions are necessary, whereas a single modifying PCR step is sufficient for the VH specific fragments (see Figure 1).

1. Subject about 1/20 of the purified PCR products obtained after the initial round of VL and VH specific PCRs to the second round of PCR amplifications in individual reactions. Use the primer combinations as shown in Figure 1.

   Note: Primer b4J2-link is employed using the first round VL PCR product generated with reverse primer VL-κ-b4J2, etc., and primer link-VH1-3-Xho is used for the modifying PCR of VH specific products generated with primers VH1, VH2, and VH3, respectively.

2. Perform 25 cycles of PCR amplification using identical conditions as described above. Separate the resulting VL specific PCR products by agarose gelelectrophoresis and subsequently extract the DNA fragments from the gel matrix.

3. Perform a second modifying PCR amplification for the VL specific DNA fragments, using J2-link-Xho as the only reverse primer (see Figure 1).

4. Purify the final VL and VH specific PCR products using any commercially available PCR product purification kit (e.g. High Pure PCR Product Purification Kit, Roche Diagnostics).

5. Digest the purified PCR products with either restriction endonucleases *Sac*I and *Xho*I (VL) or with *Xho*I and *Nhe*I (VH) for several hours. Separate the cleaved PCR fragments by agarose gel electrophoresis

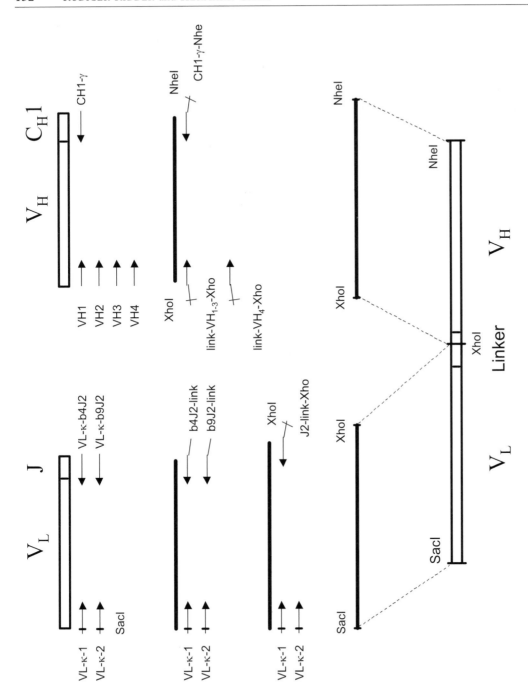

Fig. 1. General strategy for the PCR amplification of the expressed VL and VH immunoglobulin gene repertoire from rabbits. Primers used for the subsequent modifying PCR amplifications are listed in Tables 1 and 2, respectively. Restriction enzymes used for the two-step cloning of the VL and VH domains are indicated.

and subsequently extract the fragments from the gel. The VL and VH fragments are now ready for cloning into a vector suitable for the expression of the rabbit scFv fragments translationally fused to a bacterial leader sequence and to sequences coding for parts of the minor or major coat proteins of filamentous phage for surface display (e.g. pGEM3-gIII, Ridder et al., 1995).

Comments

- The restriction enzymes selected for the two-step cloning of the rabbit variable antibody domains (*Sac*I, *Xho*I, and *Nhe*I), which show no cleavage site within the framework sequence of a known rabbit variable domain, may be substituted by other cutting enzymes according to the individual cloning parameters of other plasmid vectors allowing phage display (e.g. by eight-cutters *Sfi*I and *Not*I).

  Note: It may be necessary to adjust the reading frame at the 3'-end of the scFv fragment when using different cloning sites and phage display vectors.

- For the generation of recombinant Fab fragments from rabbit, the VL and VH forward primers VL–1, VL–2, and VH1 to VH4 may be used in combination with reverse primers specific for the 3'-end of the light chain coding sequence (C–1: 5'- ACA GTC ACC CCT ATT GAA GCT CTG GAC -3'; C–2: 5'- ACA GTT CTT CCT ACT GAA GCT CTG GAC -3') and for the 3'-end of the first constant domain of the heavy chain (FAB-CH1-Cys230: 5'- GCA TGT CGA GGG TGC AAC GGT CTT GTC CAC TTT G -3'), comprising the cys residue in codon position 230 (underlined, codon position according to Kabat et al., 1991) of the rabbit heavy chain.

References

Akimenko MA, Mariame B, Rougeon F. (1986) Evolution of the immunoglobulin light chain locus in the rabbit: evidence for differential gene conversion events. Proc. Natl. Acad. Sci. USA, 83: 5180-5183

Becker RS, Knight KL. (1990) Somatic diversification of immunoglobulin heavy chain VDJ genes: evidence for somatic gene conversion in rabbits. Cell, 63:987-997

Brodeur PH, Riblet R. (1984) The immunoglobulin heavy-chain variable region (Igh-V) locus in the mouse: Part I. One hundred Igh-V genes comprise seven families of homologous genes. Eur. J. Immunol., 14:922-930

Chomczynski P, Sacchi N. (1987) Single-step method of RNA isolation by acid gua-nidinium thiocyanate-phenol-chlorofom extraction. Anal. Biochem., 162:156-159

Colcher D, Bird R, Roselli M, Hardman KD, Johnson S, Pope S, Dodd SW, Pantoliano MW, Milenic DE, Schlom J. (1990) In vivo tumor targeting of a recombinant single-chain antigen-binding protein. J. Natl. Cancer Inst., 82:1191-1197

Eldin P, Pauza ME, Hieda Y, et al. (1997) High-level secretion of two antibody single chain Fv fragments by *Pichia pastoris*. J. Immunol. Methods, 201:67-75.

Emorine L, Max EE. (1983) Structural analysis of a rabbit immunoglobulin kappa 2 J-C locus reveals multiple deletions. Nucleic Acids Res., 11:8877-8890

Emorine L, Dreher K, Kindt TJ, Max EE. (1983) Rabbit immunoglobulin genes: struc-ture of a germline *b4* allotype J-C locus and evidence for several *b4*-related se-quences in the rabbit genome. Proc. Natl. Acad. Sci. USA, 80:5709-5713

Foti M, Granucci F, Ricciardi-Castagnoli P, Spreafico A, Ackermann M, Suter M (1998) Rabbit monoclonal Fab derived from a phage display library. J. Immunol. Methods, 213:201-212

Gram H, Schmitz R, Ridder R. (1998) Secretion of scFv antibody fragments. Methods Mol. Biol. 103:179-192

Heidmann O, Rougeon F. (1983) Diversity in the rabbit immunoglobuline kappa chain variable regions is amplified by nucleotide deletions and insertions at the V-J junc-tion. Cell, 34:767-777

Hole NJ, Young CG, Mage RG. (1991) Mapping of the duplicated rabbit immunoglo-bulin kappa light-chain locus. Eur. J. Immunol., 21:403-407

Kabat TA, Wu TT, Perry HM, Gottesmann KS, Foeller C (1991) Sequences of proteins of immunological interest. US Dept. of Health and Human Services, Public Health Service, National Institutes of Health, Bethesda, MD.

Knight KL. (1992) Restricted VH gene usage and generation of antibody diversity in rabbit. Annu. Rev. Immunol., 10:593-616

Knight KL, Becker RS. (1990) Molecular basis of the allelic inheritance of rabbit im-munoglobuline VH allotypes: implications for the generation of antibody diversity. Cell, 60:963-970

Knight KL, Crane MA. (1994) Generating the antibody repertoire in rabbit. Adv. Im-munol. 56:179

Lang IM, Barbas CFIII, Schleef RR (1996) Recombinant rabbit Fab with binding activity to type-1 plasminogen activator inhibitor derived from a phag-display library against human -granules. Gene, 172:295-298

Mage RG (1998) Diversification of rabbit VH genes by gene-conversion-like and hy-permutation mechanisms. Immunol. Rev., 162:49-54

Raman C, Spieker-Polet H, Yam PC, Knight KL (1994) Preferential VH gene usage in rabbit Ig-secreting heterohybridomas. J. Immunol. 152:3935-3945

Ridder R, Schmitz R, Legay F, Gram H. (1995) Generation of rabbit monoclonal anti-body fragments from a combinatorial phage display library and their production in the yeast *Pichia pastoris*. Biotechnology, 13:255-260

Spieker-Polet H, Sethupathi P, Yam PC, Knight KL. (1995) Rabbit monoclonal anti-bodies: generating a fusion partner to produce rabbit-rabbit hybridomas. Proc. Natl. Acad. Sci. USA, 92:9348-9352

Suter M, Blaser K, Aeby P, Crameri R (1992) Rabbit single domain antibodies specific to protein C expressed in prokaryotes. Immunol. Lett. 33:53-60

Isolation of Antibody Fragments
from Combinatorial Libraries

Immunotube Selections

ROLAND E. KONTERMANN

Introduction

Selection of antibody fragments by "biopanning" on antigen-coated immunotubes is the most commonly used method for the isolation of antibody fragments from phage display libraries (Marks et al., 1991; Harrison et al., 1996; for review see Winter et al., 1994). The protocol shown here describes selections from a phagemid library (Vaughan et al., 1996) (see also Chapter 6) based on phagemid vectors such as pCANTAB6 (McCafferty et al., 1994). In these vectors the scFv fragment is cloned in front of the gene 3 with an amber stop codon located C-terminal of the scFv. For purification and detection, the scFv also contains a Myc-tag and a hexahistidyl-tag at the C-terminus preceding the amber stop codon. For the production of phage, the scFv fragment is expressed in fusion with the gene3 protein propagating the phagemids in suppressor strains such as TG1 (Gibson, 1984). The selections are performed by immobilising the antigen on a plastic surface and subsequent incubation with phage from the antibody library. Unbound phage are eliminated by a washing step and bound phage are eluted and used to infect bacteria. Phage are then rescued by co-infection with helper phage. The resulting phage particles are precipitated from the culture supernatant and are subjected to the next round of selection. Normally, three to four rounds of selections are performed to enrich for specific phage. The success of selection is monitored by polyclonal phage ELISA and monoclonal antibodies are identified by screening of soluble antibody fragments expressed by non-suppressor strains such as HB2151 (Carter et al., 1985) resulting in secretion of the antibody fragments into the culture medium.

Roland E. Kontermann, Philipps-Universität, Institut für Molekularbiologie und Tumorforschung, Emil-Mannkopff-Straße 2, 35033 Marburg, Germany
(*phone* +49-6421-286-6727; *fax* +49-6421-286-8923; *e-mail* rek@imt.uni-marburg.de)

Outline

The selection and screening procedures are outlined in Fig. 1.

Materials

Equipment   – 37°C incubator and shaker

– 30°C incubator and shaker

– ELISA plate reader

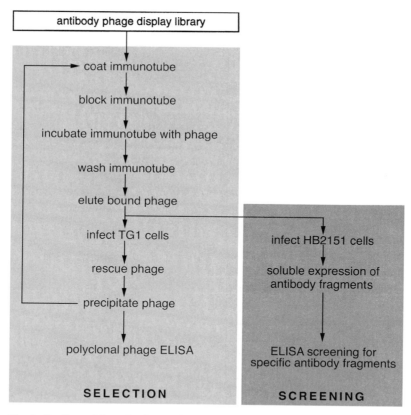

Fig. 1. Outline of the selection and screening procedures

- 5 ml immunotubes (Nunc MaxiSorp^{TM}; catalogue # 444202) **Plastic ware**

- 96 well microtitre plates for ELISA (Nunc MaxiSorp^{TM}; Falcon Microtest III, etc.)

- 96 well round bottom cell culture plates

- 1.5 ml and 50 ml tubes

- TG1 (Gibson, 1984) **Bacteria and bacteriophage**

- HB2151 (Carter et al., 1985)

- helper phage (M13KO7 from Pharmacia or M13 VCS from Stratagene)

- ampicillin-stock solution (1000 x): 100 mg/ml in H_2O **Antibiotics**

- kanamycin-stock solution (1000 x): 25 mg/ml in H_2O

- 2xTY medium: 16 g bacto-tryptone, 10 g yeast extract, 5 g NaCl per litre **Media and plates**

- TYE/amp/glucose plates (9 cm round plates and 24 cm square plates): per litre 2xTY medium containing 100 µg/ml ampicillin and 1% glucose add 15 g agar

- Minimal plates: autoclave 15 g agar in 880 ml water. Cool, then add 10 ml 20% glucose, 1 ml 1 M $MgSO_4$, and 1 ml vitamine B1 (2 mg/ml), 100 ml M9 stock (per litre: 58 g Na_2HPO_4, 30 g KH_2PO_4, 5 g NaCl, 10 g NH_4Cl, pH 7.2)

- horseradish peroxidase-conjugated goat anti-M13 (Pharmacia; catalogue # 27-9421-01) **Antibodies**

- anti-Myc antibody 9E10 (Genosys; catalogue # OM-11-908)

- horseradish peroxidase-conjugated goat anti-mouse (Sigma; cataloque # A 2554)

- Phosphate-buffered saline (PBS) **Buffers**

- PBS, 0.1% Tween-20

- 50 mM carbonate buffer pH 9.6

- 1 M Tris-HCl pH 7.5

Other solutions   – 20% glucose in H_2O

– 100 mM triethylamine in H_2O (freshly prepared)

– PEG-NaCl: 20% polyethyleneglycol 6000, 2.5 M NaCl

– Glycerol

– TMB substrate solution: 100 µg/ml 3'3',5,5'-tetramethylbenzidine (diluted from a 10 mg/ml stock in DMSO) in 100 mM sodium-acetate buffer pH 6.0, add 2 µl 30% H_2O_2 to 10 ml of the solution prior to use

Oligonucleotides   – LMB3: 5'- CAG GAA ACA GCT ATG ACC-3'

– fdSeq1: 5'-GAA TTT TCT GTA TGA GG-3'

Procedure

Selection from phagemid libraries

1. Coat immunotubes with 1 or 4 ml of antigen in a suitable coating buffer (normally we use PBS or carbonate buffer) at 1-100 µg/ml overnight at 4°C. For standard protocols we use 10 µg/ml. The volume used depends mainly on the amount of antigen available.

2. Inoculate an overnight culture of TG1 in 2xTY. Use a fresh TG1 culture grown on a minimal plate for inoculation. Since the genes for production of the pili are encoded by an episome also containing the genes for proline synthesis, growing the strain on a minimal plate ensures production of pili. From round 2 on, inoculate also an overnight culture of HB2151.

3. Next day, wash the immunotube 2-3 times with PBS and block remaining binding sites with PBS, 2% skimmed milk powder (2% MPBS) for approximately 2 hrs at RT. Fill tube to the brim (approximately 5 ml), even if you used only 1 ml for coating.

4. Preincubate phage (e.g from the library stock or from a previous selection round) for approximately 30 min with 1-4 ml 2% MPBS (depending on whether you used 1 or 4 ml for coating).

5. Empty the immunotube, add the preblocked phage and seal the tube with parafilm. Incubate for 2 hrs at RT. If you use 4 ml you can incubate the tube rotating for 30 min and then standing for 1.5 hrs.

6. Per selection, inoculate 11-12 ml of 2xTY with 100 µl of the TG1 overnight culture. Incubate at 37°C until an OD_{600} of 0.4-0.5 is reached (takes approximately 2 hrs). The OD is very critical. The bacteria should be in an early log-phase to ensure presence of pili.

   Note: The bacteria will not produce pili if grown below 33-34°C. Beginning with round 2, perform the same with HB2151. After having reached an OD_{600} of 0.4-0.5 the bacteria can be kept on ice. However, do not keep them longer than 30 min on ice since the bacteria will start to loose their pili.

7. Wash immunotubes 5-20 times with PBS, 0.1% Tween-20 and subsequently 5-20 times with PBS. Each washing step is performed by pouring the wash solution in and out, e.g. using a wash bottle. In the first two rounds we normally wash only 5-10 times with each solution so as not to lose rare binders, while in later rounds stringency can be increased by washing 10-20 times.

8. Remove the remaining wash buffer, add 1 ml of 100 mM triethylamine and incubate for 8-10 min (longer incubation will kill the phage). If you used 4 ml antigen, seal tube with parafilm and incubate while rotating.

9. Transfer solution to a sterile reaction tube containing 500 µl 1 M Tris-HCl pH 7.5 to neutralise the solution.

10. Take 1 ml of neutralised phage and add them to 10 ml of log-phase TG1. Incubate 30 min at 37°C standing and then additional 30 min shaking. In round 2-4, take also 10 µl of the neutralised phage and add them to 1 ml of log-phase HB2151 to get single colonies for soluble expression (see Section Screening of soluble scFv by ELISA).

11. Make serial dilutions of the TG1 and HB2151 culture (e.g. 1:10, 1:100, 1:1000, etc.) and plate 100 µl from each dilution as well as from the undiluted HB2151 culture onto a TYE plate containing 100 µg/ml ampicillin, 1% glucose. Incubate overnight at 37°C. The TG1 cultures are used to determine the number of eluted phage. Normally, in the first two rounds you will get between 10^4 - 10^6 eluted phage, in later rounds this number can increase to the number of bacteria used for infection ($> 10^9$) (Fig. 2). However, it does not necessarily mean that the selection has failed if the number of eluted phage is not increasing.

12. Take the remaining TG1 culture and spin for 10 min at 3500 rpm. Resuspend cell pellet in 1 ml of 2xTY. Plate onto a large 24x24 cm TYE plate containing 100 µg/ml ampicillin, 1% glucose. Incubate overnight at 30°C.

Rescue of phage with helper phage

1. Next day, add 10 ml of 2xTY to the large plate and scrape the bacteria from the plate with a glass spreader. Transfer to a 15 ml tube.

2. Make a glycerol stock from 1 ml of the bacteria.

3. Add approximately 50-200 μl of the bacteria to 50 ml of 2xTY, 100 μg/ml ampicillin, 2% glucose to get a starting OD_{600} of 0.1. Incubate shaking at 37°C until an OD_{600} of 0.4-0.5 is reached ($1-2 \times 10^{10}$ bacteria). This will take approximately 1 hr.

4. Add approximately 10 fold excess (10^{11} t.u.) of helper phage (M13 KO7 or VCS M13) to the bacteria. Incubate for 30 min at 37°C standing and then additional 30 min shaking.

5. Centrifuge for 10 min at 3500 rpm and resuspend pellet in 50 ml 2xTY, 100 μg/ml ampicillin, 25 μg/ml kanamycin (**NO glucose !**).

   Note: The expression of scFv-g3 fusion proteins without induction with IPTG and in the absence of glucose is sufficient for production of phage displaying antibody fragments.

6. Incubate overnight at 30°C.

7. Centrifuge culture for 15 min at 3500 rpm and transfer 40 ml of the supernatant to a 50 ml tube.

8. Precipitate phage by adding 8 ml (1/5 volume) 20% PEG, 2.5 M NaCl and incubate on ice for 1 h.

9. Centrifuge for 10 min at 3500 rpm and discard supernatant.

10. Resuspend pellet in 1 ml PBS. Respin to remove remaining bacteria and transfer supernatant to a fresh tube.

11. To determine the phage titre, make serial dilutions of the phage and incubate with 1 ml of log-phase TG1 as described. Plate 100 μl onto a TYE, amp, glucose plate and incubate at 37°C overnight. Count colonies to calculate phage titre. Phage titre should be around 10^{12}-10^{13} t.u./ml.

12. Repeat selection as described using 100 μl (10^{11}-10^{12} t.u.) of rescued phage for the next round. We normally perform 3-4 rounds of selection. Further rounds can be performed (and might sometimes be necessary) but this can lead to the loss of diversity of the selected antibodies and the enrichment of those having growth advantages or slow dissociation rates.

Polyclonal phage ELISA

1. Coat microtitre plate with your antigen at 1-10 µg/ml overnight in a suitable buffer (PBS or carbonate buffer pH 9.6) at 4°C. Use one or more appropriate proteins as negative controls. For each round and antigen coat two wells, plus two wells as blank.

2. Block remaining binding sites with 2% MPBS for 2 hrs.

3. Pipette 90 µl of 2% MPBS per well and add 10 µl of rescued phage. Incubate for 1 hr at RT.

4. Wash plate 6 times with PBS.

5. Add 100 µl of HRP-conjugated anti-M13 antibody diluted 1/5000 in 2% MPBS. Incubate for 1 hr at RT.

   Note: If you use an unconjugated anti-M13 antibody you have to include an additional incubation step with an HRP-conjugated secondary antibody.

6. Wash plate 6 times with PBS.

7. Add 100 µl of TMB/H_2O_2 per well and incubate until blue color has developed. Stop reaction by adding 50 µl of 1 M sulfuric acid. Read plate at 450 nm in a microtitre plate reader (see Fig. 2 for a typical result). If you have successfully enriched positive phage you should see an increase in signal (normally at round 2-3).

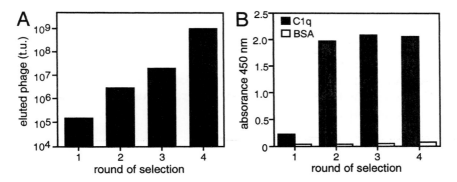

Fig. 2. A) Titre of phage eluted from immunotubes coated with C1q and selected using a naive antibody library (Vaughan et al., 1996; Kontermann et al., 1997). B) Polyclonal phage ELISA with phage selected on C1q-coated immunotubes.

Screening of soluble scFv by ELISA

1. Pipette 100 µl of 2xTY, 100 µg/ml ampicillin, 1% glucose into each well of a sterile microtitre plate. The high concentration of glucose in the medium ensures that expression of the antibody fragments which is driven by the lacZ promoter is completely inhibited.

2. Pick single colonies from the HB2151 plate of Step 11 (see Section Selection from phagemid libraries) using sterile toothpicks and transfer them to the microtitre plate (we normally use colonies from round 3 and 4). This will be your **master plate**. Leave one or two wells free as blank. Seal plate with parafilm and incubate shaking overnight at 37°C.

3. Transfer an aliquot of the culture to a fresh plate (**induction plate**) containing 125 µl of 2xTY, 100 µg/ml ampicillin, 0.1% glucose. This can be done for example with a 96-well transfer device (Sigma). Seal plate with parafilm to avoid cross-contamination.

4. Add 50 µl of 2xTY, 45% glycerol to each well of the master plate, mix well, seal with parafilm and store plate at -20 or -80°C.

5. Incubate induction plate at 37°C in a shaker until an OD_{600} of 0.8-1 is reached. This takes approximately 2-3 hrs. The low concentration of glucose will be metabolised by the time IPTG is added to induce expression.

6. Add 25 µl of 2xTY, 100 µg/ml ampicillin, 6 mM IPTG and incubate overnight at 30°C.

7. Coat a microtitre plate with your antigen as described. Use a second plate with a negative control protein.

8. Next day, block remaining binding sites with 2% MPBS for 2 hrs at RT.

9. Remove blocking solution and add 50 µl of 4% MPBS to each well.

10. Centrifuge induction plate for 10 min at 3500 rpm and transfer 50 µl of the supernatant from each well to the corresponding well of the ELISA plate. Incubate for 1 hr at RT.

11. Wash 6 times with PBS.

12. Add anti-Myc-tag antibody 9E10 diluted to 10 µg/ml in 2% MPBS and incubate for 1 hr at RT.

13. Wash 6 times with PBS.

14. Add HRP-conjugated goat-anti-mouse antibody diluted in 2% MPBS and incubate for 1 hr at RT.

15. Wash 6 times with PBS.

16. Add 100 µl TMB/H_2O_2 substrate solution and proceed as described under polyclonal phage ELISA.

Characterisation of soluble scFv

1. For expression and purification of scFv by immobilised metal affinity chromatography see Chapter 17. Purified scFv can then be directly applied to various immunological tests, such as ELISA, immunoblotting experiments, immunocyto- and histochemistry. In these experiments, bound antibody fragments can be detected via their tag sequences or using protein L or A (Chapter 21).

2. For sequence analysis, grow cultures of positive clones overnight at 37°C in 2xTY, 100 µg/ml ampicillin, 1% glucose and isolate plasmid DNA by standard procedures. Using phagemid vectors, clones can be sequenced with primers LMB3 and fdSeq1. For sequence analysis see Chapter 30 and Chapter 31.

Preparation of helper phage

1. Melt top agar and cool down to 42°C.

2. Mix 100 µl of serial dilutions of helper phage (VCSM13 or M13KO7) in 2xTY with 100 µl of log-phase TG1 ($OD_{600} = 0.5$) and 3 ml of top agar.

3. Plate onto warm TYE plates and incubate overnight at 37°C.

4. Pick a single (small) plaque from the lawn and transfer to 3 ml of 2xTY containing 100 µl of a TG1 overnight culture to obtain a good cell density.

5. Shake for 3 hrs.

6. Inoculate 500 ml of 2xTY with the grown plaque. Incubate for 1 h shaking at 37°C and then add 25 µg/ml kanamycin. Incubate shaking overnight at 37°C.

7. Transfer culture to a 500 ml centrifuge beaker and centrifuge at 6000 rpm for 30 min.

8. The supernatant contains the helper phage. Filter through a 0.45 μm Nalgene membrane (use several 150 ml filters instead of one 500 ml). Check an aliquot of the filtrate on TYE plates for remaining bacteria.

9. Freeze phage in liquid nitrogen or dry ice and transfer to -20°C. No glycerol is needed.

10. Titre phage (thawed from the frozen stock) as described above in steps 1-3 to quote actual survival titre of phage (titre should be 10^{11} - 10^{12} t.u./ml). If necessary, phage can be concentrated by PEG-NaCl precipitation as described.

Troubleshooting

- No enrichment of positive phage
  - Check the library with another antigen (e.g BSA) to ensure that the phage library you are using and your selection procedure is working.
  - Check TG1 cells used for infection of eluted phage are already infected by plating them on ampicillin and kanamycin plates. If you find colonies on these plates your TG1 stock is contaminated. Use a fresh stock of TG1.

- Only few antibodies are enriched after 3-4 rounds
  - Reduce washing stringency (i.e. less washing steps) or increase antigen density coated onto immunotubes. Check also immobilisation conditions, e.g. by detecting your antigen with available antibodies. Perform additional rounds and check for further enrichment.

- Selected antibodies have low affinities
  - If you are using a naive antibody library of low or moderate diversity (10^6-10^7), no high affinity antibodies may be present in the library. Try to use a library with greater diversity or generated by a different method (e.g. a semi-synthetic library).
  - Your protein might possess low antigenicity, e.g. peptides and haptens quite often produce only antibodies with low or moderate affinity.

Applications

Immunotube selections are generally applied for the selection of antibody fragments directed against purified proteins or other antigens which can be immobilised on polystyrol surfaces. For selection of antibodies recognising small molecules such as peptides or haptens, these molecules can be coupled to a carrier protein (e.g. BSA or keyhole limpet haemocyanin (KLH)). In that case it is advisable to perform alternating selections using two different carrier proteins to avoid enrichment for antibodies against the carrier. Applications might be limited by impure preparations of antigen in which case indirect methods for immobilisation might be used (e.g. using specific ligands such as antibodies). However, in this case libraries have to be pre-adsorbed on the ligand in order to avoid selection of antibodies directed against the ligand. Often, selections are also limited by the amount of antigen available. Selections with antigens coated at 1-10 µg/ml are routinely used with success, although lower concentrations have been applied for immunotube selections. Another problem might arise from the fact that proteins often unfold during immobilisation which might result in the selection of antibodies recognising (only) epitopes of the denatured protein. In that case other selection methods, for example selections on biotinylated antigens bound to streptavidin-coated paramagnetic beads can be applied (Chapter 10).

References

Carter P, Bedouelle H, Winter G (1981) Improved oligonucleotide site-directed mutagenesis using M13 vectors. Nucl Acid Res 13:4431-4443

Gibson TJ (1984) Studies on the Epstein-Barr virus genome. PhD. thesis, University of Cambridge

Kontermann RE, Wing MG, Winter G (1997) Complement recruitment using bispecific diabodies. Nat Biotechnol 15:629-631

Marks JD, Hoogenboom HR, Bonnert TP, McCafferty J, Griffiths AD, Winter G (1991) By-passing immunization. Human antibodies from V-gene libraries displayed on phage. J Mol Biol 222:581-597

Harrison JL, Williams SC, Winter G, Nissim A (1996) Screening of phage antibody libraries. Methods Enzymol 267:83-109

McCafferty J, FitzGerald KJ, Earnshaw J, Chiswell DJ, Link J, Smith R, Kenten J (1994) Selection and rapid purification of murine antibody fragments that bind a transition-state analog by phage display. Appl Biochem Biophys 47:157-173

Vaughan TJ, Williams AJ, Pritchard K, Osbourn JK, Pope AR, Earnshaw JC, McCafferty J, Hodits RA, Wilton J, Johnson KS (1996) Human antibodies with sub-nanomolar affinities isolated from a large non-immunized phage display library. Nat Biotechnol 14:309-314

Winter G, Griffiths AD, Hawkins RE, Hoogenboom HR (1994) Making antibodies by phage display technology. Annu Rev Immunol 12:433-455

▨ Suppliers

Genosys
London Road, Pampisford
Cambridgeshire CB2 4EF
UK

Nunc
Hagenauer Straße 21a
D-65203 Wiesbaden
Germany

Pharmacia
Munzinger Straße 9
D-79111 Freiburg
Germany

Sigma
Grünwalder Weg 30
D-82041 Deisenhofen
Germany

Stratagene Europe
Gebouw California
Hogehilweg 15
1101 CB Amsterdam Zuidoost
The Netherlands

Selections on Biotinylated Antigens

PATRICK CHAMES, HENNIE R. HOOGENBOOM, and PAULA HENDERIKX

Introduction

Phage antibody library selections on peptides or proteins are usually carried out using antigens directly coated on a plastic surface (e.g. Petri dishes, microtiter plate well, immunotubes, Chapter 9). This straightforward method is easy to perform and has been shown to be very successful for a diverse set of antigens (for review see Winter et al. 1994). However, phage-antibody selections on some proteins and especially on peptides are not always successful, often due to by immobilization-associated features. The main problem observed for selection on peptides is the very poor coating efficiency of some peptides and the altered availability of epitopes on plastic-coated peptides. The direct coating of proteins on plastic is usually more efficient but can also be problematic, because the passive adsorption on plastic at pH 9.6 is a mechanism of protein denaturation. Under these conditions, 95% of adsorbed proteins are non-functional (Butler et al. 1992; Davies et al. 1994). This problem is not very important for a classical ELISA because mostly a small fraction of proteins having a native conformation is still detectable. However, this phenomenon can be very troublesome for phage antibody library selections, because phage antibodies binding to epitopes only present in denatured molecules may be selected.

Several methods have been developed to increase peptide coating, including coupling to bigger proteins (Oshima and Atassi 1989), to amino acid linkers binding plastic (Loomans et al. 1998; Pyun et al. 1997), or to

Patrick Chames, University Hospital Maastricht, Department of Pathology, P.O. Box 5800, Maastricht, 6202 AZ, Netherlands

✉ Hennie R. Hoogenboom, University Maastricht, Department of Pathology, P.O. Box 5800, Maastricht, 6202 AZ, Netherlands (*phone* +31-43-387-4630; *fax* +31-43-387-6613; *e-mail* hho@lpat.azm.nl)

Paula Henderikx, University Hospital Maastricht, Department of Pathology, P.O. Box 5800, Maastricht, 6202 AZ, Netherlands

multiple antigen peptide (Tam and Zavala 1989). The most successful method had been the indirect coating of biotinylated antigens via streptavidin: biotinylation of the peptide and immobilisation via streptavidin improves the sensitivity in ELISA (Ivanov et al. 1992) and allows more efficient selection of anti-peptide phage-antibodies (de Haard et al. 1999; Henderikx et al. 1998).

In the case of phage library selection against proteins, the indirect coating via streptavidin results in higher density coating, more uniform distribution of antigens on the well surface, and above all, 60 to 70% of active molecules (Butler et al. 1992; Davies et al. 1994).

Most importantly, however, the use of biotinylated peptide or protein allows the use of paramagnetic streptavidin-coated microbeads to capture the biotinylated antigens with the phage bound to them. The interaction between the phage particle and the antigen therefore takes place in solution; antigen-bound phage is retrieved via a short incubation with the beads. This technique permits the precise control of the antigen concentration and the time of exposure of the antigen to the phage-antibody library, two parameters that are very useful in affinity selection, for example during affinity maturation protocols (Hawkins et al. 1992; Schier and Marks 1996). This interaction between antigen and phage antibody in solution leaves a maximum of epitopes available for binding, and avoids the selection of scFv fragments with low affinity but a high tendency to form dimers (Schier et al. 1996). The latter will be preferentially selected on antigen-coated surfaces due to their avid binding.

This chapter contains protocols for chemical or enzymatic biotinylation, as well as phage library selection in solution and sensitive ELISA procedures for using indirectly coated biotinylated antigen. The advantages and drawbacks of each method are discussed.

Subprotocol 1
Biotinylation of Proteins/Peptides with NHS-ss-Biotin

Purpose   Chemical biotinylation is the most common way to obtain a biotinylated antigen. There are many commercially available reagents which can be used for biotinylation using a variety of chemistries. For most of our biotinylations, we prefer to use the chemical reagent NHS-SS-Biotin (Sulfo-Succinimidyl-2-(biotinamido)ethyl-1,3-dithiopropionate MW= 606.70). This molecule is a unique biotin analogue with an extended spacer arm of approximately 24.3 Å in length, capable of reacting with primary amine groups (lysines and NH_2 termini). The long chain reduces steric hindrances associated with binding of biotinylated molecules to avidin or streptavidin and should not interfere with the structure of the protein/peptide involved.

The presence of the S-S linker in NHS-S-S-Biotin enables the use of a reducing agent (DTT, DTE, β-mercaptoethanol) to separate the antigen and all phage-antibodies bound to it from the beads. This feature allows a more specific elution, which is very useful when undesired streptavidin binders are preferentially selected from a phage antibody repertoire. The following method is modified from (Hnatowich et al. 1987) and the Pierce instruction manual.

Materials

- 50 mM Sodiumbicarbonate pH=8.5: 4.2 g $NaHCO_3$ adjust pH with 6 M NaOH to pH 8.5, adjust to 1 liter with H_2O, filter-sterilise, store at RT

- NHS-SS-Biotin (Pierce 21331, Rockford, Illinois)

- Peptide/Protein of interest

- The best molar ratio of biotin to the protein has to be determined empirically. Try different molar ratios if possible. NHS-SS-Biotin MW= 606.70 NHS-LC-Biotin MW= 556.58

- Alternatively if no dialysis is used: Centricon 30 or Centricon 10 (Amicon, 4306, 4304; Beverly, Massachusetts)

- Dialysis tubing (tubing with MW-cut-off of 1,000 - 50,000 kD can be found at Cellu.Sep, Waterloo, Belgium)

- 5 × PBS pH= 7.4: 43.8 g NaCl (750 mM), 7.1 g Na_2HPO_4 (40 mM), 1.08 g KH_2PO_4 (7.8 mM), adjust volume to 1 liter. Before the test, add 800 ml H_2O to 200 ml of 5 × PBS and check the pH (7.4)

Avoid buffers containing amines (such as Tris or glycine) since these compete with peptide/protein the biotinylation reaction. Also, reducing agents should not be included in the conjugation step to prevent cleavage of the disulphide bond within NHS-SS-Biotin.

Precautions

▉ ▉ Procedure

1. Dissolve 1-10 mg/ml of the peptide/protein of interest in 50 mM NaHCO$_3$ (pH=8.5). If the peptide/protein is already in another solvent, dialyse against 50 mM NaHCO$_3$ (2-3 x for at least 4 h in 1 l).

2. Calculate the amount NHS-SS-Biotin required taking into account a Molar ratio Biotin: Protein = 5:1- 20:1. Although the amount NHS-SS-Biotin required depends on the number of lysines present within the protein, usually a ratio of 5:1 works fine. When enough protein is available it is advised to test different ratios Protein: Biotin. Ideally 1-2 biotinylated residues per molecule are present. Overbiotinylation often results in non-functional protein (aggregation etc.).

3. Dissolve the required amount of NHS-SS-Biotin in distilled water and immediately add this to your protein sample, or alternatively, when using larger amounts of protein add NHS-SS-Biotin directly to the protein solution.

4. Place the tube on ice for 2 h or 30 min at room temperature.

5. Add 1 M Tris (pH=7.5) to a final concentration of 50 mM, and incubate 1 h on ice to block free biotin.

6. To remove free biotin, dialyse against the biotinylated protein in PBS at 4°C (3x for at least 4 h in 1 l PBS) overnight or alternatively: do steps 7 to 9. For small peptides (< 20 amino acids) alternative separation protocols (e.g. affinity chromatography, HPLC) should be followed.

7. Alternatively for 6: Spin at 3000 rpm a pre-treated ultra-filtration device (e.g. Centricon 10 or 30 for 15-30 min) to concentrate the sample.

8. Dilute sample in PBS to dilute out free biotin left after concentration.

9. Repeat steps 6 and 7 two more times.

10. Add sodiumazide to a final concentration of 0.1%.

11. Store in small aliquots at - 20°C or at 4°C. (Storage conditions should be tested for each type of protein).

Subprotocol 2
Enzymatic Biotinylation of Recombinant Proteins

E. coli possesses a cytoplasmic enzyme, BirA, which is capable of specifically recognising a sequence of 13 amino acids and adding a biotin on a unique lysine present on this sequence (Schatz 1993). If this sequence is fused as a tag to the N or C-terminal part of a protein, the resulting fusion will be biotinylated as well. The main advantage of this system is that the protein by itself remains fully intact. Conversely, the chemical biotinylation randomly modifies any accessible lysine. An overbiotinylation often leads to inactivation of the protein of interest, especially if a lysine is present in the active site of the protein. The use of a low ratio biotin/protein aimed to reduce this problem, may lead to poor yield of biotinylation. The enzymatic biotinylation avoids this drawback, leading to a 100% active protein, but also to a high yield of biotinylation (typically 85-95%).

Purpose

The specific biotinylation via the tag presents another important advantage: it allows an ideal orientation of the protein during the selection or the ELISA analysis. In both instances the tag will be bound to streptavidin and will thus be directed toward the solid surface (beads or plastic); the rest of the molecule is perfectly oriented, available for interaction with the phage-antibody. This allows also a very uniform presentation of the antigen whereas the chemical biotinylation will lead to a number of antigens having the epitope of interest directed toward streptavidin and thus not being available for phage-antibody binding.

The main drawbacks of this method are that it cannot be applied on non-recombinant proteins, and that the link between biotin and the antigen cannot be broken using DTT (see selection Subprotocol 6).

Procedure

The enzymatic biotinylation can be done in vivo if the antigen is produced in the cytoplasm of *E. coli*. In this case, the only requirement is to overexpress birA and add free biotin to the culture medium. Surprisingly, the biotinylation is also very efficient on intracellularly expressed proteins which will form inclusion bodies. However, if the antigen has to be produced in the periplasm of *E. coli*, the biotinylation yield is very poor (0.1-1%) (Chames et al., unpublished). In this case and in the case where the antigen is produced in another expression system, the biotinylation of the tag can still be performed in vitro, on the purified protein, using purified commercially available BirA. Both protocols are described here.

Subprotocol 3
In vivo Enzymatic Biotinylation

Materials

- Plasmid pBirCm (Avidity, Denver, USA, www.avidity.com)
- d-Biotin, SIGMA (B4501)
- 2xTY, Ampicillin, Chloramphenicol, IPTG

Procedure

1. Using PCR and classical molecular biology protocols, reclone the gene of interest fused at its 5' or 3' end with the sequence coding for the biotinylation tag (see Figure 1).

2. Once a recombinant clone is obtained, co-transform E. coli with the recombinant plasmid coding for ampicillin resistance, and pbirACm, a plasmid coding for Chloramphenicol resistance, compatible with pUC-based vectors and used to overexpress birA. Transformed bacteria have to be plated on plate containing 100 µg/ml ampicillin and 15 µg/ml chloramphenicol.

3. Dilute 10 ml of an overnight culture of one of the double-transfectants in 1L of 2xTY containing 100 µg/ml ampicillin and 15 µg/ml chloramphenicol. Incubate at 37°C with shaking until OD_{600} of 0.5.

4. Add 1 mM IPTG to induce BirA and the protein of interest expression, 50 µM biotin from a 500 mM stock in DMSO and incubate 3 h at 37°C (if the gene of interest is not under the control of tac or lac promoter, add the required inducer as well). The production can alternatively be done at 30°C.

5. Spin the culture for 20 min at 4000 rpm.

```
 G   S   L   H   H   I   L   D   A   Q   K   M   V   W   N   H   R
GGA TCC CTG CAT CAT ATT CTG GAT GCA CAG AAA ATG GTG TGG AAT CAT CGT
 BamHI
```

Fig. 1. Sequence specifically biotinylated by the *E. coli* enzyme BirA. (The modified lysine is shown in bold.)

6. Resuspend the pellet in 10 ml of sonication buffer 20 mM Tris-Hcl pH 7.5, 100 mM NaCl.

7. Lyse the cells by sonication. Keep the solution on ice during the whole process.

8. Purify the protein by affinity chromatography. If the protein possesses a his tag, use IMAC procedures (Chapter 17). An interesting alternative is to purify the protein using avidin monomers (allowing soft elution condition unlike tetramers) fused to sepharose (Softlink soft release avidin resin, Promega, V2011). This method allows to recover only the biotinylated fraction of the recombinant but will also purify the only endogenously biotinylated of *E. coli* (although present in very low amount in *E. coli*, in the µg/l range).

Subprotocol 4
In vitro Enzymatic Biotinylation

Materials

birA enzyme and buffers (**biomix A and B**) (Avidity, Denver, USA, www.avidity.com) Centricon (Amicon, Beverly, MA).

Procedure

1. Using PCR and classical molecular biology protocols, reclone the gene of interest fused at its 5' or 3' end with the sequence coding for the biotinylation tag (see Figure 1, Subprotocol 3).

2. Produce and purify the recombinant protein using the appropriate method.

3. Use ultracentrifugation devices (Centricon, Amicon) to exchange the buffer for Tris-HCl 20 mM pH 7.5 and concentrate the protein to 40 µM (the birA enzyme is inhibited by NaCl and glycerol).

4. Mix one volume of recombinant protein at 40 µM with 0.125 volume of Biomix A and Biomix B plus 2.5 µl of birA for each 10 nmol of protein, in less than 500 µl.

5. Incubate for 1 h at 30°C.

6. Remove the excess of biotin by gel filtration (for example on Superdex S75 column, Pharmacia), collect fractions of 500 µl, pool fractions containing the protein and estimate the protein concentration by UV measurement at 280 nm.

Subprotocol 5
Determination of Biotinylation Efficiency

Purpose   It is important to determine the percentage of protein that has actually been biotinylated. If the antigen has to be used for selection in solution, the non-biotinylated part of the preparation will be very detrimental, blocking specific phages and impairing their binding to the biotinylated fraction. Hence, this non-biotinylated fraction must represent less than 10-15%. The following protocol permits the determination of this percentage. This protocol is also used to determine the amount of biotinylated peptide captured by a certain amount of magnetic beads. Extrapolation of the results can be used for determining the concentration of antigen and amount of beads to be used during phage library selection.

■ ■ Materials

- Streptavidin dynabeads (Dynal, M280, Oslo, Norway), magnetic separation device (Dynal)

- Biotinylated peptide/protein, (5 dilutions between 5-50 nM in PBS, in 200 µl)

- PBS: see Subprotocol 1 of this chapter

- T-PBS: 1 ml Tween 20 in 1 liter PBS

- 0.05 M Tris-buffer pH=7.6: 0.6 g Tris/ 100 ml H_2O. Adjust pH to 7.6 with HCl

- SDS-page gel (12%, 10% or 7.5 dependent on the size of the peptide/ protein)

- Coomassie blue 0.1% in 40% methanol, 10% acetic acid, 50% H_2O

- Destainer: 40% methanol, 10% acetic acid, 50% H_2O

– Non-reducing protein sample buffer: 30% glycerol, 0.025% bromophenol blue, 0.05 M Tris pH 6.8 in H_2O

– End-over-end rotator

– UV spectrophotometer (280 nm), quartz cuvettes

▪▪ Procedure

1. Resuspend Dynabeads by gentle shaking.

2. Transfer 10 µl per peptide/protein dilution (e.g. 50 µl for 5 dilutions of peptide/protein) to a tube that fits into the magnetic separator and add an excess of PBS, shake gently.

3. Put the tube in the separation device for 2 min and pipette off the PBS.

4. Add 0.5 ml T-PBS and incubate for 60 min.

5. Remove T-PBS and resuspend the beads in 50 µl T-PBS.

6. Transfer 10 µl of beads to appropriate tubes.

7. Add 100 µl of peptide/protein dilution to each tube.

8. Incubate for 30 min at room temperature in an end-over end rotator.

9. Remove 100 µl of incubated peptide/protein fractions by means of the magnet and store them (fractions 1).

10. Wash Dynabeads 5x with T-PBS and add finally 100 µl SDS non-reducing protein sample-buffer (fractions 3) if protein measurements will be done by SDS page, incubate for10 min.

11. Load two volumes (e.g. 10 µl and 50 µl) of each sample on SDS-PAGE (use non-reducing sample buffer). (Unbiotinylated protein = fraction 0. Input biotinylated protein = fraction 1. Supernatant = fraction 2. Dynabeads = fraction 3.).

12. Perform SDS-page electrophoresis.

13. Stain with Coomassie blue, destain.

14. Compare protein bands.

15. Alternatively, measure UV_{280} absorption of fraction 0 and fraction 1.

16. Check whether 100% of the protein is biotinylated by comparing the fractions of the lowest concentrations used (protein should be seen in

fraction 0 and fraction 3) and check the maximum amount of protein able to bind 10 µl of Streptavidin-Dynabeads. Extrapolate this amount to phage-selection conditions (e.g. a maximum of 30 nM can be bound at about 100% to 10 µl of beads: for 500 mM used during the selections, 166 µl of magnetic beads should be used).

17. To determine the number of biotin molecules per protein/peptide the HABA method can be used (see Pierce, http:/www.piercenet.com).

Troubleshooting

NHS-ss-Biotin biotinylation

For very low amounts of protein/peptide: keep them as concentrated as possible, because hydrolysis of the NHS-ester does compete for reaction with the amine, and use a high amount of NHS-ester (without exceeding a 30 molar excess) (Pierce, personal communication).

Enzymatic biotinylation

- Failure to obtain good yields of biotinylation using these techniques can be explained most of the time by a degradation of the biotinylation tag due to the presence of protease co-purified with the protein of interest. This problem can occur during biotinylation for the in vitro protocol during the 1h incubation at 30°C. To avoid this, one can add a cocktail of protease inhibitor to the reaction mix. However, EDTA, a common metalloprotease inhibitor, must not be added (magnesium is needed for enzymatic activity). Use for example Complete™ EDTA-free (Boehringer Mannheim, Germany, 1 836 170).

- For in vivo biotinylation, tag degradation can occur during storage. In this case, a complete protease inhibitor cocktail can be used (e.g. Complete™, Boehringer Mannheim, Germany, 1 697 498)

- For in vitro biotinylation, a common problem is a loss of activity of the purified BirA. The enzyme is delivered frozen and can be kept several months at -70°C. However, several cycles of freeze and thaw rapidly inactivate the enzyme. After the first thawing of the enzyme, it is thus strongly recommended to store the remaining enzyme in small frozen aliquots.

Subprotocol 6
Selection of Phage-Antibodies Using Biotinylated Peptide/Protein

The aim of this procedure is to select phage antibodies directed to a bio-   **Purpose**
tinylated antigen. The principle is to incubate the phage antibody repertoire in solution with the biotinylated antigen. Once specific phages are
bound to the antigen, paramagnetic beads coupled to streptavidin are
added into the solution. The biotinylated antigens with bound phages
are captured and the whole complex is drawn out from the suspension
by applying a magnet on the side of the tube. Beads are washed several
times and specific phages are eluted from the beads (see Figure 2).

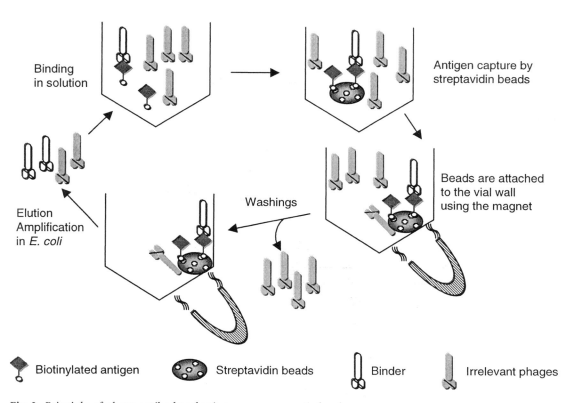

Fig. 2. Principle of phage antibody selection on paramagnetic beads.

▨▨ Materials

- DMSO

- Skimmed dry milk (marvel)

- Dynabeads M280 (Dynal, Oslo, Norway)

- 5 × PBS and PBS see Subprotocol 1

- 2% M-PBS: 2 g dried skimmed milk (marvel) in100 ml PBS

- 4% M-PBS: 4 g dried skimmed milk powder (marvel, Sainsbury, UK)/ 100 ml PBS

- MT-PBS: 2% Marvel, 2% Tween 20 in PBS

- MT-PBS (DMSO): MT-PBS 5% DMSO (for peptides difficult to disolve)

- 100 mM TEA: 140μl triethylamine in10 ml water (pH 12)

- 1 M Tris-HCl, pH=7.4

- rotator

Precautions Check whether the antigen is water soluble in the selection buffers used. If the antigen (peptide) is too hydrophobic, you have to find alternative buffer conditions in which it remains in solution and use these conditions for the selection. We have for example successfully used 5% DMSO in all solutions.

▨▨ Procedure

1. Mix equal volumes of a phage library and 4% M-PBS. During the 1st selection, the number of phage particles should be at least 100 times higher than the library size.

2. Incubate on a rotator at room temperature for 60 min.

3. While preincubating the phage, wash 100 μl (typically 200 μl for the first round) streptavidin-dynabeads per peptide in a tube, fitting in the magnetic separation device, as described in Subprotocol 5. The minimal required amount of beads for selection can be calculated as described in Subprotocol 5 (step 16).

4. Remove PBS and resuspend beads in 2% M-PBS.

5. Equilibrate on rotator at room temperature for 1-2 h.

6. Add biotinylated antigen (100-500 nM; see Subprotocol 5) diluted in PBS (DMSO) directly into equilibrated phage mix.

7. Incubate on rotator at RT for 30 min to 1 h.

8. Draw equilibrated beads (from Step 5) to one side with magnet and remove M-PBS.

9. Resuspend Dynabeads in 250 µl per antigen in MT-PBS(DMSO).

10. Add Dynabeads to phage-antigen mix and incubate on a rotator at room temperature for 15 min.

11. Place tubes in the magnetic separator and wait until all beads are bound to the magnetic site (20 sec).

12. Tip rack upside down and back again, with caps closed. This will wash down the beads from the cap.

13. Leave tubes in the rack for 2 min.

14. Aspirate the tubes carefully, leaving the beads on the side of the tube.

15. Wash the beads carefully 6 times with 1 ml MT-PBS.

16. Transfer beads to a new eppendorf tube.

17. Wash the beads 6 times with 1 ml MT-PBS.

18. Transfer the beads to a new eppendorf tube.

19. Wash the beads 2 times with 1 ml PBS.

20. Transfer to a new tube.

21. Elute page from beads with 1ml 100 mM TEA, or in case of an -SS-biotin labelled protein/peptide use 200 µl 10 mM DTT, rotate 5 min at room temperature.

22. In case of elution with TEA: transfer the solution to an eppendorf tube containing 0.1 ml Tris-HCl, 1M, pH=7.4 and mix by inversion. It is necessary to neutralise the phage eluate immediately after elution. In case of DTT elution, transfer solution to new tube.

23. Titrate in and output by infection of bacterial cells (TG1 cells) (Chapter 9).

24. Re-infect bacterial cells (TG1 cells) with selected phages for another round of selection (Chapter 9).

25. Store the remaining beads or eluate at 4°C as a backup.

▨▨ Troubleshooting

- If a significant proportion of the peptide/protein is not labelled, one can incubate the antigen with the avidin-dynabeads first with the streptavidin beads, taking into account the molarity of the biotinylated peptide/protein, and wash away the non-biotinylated peptide. The beads are then used directly for the selection.

- For the selection of high affinity antibodies, it is advised to select with a decreasing antigen concentration. For example with: 100 nM biotinylated peptide during the first round, with 20 nM for the second round, with 5 nM for the third round and with 1 nM for the 4th round.

- If streptavidin binders are preferentially selected (which may be the case when using non-immunized or synthetic antibody libraries) the following steps can be undertaken:
  - If the peptide/protein is biotinylated with NHS-L-C-biotin, it is advised to deplete the library by incubating for 1 h (from round 2 on and later) with 100 μl streptavidin-Dynabeads before adding the biotinylated antigen to the depleted library.
  - If the peptide is biotinylated with NHS-SS-biotin, 10 mM DTT should be used to elute the antigen-binding phage specifically.

Subprotocol 7
(Inhibition) ELISA with Indirectly Coated Biotinylated Antigen

Purpose   This very sensitive ELISA aimed to screen monoclonal phage uses the same biotinylated antigen already used during the selection step. The indirect coating via streptavidin ensures maintenance of the native structure of the antigen. A precoating of the plastic surface with biotinylated BSA is used to circumvent the low adsorption properties of streptavidin (see Figure 3).

▨▨ Materials

- Supernatant of induced fractions coming from colonies picked from the selections with Dynabeads

- Biotinylated antigen of interest, at a concentration of 1-5 μg/ml M-PBS(DMSO). For inhibition ELISA the concentration should be 1 μg/ml

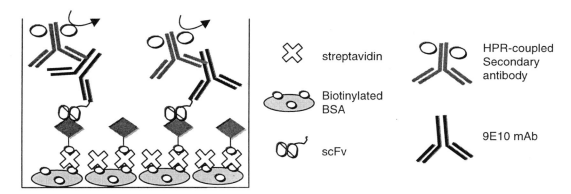

Fig. 3. ELISA using biotinylated antigen and soluble antibody fragments.

- *For inhibition ELISA (IE) Non-biotinylated antigen at a concentration of 1 mg/ml in 2% M-PBS. Steps involved in inhibition ELISA are printed in italics.*

- PBS, PBS (DMSO), M-PBS, T-PBS see Subprotocol 6.

- T-PBS: 2% Tween 20 in PBS

- Biotinylated BSA, stock solution: 2 mg/ml PBS, working solution: per microtiter plate add 10 µl of stock solution to 10 ml PBS

- PBS/0.5% gelatine: 0.5 g gelatin/100 ml PBS. Store at 4°C. Pre-warm this solution before use (60°C or microwave) to dissolve the gelatine. Use at room temperature.

- Streptavidin solution: stock solution: 1 mg/ml H_2O, working solution: Add per plate 100µl of the stock solution to 10 ml PBS/0.5% gelatine

- 2% M-PBS: see Subprotocol 6

- 4% M-PBS: see Subprotocol 6

- 9E10 monoclonal antibody (Pharmacia, Upsalla, Sweden): used for Myc-tagged antibodies, dilute with 2% M-PBS at a dilution advised by the supplier

- Rabbit Anti Mouse PerOxidase (RAMPO) working solution: dilute the antigen at a concentration advised by the supplier in 2% M-PBS

- 10 times Tetra-Methyl-Benzidine-buffer (TMB): dissolve 37.4 g NaA-cetate.3H_2O in 230 ml H_2O. Adjust pH with saturated citric acid (92.5 g citric acid/50 ml H2O) and adjust volume to 250 ml

- TMB stock: Dissolve 10 mg of TMB in 1 ml DMSO

- TMB-staining solution: Mix 1 ml 10 times TMB with 9 ml of H_2O per microtiterplate. Add, per plate, 100 µl of the TMB to 10 ml of TMB-buffer. Add 1 µl of 30% hydrogen peroxidase. Make this solution fresh and keep it in the dark.

- 1 M H_2SO_4: 53.2 ml of concentrated H_2SO_4/ liter H_2O

- 96-well flat-bottomed ELISA microtiter plates (2 plates to screen 96 colonies) (Falcon 3912)

- *Only for IE: + microtiter plates with low coating efficiency (2/96 colonies)*

- Microtiterplate reader (for OD_{450} measurements)

▨▨ Procedure

1. Add 100 µl biotinylated BSA to each well of the microtiter plate. For 96 induced fractions, 2 plates should be coated (negative control and positive plate).

2. Incubate for 1 h at 37°C or overnight at 4°C.

3. Discard the solution of step 2. Wash plates 3 times for 5 min with T-PBS by submersing the plate into the buffer and removing the air bubbles by rubbing the plate.

4. Add 100 µl/well streptavidin to negative and positive plates.

5. Incubate for 1 h at room temperature while shaking.

6. Wash plates as described in Step 3.

7. Add 100 µl biotinylated antigen to each well of the positive plate and add 100 µl 2% M-PBS to the wells of the negative control plate. (For IE: *add the biotinylated antigen to both plates*).

8. Incubate for 1 h at room temperature.

9. Wash plate 3 times with T-PBS (DMSO) as described in Step 3.

10. Block with 2% M-PBS/DMSO. Use 120 µl/well incubate for at least 30 min at room temperature.

11. Empty wells, add 50 µl/well of 4% M-PBS (DMSO) to all wells of both plates (for IE: *use 2 non-coated plates with low coating efficiency*).

12. Add 50 µl/well of culture supernatant containing soluble antibody fragment.

13. Only for IE: *add 10 µl/well of M-PBS to the positive plate, add 10 µl/well of non-biotinylated antigen to the negative plate. Mix by pipetting and incubate for 30 min. Transfer 100 µl to microtiter plates coated with biotinylated antigen.*

14. Mix by pipetting and incubate for 1.5 h at room temperature.

15. Wash 3 times with T-PBS as described in Step 3.

16. Add 100 µl/well of 9E10 solution to all wells and incubate for1 h RT.

17. Wash as in Step 3.

18. Add 100 µl/well of RAMPO solution to all wells and incubate for 1 h at room temperature.

19. Wash as in Step 3.

20. Stain with TMB by adding 100 µl/well of TMB staining solution. Incubate for 10-30 min in the dark.

21. Add 50 µl/well 1 M H_2SO_4

22. Read plate at 450 nm.

23. If the optical density of a clone on the positive plate is higher than 2 times the optical density of the same clone on the negative plate, it can be considered positive and should be tested further.

References

Butler, J. E., Ni, L., Nessler, R., Joshi, K. S., Suter, M., Rosenberg, B., Chang, J., Brown, W. R., and Cantarero, L. A.: The physical and functional behavior of capture antibodies adsorbed on polystyrene. J Immunol Methods 150: 77-90, 1992

Davies, J., Dawkes, A. C., Haymes, A. G., Roberts, C. J., Sunderland, R. F., Wilkins, M. J., Davies, M. C., Tendler, S. J., Jackson, D. E., and Edwards, J. C.: A scanning tunnelling microscopy comparison of passive antibody adsorption and biotinylated antibody linkage to streptavidin on microtiter wells. J Immunol Methods 167: 263-9, 1994

de Haard, H. J., van Neer, N., Reurs, A., Hufton, S. E., Roovers, R. C., Henderikx, P., de Brune, A. P., Arends, J. W., and Hoogenboom, H. R.: A Large Non-immunized Human Fab Fragment Phage Library That Permits Rapid Isolation and Kinetic Analysis of High Affinity Antibodies. J Biol Chem 274: 18218-18230, 1999

Hawkins, R. E., Russell, S. J., and Winter, G.: Selection of phage antibodies by binding affinity. Mimicking affinity maturation. J Mol Biol 226: 889-96, 1992

Henderikx, P., Kandilogiannaki, M., Petrarca, C., von Mensdorff-Pouilly, S., Hilgers, J. H., Krambovitis, E., Arends, J. W., and Hoogenboom, H. R.: Human single-chain Fv antibodies to MUC1 core peptide selected from phage display libraries recognize unique epitopes and predominantly bind adenocarcinoma. Cancer Res 58: 4324-32, 1998

Hnatowich, D. J., Virzi, F., and Rusckowski, M.: Investigations of avidin and biotin for imaging applications. J Nucl Med 28: 1294-302, 1987

Ivanov, V. S., Suvorova, Z. K., Tchikin, L. D., Kozhich, A. T., and Ivanov, V. T.: Effective method for synthetic peptide immobilization that increases the sensitivity and specificity of ELISA procedures. J Immunol Methods 153: 229-33, 1992

Loomans, E. E., Gribnau, T. C., Bloemers, H. P., and Schielen, W. J.: Adsorption studies of tritium-labeled peptides on polystyrene surfaces. J Immunol Methods 221: 131-9, 1998

Oshima, M. and Atassi, M. Z.: Comparison of peptide-coating conditions in solid phase plate assays for detection of anti-peptide antibodies. Immunol Invest 18: 841-51, 1989

Pyun, J. C., Cheong, M. Y., Park, S. H., Kim, H. Y., and Park, J. S.: Modification of short peptides using epsilon-aminocaproic acid for improved coating efficiency in indirect enzyme-linked immunosorbent assays (ELISA). J Immunol Methods 208: 141-9, 1997

Schatz, P. J.: Use of peptide libraries to map the substrate specificity of a peptide- modifying enzyme: a 13 residue consensus peptide specifies biotinylation in Escherichia coli. Biotechnology (N Y) 11: 1138-43, 1993

Schier, R., Bye, J., Apell, G., McCall, A., Adams, G. P., Malmqvist, M., Weiner, L. M., and Marks, J. D.: Isolation of high-affinity monomeric human anti-c-erbB-2 single chain Fv using affinity-driven selection. J Mol Biol 255: 28-43, 1996

Schier, R. and Marks, J. D.: Efficient in vitro affinity maturation of phage antibodies using BIAcore guided selections. Hum Antibodies Hybridomas 7: 97-105, 1996

Tam, J. P. and Zavala, F.: Multiple antigen peptide. A novel approach to increase detection sensitivity of synthetic peptides in solid-phase immunoassays. J Immunol Methods 124: 53-61, 1989

Winter, G., Griffiths, A. D., Hawkins, R. E., and Hoogenboom, H. R.: Making antibody by phage display technology. Ann. Rev. Immunol. 12: 433-455, 1994

Guided Selection on Cells

SILVANA CANEVARI and MARIANGELA FIGINI

Introduction

Need for guided selection

Routine screening of phage display libraries for antibodies has been and continues to be successfully applied, exploiting the affinity interaction with a large variety of soluble antigens to identify antibodies. Standard procedure is based on the use of:

a. antigens coated directly or indirectly (through biotin-avidin interaction) on plastic surfaces (plates or immunotubes) or on different solid matrices (affinity chromatography), see Chapter 9;

b. antigens that are biotinylated and coupled to streptavidin-coated paramagnetic beads (magnetic selection), see Chapter 10.

However in principle, any target that is suitable for affinity interaction with phage can be used. Thus, selection has also been successfully carried out with whole cells (Figini et al., 1998; Cai and Garen, 1995; Noronha et al., 1998; Hoogenboom et al., 1999) and even with live organisms (Yip et al., 1999).

Selection format can be a limiting factor, particularly when:

a. molecules against which to select phage antibodies are not available in purified and soluble form;

b. molecules/determinants are poorly immunogenic in humans;

✉ Silvana Canevari, Istituto Nazionale Tumori Milano, Experimental Oncology Dept., Via Venezian 1, Milano, 20133, Italy (*phone* +39-02-2390567; *fax* +39-02-2362692; *e-mail* canevari@institutotumori.mi.it)
Mariangela Figini, Istituto Nazionale Tumori Milano, Experimental Oncology Dept., Via Venezian 1, Milano, 20133, Italy

c. the phages of interest are extremely rare in a large library.

Thus, novel selection formats have recently been envisaged and shown to be useful.

Selection on cells

Although conventional biochemistry and advanced biotechnological approaches have led to the availability of a large variety of molecules in purified and/or recombinant form, such molecules in many cases do not maintain the correct conformation in soluble form. This is particularly evident in molecules that are deeply embedded in the membrane lipid bilayer and/or whose conformation/activity depends not only on post-translational modification (GPI-anchorage, acetylation, farnesylation), but also on complex interactions with other cell surface components in specialized membrane micro-domains (Harder and Simons, 1997). Membrane micro-domains and their associated molecules are good candidates as targets for antibody generation since they are relevant in the transduction/regulation of the mitogenic signal (Shaul ans Anderson, 1998); however these structures are easily disrupted by biochemical manipulation. Thus, the use of whole living cells as a direct source of target antigen is advisable to retain the physiological status of the molecule as much as possible. However the selection of phage antibodies against unpurified cell surface markers by panning on whole cells has proved very difficult due to the huge number of different antigens with different expression levels.

Guided selection

Guided selection involves the use of a mouse template antibody chain (VHCH$_1$ or VLCL) to help fashion the antigen binding site of a human antibody in combination with a complementary human chain from a repertoire (see Chapter 39). Guided selection is particularly useful in overcoming the limitations associated with molecules that are poorly immunogenic in humans. In fact, molecules such as self-antigens, for which very few natural antibodies are likely to be present, frequently exhibit high immunogenicity in mice and so a large number of mouse monoclonal antibodies recognizing distinct determinants can easily be derived. The novel format of guided selection on cells not only circumvents problems of poor immunogenicity and unavailability of purified antigen, but can also increase the probability of identifying extremely rare phages in a large library. Guided selection on cells provides a simple means of generating human antibody against cell surface antigens for which a rodent monoclonal antibody is available but the purified antigen is not. Potential candidates for guided selection on cells include several tumor-associated antigens (TAA) on malignant cells and the cluster differentiation (CD) antigens on human lymphocytes.

Advantages and limitations

Practical and theoretical advantages of the method are:

a. the possibility to avoid cumbersome purification of antigens;

b. the ability to raise antibodies against molecules not available in purified form;

c. the ability to obtain completely human antibody against a target of interest for which murine monoclonal antibodies are available and have given promising results.

On the other hand, potential limitations of the method are represented by:

a. the topobiology of the target antigen (generic expression level, local density, accessibility/crypticity);

b. the high phagocytic activity of some cells;

c. lack of precise homologues to the mouse templates in the human repertoire.

Limitations associated with topobiology can only be partially overcome by applying alternative preparation procedures that reduce the turnover of the target antigen and stabilize its expression on the cell membrane. Similar procedures could be applied to reduce the impact of target antigen internalization (see comments). Finally, the limit c) is mainly due to the fact that the majority of murine monoclonal antibodies uses k light chain and some of them have no homologue in the human repertoire. The use of a less precise homologue could account for a lower affinity of the human antibody compared to the murine template. It should be possible to improve the affinity by in vitro mutagenesis (Hawkins et al., 1992; Chowdhury and Pastan, 1999) or in bacterial mutator strains (Low et al., 1996) (see Chapter 37).

Outline

The protocol described (outlined in Figure 1 and Figure 2) was initially set up at the MRC Center for Protein Engineering in the laboratory of G. Winter (Figini et al., 1994; Jespers et al., 1994) and further developed at the Department of Experimental Oncology of Istituto Nazionale Tumori (Figini et al., 1998) for adherent cells. Modification of this protocol might be necessary for special applications, for example, when using cells grown in suspension. Suggestions for modified protocols are given in comments at the end of the chapter to overcome the more frequently encountered difficulties.

Materials

- Culture medium RPMI-1640 (SIGMA)
- Fetal Calf Serum (FCS) (Gibco)
- Glutamine (SIGMA)
- CO_2 incubator
- FACScan (Becton Dickinson)
- Trypsin-EDTA (BioWhittaker)
- Benchtop centrifuge
- Cell culture petri dishes (Costar)
- 96-well plates flat-bottom for cell culture (Costar)
- MPBS=Marvell PBS (Marvell is a Non fat dry milk)
- *E. coli* TG1
- Large plates (243 mm x 243 mm Nunc)
- 2 liter shake flask
- Sterile glycerol
- Polypropylene freezing tubes
- ELISA multichannel reader (Biorad)
- Shaking incubator at 37°C
- anti-M13 HRP conjugate antibody (PHARMACIA Biotech Cat. No. 27-9411-01)

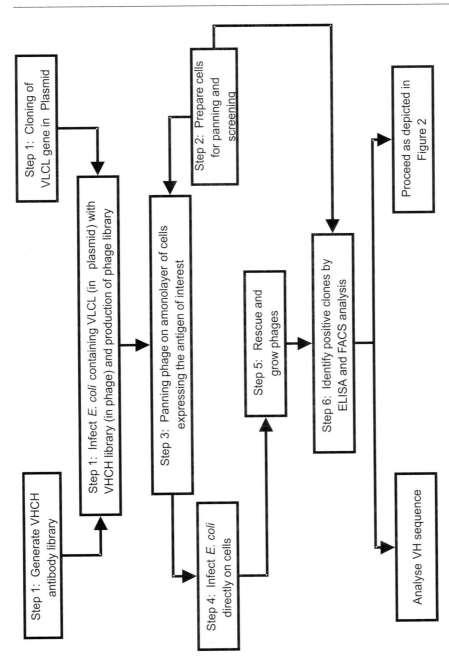

Fig. 1. Flow-chart of the guided selection on cells: identification of human VH

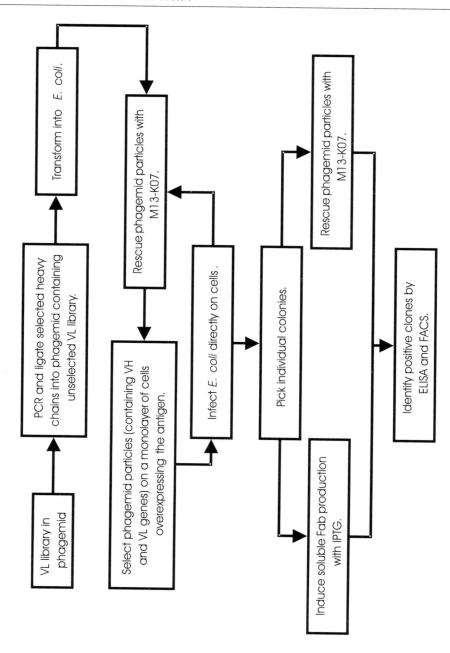

Fig. 2. Flow-chart of the guided selection on cells: identification of the completely human Fab

- FITC-conjugate anti-sheep antibody (SIGMA)

- Tetramethyl-Benzidine Dihydrochloride (TMB) (SIGMA)

- anti-M13 (PHARMACIA Biotech Cat. No. 27-9410-01)

- FITC-conjugate anti-sheep antibody (SIGMA)

- poly-L-lysine

- Microcentrifuge

- Sterile microcentrifuge tubes

- Sterile pipette tips

- Vortex mixer

- High-speed centrifuge (Sorvall or equivalent) and GSA rotor

- Sterile 50 ml Falcon tubes

- 15 ml polypropylene culture tubes

- Glutaraldehyde (Polysciences.inc)

- 2xTY broth, per liter:
  - 16 g tryptone
  - 10 g yeast extract
  - 5 g NaCl
  - Sterilize by autoclaving

- Minimal agar, per liter:
  - 12.8 g $Na_2HPO_4 \cdot 7 H_2O$
  - 3 g KH_2PO_4
  - 0.2 g $MgCl_2$
  - 4 g glucose
  - 0.5 g NaCl
  - 5 mg thiamine hydrochloride
  - 1 g NH_4Cl
  - 15 g agar

- TYE plates, per liter:
  - 2xTY
  - 100 µg/ml ampicillin
  - 12.5 µg/ml tetracyclin
  - 1% glucose
  - 15 g agar

Procedure

STEP 1   The procedures of Step 1 are fully described in other chapters of this manual to which we refer.

Preparation of the library

Protocols for the preparation and growth of the phage library are described in Chapters 5 to 7 and in Figini et al., 1994.

Cloning monoclonal antibody VLCL gene

For the preparation of the murine VL gene, see Chapter 2 and Figini et al., 1994.

Note: Experimental evidence indicates that many hybridomas also express non-functional mRNAs that encode abortive immunoglobulin variable regions and that can be amplified during PCR. It is therefore advisable to construct scFv from the hybridoma before proceeding in order to ensure the identification of functional VL.

Infection of E. coli (containing VLCL) with VHCH1 library

Described in Chapters 5 and 6.

Rescue and grow phage library

Described in Chapters 5 and 6.

STEP 2   **Preparation of cells**

1. Maintain the cells expressing the target antigen in complete culture medium (RPMI-1640 medium supplemented with 10% FCS, 2 mM glutamine) in a humidified 5% CO_2 atmosphere at 37°C.

    Note: Routinely confirm cell surface expression of the antigen by FACS analysis.

2. Recover the cells, depending on their growth characteristics (see Comment 1 below), by detachment with trypsin-EDTA or by centrifugation.

3. For panning purposes, seed cells in 100-mm cell culture petri dishes at a concentration of 4×10^5/ml (depending on cell size and growth rate) and grow as a monolayer (80% confluent).

4. For purposes of phage ELISA selection, seed cells in flat-bottom 96-well plates at a concentration of $1-3 \times 10^5$ cells/ml and grow as a monolayer.

5. If necessary, fix the cells using the following procedure:

Cell fixation

1. Start from Step 2.3 or 2.4.

2. Gently pour off the cell culture supernatant.

3. Add PBS+0.1% glutaraldehyde (200 µl to each well of a 96-well plate or 10 ml per culture dish).

4. Incubate for 5 min (longer incubation could damage the cells).

5. Wash 5 times with PBS.

6. Add PBS+0.75% glycine (0.1 M) + 0.001% phenol red to block the free aldehyde groups.

7. Incubate for 5 min at room temperature.

8. Proceed directly with the ELISA or panning on cells by washing 5 times with MPBS, or for the ELISA plate add 200 µl of PBS+1% BSA+0.2% NaN$_3$ and store the plate.

   Note: ELISA plates can be kept at 4°C up to 6 months in a closed moist box (to avoid dehydration) with no significant loss of antigenic activity and no cell detachment.

Panning on cell monolayers   STEP 3

1. Incubate the cells in 20 ml of 2% MPBS for 2 hr at 37°C to block any remaining unsaturated binding sites on the plastic.

2. Remove the MPBS.

3. Add ~10^{12} t.u. of phage library displaying hybrid Fab fragments in 10 ml of 2% MPBS; shake slowly for 1 hr at room temperature.

Note: Culture plates are treated to ensure cell adhesion even in presence of excess protein so that shaking is necessary to reduce non-specific phage adherence to plates.

4. Wash 5 times with 10 ml of 0.1% Tween-20/PBS, and then 10 times with 10 ml of PBS alone. In each washing step, the buffer is gently added, briefly swirled in the plate, and immediately removed.

Note: If the cells being used are detergent-sensitive, wash further with PBS alone or use fixed cells (see Comment 2).

STEP 4   Infection

1. Prepare 10 ml of log phase *E.coli* TG1 bearing the plasmid that contains the light chain and add it directly to the plates from Step 3; incubate for 30 min at 37°C without shaking.

   Note: Infection of bound phages by direct addition of the E. coli strain containing the VLCL to the cells generally leads to an increase in the number of phage rescued without an increase in background. When direct infection is not possible (see Comment 4), elution of the phage from the cells is recommended in either of the following ways:
   a) Add 5 ml of 100 mM triethylamine (diluted immediately before use) and keep under agitation for 10 min; add 2-3 ml of 1 M Tris, pH 7.4, for quick neutralization; use this phage for infection.
   b) Add 5 ml of 50 mM glycine-HCl, pH 2.7, plus 150 mM NaCl and incubate with shaking for 10-20 min at room temperature; neutralize immediately with 2 M Tris-base until physiological pH is reached; use this phage for infection.

2. Use 0.5 ml to make serial dilutions in 2X TY broth; place 100 ml of the dilutions (from 10^{-4} to 10^{-8}) on TYE Amp-Tet plates to determine the titer of the library.

3. Spin the remaining 9.5 ml at 3,300g for 10 min and resuspend the pellet in 0.6 ml of 2xTY.

4. Plate the resuspended pellet in a 243x243-mm square plate containing TYE-Amp-Tet.

5. Grow overnight at 30°C.

Phage production   STEP 5

1. Add 5 ml of 2X TY containing 12.5 µg/ml tetracycline and 15% glycerol to each plate and rescue the colonies by scraping with a sterilized glass spreader.

2. Use 50 µl of the phage culture to inoculate a 2-liter conical flask containing 500 ml of 2X TY and 12.5 µg/ml tetracycline and grow for 16-24 hr at 30°C.

   Note: Starting OD at 600 nm should be less than 0.1.

3. Find the titer of the rescue. Use 10^{10} clones for further phage preparations.

4. Store the remaining culture at -70°C.

5. Spin the 500-ml overnight culture at 10,800 g for 10 min or at 3,300 g for 30 min.

6. Add a 1/5 of 20% polyethylene glycol 6000 and 2.5 M NaCl (PEG/NaCl) to the supernatant; mix well and incubate for 1 hr or more at 4°C.

7. Spin at 10,800 g for 10 min or 3,300 g for 30 min and carefully aspirate the supernatant.

8. Resuspend in 10 ml of PBS and add 2.5 ml of PEG/NaCl. Mix and incubate for 20 min or more at 4°C.

9. Respin briefly and aspirate any remaining traces of PEG/NaCl.

10. Resuspend in 2 ml of PBS and spin at 11,600 g for 10 min to remove most of the remaining bacterial debris.

11. Store the phage at 4°C for short-term storage or in PBS and 15% glycerol for longer term storage at -70°C.

12. Titrate the phage stock (see Chapter 6).

13. Check phage binding by polyclonal phage ELISA.

After every round of infection, it is necessary to check the titration of the phage to monitor the extent of enrichment, and to check cell binding by polyclonal ELISA.

Note: Repeat Step 2 through Step 5 until a positive signal is detected, and grow single colonies in a 96-well plate to isolate single positive clones.

STEP 6   **Screening: Single clone ELISA**

1. Gently touch the top of individual colonies with a sterile toothpick. Transfer the toothpick into 200 ml of 2X TY medium with the appropiate antibiotic in a 96-well plate and grow with shaking (300 RPM) for 16-20 hr at 30°C.

   Note: At this stage, the culture is saturated and should yield 10^{10} t.u phage/ml. Also, aeration is very important at this stage of growth, increased aeration and higher yields can be obtained by placing the 96-well plate without a lid into a box.

2. Spin the 96-well plate with growing bacteria at 3,300 g for 10 min.

3. Rinse the ELISA plate containing the cell monolayer (Step 2.4) 3 times with PBS and block unsaturated sites on the plastic by adding 200 µl of 2% MPBS to each well; incubate for 2 hr at 37°C.

4. Remove the 2% MPBS by aspiration and add 20 µl of 10% MPBS to each well.

5. Using a multichannel pipette, transfer 80 µl of bacterial culture supernatant to the ELISA plate, mix, and leave for 1 hr at room temperature.

6. Wash the wells 3 times with 0.1% Tween-20/PBS and then 3 times with PBS.

7. Add 100 µl of a 1:5000 dilution of anti-M13 HRP conjugate antibody to each well.

8. Wash the wells 3 times with 0.1% Tween-20/PBS and then 3 times with PBS.

9. Add 100 µl of TMB to each well and leave at room temperature for 10-20 min.

10. Block the reaction with 1 M H_2SO_4 and read absorbance at 450 nm in an ELISA multichannel reader.

Screening: FACS analysis

1. For each sample, add 100 µl of PBS+1% of BSA containing approximately 5×10^9 antibody-bearing phage particles to a pellet of 3×10^5 cells expressing the antigen of interest and incubate for 1 hr at room temperature.

Note: Incubation temperature can be modified depending on the characteristics of the antigen.

2. Wash 3 times with PBS.

3. Add 100 µl of 1:5000-diluted anti-M13 and incubate for 30 min on ice.

4. Wash 3 times with PBS.

5. Add 100 µl of 1:1000-diluted FITC-conjugate anti-sheep antibody (SIGMA) and incubate for 30 min on ice.

6. Wash 3 times with PBS.

7. Analyze the cells with a FACScan.

At the end of 3-6 rounds of positive selection and screening, at least one human VHCH1 should be identifiable, which in combination with the original murine light chain used as the guide probe, can bind the target antigen. To obtain a completely human antibody, the VHCH1 selected must be subcloned into a plasmid containing a repertoire of human light chains. This procedure is outlined in Figure 2.

Note: Because a phagemid instead of a phage is being used at this point, different protocols for the rescue must be followed (see Figini et al., 1994). Also, all caveats that apply to the use of cell monolayers as target antigen must also be considered.

Elution of bound phages from cell monolayers

Comments

1. Cell fixation

All the Steps described above that involve use of cells are performed on live cells. If the cells are detergent-sensitive (see Comment 3) or if the target antigen shows rapid turnover (see Comment 4), fixation of the cells is recommended. Fixation should also be considered when cells do not adhere well to the plastic and detachment during rinsing is possible. However fixation must be applied with caution since it has been reported to alter the conformation of several cell surface molecules. Preservation of the determinant of interest after fixation should be tested, using the monoclonal antibody used to guide selection as positive control.

2. Use of cells in suspension

The procedures above are described for adherent cells. Similar procedures can be applied to cells growing in suspension, using centrifugation for each wash step. The major risk with suspended cells is their loss, and consequently that of the bound phage, during centrifugation. Thus, a centrifugation speed must be selected that is appropriate to the size and sensitivity of the cells, balancing the risk of cell loss due to flotation (low speed) and due to centrifugation damage (squashing) (high speed). Alternatively, the cells can be attached to the plastic substrate using polylysine.

1. Wash the plate twice with PBS.

2. Add 50 μl/well or 10 ml/petri plate of poly-L-lysine at a concentration of 1 mg/ml in PBS and incubate for 15 hr at room temperature.

3. Wash 3 times with PBS.

4. Seed the cells at 3×10^4 cells/well or $5-6 \times 10^6$ in a petri plate and incubate for 1 hr at room temperature.

   Note: The number of cells depends on their size.

5. Wash another 3 times with PBS and proceed with panning or with fixation.

3. Sensitivity of cells to detergent lysis

In detergent-sensitive cells, cellular fragments and the bound phage are washed away and lost. Although the use of detergent can be avoided, it is then necessary to perform more washes with PBS alone, which frequently results in increased background. In that case, cell fixation might be preferable.

4. Topobiology of the target antigen

The topobiology of a molecule is comprised of a complex series of parameters such as antigen expression level, local density of antigen expression in membrane microdomains, and the local environment that controls antigen/determinant accessibility/crypticity. When applicable, such parameters should be optimized or at least the cells with the most favorable parameters should be identified. Identification of cells with an appropriate topobiology of the target antigen is a critical step in the procedure for guided selection on cells and strongly influences the final outcome of the selection. In particular, cells frequently exhibit a lower antigenic density as compared to immunotubes.

5. Internalization of the target antigen

Phage bound to a target antigen expressed on live, metabolically active cells, might be removed from the cell surface through the specific process of cellular internalization or through the passive intake of larger membrane areas. Either process can drastically reduce the number of phage available for infection and seriously impair the selection of rare phage. To obviate the internalization activity of some specific target antigens (such as membrane receptors) and/or the inherent phago/potocytic activity of some specialized cells, selection can be carried out on fixed cells or on cells maintained at temperatures below the transition state of the lipid bilayer of the cell membrane (16 °C). However these procedures should be applied with caution since fixation can induce conformational alterations of some cell surface molecules, while selection at low temperatures can lead to bias favoring antibodies with a rapid Koff under physiologic conditions.

6. Positive versus negative selection

The described protocol is based on several sequential rounds of positive selection, i.e. recovery of the phages bound to the cells. This strategy is generally preferable, at least in the first one/two rounds, to negative selection, i.e. recovery of unbound phages, for two main reasons. On the one hand, all phages are quite sticky and so tend to bind nonspecifically as well as specifically; on the other hand, the specific phage in a large library are quite rare and can be easily lost in negative selection. With cells of the same line, with or without the target antigen of interest, it is possible to intercalate 1 round of negative selection for every 2 rounds of positive selection. However, this complex strategy is generally unnecessary since positive selection per se enables a rapid specific phage number enrichment (see Comment 7).

7. Enrichment

The first step is the most critical. Due to the nature of the enrichment process, any abnormalities or mistakes selected at this point will be amplified during panning. Figure 3 shows two examples of phage enrichment. Usually starting with 10^{12} phage, the first round of selection should yield at least 10^4 phage. It is advisable to perform initial rounds of selection under low stringency (using cells with high antigen density, and minimizing the number and duration of washes) so as not to lose rare binders, and then to employ more stringent conditions in later rounds. The stringency conditions should be fine-tuned by the operator according to the topobiology of the target antigen. All rescues should be checked for the presence of insert (by PCR), and the number of positive clones should scored. If no enrichment is obtained after 4-6 pannings, it is advisable to: check the length of the insert; change the conditions of panning (temperature, incubation time, fixation of cells); if possible, use an alternate cell line; start a new selection from the beginning.

Acknowledgements. This work was supported by Italian Health Ministry, finalized program and Italy-USA program "therapy of tumors".

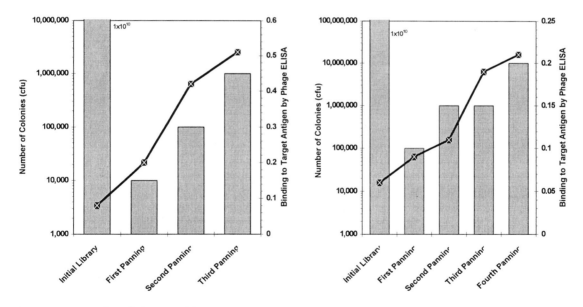

Fig. 3. Two examples of phage enrichment

References

Cai X, Garen A (1995). Anti-melanoma antibodies from melanoma patients immunized with genetically modified autologous tumor cells: Selection of specific antibodies from single-chain Fv fusion phage libraries. Proc Natl Acad Sci USA 92: 6537-6541

Chowdhury PS, Pastan I (1999). Improving antibody affinity by mimicking somatic hypermutation in vitro. Nature Biotechnol 17: 568-572

Figini M, Marks JD, Winter G, Griffiths AD (1994). In vitro assembly of repertoires of antibody chains on the surface of phage by renaturation. J Mol Biol 239: 68-78

Figini M, Obici L, Mezzanzanica D, Griffiths AD, Colnaghi MI, Winter G, Canevari S (1998). Panning phage antibody libraries on cells: isolation of human Fab fragments against ovarian carcinoma using guided selection. Cancer Res 58: 991-996

Harder T, Simons K (1997). Caveolae, DIGs, and the dynamics of sphingolipid-cholesterol microdomains. Curr Opin Cell Biol 9: 534-542

Hawkins RE, Russell SJ, Winter G (1992). Selection of phage antibodies by binding affinity: mimicking affinity maturation. J Mol Biol 226: 889-896

Hoogenboom HR, Lutgerink JT, Pelsers MM, Rousch MJ, Coote J, van Neer N, De Bruine A, Van Nieuwenhoven FA, Glastz JF, Arends JW (1999). Selection-dominant and nonaccessible epitopes on cell-surface receptors revealed by cell-panning with a large phage antibody library. Eur J Biochem 260: 774-784

Jespers LS, Roberts A, Mahler SM, Winter G, Hoogenboom HR (1994). Guiding the selection of human antibodies from phage display repertoires to a single epitope of an antigen. BioTechnology 12: 899-903

Low N, Holliger P, Winter G (1996). Mimicking somatic hypermutation: affinity maturation of antibodies displayed on bacteriophage using a bacterial mutator strain. J Mol Biol 260: 359-368

Noronha EJ, Wang X, Desai SA, Kageshita T, Ferrone S (1998). Limited diversity of human scFv fragments isolated by panning a synthetic phage-display scFv library with cultured human melanoma cells. J Immunol. 161: 2968-2976

Shaul PW, Anderson RG (1998). Role of plasmalemmal caveolae in signal transduction. Am J Physiol 275,: L843-L851

Yip YL, Hawkins NJ, Smith G, Ward RL (1999). Biodistribution of filamentous phage-Fab in nude mice. J Immunol Methods 225: 171-178

Proximity (ProxiMol) and Step-Back Selections

JANE K. OSBOURN

Introduction

Other chapters in this volume describe the selection of antibodies specific to cell surface molecules in a non-guided manner. Cell surfaces are antigenically complex and this results in difficulties in specifically selecting antibodies which bind to a given cell surface molecule without selecting for a host of unwanted specificities in parallel. The desired antigen specificity can in many cases be obtained, but only after an intensive screening programme has been undertaken. One of the rationales for the development of ProxiMol selection (previously known as Pathfinder selection) was to provide an efficient method for the generation of a population of antibodies enriched for binders which bind in close proximity to a particular desired cell surface antigen alleviating the necessity for labour intensive screening programmes. This selection regime greatly narrows the scope of the selection and increases the probability of selecting antibodies of interest.

Principles and applications

ProxiMol selection is a development of catalysed reporter enzyme deposition (CARD), which has previously been described as a method of signal amplification in ELISA, immuno-cytochemistry and blotting formats (Bobrow et. al., 1989, Bobrow et. al., 1992, Merz et. al., 1995, Adams, 1992). CARD is based on the use of horseradish peroxidase (HRP)-conjugated reagents, such as antibodies, in conjunction with a biotin tyramine substrate. When hydrogen peroxide is added, the enzyme catalyses formation of biotin tyraminefree radicals, which then react with proteins in the immediate vicinity of the enzyme, hence biotinylating them. By then adding streptavidin-HRP complex, the number of molecules of HRP at a specific

Jane K. Osbourn, Cambridge Antibody Technology, The Science Park, Melbourn, Cambridgeshire, SG8 6JJ, United Kingdom (*phone* +44-1763-269237; *fax* +44-1763-263413; *e-mail* jane.osbourn@camb-antibody.co.uk)

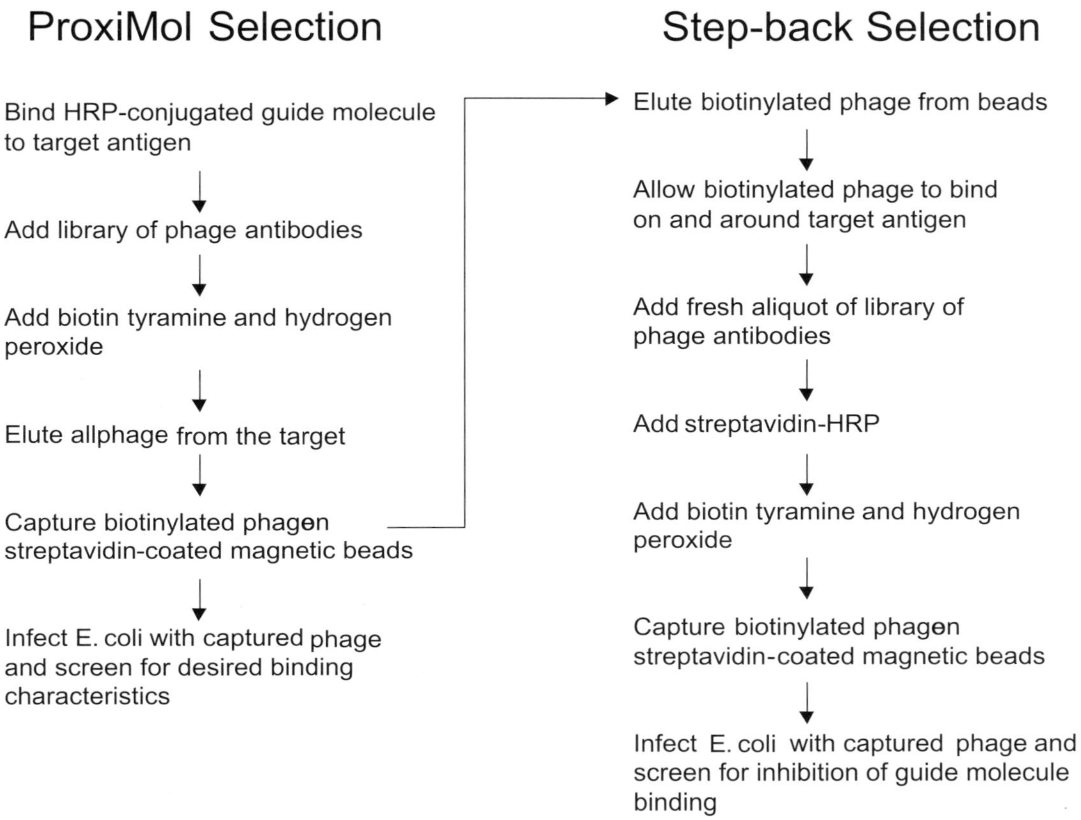

ProxiMol Selection

Bind HRP-conjugated guide molecule
to target antigen

↓

Add library of phage antibodies

↓

Add biotin tyramine and hydrogen
peroxide

↓

Elute allphage from the target

↓

Capture biotinylated phagen
streptavidin-coated magnetic beads

↓

Infect E. coli with captured phage
and screen for desired binding
characteristics

Step-back Selection

Elute biotinylated phage from beads

↓

Allow biotinylated phage to bind
on and around target antigen

↓

Add fresh aliquot of library of
phage antibodies

↓

Add streptavidin-HRP

↓

Add biotin tyramine and hydrogen
peroxide

↓

Capture biotinylated phagen
streptavidin-coated magnetic beads

↓

Infect E. coli with captured phage and
screen for inhibition of guide molecule
binding

Fig. 1. Flow chart of the procedure.

site is increased resulting in signal enhancement when the enzyme is detected colorimetrically. When CARD is used as an amplification system in immunocytochemistry no detectable loss of image resolution is apparent, demonstrating that the deposition occurs in close association with the catalytic enzyme. The ProxiMol selection approach is a modification of the CARD methodology that uses the ability of HRP, in conjunction with biotin tyramine, to biotinylate phage binding around the site of enzyme activity at a specific site on a cell surface. Biotinylated phage can be recovered on streptavidin-coated magnetic beads. ProxiMol selections can be carried out on any cell type, but also on antigen in many other forms, such as purified antigen, cell extracts or membrane preparations. The technique relies on the availability of a ligand to the target antigen or cell type that can be HRP conjugated and used to guide the selection process. This ligand could be a naturally occurring ligand such as a growth factor, chemokine, existing antibody, or any other type of molecule known to bind the cell type of interest. A range of example selections using different guide molecules has been carried out (Osbourn et. al., 1998a). The method directs biotinylation of phage particles that bind up to an approximate radius of 25 nm from the original binding site and results in very low numbers of phage particles being recovered (in some cases only a few hundred). The low number of recovered phage particles is due to the highly specific nature of the selection procedure and often means that it is only necessary to carry out a single round of selection.

ProxiMol selections using either existing antibodies or natural ligands as guide molecules result in panels of antibodies that bind close to the site of antibody or ligand binding, but which do not overlap with it. This is because the selection method relies on the presence of the guide molecule on the cell surface to direct the selection. It is possible, however, to generate antibodies which block ligand or antibody binding to the cell by carrying out a second round of ProxiMol selection using the output from the first round of selection to guide a further round. This second stage of selection is referred to as a "step-back" selection. Antibodies which block MIP-1α binding to CD4$^+$ cells have been generated using this approach (Osbourn et. al., 1998b).

Materials

Equipment
- magnetic rack (Dynal, Oslo, Norway)
- 16 well chamber slides (Nunc)
- rotating wheel to which eppendorf tubes can be attached

Reagents and solutions
- biotin tyramine (NEN, available as part of the Renaissance TSA kit)
- streptavidin-coated magentic beads (M-280 Dynal, Oslo, Norway)
- streptavidin-horseradish peroxidase conjugate (Amersham, Bucks, UK)
- hydrogen peroxide
- 100 mM triethylamine
- 1 M Tris-HCl pH 7.4

– PBS (phosphate buffered saline) made using tablets supplied by Oxoid Ltd, Basingstoke, UK)

– PBSM (PBS containing 3% Marvel)

– PBST (PBS containing 0.1% Tween)

– TE (10 mM Tris pH 8.0, 1 mM EDTA)

Procedure

ProxiMol selection using a guide antibody on an adherent cell line

In some situations an antibody may already exist which binds the target antigen, but may not recognise the desired epitope. In such cases it is possible to use the existing antibody as a guide molecule in a ProxiMol selection to generate a population of antibodies which binds very close to, but not at the same epitope as, the original antibody. Such an approach has been used to generate a panel of human anti-carcinoembryonic antigen (CEA) antibodies using a commercially available mouse mAb as the guide molecule (Osbourn et. al., 1998a). In this case the HRP was directed to the target site using an anti-species HRP conjugate. The protocol described below assumes the target antigen is expressed on the surface of an adherent cell line, but target antigens expressed on non-adherent cells, or purified antigens can also be used in ProxiMol selections.

1. Grow the cell line under normal culture conditions on 16 well chamber slides to approximately 80% confluence (1×10^5 to 1×10^6 cells per well). It is preferable to carry out selections on fixed cells, but if it is necessary to use unfixed cells all the following steps should be carried out at 4°C. A suggested fixation regime is treatment with 0.1% gluteraldehyde for 15 min at room temperature.

2. Block the cells in 100 μl PBSM for 1-2 hours at room temperature.

3. Gently wash the cells with three changes of PBS.

4. Add the primary guide antibody at an appropriate dilution (i.e. one that gives well-defined staining by ICC) in PBSM. Incubate for 1 hour at room temperature. If a HRP-conjugated guide antibody is available this can be used directly and step 8 of this protocol can be omitted.

5. Wash the cells as in step 3.

6. Add an aliquot of phage library (the number of phage added should be 100-fold greater than the library size i.e. for a 1×10^{10} library 1×10^{12} phage particles should be added per selection) in 100 μl of PBSM and incubate for 1 hour at room temperature.

7. Wash as before.

8. Add the secondary anti-species-HRP conjugate at an appropriate dilution (normally 1:1000 to 1:5000) in PBSM and incubate for 1 hour at room temperature.

9. Wash, then resuspend the cells in 100 μl 50 mM Tris-HCl pH 7.4 containing 0.03% H_2O_2 and 0.4 μl of biotin tyramine (at 1 mg/ml) and incubate at room temperature for 10 min.

10. Wash the cells as before and elute the bound phage by adding 100 μl of freshly made 100 mM triethylamine and incubating at room temperature for 10 min.

11. Transfer the eluted phage to a 1.5 ml eppendorf tube and neutralise immediately with 50 μl 1M Tris-HCl, pH 7.4.

12. Preblock 20 μl of streptavidin-coated magnetic beads by incubation in PBSM for 1 hour at room temperature. Add the blocked beads to the eluted phage and rotate for 15 min at room temperature.

13. Pellet the beads using the magnetic rack and wash three times in 1 ml of PBST, then three times in 1 ml of PBS.

14. Resuspend the beads in 100 μl of PBS and use half of this to infect E coli and to titre the phage as described in previous chapters. There is no need to elute the phage off the magnetic beads for successful infection of the bacteria to occur. Resultant colonies can be screened by cell ELISA, flow cytometry, or any other appropriate assay.

ProxiMol selections using natural ligands

In many situations it may not be possible or appropriate to use an existing antibody as a guide molecule to direct the selection, but another type of molecule may be available. A number of natural ligands have been used as guide molecules for ProxiMol selections such as; sialyl Lewis X to generate antibodies to P- and E-selectins (Osbourn et. al., 1998a), and macrophage inflammatory protein-1α (MIP-1α) to generate antibodies to the CCR5 receptor (Osbourn et. al., 1998b). In these examples the natural ligand was biotinylated and streptavidin-HRP was added to the system to provide HRP localisation in the region of ligand-cell interaction. The only limitation is that the ligand must be either HRP-conjugated directly or tagged in some way to allow indirect conjugation to HRP, e.g. by biotinylation followed by the addition of streptavidin-HRP. It must be ensured that biotinylation of the ligand does not affect the receptor recognition site. The following protocol describes the use of a biotinylated ligand to guide selection of antibodies to a cell surface receptor on cells in suspension. It is possible to modify this method for use with adherent cells, or for other methods of antigen presentation, and for use with directly conjugated ligands.

1. Prepare approximately 1×10^5 to 1×10^6 cells in solution in an approximate volume of 100 µl PBSM and incubate for 1 h at room temperature. If receptor internalisation is known to occur all steps except the biotin tyramine reaction (step 8) should be carried out on ice.

2. Add biotinylated ligand at an appropriate concentration and incubate for 1 h at room temperature. The chosen concentration of ligand added may be affected by receptor abundance and ligand availablilty (e.g. for selections on $CD4^+$ cells using biotinylated MIP-1α the ligand was added to a final concentration of 375 nM).

3. Gently wash the cells by low speed centrifugation with three changes of 1 ml PBS.

4. Add streptavidin-HRP conjugate at a dilution of 1:1000 in PBSM and incubate at room temperature for 1 h.

5. Wash the cells as in step 3.

6. Add 1×10^{12} phage in 100 µl of PBS containing 3% (w/v) skimmed milk and incubate for 1-2 h at room temperature.

7. Gently wash the cells as in step 3.

8. Add 0.4 µl of biotin tyramine (approximately 1 mg/ml) in 100 µl 50 mM Tris-HCl pH 7.4 containing 0.03% H_2O_2 and incubate at room temperature for 10 min.

9. Wash the cells as before and resuspend the cells in 100 µl Tris-EDTA (TE) containing 0.5% Triton X 100. This is an alternative to TEA elution which lyses the cells to release the phage.

10. Add 20 µl of streptavidin-coated magnetic beads (which have been preblocked in PBS containing 3% skimmed milk) to the cell lysates and rotate for 15 min at room temperature.

11. Pellet the beads using the magnetic rack and wash three times in 1 ml of PBST, followed by three washes in 1 ml PBS.

12. Resuspend the beads in 100 μl of PBS and use 50 μl of this to infect E. coli and titre the phage as described in previous chapter.

Step-back selections

ProxiMol selections result in the selection of panels of antibodies which bind in close proximity to the guide ligand, but at a site which does not overlap with the ligand binding site, since that site is blocked by the presence of the ligand. It is possible, however, to generate antibodies which block ligand binding by carrying out a second round of ProxiMol selection using output from the first round of selection as guide molecules. The biotinylated phage particles which have been captured from the first stage of the selection are recovered and used directly, without amplification, as guide molecules which bind around the original ligand binding site. Streptavidin-HRP is then added, along with a new aliquot of library and biotin tyramine treatment is carried out. Such a selection is termed a "step-back" selection and will generate a panel of antibodies, some of which will bind at the original ligand binding site, and some of which will bind at sites around the original site. Antibodies which inhibit MIP-1α binding to CD4$^+$ cells have been generated using this approach (Osbourn et. al., 1998b).

1. Carry out a ProxiMol selection (either antibody- or ligand-directed) exactly as described in the previous protocol to stage 12 at which point biotinylated phage particles have been recovered on the magnetic beads.

2. Elute the biotinylated phage from the beads by pelleting the beads using the magnet then resuspending the beads in 100 μl of 100 mM triethylamine (TEA) and leaving for 15 min at 37°C. Neutralise the TEA by the addition of 50 μl of 1M Tris-HCl pH 7.4. Re-pellet the beads and retain the supernatant. This process is not 100% efficient, but the majority of the phage appear to be released from the beads, and provide a sufficient population for use in the next stage of the selection.

3. Prepare approximately 1×10^5 to 1×10^6 target receptor-expressing cells in solution in an approximate volume of 100 μl PBSM and incubate for 1 h at room temperature.

4. Add 75 μl of the biotinylated phage, along with a fresh aliquot of library consisting of 1 x 10^{12} phage in 100 μl of PBSM, directly to the cells and incubate for 1-2 hour at room temperature.

5. Gently wash the cells by low speed centrifugation with three changes of 1 ml PBS.

6. Resuspend the cells in 200 μl of PBSM containing streptavidin-HRP at a dilution of 1: 1000 and incubate for 1 hour at room temperature, then wash the cells as in step 5.

7. Resuspend the cells in 100 μl 50 mM Tris-HCl pH 7.4 containing 0.03% H$_2$O$_2$ and 0.4 μl of biotin tyramine (approximately 1 mg/ml) and incubate at room temperature for 10 min.

8. Wash the cells as before and resuspend the cells in 100 μl Tris-EDTA (TE) containing 0.5% Triton X 100.

9. Add 20 μl of streptavidin-coated magnetic beads (which have been pre-blocked in PBSM) to the cell lysates and rotate for 15 min at room temperature.

10. Pellet the beads using the magnetic rack and wash three times in 1 ml of PBST, followed by three washes in 1 ml PBS.

11. Resuspend the beads in 100 μl of PBS and use 50 μl of this directly to infect E coli. Resultant single colonies can then be picked and screened to identify antibodies which inhibit ligand binding.

Results

ProxiMol and step-back selection are highly specific, directed selections which often result in the recovery of very low numbers of phage particles compared with standard selection methods. Because of the highly specific nature of the selection and the low numbers of recovered phage it is often necessary to carry out only one round of selection.

The protocols described have been written for cell surface selections, but ProxiMol selections can also be carried out on purified antigen to select for antibodies which bind to differing epitopes from that recognised by the guide molecule. An example of this is the selection of anti-tranforming factor β1 antibodies (Osbourn et. al. 1998a). In theory there is no limit to the type of target antigen which can be used in the selections; membrane preparations, cell extracts, or tissue sections provide other possible starting materials.

References

Adams JC. (1992). Biotin amplification of biotin and horseradish peroxidae signals in histochemical stains J Histochem Cytochem 40 1457-1463

Bobrow MN, Harris TD, Shaughnessy KJ, Litt GJ (1989) Catalyzed reported deposition, a novel method of signal amplification. J Immunol Methods 125 279-285

Bobrow MN, Litt GJ, Shaughnessy KJ, Mayer PC and Colon J (1992). The use of catalysed reported deposition as a means of signal amplification in a variety of formats. J Immunol Method 150 145-149

Merz H, Malisius R, Mannweiler S, Zhou R, Hartmann W, Orscheschek K, Moubayed P, Feller AC (1995). A maximised immunohistochemical method for the retrieval and enhancement of hidden antigens. Lab Invest 73 149-156

Osbourn JK, Derbyshire EJ, Vaughan TJ, Field AW, Johnson KS (1998a). Pathfinder selection; in situ isolation of novel antibodies. Immunotechnology 3 293-302

Osbourn JK, Earnshaw JC, Johnson KS, Parmentier M, Timmermans V, McCafferty J (1998b). Directed selection of MIP-1α neutralizing CCR5 antibodies from a phage display human antibody library. Nature Biotech 16 778-781

Selections on Tissue Sections

JESPER MAX TORDSSON, THOMAS NIKLAS BRODIN,
and PIA JASMINE KARLSTRÖM

Introduction

Antibody reagents can be generated to virtually any antigen using phage display technology provided that a large phage library of high diversity and quality and a purified antigen is available. Thus, a number of research groups have invested in the construction of phage libraries as a general supply of antibody reagents (see other sections of this book).

To promote rapid progression in the discovery and knowledge of novel biological target molecules, including their expression and function, general and efficient technology for target discovery and characterisation is desired. Efficient production of antibody reagents useful for characterisation and purification of novel targets is critical for this development. However, the majority of molecules expressed in animal tissues are neither available nor simple to purify without a specific reagent. In addition, the purification process itself may alter the original in vitro/in vivo structure of an antigenic molecule/complex.

In these cases, the use of the "crude" antigen source as the materials for phage selection, such as sections of tissues, could be the most efficient method for identification of a novel antigen and for cloning of a useful reagent. Tissue samples can be stored deep frozen with preserved integrity for years and cryosectioned at any time. If mildly treated, tissue sections should represent a generally applicable material that very closely resembles the original in vivo phenotype and that in theory makes all tissue components available for phage selections. The phage library added to

✉ Jesper Max Tordsson, BioInvent Therapeutic AB, Box 724, Lund, 223 70, Sweden
(*phone* +46-46-2868575, 191138; *fax* +46-46-2110806;
e-mail jesper.tordsson@bioinvent.se)
✉ Thomas Niklas Brodin, Active Biotech Research AB, Box 724, Lund, 220 07, Sweden
(*phone* +46-46-191255, 191138; *fax* +46-46-191134;
e-mail thomas.brodin@activebiotech.com)
Pia Jasmine Karlström, Active Biotech Research AB,

the tissue sections could be thought of as a tool for molecular dissection and identification of the antigens expressed. Tissues in certain developmental or diseased states exemplify biological materials of particular interest to be used with this technology.

In the present chapter we will give a detailed description on how to select phage antibodies using frozen tissue sections. Following phage antibody selections towards complex antigens, it is also important that discriminating screening assays are used. Thus, protocols for production of soluble antibodies and for immunohistochemical analysis that supplies qualitative information regarding the antigenic tissue expression patterns are included.

Materials

Equipment

– Tissue sample transported in 0.15 M NaCl on ice before use

– Scissors, forceps, scalpel, 1.8 ml sterile cryotubes with screw cap

– Freezer for -70°C or lower storage, cryostat, microscope slides, we use multitest slides (ICN Biomedicals,Inc.USA) for immunohistochemistry, microscope including low power and up to 100× magnifying lenses (we use a Leitz Aristoplan research microscope)

– PAP-PEN from Histolab (Gothenburg, Sweden)

– Library phage stock solution (titer of ~10^{13} T.U. (transducing units)/ ml) stored in aliquotes at -20°C

– Sterile filtertips for micropipette

– Belly Dancer Shaker (Stovall Life Science, Inc. USA)

– Safe Seal Microcentrifuge Tubes, pre-lubricated and certified Rnase/ Dnase-Free (Bioscience, Inc. USA)

– Autoclave

– 90 mm petridishes containing 15 g/l luria agar diluted in 2×YT medium supplemented with the appropriate antibiotic (LB agar plates)

– *E. coli* DH5αF', amber stop codon suppressor strain for phage antibody production and *E. coli* HB 2151, non-suppressor strain for soluble antibody production, both grown in 2×YT at +37°C in culture flasks on a rotating shaking plate at 250 rpm to $OD_{600} \approx 1.0$ and 0.5 respectively.

Cultures originate from single colonies grown on minimal agar in a petri dish. (ref. Molecular Cloning, Sambrook, Frisch and Maniatis)

- Sterile 96U (U-shaped) microwell plates (Nunc Brand Products, Denmark) with lid for soluble antibody production

- Autoclaved toothpicks, used to pick single bacterial colonies

- Centrifuge for microtiter plates (we use Hettich Universal 30RF)

- 2-Methylbutane, liquid nitrogen, acetone at -20°C, 2% methyl green in water   **Solutions**

- Tris buffer (TBS): 50 mM Tris (hydroxymethyl)-aminomethane, 150 mM NaCl, pH 7.6, adjusted with HCl

- 20% Foetal Bovine Serum (FBS) diluted in TBS

- 1×Dulbecco's phosphate buffered saline (DPBS) with calcium and magnesium, pH 7.4

- High-salt washing buffer, 0.05 M Tris, 1.0 M NaCl, pH 7.6, adjusted with HCl

- Elution buffer : 0.1 M Triethylamine diluted in distilled water

- 5% bovine serum albumin (BSA; Fraction V, (Boehringer-Mannheim, Mannheim, Germany) diluted in DPBS

- Neutralization solution: 80 µl 1 M Tris pH 7.4 and 70 µl 5% BSA (for each slide)

- 2×YTmedium: 16 g select peptone (Life Technologies, Paisley UK), 10 g select yeast extract (Life Technologies, Paisley UK), 5 g sodium chloride per litre destilled water. Adjust pH to 7.0 with NaOH, autoclave for 20 minutes at 15 lb/sq. in on liquid cycle

- 1 l 10×MOPS (3-(N-Morpholinol-propanesulfonic acid) medium:
    - 400 ml 1 M Mops, adjust the pH to 7.4 with KOH
    - 40 ml 1 M Tricine, adjust the pH to 7.4 with KOH
    - 10 ml 0.01 M $FeSO_4$
    - 50 ml 1.9 M NH_4Cl
    - 10 ml 0.276 M K_2SO_4
    - 10 ml 5×10^{-4} M $CaCl_2$
    - 10 ml 0.528 M $MgCl_2$
    - 100 ml 5 M NaCl
    - 10 ml trace elements as follows:

- 3 μM $(NH_4)_6 (MO_7)_{24}$
- 400 μM H_3BO_3
- 30 μM $CoCl_2$
- 10 μM $CuSO_4$
- 80 μM $MnCl_2$
- 10 μM $ZnSO_4$

Dilute to 1L with distilled water and sterilize by filtration through a 0.22-micron filter. Store at -20°C in 100 ml aliquots.

- 1 l low phosphate containing Neidhart medium (Neidhardt 1974):
  - 100 ml MOPS-medium
  - 0.2% Glucose
  - 1.5 mM Thiamine
  - 20 mg/ml Adenine
  - 0.2% Caseine
  - 0.05 mM NaH_2PO_4
  - 0.05 mM Na_2HPO_4

Dilute to 1 l with distilled water and autoclave for 15 minutes at 15 lb/sq.in on liquid cycle

Procedure

Preparation of tissue sections for selection

1. Keep surgically resected tissue samples in 150 mM NaCl on ice not more than two hours before use.

2. Remove fat, necrotic tissue and irrelevant tissue components.

   Note: A very heterogeneous sample might be difficult to section due to variations in freezing points (lipid content) and structures.

3. Cut the tissue sample in pieces of approx. 5 mm (2-15) in diameter each.

4. Chill 2-Methylbutane in a glass vessel in liquid nitrogen in an isolated container to a temperature just above its freezing point. Label empty cryotubes and prechill them in the nitrogen with the lids tightly screwed.

5. Freeze each piece of tissue for 30 seconds in the 2-Methylbutane and then put it on a Cleanex paper cloth. Let the excess 2-Methylbutane

evaporate. When the tissue sample is frosty on its surface put it in a prechilled cryotube and return it to liquid nitrogen.

6. Store the tissue samples in a freezer at a temperature of -70°C or lower.

7. Collect tissue samples from the freezer and mount them in a cryostat.

8. Prepare 4-8 μm thick cryosections.

9. Stain a cryosection thawed on a glass slide with 2% methyl green (cell nuclei staining that facilitate tissue evaluation) for 15 seconds and then wash it quickly in tap water. Remove the water from the backside of the slide but keep a film of water over the section (e.g. using a coverslip).

10. Study the section in a microscope to check tissue orientation, tissue quality and quantity and proportion of the structure of interest.

    Note: For some tissue structures this evaluation requires immunohistochemical staining.

11. Cover the surface of a glass slide with cryosections, but leave a frame of at least 3 mm of free glass surface around the sections.

    Note: If high similarity between slides is required (e.g. in model experiments), place the first section on the first slide, the second section on the second slide, and so on until the last slide and then repeat this until all slides are covered with sections. This will minimise variations caused by differences in section size and tissue composition at varying depths of the tissue.

12. Let the tissue sections air-dry on the slides for at least 12 hours at room temperature (RT).

Phage selection on tissue sections

1. Thaw an aliquote of the phage stock.

2. Fix the tissue sections in a glass vessel containing cold (-20°C) acetone for 10 minutes.

3. Let the acetone evaporate from the slides on a ventilated surface at RT for 10 minutes.

4. Draw a line around the section area using a PAP-PEN (helps to keep all following solutions on the slide).

5. Add 0.5 ml 20% foetal bovine serum in TBS (20% FBS) to the slides such that it covers all the sections. Let the solution incubate at RT for 1 h. This step rehydrates the tissue and blocks non-specific binding sites.

6. Add 10^{12} and 10^{11} T.U. of library phage to safe seal microcentrifuge tubes with 0.4 ml 20% FBS and let them incubate at RT for 1 h.

   Note: We include an internal negative control phage at 10^{11} and 10^{10} T.U. in the library solutions. This non-binding phage clone contains a different antibiotic resistance marker and allows monitoring of specific enrichment and non-specific background during selections.

7. Remove the blocking/rehydration solutions from the slides by aspiration using a pasteur pipette and then immediately add the phage solutions. Incubate at 4°C for at least 12 h in a humid atmosphere.

8. Wash the sections with 10 ml TBS using a 10 ml pipette to remove excess non-bound phage.

9. Put the slides in 50 ml Falcon tubes with 40 ml TBS. Let the tubes rock horizontally on a belly dancer platform such that the solution slowly moves over the sections within the tubes. Incubate (wash) at RT for at least 10 minutes and then change to a fresh solution. Repeat this five times.

10. Wash two times for 5 minutes each using the high-salt washing buffer.

11. Wash rapidly two times with DPBS. Keep the slides in DPBS. These washes are performed to remove the high-salt buffer that would otherwise reduce the elution efficiency in the next step.

12. Wipe the slide surface edge with a Cleanex paper cloth close to the sections and then immediately add 0.8 ml DPBS to cover the section area.

    Note: Perform this step rapidly to avoid drying of the sections.

13. Remove the DPBS from the sections by aspiration and add 0.4 ml of the elution solution to each slide. Incubate at RT for 15 minutes.

14. Add 150 µl neutralization solution to empty safe seal microcentrifuge tubes.

15. Repeatedly (approx. 10 times) dispense and aspirate the elution solution using a 1 ml pipette while moving the pipette over the entire section area.

16. Transfer the eluate to the safe seal microcentrifuge tube.

   Note: For library selections try to remove as much as possible of the eluate solution. For model experiments remove only a fixed volume (e.g. 300 µl) to reduce errors between parallels.

17. For analysis of the phage yield make dilutions of the eluted phage for titration in 2×YT and mix (in Falcon 2059 tubes) 100 µl of the diluted phage with 250 µl *E.Coli* DH5αF' bacteria grown to OD_{600}=1.0. Let the phage infect the bacteria for 5 min. at RT (no motion). Incubate at 37°C on a shaking platform (200 rpm) for 75 minutes and then spread the bacteria on antibiotic containing LB agar plates.

   Note: The appropriate dilution that will result in more than 50 colonies on an agar plate varies from $1-10^6$ times dilution depending on the fraction of the specific phage and the adsorption, washing and elution efficiency.

18. For phage rescue and amplification, refer to sections of this manual covering these steps.

Screening of soluble scFv for tissue reactivity and specific binding pattern: Production of soluble scFv in non-suppressor strain E. Coli HB 2151

1. Make 10-fold dilutions of the amplified phage stock in 2×YT medium. Infect 100 µl *E. Coli* HB2151 with phage stocks diluted 10^9 - 10^{12} fold. Spread on LB agar (for **lacZ** phagemids, include 2% glucose) plates. Incubate overnight at +30°C.

2. Pick bacterial cells from 200-300 well isolated colonies using auto-claved toothpicks and put into 96U-well plates containing 200 µl 2×YT medium per well and appropriate antibiotic.

   Note: The U-shaped well will pellet bacterial cells better in the following centrifugation steps. 96-well plates with deeper wells holding larger volumes could be used as well as plates with fewer wells holding larger volumes, if high throughput or sensitivity of the assay are important issues.

3. **Alternative 1,** we have used for phoA promoter driven scFv amber mutation expression phagemids: Incubate with lid for 7 h at +37°C on a rotating shaking plate.

4. Transfer 5 µl of the bacterial cell suspension to a well in a corresponding position in a new microtiter plate containing 200 µl Neidhart medium plus the appropriate antibiotic. Store the remainder of the culture as a master plate at +4°C. For longer storage, add to 15 % final conc. of sterile glycerol and freeze at -20°C. Culture overnight at +30°C on a rotating shaking plate.

5. Spin the plate at 2200 rpm for 7 min. Transfer 100 µl supernatant to a well in a corresponding position in a new microtiter plate containing 100 ml 1% BSA in DPBS.

6. **Alternative 2**, used for lacZ promoter driven scFv amber mutation phagemids or expression plasmids (including those containing the lac I^q suppressor): Include 2% glucose in all incubation steps in flasks or on plates to further suppress expression. From step 2 above, incubate plates overnight at +30°C shaking at 220 rpm. To prepare a master plate, transfer 5 µl of the bacterial cell suspension to a well in a corresponding position in a new microtiter plate containing 200 µl 2×YT medium, 2% glucose plus the appropriate antibiotic. Incubate this plate for 3-5 hours to establish a master plate culture for further use of the clones. Store at +4°C or add to 15% final conc. of sterile glycerol and freeze at -20°C.

7. The remainder of the plate is used for expression: Spin this plate at 4000 rpm for 7 min. By a flicking wrist movement, empty the plate of supernatants. Excess fluid on plate edges is removed using a paper towel. Add 200 µl 2×YT medium using a multipipette. This will resuspend the cells. Spin again and remove the supernatant similarly. Add 200 µl 2×YT medium including antibiotics. Incubate overnight at +30°C shaking at 220 rpm.

    Note: Alternatively to this low-grade induction of expression (by removing glucose) we have used addition of 2 mM IPTG. Individual clones will yield more or less of soluble scFv material in the supernatant with these two protocols, dependent on e.g. the presence of the lac I^q suppressor, the leakage caused by toxic effects on the bacterial cell, and expression levels.

8. Centrifuge the plate at 2200 rpm for 7 min. Harvest 100-150 µl of the supernatant and transfer to a plate containing 100 µl 1% BSA in DPBS.

9. Use the supernatants for analysis within 1-2 days or store at -20°C.

Immunohistochemical analysis of soluble scFvs

1. Prepare frozen tissue sections on multitest slides, and fix and rehydrate the sections as described above.

   Note: It is useful to include at least one control tissue section together with the tissue used for selection in the same test area on the multitest slide.

2. Add 50 μl of soluble scFv antibody culture supernatants diluted 1:2 to 1:10. Incubate at RT for 1 h in a humid athmosphere.

3. Wash three times with TBS preferentially using a multipipette and aspiration using a pasteur pipette or similar connected to a vacuum line. Add 5 μg/ml (diluted in 1% BSA in DPBS) of the second antibody, affinity purified rabbit polyclonal antibodies directed to the tag peptide, ATPAKSE, at the C-terminus of the scFv. Incubate for 1 h at RT.

4. Wash as above. Add the third step, commercially available biotinylated anti-rabbit polyclonal antibodies (we use goat anti-rabbit Ig (eg. from DAKO A/S Copenhagen Denmark) diluted to 1-5 μg/ml in 1% BSA in DPBS. Incubate for 1 h at RT.

5. Wash as above. Add commercially available avidin - biotinylated peroxidase complex (e.g. from DAKO A/S Copenhagen, Denmark) diluted according to the manufacturer in 1% BSA in DPBS. Incubate for 1 h at RT.

6. Wash as above. Develop using e.g. diaminobenzidine substrate (Sigma, St. Louis, MO) at 0.5 mg/ml in 50 mM tris buffer pH 7.6. Include optional counterstain using methyl green. Mount in DPX or similar.

7. Evaluate both by macroscopic inspection and microscopy at low and high magnification.

   Note: The choice of detection tag system could be critical. Preferentially the tag sequence or any crossreactive structure should not be expressed by the mammalian species used, and the available tag reagents should be of high affinity. We have not identified any optimal tag among known examples. In cases where the ATPAKSE tag has not worked appropriately for detection due to endogenous expression of crossreacting structures, we have recloned the entire enriched phage population into a plasmid fusion protein expression vector. This employs conventional cassette subcloning techniques of relatively small libraries (10^3- 10^5 members) that can be found elsewhere in this manual. The fusion partner we have used, staphylococcal enterotoxin A (D227A) promote expression of the recombinant fusion protein and can be detected in sensitive immunoassays using affinity purified antibody reagents that do not crossreact with mammalian tissue structures.

Troubleshooting

No enrichment is seen even after 4-5 rounds of repeated selection-amplification

- The diversity/quality of the library is too low. Check the library by selecting against purified antigens.

- The power of enrichment is too low.
  - The background is too high. Extend the washing procedure. Monitor by quantification of a non-specific negative control phage clone in the library or in a model system. The total washing time could be increased or possibly the salt concentration. Salt concentrations of up to 3 M NaCl in the washing buffers reduced the background of non-specifically bound phage more than the yield of the specific phage in our model experiments (unpublished results). However, other specificities and tissue antigens may be much more sensitive than our model phage.
  - The representation of appropriate tissue antigens are too poor. The total area or expression of antigenic area is too low. Use microscopic monitoring to optimize the tissue dissection and cryosectioning steps.

- The input of phage T.U number is too low relative to the power of enrichment and the diversity of the library. Try to increase phage input without increasing output of non-specific control phage (monitor and if possible improve by washing more extensively). The yield of specific and non-specific phage from tissue sections increases equally with an increased concentration of added phage particles up to approximately 5×10^{12}/ml. At even higher concentrations the viscosity of the phage solution gradually increases. This probably impairs the diffusion and concequently the phage yield. For our model phage only a small fraction of all potential binding sites could be saturated at the highest possible phage concentration. Increase of incubation time or incubation temperature (provided preservation of the antigenic structures) should also be considered. As a rule-of-thumb we aim for an output of more than 10^4 T.U. in the first selection round. Calculated from an average enrichment of 100-fold, and a frequency of tissue antigen binding phage of 10^{-6}, this would yield at least one specific phage clone for amplification. However, for some clones, enrichment of several thousand fold were seen.

Enrichment of too few specificities

- Too low diversity/quality or representation of the library. Check the library and adjust input numbers as described above.

- The optimal selection round has not been analyzed. Screen for specificities in earlier (less biased) or later (higher frequency of binders) selection rounds.

Enrichment of specificities only for non-relevant phenotypes

- The area representing the relevant structure is a too small fraction of the sections.
  - Use microdissection of the tissue before freezing/sectioning.
  - Use an immune library biased to the tissue components of interest.
  - If an appropriate guiding molecule is available, a pathfinder selection principle (described elsewhere in this book) could be employed to direct the selection to relevant areas in the sections.

- Binding clones recognize the relevant structures but are not phenotype-specific. A subtractive procedure employing negative selection using an appropriate control tissue or cell suspension should be explored. We have recently established protocols with sufficient efficiency for negative selection demonstrated in model and library selections (Tordsson et al., manuscripts in preparation).

Only low affinity antibodies are selected

- High-affinity scFv clones resist elution by triethylamine. Try elution using proteolytic cleavage, we have successfully used a mutated version of subtilisin (Tordsson et al., 1997). This should potentially also improve selectivity of the elution procedure.

- Multivalent phage binding predominate. If it is possible, reduce the level of displayed antibodies to obtain a monovalent library stock, e.g. by suppressing the promoter during the phage culture.

Applications

We constructed a large scFv antibody phage library from lymph nodes of a monkey (*Macaca fascicularis*) repeatedly immunized with a suspension of pooled human malignant melanoma metastatic tissue mixed with alun adjuvant. Following three selection rounds of this library on tissue sections of malignant melanoma, the library phage yield increased compared to an internal reference phage. The library scFv genes from the second selection round were recloned in an expression vector in fusion with an immunostimulatory effector molecule, staphylococcal enterotoxin A (SEA). The ligated expression vector was transformed into *E. coli* UL635 for production of scFv-SEA fusion proteins. Individual clones of this selected and recloned library were screened for tissue reactivity using immunohistochemistry. Examples of tissue reaction patterns of four fusion protein clones that could be identified by immunohistochemical stainings of human melanoma tissue sections or other illustrative tissues are shown in Figure 1. Furthermore, tissue reactive scFv antibodies have also been successfully identified using tissue section selection of immune primate colorectal cancer libraries and a non-immune library combined with immunohistochemical screening.

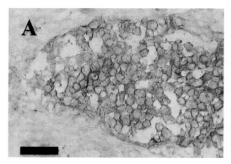

Fig. 1. Microphotographs showing immunohistochemical stainings of tumor cells in tissue sections of human malignant melanoma by the scFv K373-SEA fusion protein in (A) compared to no primary antibody control (B).

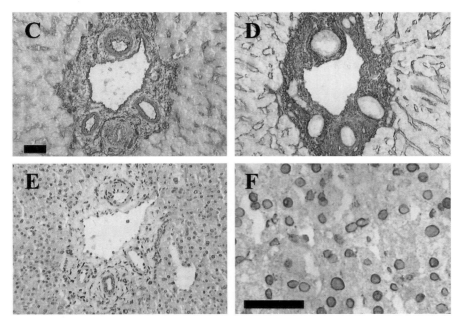

Fig. 1 (cont.) The fusion proteins, scFv K378-SEA in (C) and scFv K578-SEA (D) both bind to a component of the sinusoids and to an extracellular component in human liver, but differ in their binding to blood vessels and to the bile duct. Intracellular staining with a nuclear membrane associated pattern in human liver by the scFv K638-SEA in (E and F). Scale bar in A and F corresponds to 50 μm and in C to 100 μm. (This figure was reproduced from J. Tordsson et al. Efficient selection of scFv antibody phage by adsorption to in situ expressed antigens in tissue sections. Journal of Immunological Methods 210 (1997) 11-23, with kind permission from Elsevier Science B. V., PO Box 1527 1000 BM Amsterdam, Netherlands, +31-20-485-2431).

Acknowledgements. We would like to thank Dr. Lennart Ohlsson, Elisabeth Folkesson and Charlotte Nordenberg for their excellent immunohistochemical analysis, Dr. Christian Ingvar for supplying human tissue samples, and Dr. Lars Abrahmsén, Åsa Gahne and Catherine Ljung for constructing the antibody phage library. We are also grateful to Elsevier Verlag for permitting the reproduction of the results shown in Figure 1.

References

Neidhardt FC (1974) Culture medium for enterobacteria. Journal of Bacteriology Sept:736-747

Tordsson MJ (1997) Efficient selection of scFv antibody phage by adsorption to in situ expressed antigens in tissue sections. Journal of Immunological Methods 210:11-23

Selection of Human Antibody Fragments by Anti-Idiotypic Interaction

PETER FISCHER

Introduction

The direct biopanning of specific antibody presenting phage may be difficult or impossible when the specificity or affinity for the antigen is too low compared to the background phages or the antigen is not available in a suitable form, resulting in nonspecific adherence of the phages. For this reason researchers used anti-idiotypic antibodies for the panning procedure. These were generated by immunization of rabbits with human IgG Fab (idiotype) that were affinity purified with the target antigen (Ishida et al. 1995). Alternatively, anti-idiotypic antibodies mimicking the target antigen were used for isolating the specific phage (Hombach et al. 1998). In addition, the cloning of anti-idiotypic antibodies themselves may be required to isolate antigen-mimicking antibody fragments, e. g. for vaccination, or to derive a tool for targeting clonal myeloma cells (Willems et al. 1998). On the other hand, the cloning of unknown antibodies that are the target of regulatory anti-idiotypic antibodies may be the primary goal. This was the case when autoantibodies from patients with autoimmune disease where selected by anti-idiotypic panning using intravenous IgG preparations (IVIG). This technique did not only enable to determine the genetic origin of those antibodies (Jendreyko et al. 1998) but also allowed for the first time to clone platelet reactive IgG antibodies from patients with autoimmune thrombocytopenia (Fischer et al. 1999). The latter procedure will be described below.

Peter Fischer, Humboldt-University, Charité Children's Hospital, Molecular Biology Laboratory, Ziegelstr. 5-9, 10117 Berlin, Germany (*phone* +49-30-2802-6583; *fax* +49-30-2802-6528; *e-mail* peter.fischer@charite.de)

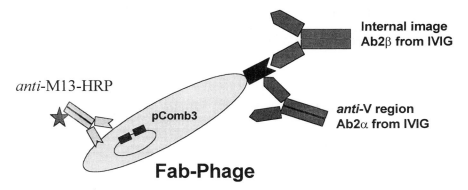

Fig. 1. Principle of anti-idiotypic interactions used for selection of human Fabs. Anti-idiotypic (IVIG) molecules are shown as either anti-V region (e.g. 3-23 or 3-30/3-30.5 origin) Ab2α or internal image Ab2β. In case of an internal image interaction (molecular mimicry of platelet antigens, for example), the Fab displayed by the selected phage may have the same antigen specificity as the antibody that generated the anti-idiotype. These modes of interaction can be used for biopanning with IVIG or other anti-idiotypes immobilized to immunotubes. In a comparative manner, selected Fab-phages can be tested in ELISA with peroxidase (HRP) labeled anti-M13 antibodies.

Materials

SB Super Broth Medium

- 60 g Tryptone
- 40 g yeast extract
- 20 g MOPS
- aqua dest. to 2 l
- adjust pH to 7.0
- autoclave

TBS, Tris-buffered saline

- 50 mM Tris
- 150 mM NaCl
- adjust to pH 7.5 with HCl
- autoclave

Procedure

Generation of Fab displaying phage libraries

The general procedure of generating human phage display libraries is described elsewhere in the book. The following protocols were developed by using the pComb3 and pComb3H cloning vectors (Barbas CFI et al. 1991) as described before (Barbas and Burton 1993, Barbas and Lerner 1991, Duchosal et al. 1992, Fischer et al. 1999, Jendreyko et al. 1998, Watkins et al. 1996).

Enrichment of Fab-phages by anti-idiotypic panning

1. Coat Maxisorp immunotubes (NUNC) with 300 μl IVIG or other anti-idiotypic immunoglobulin preparations at a concentration of 30 μg/ml in PBS over night at 4°C.

2. Block the plates with 1% casein for at least 1 h at RT.

3. Select IVIG-binding Fab-phages by panning of approximately 10^{12} recombinant phages. Incubate for 1-2 h at RT on a turning wheel.

4. Unbound phages are removed and the tubes are washed vigorously up to 10 times with 0.2% casein in PBS.

5. Then, elute the bound phages with 300 μl of 0.1 M HCl/glycine (pH 2.2), and immediately neutralize with 60 μl 2 M Tris/HCl at pH 9.

6. Infect 3 ml of a fresh E. coli XL1 Blue culture grown in SB with tetracycline (10 μg/ml) with the eluted phages.

7. After a 20 min incubation at 37°C on a shaker at 240 rpm, dilute the transformed bacteria in 10 ml SB supplemented with 10 mM $MgCl_2$ and a low concentration of carbenicillin (20 μg/ml). Determine the number of eluted phages ("Output") by plating appropriate dilutions of infected bacteria on carbenicillin plates.

8. Grow the bacteria for 1 h as above and then add 90 ml SB with a high concentration of carbenicillin (50 μg/ml) and 10 mM $MgCl_2$.

9. After 1 h incubation, add 10^{11} VCSM13 helper phages.

10. Following another 2 h incubation, add kanamycin to 70 μg/ml (final concentration) and grow the culture overnight.

11. The next morning, remove the bacteria by centrifugation. After the final panning round, keep the pellet to purify plasmid DNA.

12. Transfer the supernatant into new centrifuge tubes and incubate with 3 g NaCl and 8 g PEG 8000 for 20 min on ice. Precipitate the phages for 20 min at 4°C with 12000 xg. Resuspend the phage pellet in 2 ml TBS with 1% casein and 0.02% NaN_3. For some days, the phages can be stored at 4°C until further use.

13. Repeat the panning procedure at least three times.

Simultaneous growth of Fab-phages from single colonies

1. After the last panning step, pick single colonies from the plates used to determine the "Output" with sterile pipette tips or toothpicks. Inoculate the wells of a 48-well plate filled with 300 µl SB containing carbenicillin (50 µg/ml) and 10 mM $MgCl_2$. Leave blank wells to control for cross-over contamination during the following incubation steps.

2. Seal the plate and lid with tape and incubate on a shaker at 37°C for 5.5 h. Add 50 µl VCSM13 helper phages diluted in SB (final titer 1011 per ml) with a multistepper pipette to each well.

3. After 1 h of further incubation, add 50 µl kanamycin (final concentration 70 µg/ml) and grow the cultures over night.

4. Collect phage supernatants for ELISA after precipitating the bacteria for 30 min at 2000 rpm using a plate rotor.

5. The bacterial pellets may be used for analytical DNA preparations directly in the plate as described (Fischer P et al. 1994).

ELISA of recombinant Fab-phages on IVIG

1. Coat microtiter plates (Maxisorp, NUNC or #3690, COSTAR) with equal concentrations of IVIG (or respective immunoglobulins), human Fc-fragments (#009-000-008, DIANOVA) as control and other test-antigens at 600 ng/well each or PBS for the blanks at 4°C overnight.

2. After 1x washing with PBS containing 0.2% casein, block the plates with 1% casein in PBS for 1 h at 37°C.

3. After 1x washing with PBS/0.2% casein add 50 - 100 µl/well of the phage supernatants collected from the 48-well cultures and incubate at 37°C for 1 hour.

4. After 5x washing with PBS/casein, incubate the plate with peroxidase-labeled anti-M13 phage antibody (#27-9421, PHARMACIA) diluted 1:3000 – 1:5000 in PBS/casein at 37°C for 1 hour.

5. After 5x washing with PBS bound phages are then detected by addition of an appropriate staining solution such as TMB (3,3',5,5'-tetramethyl-benzidine) substrate.

6. The reaction can be stopped with 1 M H_3PO_4. This additionally enhances the staining intensity. The absorption may be quantified at 450 nm in an ELISA reader.

Results

Anti-idiotypic biopanning of human phage display libraries with IVIG enabled for the first time the selection of many IgG anti-platelet antibodies from patients with autoimmune thrombocytopenia (Fischer et al. 1999). In addition, we could show that IVIG preferentially selected Fab fragments that were derived from the V_H germ-line gene segments 3-23 or 3-30/3-30.5 (Fischer et al. 1998). This favorite interaction of IVIG was observed with IgG libraries from autoimmune patients (Jendreyko et al. 1998) as well as with IgG and IgM libraries from a healthy donor (Hoffmann et al. 2000). Our combined results suggested a specific interaction of a subset of IVIG, respectively normal immunoglobulin repertoires, with B cells that present B cell receptors derived from these two germ-line genes. Because 3-23 and 3-30/3-30.5 are the most frequently rearranged VH germ-line gene segments among human B cells, this restricted anti-idiotypic interaction may have an important role for the development and control of the normal B cell repertoire in health and disease (Fischer et al. 1998, Hoffmann et al. 2000).

Troubleshooting

- It is important that the recombinant Fab-phages used for panning are fresh. Therefore we purify recombinant pComb3 DNA from the bacterial pellet of a new library using DNA purification kits (QUIAGEN) and freeze it. Before starting a panning procedure, one µg of this DNA is then used to transfect E. coli by electroporation, and new Fab-phages are rescued by addition of helperphages as described.

- To identify false negative or false positive clones, it is important to control for comparably high Fab-phage titers in the ELISA. Coat additional wells or a separate plate with unlabeled anti-M13 antibodies, incubate with 1:10 dilutions of the investigated Fab-phage supernatants and controls and develop as described with peroxidase labeled anti-M13 antibody from the same source (e.g. PHARMACIA). All A_{450} should be comparatively high, normally above 1.7 after 5 min development with TMB.

Comments

- Using Fab-phages instead of soluble Fab in the ELISA allows for antigen-specific detection of bound Fab-phages with labeled anti-M13 antibodies which do not cross-react with human anti-idiotypes such as IVIG.

- It is recommended to check after the final panning step if the total plasmid DNA still contains both heavy and light chain inserts. In case of panning with IVIG, phages expressing the heavy chain only without functional light chain may be specifically selected. In case of other "antigens", a loss of the light chain insert or both inserts indicates a nonspecific panning procedure.

Acknowledgements. I thank Prof. Dr. Gerhard Gaedicke, Dr. Martina Uttenreuther-Fischer, and Ms. Heike Lerch for discussion. Supported by the DFG.

References

CFI Barbas DR Burton (1993): 1993 Cold Spring Harbor Laboratory Course on: Monoclonal Antibodies from Combinatorial Libraries. The Scripps Research Inst., La Jolla

Barbas CFI, Kang AS, Lerner RA, Benkovic SJ (1991) Assembly of combinatorial antibody libraries on phage surfaces: the gene III site. Proc Natl Acad Sci USA 88: 7978-7982

Barbas CFI Lerner RA (1991) Combinatorial immunoglobulin libraries on the surface of phage (Phabs): rapid selection of antigen-specific Fabs. Methods Companion Methods Enzymol 2: 119-124

Duchosal MA, Eming SA, Fischer P, Leturcq D, Barbas CFI, McConahey PJ, Caothien RH, Thornton GB, Dixon FJ, Burton DR (1992) Immunization of hu-PBL-SCID mice and the rescue of human monoclonal Fab fragments through combinatorial libraries. Nature 355: 258-262

Fischer P, Jendreyko N, Lerch H, Uttenreuther-Fischer MM, Gaedicke G (1998) Genetic origin of IgG antibodies cloned by phage display and antiidiotypic panning from three patients with autoimmune thrombocytopenia. Immunobiol. 199:510-511

Fischer P, Jendreyko N, Hoffmann M, Lerch H, Uttenreuther-Fischer MM, Chen PP, Gaedicke G (1999) Platelet reactive IgG antibodies cloned by phage display and panning with IVIG from three patients with autoimmune thrombocytopenia. Br J Haematol 105: 626-640

Fischer P, Leu SJC, Yang YY, Chen PP (1994) Rapid simultaneous screening for DNA integrity and antigen specificity of clones selected by phage display. Biotechniques 16: 828-830

Hoffmann M, Uttenreuther-Fischer MM, Lerch H, Gaedicke G, Fischer P (2000) IVIG-bound IgG and IgM cloned by phage display from a healthy individual reveal the same restricted germ-line gene origin as in autoimmune thrombocytopenia. Clin Exp Immunol 121: 37-46

Hombach A, Pohl C, Heuser C, Sircar R, Diehl V, Abken H (1998) Isolation of single chain antibody fragments with specificity for cell surface antigens by phage display utilizing internal image anti- idiotypic antibodies. J Immunol Methods 218: 53-61

Ishida F, Gruel Y, Brojer E, Nugent DJ, Kunicki TJ (1995) Repertoire cloning of a human IgG inhibitor of alpha(IIb)beta(3) function. The OG idiotype. Mol Immunol 32: 613-622

Jendreyko N, Uttenreuther-Fischer MM, Lerch H, Gaedicke G, Fischer P (1998) Genetic origin of IgG antibodies cloned by phage display and antiidiotypic panning from three patients with autoimmune thrombocytopenia. Eur J Immunol 28: 4236-4247

Watkins BA, Davis AE, Fiorentini S, di Marzo Veronese F, Reitz MS (1996) Evidence for distinct contributions of heavy and light chains to restriction of antibody recognition of the HIV-1 principal neutralization determinant. J Immunol 156: 1676-1683

Willems PMW, Hoet RMA, Huys ELPG, Raats JMH, Mensink EJBM, Raymakers RAP (1998) Specific detection of myeloma plasma cells using anti-idiotypic single chain antibody fragments selected from a phage display library. Leukemia 12: 1295-1302

Suppliers

Amersham-Pharmacia Biotech Europe GmbH, Munzinger Str. 9, 79111 Freiburg, phone: 0761-49030, FAX: 0761-4903247

Nunc GmbH, Hagenauer Str. 21 a, 65203 Wiesbaden-Biebrich, phone: 0611-67095, FAX: 0611-607348

Costar *via* Corning Inc., Corning, NY 14831, USA, phone: (800)-492-1110, FAX: (978)-635-2476 or via Multilab, Reichenberger Str. 153, 10999 Berlin, Germany, phone: 030-6112005, FAX: 030-6116028

Dianova *via* Jackson Immuno Research, Klein Fontenay, 20354 Hamburg, phone: 040-450670, FAX: 040-45067390

Qiagen GmbH, Max-Volmer-Str. 4, 40724 Hilden, phone: 0210-3892230, FAX: 0210-3892233

Selecting Intracellular Antibodies Using the Two-Hybrid System

MICHELA VISINTIN and ANTONINO CATTANEO

Introduction

The ectopic expression of functional antibodies can be utilized to specifically interfere with the function of selected antigens in cells, tissues or in organisms [1]. The efficacy of the ectopic antibody expression has been already demonstrated in several cases and its use has potential applications to disease therapy and in functional genomic studies.

Antibodies can be directed to many intracellular compartments by the fusion of suitable targeting signals [2] [3]. Among the different intracellular locations, expression in the cytoplasm represents a "worst case", since the reducing environment [4] presents the formation of the conserved intradomain disulphides in antibody variable regions [5]. The intradomain disulphide contributes about 4-5 kcal/mol to the stability of antibody domains. Therefore, antibody fragments expressed in a reducing environment are strongly destabilized. Nevertheless, a number of cytoplasmically expressed antibody fragments are active [1], because they are intrinsically non-stable, and tolerate the absence of the disulphide bond (see Fig. 1).

Natural repertoire contain many stable antibodies that could be rescued with an appropriate selection method without the need for molecular evolution strategies [6].

We have therefore developed an *in vivo* assay for functional intracellular antibodies using a two-hybrid approach [7].

The two-hybrid procedures described in this chapter have been extensively used for detecting interactions between proteins in the yeast *Sac-*

Michela Visintin, International School for Advanced Studies (SISSA), Neuroscience Programme, Via Beirut 2/4, Trieste, 34013, Italy
✉ Antonino Cattaneo, International School for Advanced Studies (SISSA), Neuroscience Programme, Via Beirut 2/4, Trieste, 34013, Italy (*phone* +39-40-3787247; *fax* +39-40-3787249; *e-mail* cattaneo@sissa.it)

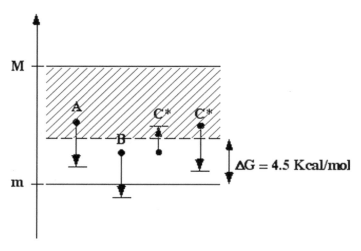

Fig. 1. The folding stability of antibody domains has a range between a minimum (m) and a maximum (M) value and is contributed by many residues in the frameworks [6]. Different scFv fragments have different overall stabilities. Some scFv could tolerate the removal of intrachain disulphide bond and remain folded because they are in the upper range of folding stability (A), while those that are in the lower range will not (B). Moreover, there are some mutations that could stabilize the antibody fragments rendering them tolerant to the absence of the disulphide bonds (C*).

charomyces cerevisiae. We have adapted this system for expression of scFv fragments and their corresponding antigens, to select recombinant antibodies on the basis of their binding activity, under conditions of intracellular expression [7] (see Fig. 2).

Yeast strain

For intracellular antigen-antibody interactions, we found that the choice of the yeast strain and the relative vectors was of critical importance.

We used the yeast strain L40 [8] which contains lexA operator-responsive reporters chromosomally integrated; the genotype of L40 is: Mata his3Δ200 trp1-901 leu2-3,112 ade2 LYS2::(lexAop)$_4$-HIS3 URA3:: (lexAop)$_8$-lacZ GAL4

The expression of the HIS3 and lacZ coding sequences are driven, respectively, by minimal HIS3 and GAL1 promoters fused to multimerized lexA binding sites. The expression of HIS3 provides a growth selection for interaction while the expression of lacZ, which encodes the enzyme β-ga-

A)

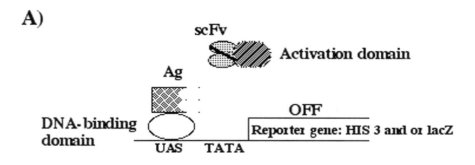

B)

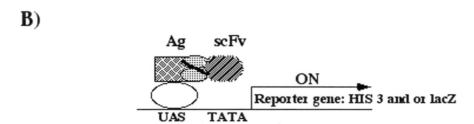

Fig. 2. Dual selection/screen by using the two-hybrid system. **A)** The antigen (Ag) fused to the DNA-binding domain of the *E.coli* repressor LexA is expressed in the yeast strain L40 carrying two reporter genes, lacZ and HIS3, under control of the LexA regulatory sequence. This hybrid protein can bind to its operator site but will not activate transcription. A second plasmid which expresses the scFv fragments fused to the herpes simplex virus VP16 transcriptional activation domain is introduced into the same strain; this hybrid protein does not bind to the upstream activation sequence (UAS) because it lacks the DNA-binding domain. **B)** The interaction of the activation domain fusion scFv with the DNA-binding fusion Ag results in the expression of the reporter construct.

lactosidase, provides a colorimetric assay based on the activity of β-galactosidase: the lacZ⁺ yeasts form blue colonies in the presence of the chromogenic substrate 5-bromo-4-chloro-3-indolyl-β-D-galactoside (X-gal).

The L40 strain is deficient for TRP and LEU (auxotrophic phenotype) and cannot grow on minimal medium lacking those nutrients unless functional TRP1 and LEU2 genes are introduced. Moreover, this strain carries the ade2 mutation, which confers a red color (due to a red pigment accumulation) on medium containing limiting amounts of adenine, that turns darker as the colony age.

Plasmids

There are several vectors for use with the two-hybrid system and in general they have a number of features in common. In our experience we have seen that not all the vectors in use can be utilized for our purpose. In particular, the choice of terminal fusion for the scFv fragment can be crucial because of the steric hindrance of the molecule.

The two-hybrid vectors described here contain replicators and genetic markers that allow their selection and mantenance in *S. cerevisiae* and in *E. coli*. The drug resistance markers and replicators that allow selection and maintenance in *E.coli* are standard (*bla* and *ori* sequences respectively), whereas those that allow use in yeast are more specialized (see Fig. 3 and Fig. 4).

Both plasmids contain ADH1 transcription terminators, downstream of the promoter and the cloning-fusion sites, to cause RNA polymerase II to cease transcribing. Their presence in transcribed regions causes formation of the 3'-ended polyadenylated transcripts [9] [10] [11]. The presence of these sequences downstream of the promoter increases the total amount of messenger RNA as well as the total amount of protein expression.

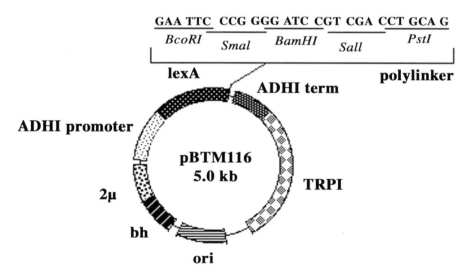

Fig. 3. pBTM116 [22] 5-kb DNA-binding domain vector is used to generate fusions of the target antigen with the complete lexA coding sequence, with a downstream polylinker to generate in-frame protein fusions to lexA, expressed from the yeast ADH1 promoter. It contains TRP1 gene, a selectable marker for yeast and the yeast 2 μ origin of replication.

BcoRI **stop VP16** NL5 *NotI* **scFv** library *SAI* **NLS ATG** *HindIII*

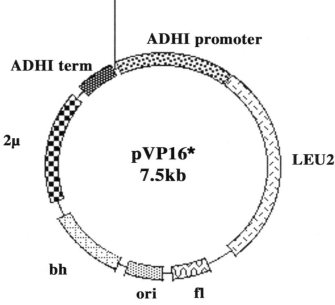

Fig. 4. pVP16* [8] 7.5-kb activation domain vector is used to generate fusions of a known scFv or a scFv library with the VP16 activation domain, with an upstream polylinker to generate in-frame fusions to VP16, expressed from the yeast ADH1 promoter. It contains LEU2 gene, a nutritional marker which must be used in conjunction with special host strains, as L40, that carry the appropriate complementable mutations. It contains also two SV40 large antigen nuclear localization sequences (NLS) fused to the VP16 acidic activation domain and the yeast 2 μ origin of replication. The pVP16 polylinker allows the cloning of scFv libraries containing SfiI-NotI sites, as for example Vaughan et al. [23] and Sheets et al. [24].

Construction and characterization of the "bait"

The first step in a two-hybrid assay is the construction of the "bait". A series of control experiments must be performed to establish whether the fusion protein is suitable as such or whether it must be modified; a "well behaved" bait needs

a. testing for activity

b. testing for expression

c. testing for toxicity

Materials

In this chapter we will discuss only the protocols that we have actually adjusted and optimized for scFv/antigen intracellular interactions and will not discuss other possible protocols that already exist for two-hybrid approach.

Preparation of media and reagents

Yeast strains, which employ auxotrophic mutations as markers, are grown on nutrient-rich media at 30°C to minimize selection of revertants.

Strains to be preserved are grown to logarithmic phase on YPD plates; the yeast is then scraped up with a sterile inoculation loop and suspended in a 15-30% (v/v) glycerol-YPD medium; yeast can be stored now indefinitely at -70°C.

Note: Vortex briefly the vials before freezing at -70°C to avoid cells settling to the bottom of the tube.

Yeast strain can be revived by transferring a small portion of the frozen sample onto YPD plates; yeast can also be stored at 4°C on YPD plates for up to 2 months.

Note: Plates must be stored sealed with parafilm to avoid contamination and drying up of medium.

Note: Yeast L40 colonies should appear slightly pink onto YPD plates and grow to ›2 mm in diameter. To verify the phenotype of the yeast strain provided, streak few colonies from the working stock onto separate YC plates (table 1) and incubate at 30°C.

YPD medium

- 10 g yeast extract (Difco # 0127-17)

- 20 g peptone (Difco # 0118-17)

- 20 g bacto-agar (Difco # 0140-01)

- 2% glucose

YPAD medium

- 10 g yeast extract

- 20 g peptone

- 20 g bacto-agar

- 0.1 g Adenine hemisulfate salt (Sigma # I-7383)

- 2% glucose

- Add H_2O to 900 ml. Adjust pH to 5.8, autoclave 121°C for 15 min then cool to 50-60°C. Add 100 ml of sterile 20% glucose stock solution.

YPA medium

- 10 g yeast extract

- 20 g peptone

- 20 g bacto-agar

- 0.1 g Adenine hemisulfate salt

- Add H_2O to 950 ml. Adjust pH to 5.8 then adjust to 1 liter. Autoclave 121°C for 15 min.

YC medium

- YNB w/o aa & $(NH_4)_2SO_4$:
  - 1.2 g yeast nitrogen base, w/o amino acids and ammonium sulfate (Difco # 0335-15-9)
  - 20 g bacto-agar
  - Add H_2O to 800 ml. Autoclave 121°C for 15 min.
- Salts:
  - 5.4 g NaOH
  - 10 g succinic acid (Sigma # S-7501)
  - 5 g ammonium sulfate (Sigma # A-3920)
- Glucose:
  - 22 g D-glucose
  - Dissolve in 16 ml of H_2O. Add H_2O to 100 ml and dissolve all components one by one. Add glucose and H_2O to make a final volume of 150 ml.

aa MIX

- 5.8 g NaOH

- 1 g Adenine hemisulfate salts

- 1 g L-Arginine HCl (Sigma # A-5131)

- 1 g L-Cysteine (Sigma # C-8277)

- 1 g L-Threonine (Sigma # T-8625)

- 0.5 g L-Aspartic acid (Sigma # A-4534)

- 0.5 g L-Isoleucine (Sigma # I-7383)

- 0.5 g L-Methionine (Sigma # M-9625)

- 0.5 g L-Phenylalanine (Sigma # P-5030)

- 0.5 g L-Proline (Sigma # P-4655)

- 0.5 g L-Serine (Sigma # S-5511)

- Dissolve in 80 ml H_2O

L-Tyrosine

- 0.5 g L-Tyrosine (Sigma # T-3754)

- 0.2 g NaOH

- Dissolve in 10 ml by heating

Omitted nutrients

- L-Histidine (Sigma # H-9511): 5 g/l H_2O

- Uracil (Sigma # U-0750): 10 g/l H_2O (+2 pellets NaOH)

- L-Leucine (Sigma # L-1512): 10 g/l H_2O

- L-Lysine HCl (Sigma # L-1262): 10 g/l H_2O

- L-Tryptophan (Sigma # T-0254): 10 g/l H_2O

Add to aa MIX the L-Tyrosine solution and H_2O to make a final volume of 100 ml. Filter sterilize, aliquot and store at -20°C for up to 1 year.

Filter-sterilize and aliquot individually omitted nutrients (H, W, L, K, U) and store at -20°C for up to 1 year.

Before preparing YC plates or media by mixing appropriate solutions, adjust pH to 5.8 with NaOH 10 M and sterilize by filtering salts + aa final mix. Then mix all the components as suggested in Table 1.

Table 1. Mix of components (Note: All the suggested volumes are expressed in ml)

	YNB w/o aa & $(NH_4)_2SO_4$	Salts	aaMIX	W	H	U	L	K	H_2O
-W	800	150	10		10	10	10	10	
-L	800	150	10	10	10	10		10	
-WL	800	150	10		10	10		10	10
-UW	800	150	10		10		10	10	10
-WHUK	800	150	10				10		30
-WHULK	800	150	10						40

Subprotocol 1
Activity – LiAc Transformation (Small-Scale)

A well-behaved bait should not transactivate in a non-specific way the reporter genes in the L40 strain and should not interact with either the nuclear localization signals or with the VP16 activation domain. Therefore, the L40 strain transformed with the bait alone or together with the pVP16 plasmid should not grow in the absence of histidine and should not contain any detectable β-galactosidase activity. Note: the bait must be modified by using a particular domain or a deletion mutant of the same protein in frame to lexA, if it does not fulfil all the desired features.

Test the extent of non specific activation of the reporter gene by the bait plasmid with the following protocol:

1. transform the bait plasmid into L40 strain using the small-scale yeast transformation protocol

2. select for transformants on appropriate YC plates as described in Table 2

3. assay the bait construct for activation of HIS3 reporter gene as described in Table 2 and for activation of the lacZ reporter gene using the β-galactosidase colony-filter assay as described below.

This protocol follows a modification of published methods made by Clontech laboratories [12], [13], [14]. The expected transformation efficiency is 10^3-10^4/μg plasmid DNA.

Materials

- 10X LiAc buffer: 1M LiAc, pH 7.5 adjusted with dilute glacial acetic acid filter-sterilized (lithium acetate dihydrate "Sigma # L-6883)

- 50% (w/v) filter-sterilized PEG 4000: polyethylene glycol, avg. mol. wt.=3350-Sigma # P-3640.

  Note: Store the solution in a tightly sealed glass bottle to avoid evaporation

- 10X TE buffer: 100mM Tris, 10mM EDTA, pH7.5, filter-sterilized

- 100% DMSO: dimethyl sulfoxide

- 10 mg/ml denatured herring testes carrier DNA (Clontech # K1606)

Procedure

Day 1   Inoculate few colonies of L40 in 50 ml of YPD and incubate for 16-18 hr with shaking at 250 rpm at 30°C to place the culture at mid log phase the next day ($OD_{600} > 1.5$). Note. Use only glass flasks carefully washed with ultrapure, pyrogen-free water and sterilized by autoclaving 15 min at 121°C.

Day 2   1. Dilute the overnight culture to OD_{600} 0.2-0.3 in 300 ml of YPD prewarmed to 30°C. Grow at 30°C for 3 hours with shaking (230 rpm).

2. Pellet the cells by centrifugation (1000 X g for 5 min) at room temperature, discard the supernatant and resuspend the pellet in 50 ml of H_2O.

3. Centrifuge the cells again as in Step 2, decant the supernatant

4. Resuspend the pellet in 1.5 ml of freshly prepared 1X TE/LiAc (10 mM TE, 0.1M LiAc).

5. Prepare in a tube a mixture of:
   - 0.1 µg lexA-Ag vector construct
   - 0.1 µg scFv-VP16 vector construct (if you need to test specific antigen-antibody partners)
   - 0.1 mg denatured herring testis carrier DNA

6. Add 0.6 ml of a sterile PEG/LiAc (0.1 M LiAc, 10 mM TE, PEG 4000 40%) to the tube and vortex to mix.

7. Incubate 30 minutes at 30°C with shaking (230 rpm).

8. Add 70 µl of DMSO, mix gently by inversion and heat shock for 15 min in a 42°C water bath.

9. Chill cells on ice.

10. Pellet cells by centrifugation (20 sec at maximum speed).

11. Remove supernatant and resuspend cells in 0.5 ml of sterile 1X TE; spread 100 µl for single transformation or 250 µl for a co-transformation on each 100 mm plate.

Subprotocol 2
Activity – LiAc Transformation (Maxi-Scale)

The following transformation protocol is a modification of published methods made by Invitrogen laboratories [12], [13], [14]. This protocol should be adapted for cDNA library transformation, when a higher efficiency of transformation ($\cong 10^6$/µg DNA) is needed.

▪▪ Materials

- 500 µg scFv/VP16 DNA
- 150 ml YC-UW
- 2 l YPAD
- 1 l YPA
- 1.5 l YC-UWL + 10 YC-UWL plates (100 mm)
- 1.5 l YC-WHULK + 100 YC-WHULK plates (100 mm)
- 100 ml 10X TE
- 20 ml 10X LiAc
- 1 ml of 10 mg/ml denatured herring testes carrier DNA
- 150 ml 50% PEG 4000
- 20 ml DMSO

▪▪ Procedure

Day 1   1. Grow L40 yeast containing bait plasmid in YC-UW O/N.

Day 2   2. Inoculate 100 ml of YC-WU with an aliquot of the overnight culture in order to find a dilution that places the 100 ml culture to logarithmic phase the next day.

Day 3   3. Transfer enough overnight culture in 1 l of prewarmed to 30°C YPAD to produce an $OD_{600} = 0.3$.

4. Grow at 30°C for 3 hours.

5. Centrifuge the cells at 1500 X g for 5 min at room temperature.

6. Wash pellet in 500 ml of 1X TE then centrifuge again the cells at 1500 X g for 5 min at room temperature.

7. Resuspend pellet in 20 ml 1X LiAc, 0.5X TE and transfer to a new flask.

8. Add 500 μg DNA library and 1 ml herring testes carrier DNA.

9. Add 140 ml of 1X LiAc, 40% PEG 3350, 1X TE; mix and incubate for 30 min at 30°C with gentle shaking.

10. Add 17.6 ml of DMSO; swirl to mix.

11. Heat shock for 10 minutes at 42°C in a waterbath, swirl occasionally to mix.

12. Rapidly cool cells at room temperature in a waterbath diluting with 400 ml YPA.

13. Pellet cells by centrifugation and wash with 500 ml YPA.

14. After centrifugation resuspend pellet in 1 l of prewarmed YPAD.

15. Incubate for 1 hour at 30°C with gentle shaking.

16. Pellet cells from 1 ml and resuspend in 1 ml YC-UWL; spread 100 μl of a 1:1000, 1:100, 1:10 dilutions for transformation efficiency controls.

17. Pellet cells from the remaining culture.

18. Wash pellet with 500 ml YC-UWL.

19. Resuspend in 1 l of prewarmed YC-UWL and incubate O/N at 30°C with gentle shaking.

20. Pellet cells and wash twice with 500 ml of YC-WHULK.   **Day 4**

21. Resuspend final pellet in 10 ml of YC-WHULK.

22. Spread dilutions of the total on YC-UWL plates to compare to the number of primary transformants.

23. Spread the remaining transformation suspension on YC-WHULK plates.

Subprotocol 3
Activity – β-Galactosidase Filter Assay [15]

▮▮ Materials

– Nitrocellulose filter circles (Scheicher and Schuell BA85)

– Buffer Z (60 mM Na_2HPO_4, 40 mM NaH_2PO_4, 10 mM KCl, 1 mM $MgSO_4$, pH 7.0)

– 50 mg/ml 5-bromo-4-chloro-3-indolyl-β-D-galactoside

– Whatman filter circles

– Liquid nitrogen

▮▮ Procedure

1. Patch yeast colonies to a nitrocellulose filter circle.

2. Lift filter and place colony side up on a pre-cooled aluminium boat floating upon a sea of liquid nitrogen.

3. After 20 seconds, immerse boat and filter for 5 seconds.

4. Allow the filter to come to room temperature and place on top of Whatman filter circle that had been prewet in 3 ml of buffer Z containing 30 μl of X-gal.

5. Incubate the filter for up to 5 hours. Blue coloration is indicative of β-gal activity.

Table 2. Transformation controls for a well behaved "bait"

"bait"	"prey"	YC selection medium	HIS3 phenotype	lacZ phenotype
Ag/BTM116		-W		white
Ag/BTM116		-WHUK	no growth	
Ag/BTM116	VP16	-WL		white
Ag/BTM116	VP16	-WHULK	no growth	

Subprotocol 4
Expression

To verify that the bait fusion protein is properly synthesized, an SDS-PAGE and immunoblot analysis on crude yeast lysate must be performed.

■■ Materials

- master plate with bait-containing positive and control yeasts
- antibody to lexA (Invitrogen) or monoclonal/polyclonal to fusion protein
- Laemmli sample buffer 2x

■■ Procedure

Day 1   Incubate overnight at 30°C a 5-ml culture of the bait being tested and relative controls in the appropriate YC medium.

Day 2   1. From each overnight culture start a new 5-ml culture at $OD_{600}= 0.15$. Incubate at 30°C until the culture has reached $OD_{600}= 0.5$-0.7.

2. Remove 1.5 ml to a tube and centrifuge cells 3 min at maximum speed.

3. Remove the supernatant and working rapidly resuspend in 50 µl of 2x Laemmli sample buffer.

4. Vortex and place the tube immediately on dry ice.

5. Boil the sample 5 min and centrifuge 1 min at maximum speed before loading it on SDS-PAGE.

A high level expression of the fusion protein could be toxic to the reporter strain and this could lead the transformant cells to be unable to grow. To alleviate this detrimental effect the truncation of the toxic protein or a conditional promoter on hybrid plasmid could be used.

Toxicity

Subprotocol 5
Construction and Characterization of "Prey"

The second step in a two-hybrid assay is the construction of the "prey". In our case the prey is a single scFv fused in frame with VP16 or a scFv library. A series of control experiments must be performed also with the prey constructs, in order to verify the expression level of the protein and its specificity (see Table 3).

In order to evaluate the specificity of the scFv fusion proteins, all scFv-VP16 fragments tested should interact specifically with their corresponding bait but not with a non-relevant antigen (human lamin) [16] [8].

To monitor the expression levels of the scFv fusion protein a Western blot must be performed using an anti VP16 polyclonal antibody (Santa Cruz Biotechnology); in our experience, the high background level produced by this polyclonal antibody affects the resolution of the experiment. We have seen that the myc tag present at the 3' end of the scFvs does not interfere in the fusion constructs. Therefore, to have better results to detect these tagged proteins in Western blot the monoclonal antibody 9E10 [17] should be used.

Table 3. Transformation controls for a well -behaved "prey"

"bait"	"prey"	YC selection medium	HIS3 phenotype	lacZ phenotype
Lamin/BTM116		-W		white
Lamin/BTM116		-WHUK	no growth	
	scFv/VP16	-L		white
	scFv/VP16	-WHUK	no growth	
Lamin/BTM116	scFv/VP16	-WL		white
Lamin/BTM116	scFv/VP16	-WHULK	no growth	
Ag/BTM116	scFv/VP16	-WHULK	growth	blue

Verification of a positive Ag-scFv two-hybrid interaction

The general use of yeast cells to detect antigen-scFv interaction was assessed with a panel of scFvs derived either from monoclonal antibodies or from phage display antibody libraries [7]. Some of these had been previously shown to have biological activity when assayed *in vivo* as intracellular antibodies. The final validation for a positive interaction was assessed testing the co-transformed yeast colonies for His and lacZ gene activation. In Figure 5 two examples of positive interaction are represented.

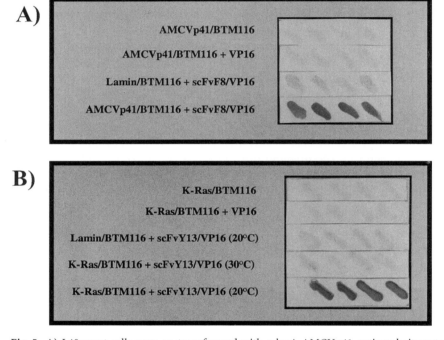

Fig. 5. A) L40 yeast cells were co-transformed with a lexA-AMCVp41 antigen bait vector (AMCVp41/BTM116) and scFvF8/VP16 fusion vector or the pVP16 vector alone. β-galactosidase activation was observed only when the lexA-AMCVp41 bait was co-expressed with scFvF8/VP16. **B)** Transformation of the K-ras bait with scFvY13/VP16 or with pVP16 vector alone. β-galactosidase and His activation was observed only when co-transformed cells were grown at 20°C and not when they were grown at 30°C; this is probably because this scFv fragment aggregates at higher temperatures [25].

Model selection

The use of the yeast two-hybrid to select those antigen-specific scFv capable of binding antigen intracellularly is demonstrated by a model selection, in which yeast L40 was co-transformed with different increasing dilutions of one scFv-VP16 DNA mixed with DNA encoding a non relevant scFv library, using a well characterized Ag-scFv pair [7]. The ability of this system to isolate intracellular antigen-specific scFv fragments from a mixture of non-relevant scFv fragments also at the dilution factor of 5×10^5 indicates that this selection procedure could be applied in the identification of candidate antibodies for intracellular antibody applications.

From the methodological point of view, a library experiment will be performed with the same experimental protocols described above, but care should be taken in planning the selection. In the following, we shall discuss the factors that need to be taken into account when planning a two-hybrid selection of scFv fragments, as well as the flow of experimental steps and the protocols required for the rescue of the selected clones for their final validation. The input of scFv fragments for a two-hybrid based selection can be twofold (Figure 6):

1. phage display library

2. immune spleen library

After the transformation of the yeast with the scFv library and the selection, the recommended approach to verify the specifity of the isolated scFv fragments and to eliminate false positives is described below.

1. rescue of His+ and lacZ+ isolated clones

2. plasmid segregation to remove original bait plasmid

3. rescue of scFv/VP16 plasmid DNA

4. co-transformation of isolated scFv/VP16 plasmid with the original bait plasmid

5. validation of positive interaction by activation of the two reporter genes

6. test for specificity of the scFv fragments using lamin as a bait

7. fingerprinting and sequence of the scFv/VP16

8. cloning of scFv fragments in phagemid vector [18] [19] [20]

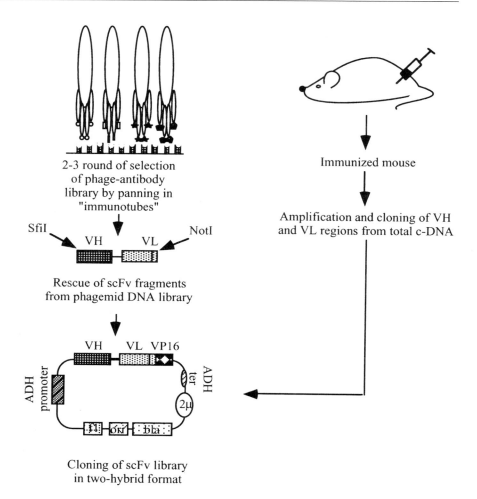

2-3 round of selection
of phage-antibody
library by panning in
"immunotubes"

SfiI   VH   VL   NotI

Rescue of scFv fragments
from phagemid DNA library

VH   VL   VP16

ADH promoter   ADH ter   2μ

Cloning of scFv library
in two-hybrid format

Immunized mouse

Amplification and cloning of VH
and VL regions from total c-DNA

Fig. 6. The scheme for generating intracellular scFv fragments might be first to use antigen-capture phage scFv libraries to "pan" for all scFv capable of binding to a particular antigen followed by shot-gun screening of this sub-library of antigen-specific scFv through the yeast antibody-antigen two-hybrid method. Another way to engineer the scFv two-hybrid library is to start from an immune spleen.

9. assay for *in vitro* binding affinity by antigen binding ELISA, surface plasmon resonance

10. cloning of scFv fragments in scFv express mammalian vector [21]

11. screening for solubility and for inhibition of intracellular functions

▢▢ Procedure

Plasmid segregation to remove bait plasmid

1. Inoculate lacZ+ His+ transformants in 5 ml of YC-L and grow at 30°C with shaking for 2 days.

2. Plate 100 µl of a 1:1000 dilution of the 2 days cultures onto YC-L plate and incubate for 2 days at 30°C.

3. Replica plate the colonies growing on YC-L first on YC-WL and then to YC-L; incubate 2 days at 30°C.

4. Isolation of colonies that grow on YC-L but not on YC-WL.

Plasmid isolation from yeast

The rapid procedure for the isolation of plasmid DNA from yeast is based on the method adapted from QIAprep Spin Miniprep Kit protocol by Michael Jones, Chugai Institute for Molecular Medicine, Ibaraki, Japan. Plasmid DNA isolated from yeast is contaminated by yeast genomic DNA: the procedure described below has been successfully used to generate plasmids for transformation of *E.coli* by electroporation. For restriction analysis, PCR screening and sequencing it is recommended to isolate plasmid DNA from transformant *E.coli* cells.

1. Inoculate single, well isolated yeast transformant colonies into 5 ml of appropriate selective media at 30°C with shaking at 250 rpm until the culture is saturated (16-24 hours).

2. Spin down the culture by centrifuging at 5000 x g and resuspend cells in 250 µl of Buffer 1 containing 0.1 mg/ml RNase A; transfer the cell suspension in a 1.5 ml microfuge tube.

3. Add 0.3 g of acid-washed glass beads and vortex for 5 min at room temperature.

4. Let stand to allow the beads to settle then transfer the supernatant to a new 1.5 ml microfuge tube.

5. Add 250 µl lysis buffer P2 to the supernatant and invert gently 4-6 times to mix. Incubate at room temperature for 5 min.

6. Add 350 µl neutralization buffer N3 to the tube and invert immediately 4-6 times.

7. Centrifuge at maximum speed for 10 min.

8. Transfer the cleared lysate to QIAprep spin column placed in a 2 ml collection tube by pipetting.

9. Centrifuge for 1 min at maximum speed; discard the flow-through.

10. Wash QIAprep spin column by adding 0.75 ml of Buffer PE and centrifuging for 1 min.

11. Discard the flow-through and centrifuge for an additional 1 min to remove residual ethanol from wash buffer.

12. Place QIAprep spin column in a clean 1.5 ml microfuge tube.

Add 50 µl of H_2O to the center of each QIAprep spin column; let stand for 1 min, and centrifuge for 1 min at maximum speed.

Comments

The described set of experimental procedures has been tested and used to demonstrate that a positive outcome of a scFv/antigen interaction in two-hybrid is diagnostic for a biological activity of the scFv as an intrabody and can be used to select and isolate scFv fragments that are intrinsically stable and hence tolerate the absence of the intrachain disulphide bond [7]. This approach therefore opens the possibility of a scaling in a high-throughput format for the generation of the scFv fragments for functional genomics projects.

References

1. Cattaneo, A. and S. Biocca, Intracellular Antibodies: Development and Applications, ed. Springer. 1997.
2. Biocca, S., M.S. Neuberger, and A. Cattaneo, Expression and targeting of intracellular antibodies in mammalian cells. Embo J, 1990. 9(1): p. 101-8.
3. Biocca, S. and A. Cattaneo, Intracellular immunization: antibody targeting to subcellular compartments. Trends Cell Biol, 1995. 5: p. 248-252.
4. Gilbert, H.F., Molecular and cellular aspects of thiol-disulfide exchange. Adv Enzymol Relat Areas Mol Biol, 1990. 63: p. 69-172.
5. Biocca, S. et al., Redox state of single chain Fv fragments targeted to the endoplasmic reticulum, cytosol and mitochondria. Biotechnology (N Y), 1995. 13(10): p. 1110-5.
6. Proba, K. et al., Antibody scFv fragments without disulfide bonds made by molecular evolution. J Mol Biol, 1998. 275(2): p. 245-53.

7. Visintin, M. et al., Selection of antibodies for intracellular function using a two-hybrid in vivo system. Proocedings of the National Academy of Sciences USA, in press.

8. Hollenberg, S.M. et al., Identification of a new family of tissue -specific basic helix-loop-helix proteins with a two-hybrid system. Molecular and Cellular Biology, 1995. 15(7): p. 3813-3822.

9. Heidmann, S. et al., Flexibility and interchangeability of polyadenylation signals in Saccharomyces cerevisiae. Mol Cell Biol, 1994. 14(7): p. 4633-42.

10. Wahle, E. and W. Keller, The biochemistry of 3'-end cleavage and polyadenylation of messenger RNA precursors. Annu Rev Biochem, 1992. 61: p. 419-40.

11. Wahle, E., The end of the message: 3'-end processing leading to polyadenylated messenger RNA. Bioessays, 1992. 14(2): p. 113-8.

12. Gietz, D. et al., Improved method for high efficiency transformation of intact yeast cells. Nucleic Acids Res, 1992. 20(6): p. 1425.

13. Hill, J. et al., DMSO-enhanced whole cell yeast transformation [published erratum appears in Nucleic Acids Res 1991 Dec 11;19(23):6688]. Nucleic Acids Res, 1991. 19(20): p. 5791.

14. Schiestl, R.H. and R.D. Gietz, High efficiency transformation of intact yeast cells using single stranded nucleic acids as a carrier. Curr Genet, 1989. 16(5-6): p. 339-46.

15. Breeden, L. and K. Nasmyth, Regulation of the yeast HO gene. Cold Spring Harb Symp Quant Biol, 1985. 50: p. 643-50.

16. Bartel, P. et al., Elimination of false positives that arise in using the two-hybrid system, in Biotechniques. 1993. p. 920-4.

17. Evan, G.I. et al., Isolation of monoclonal antibodies specific for human c-myc proto- oncogene product. Mol Cell Biol, 1985. 5(12): p. 3610-6.

18. McCafferty, J. et al., Selection and rapid purification of murine antibody fragments that bind a transition-state analog by phage display. Appl Biochem Biotechnol, 1994. 47(2-3): p. 157-71; discussion 171-3.

19. Griffiths, A.D. et al., Isolation of high affinity human antibodies directly from large synthetic repertoires. Embo J, 1994. 13(14): p. 3245-60.

20. Hoogenboom, H.R. et al., Multi-subunit proteins on the surface of filamentous phage: methodologies for displaying antibody (Fab) heavy and light chains. Nucleic Acids Res, 1991. 19(15): p. 4133-7.

21. Persic, L. et al., Targeting vectors for intracellular immunisation. Gene, 1997. 187(1): p. 1-8.

22. Bartel, P.L. et al., Using the two-hybrid system to detect protein-protein interaction, in Cellular interaction in development: a practical approach, D.A. Hartley, Editor. 1993, IRL Press: Oxford. p. 153-179.

23. Vaughan, T.J. et al., Human antibodies with sub-nanomolar affinities isolated from a large non-immunized phage display library [see comments]. Nat Biotechnol, 1996. 14(3): p. 309-14.

24. Sheets, M.D. et al., Efficient construction of a large nonimmune phage antibody library: the production of high-affinity human single-chain antibodies to protein antigens [published erratum appears in Proc Natl Acad Sci USA 1999 Jan 19;96(2):795]. Proc Natl Acad Sci USA, 1998. 95(11): p. 6157-62.

25. Cardinale, A. et al., The mode of action of Y13-259 scFv fragment intracellularly expressed in mammalian cells. FEBS Lett, 1998. 439(3): p. 197-202.

Selection of Phage Antibody Libraries for Binding and Internalization Into Mammalian Cells

ULRIK B. NIELSEN and JAMES D. MARKS

Introduction

Recently, it has proven possible to directly select peptides and antibody fragments binding cell-surface receptors from filamentous phage libraries by incubation of phage libraries with a target cell line (Andersen et al. 1996; Barry et al. 1996; Cai and Garen 1995; de Kruif et al. 1995; Marks et al. 1993). This has led to a marked increase in the number of potential targeting molecules. However, the isolation of cell type specific antibodies from naïve libraries has been difficult because selections often result in binders to common cell surface antigens (Hoogenboom et al. 1999).

The ability of bacteriophage to undergo receptor-mediated endocytosis (Barry et al. 1996; Hart et al. 1994) indicated that phage libraries might be selected not only for cell binding but also for internalization into mammalian cells. Such an approach would be particularly useful for generating ligands which could deliver drugs or toxins into a cell for therapeutic applications. Recently, a methodology was developed (Becerril et al. 1999) which allows isolation of specifically internalized phage while excluding phage merely bound to the cell surface. In the model system studied, not only could phage be selected on the basis of endocytosis, but enrichment ratios were significantly greater when phage were recovered from within the cell rather than from the cell surface.

This strategy of selection for internalization was employed to select scFv from a large naïve library (Sheets et al. 1998) capable of undergoing endocytosis into breast tumor cells upon receptor binding (Poul et al. unpublished results). Upon analysis of a large number of different antibodies

Ulrik B. Nielsen, University of California, Departments of Anesthesia and Pharmaceutical Chemistry, San Francisco, USA

✉ James D. Marks, University of California, Departments of Anesthesia and Pharmaceutical Chemistry, San Francisco, USA (*phone* +01-415-206-3256; *fax* +01-415-206-3253; *e-mail* marksj@anesthesia.ucsf.edu)

isolated in this way, distinct cell surface antigens recognized by the scFv's were identified. Several of the scFv's recognized the ErbB-2 growth factor receptor and another scFv bound the transferrin receptor. Interestingly, both the ErbB-2 and the transferrin receptor are rapidly internalized and are specific markers for a number of cancers. Both the phage antibodies and the native scFv were rapidly endocytosed into cells expressing the appropriate receptor. It is likely that selection on other cell types will identify cell specific markers, because endocytosed receptors are more likely to be associated with specific cell functions (such as growth factor and nutrient transport receptors on cancer cells or Fc receptors on cells of the immune system).

Internalizing antibodies (or fragments of antibodies) are also required for many targeted therapies, such as targeted drugs (toxins, RNases, radioisotopes), targeted liposome therapy (e.g. for delivery of chemotherapeutics) or for targeted delivery of genes (especially of non-viral vectors). It is our experience that not all antibodies which bind internalizing receptors are rapidly internalized. Rather, only antibodies to certain epitopes trigger internalization (Poul et al. unpublished data).

The strategy described below has aided us in isolating a number of cancer cell specific antibodies. We have also used the protocol to isolate antibodies to receptors stably expressed on CHO cells (T. Heitner et al. unpublished data). All of the isolated antibodies have proven to be rapidly internalized into cells expressing the target antigens, something we did not observe with antibodies to the same receptors selected on recombinant antigen using traditional panning strategies (Poul et al. unpublished data). The protocols used for this type of selection are outlined below.

Materials

- Biotinylated anti-mouse antibody (Amersham)
- Anti-myc antibody 9E10 (Santa Cruz)
- Biotinylated anti-M13 antibody (Pharmacia)
- Streptavidin-Cy3 (CALTAG)
- VECTOR shield containing DAPI (Vector Labs)

▨ Procedure

Internalization selection from phage library

1. Incubate 80% confluent target cells plated in a 20 cm plate with 1 ml of the phage library (1×10^{13} pfu total phage) in 20 ml of cold culture medium containing 10% fetal calf serum for 1.5 hours at 4°C.

2. Wash gently 6 times with 25 ml PBS/Ca^{2+}/Mg^{2+} at room temperature.

3. Add 30 ml of pre-warmed complete culture medium containing 10% fetal calf serum. Incubate 15 minutes at 37°C to allow internalization.

Removal of cell surface bound phage

4. Strip the cell surface of non-internalized phage by washing three times 10 minutes with 20 ml of glycine buffer at room temperature (50 mM glycine pH 2.5, 500 mM NaCl). Keep the glycine washes for phage titering.

5. Wash cells with 30 ml PBS to neutralize.

6. Add 5 ml of 0.05% Trypsin/ 0.5% EDTA and incubate at 37°C until cells detach. Add 10 ml of PBS to plate and transfer cells to a 15 ml centrifuge tube. Pellet cells by spinning at 1200 rpm for 5 minutes. Discard the supernatant.

7. Lyse the cell pellet by resuspending in 1 ml of 100 mM triethylamine and incubating for 5 min at room temperature. Neutralize with 1.5 ml of 0.5 M Tris, pH 7.4.

Re-infection of the internalized phage into E. coli

8. Add 2 ml of the phage stock to 18 ml of exponentially growing *E. coli* strain TG1 ($OD_{600} \sim 0.5$). (Store the remaining phage mix at 4°C).

9. Incubate at 37°C for 30 minutes without shaking.

10. Titer TG1 infection by plating 2 μL and 20 μL onto small TYE/100 μg/ml ampicillin/2% glucose plates (this is a 10^4 and 10^3 dilution, respectively).

11. Centrifuge the remaining bacteria solution at 1700 x g for 10 min. Resuspend pellet in 250 μl media and plate onto two large TYE/100μg/ml ampicillin/2% glucose plates and incubate overnight at 37°C.

12. Add 3 ml of 2xTY/100μg/ml ampicillin/2% glucose media to each plate, then scrape the bacteria from the plate with a bent glass rod. Make glycerol stocks by mixing 1.4 ml of bacteria and 0.6 ml 50% glycerol (filtered). Save stock at -70°C.

13. Phage are then rescued for the next round of selection as described in Chapter 9.

14. After two to four rounds of selection, binding phage antibodies are identified by an appropriate screening technique, such as cell ELISA.

   Note: In step 1 above, it may be necessary to use a normal cell line in suspension to deplete for phage antibodies which bind common internalizing receptors.

Detection of phage or scFv internalization by fluorescence microscopy

1. Place one sterile 15 mm round coverslip per well in a six-well tissue culture plate. Add 2 ml of appropiate tissue culture media containing cells. Grow cells to 20-50 % confluency.

   Internalization into cells grown on coverslips

2. Change media to remove dead cells and let stand at 37°C for 10 minutes.

3. If **phage** is used for the internalization, add purified phage to a final concentration of 10^{10} cfu/ml. For **scFv**, add purified scFv to a final concentration of 20 µg/ml. Incubate at 37°C, 5% CO_2 atmosphere, for two hours.

4. The coverslips are then washed six times with PBS, followed by three times for ten minutes with glycine buffer (50 mM glycine (pH 2.8), 500 mM NaCl).

   Stripping and fixation of cell surface bound scFv/phage

5. After another wash with PBS, the cells are fixed on the coverslips with PBS containing 4% paraformaldehyde for five minutes at room temperature.

6. Wash coverslips two times with 3 ml of PBS. The procedure can be interrupted at this point. The coverslips may be stored overnight in PBS at 4°C.

7. The cells are then permeabilized with cold methanol for 10 minutes.

8. Wash cells three times with PBS.

9. Place cells in a new 6 well plate and saturate with PBS containing 1% BSA for 30 minutes.

   Antibody staining for internalized scFv/phage

10. **For phage stain:** Add biotinylated anti-M13 antibody diluted 1/5000 in PBS/1% BSA.
    For scFv stain: Add 9E10 antibody at 0.2 µg/ml in PBS/1% BSA. Incubate for 1 hour on a rocker at 4°C.

11. Wash 10 times with 3 ml of cold PBS.

12. **For phage stain:** Add streptavidin-Cy3 diluted 1/1000 in PBS/1% BSA and incubate for 30 min on a rocker at 4°C.
    For scFv stain: Add biotinylated anti-mouse antibody diluted 1/200 in PBS/1% BSA. Incubate for 1 hour on a rocker at 4°C. Wash 10 times with cold PBS. Then add streptavidin-Cy3 diluted 1/1000 in PBS/1% BSA and incubate for 30 min on a rocker at 4°C.

13. Wash 10 times with cold PBS.

14. Mount coverslips onto microscope slides in 5 µL VECTOR shield containing DAPI. Slides can be stored at 4°C in the dark for up to a week. For long-term storage the slides may be kept at -20°C.

Detection of phage internalization by green fluorescent protein (GFP) reporter gene expression

Filamentous phage displaying internalizing antibody fragments can also be used to directly target genes to mammalian cells expressing the specificity of the scFv (Poul and Marks 1999). When the gene targeted to the eukaryotic cell is a reporter gene, such as green fluorescent protein (GFP), gene expression can be used as an assay for internalization of the scFv.

We have employed two approaches to deliver reporter genes packaged in phage: In one approach, helper phage are used to infect E coli containing a phagemid in which the reporter gene and eukaryotic promoter are cloned. Our lab has cloned the GFP reporter gene and promoter into pHEN1, with SfiI/NotI cloning sites for scFv subcloning (the vector, pFROG, is available upon request from JDM). Phages recovered from the culture supernatant display an average of 1 scFv-pIII fusion protein and 99% of them package the GFP reporter gene. In the other approach, the scFv gene is cloned into the fd phage genome (such as the vector fd-DOG1) for expression as a scFv-pIII fusion. fd-scFv phages are then used to infect E coli containing a reporter phagemid vector (such as pcDNA3-GFP). Phages purified from the culture supernatant display multiple scFv-pIII fusion protein and approximately 50% package the reporter gene.

Most of the phages selected for internalization enter an intracellular trafficking pathway that ultimately leads to reporter gene expression. When green fluorescent protein (GFP) is used as a reporter gene, expression can be detected with as few as 2×10^7 cfu of phage and increases with increasing phage titer up to 10% of cells (Poul and Marks 1999). The reporter gene expression form the internalized bacteriophage can be used as a rapid assay for internalization. The following protocol is used for the infection and expression of GFP reporter gene packaged in antibody displaying phage.

1. Phage preparations of scFv-phages containing a reported gene are diluted at least tenfold in complete cell culture medium. Media and phage is filtered through a 0.45 µm filter.

Bacteriophage mediated cell infection

2. Media containing the phage is added to 30 % to 50 % confluent cells.

3. After 48 hrs of incubation, the media is changed and cells are incubated for another 24-48 hrs.

4. **For fluorescence microscopy analysis:** Media is aspirated and cells are washed twice in PBS before analysis.
   For FACS analysis: Cells are trypsinized with trypsin/EDTA and washed once in PBS containing 2% FCS and analyzed for GFP expression by FACS in the FL-1 channel.

Analysis for GFP expression

References

Andersen PS, Stryhn A, Hansen BE, Fugger L, Engberg J, Buus S (1996) A recombinant antibody with the antigen-specific, major histocompatibility complex-restricted specificity of T cells. Proceedings Of The National Academy Of Sciences Of The United States Of America 93: 1820-4

Barry MA, Dower WJ, Johnston SA (1996) Toward cell-targeting gene therapy vectors: selection of cell-binding peptides from random peptide-presenting phage libraries. Nature Medicine 2: 299-305

Becerril B, Poul MA, Marks JD (1999) Toward selection of internalizing antibodies from phage libraries. Biochemical And Biophysical Research Communications 255: 386-93

Cai X, Garen A (1995) Anti-melanoma antibodies from melanoma patients immunized with genetically modified autologous tumor cells: selection of specific antibodies from single-chain Fv fusion phage libraries. Proceedings Of The National Academy Of Sciences Of The United States Of America 92: 6537-41 de Kruif J, Terstappen L, Boel E, Logtenberg T (1995) Rapid selection of cell subpopulation-specific human monoclonal antibodies from a synthetic phage antibody library. Proceedings Of The National Academy Of Sciences Of The United States Of America 92: 3938-42

Hart SL, Knight AM, Harbottle RP, Mistry A, Hunger HD, Cutler DF, Williamson R, Coutelle C (1994) Cell binding and internalization by filamentous phage displaying a cyclic Arg-Gly-Asp-containing peptide. Journal Of Biological Chemistry 269: 12468-74

Hoogenboom HR, Lutgerink JT, Pelsers MM, Rousch MJ, Coote J, Van Neer N, De Bruïne A, Van Nieuwenhoven FA, Glatz JF, Arends JW (1999) Selection-dominant and nonaccessible epitopes on cell-surface receptors revealed by cell-panning with a large phage antibody library. European Journal Of Biochemistry 260: 774-84

Marks JD, Ouwehand WH, Bye JM, Finnern R, Gorick BD, Voak D, Thorpe SJ, Hughes-Jones NC, Winter G (1993) Human antibody fragments specific for human blood group antigens from a phage display library. Bio/Technology 11: 1145-9

Poul MA, Marks JD (1999) Targeted gene delivery to mammalian cells by filamentous bacteriophage. Journal Of Molecular Biology 288: 203-11

Sheets MD, Amersdorfer P, Finnern R, Sargent P, Lindqvist E, Schier R, Hemingsen G, Wong C, Gerhart JC, Marks JD (1998) Efficient construction of a large nonimmune phage antibody library: the production of high-affinity human single-chain antibodies to protein antigens. Proceedings Of The National Academy Of Sciences Of The United States Of America 95: 6157-62

Expression and Purification
of Antibody Fragments in Bacteria

Purification and Characterisation of His-Tagged Antibody Fragments

MARKUS FIEDLER and ARNE SKERRA

Introduction

Since the production of functional recombinant immunoglobulin fragments in *Escherichia coli* was first established (Skerra, Plückthun 1988) it has become the method of choice in a variety of applications, ranging from immunochemical analysis to therapeutic treatment. The bacterial synthesis of antibody fragments provides important advantages mainly owing to the robustness and facile handling of the procaryotic organism.

Although the in vitro folding from intracellular inclusion bodies, following cytoplasmatic expression of the immunoglobulin chains, is possible the secretory strategy is most widely used and will also be described here. It leads to correctly folded and functional protein via translocation of the polypeptide chains into the bacterial periplasm, where protein folding and formation of the disulphide bonds can take place. Selective release of the proteins from this compartment furthermore facilitates protein purification because cytoplasmic constituents are efficiently removed during the cell fractionation.

Generally, the production of three different antibody fragments is of interest: The Fab fragment comprises the light chain coupled by a disulphide bond to the variable and of first constant domains (so-called Fd fragment) of the heavy chain. The smaller Fv fragment consists merely of the variable domains of the light and heavy chain, which are not covalently linked. Attachment of the two domains by a short peptide segment may enhance their stability and results in a single chain Fv fragment (scFv).

Markus Fiedler, Technische Universität München, Lehrstuhl für Biologische Chemie, 85350 Freising-Weihenstephan, Germany

✉ Arne Skerra, Technische Universität München, Lehrstuhl für Biologische Chemie, 85350 Freising-Weihenstephan, Germany (*phone* +49-8161-714351; *fax* +49-8161-714352; *e-mail* skerra@weihenstephan.de)

For the purification of bacterially produced antibody fragments by affinity chromatography one can utilize either their specific antigen affinity, their class-dependent binding to bacterial Ig receptor proteins or, most universally, one of several affinity tags which were developed during the past years. Most of these affinity tags (e. g. the *Strep*-tag or the *myc* tag) consist of a short but specific amino acid sequence that is recognized by a protein affinity reagent immobilized to a column. In contrast, the His tag simply comprises an oligo-histidine peptide, which tightly adsorbs to chelate columns charged with divalent cations of a transition metal.

The Immobilized Metal Affinity Chromatography (IMAC) of His tag fusion proteins was first introduced by Hochuli et al. (1988) as a method for the purification of recombinant proteins. Soon thereafter it was successfully adapted to the purification of scFv fragments (Skerra et al. 1991). The IMAC purification of such Ig fragments, however, which are composed of two different polypeptide chains, i. e. Fv and Fab, turned out to be more difficult.

Fv fragments can be purified when both of the variable domains are equipped with a His tag (Essen, Skerra 1993). In the case of Fab fragments, whose chain association is more stable due to the presence of the constant domains and of an interchain disulphide bond, a single tag is sufficient (Skerra 1994a). This His tag is best fixed at the C-terminus of the heavy chain because otherwise free Ig light chains, which have a natural tendency of forming Bence-Jones dimers, will be co-purified. In contrast, Fd fragments cannot normally dimerize so that soluble protein, i. e. Fab fragment, is then only isolated as a functional heterodimer with the light chain.

In practice a sequence of at least three, but optimally six consecutive His residues is fused to the C-terminus of the polypeptide chain. Choice of the N-terminus for fusion is less advisable because it is close to the combining site and antigen-binding might thus be affected. In principle, the His tag promotes binding to several types of chelating matrix, most prominently IDA-Sepharose or NTA-agarose, which can be charged with different divalent cations like Ni(II), Cu(II) or Zn(II). We prefer the combination Zn(II)/IDA-Sepharose, which leads in our hands to the highest degree of purification because of a better selectivity against host cell proteins.

High ionic strength of the chromatography buffer (e. g. 1 M NaCl) prevents unspecific binding due to ion exchange effects with the charged matrix. In the purification of Fv fragments high concentrations of NaCl may however cause dissociaton of the two chains. In this case the use of glycine betaine as a neutral electrolyte instead of NaCl is recommended because it has a stabilising effect on the recombinant immunoglobulin (Essen, Skerra 1993).

Elution of bound protein from the IMAC column can be achieved by applying a gradient of increasing imidazole concentration for competitive replacement. The use of a step gradient may serve the same purpose. Although this would seem easier than the operation of a gradient mixer one should be aware that either yield or selectivity is lost in this case, especially if an Ig fragment elutes earlier than average.

Taken together, the purification of His-tagged antibody fragments by IMAC constitutes a well established one-step strategy in order to obtain highly pure and functional protein preparations. This method was even successfully employed at preparative scale during fermenter production of Fab fragments, yielding sufficiently homogeneous protein for crystallisation or animal experiments (Schiweck, Skerra 1995; Bandtlow et al. 1996).

A convenient vector for the cloning and periplasmatic secretion of recombinant Fab fragments carrying the His tag is the expression plasmid pASK85 (Skerra 1994b), which is shown in Figure 1. The genes for the heavy and light chains of the Fab fragment are fused at their N-termini with the bacterial OmpA and PhoA signal sequences, respectively. Murine constant domains of class IgG1/κ are already encoded, whereby the heavy chain is fused with a His$_6$ tag at its C-terminus. Singular restriction sites permit the subcloning and exchange of immunoglobulin variable genes

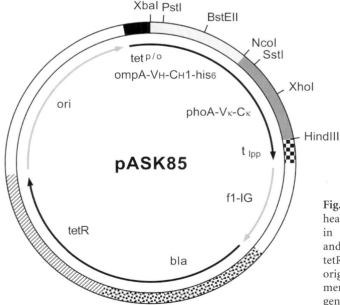

Fig. 1. Expression vector pASK85. The light and heavy chains of the Fab fragment are arranged in a dicistronic operon (OmpA-V_H-C_H1-His$_6$ and PhoA-V_κ-C_κ). tet$^{p/o}$: *tet* promoter/operator; tetR: *tet* repressor gene; ori, f1-IG, and bla denote origin of replication, intergenic region of filamentous phage f1, and ampicillin resistance gene, respectively.

(Skerra 1994a). Compatible vectors are available with different types of constant domains, including human ones, so that domain swapping is easily possible (Schiweck, Skerra 1995; Schiweck et al. 1997).

The structural genes for both polypeptide chains are arranged in a dicistronic operon under transcriptional control of the tightly regulated *tet* promoter/operator (Skerra 1994b). Foreign gene expression can be conveniently induced by anhydrotetracycline.

As an example for the production and IMAC purification of a recombinant Fab fragment the variable genes of the antibody IN-1 (Bandtlow et al. 1996) will here be used.

Outline

An overview of the procedure for production, purification, and characterisation of a recombinant Ig fragment carrying a His tag is given in Figure 2.

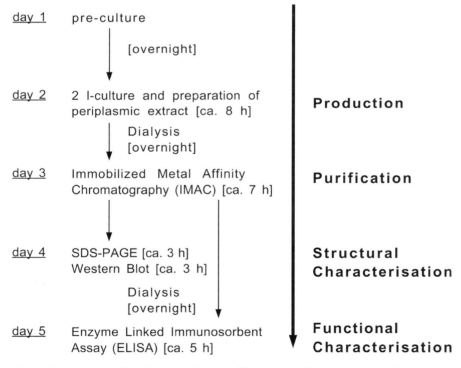

Fig. 2. Short protocol for the production, purification and characterisation of His-tagged antibody fragments.

Subprotocol 1
Production of a His-Tagged Antibody Fragment in the Shaker Flask

Materials, media and solutions should be sterilized by autoclaving or filtration.

Materials

- Incubation shaker, 22°C and 37°C (Infors or New Brunswick)

- Preparative centrifuge, rotors, tubes (Sorvall or Beckman)

- Luria-Bertani (LB) Medium
    - 10 g/l Bacto Tryptone (Difco)
    - 5 g/l Bacto Yeast Extract (Difco)
    - 5 g/l NaCl
    - adjust to pH 7.5 with NaOH

Procedure

1. A fresh single colony of *E. coli* transformed with the derivative vector pASK85-IN1 is used for inoculating 50 ml LB medium containing 100 µg/ml ampicillin. The preculture is incubated at 37 or 30°C and 200 rpm overnight.

2. 40 ml of the preculture is added to 2 liter of LB/ampicillin medium in a 5 l Erlenmeyer flask. Cells are incubated at 22°C and 200 rpm and growth should be documented by measuring OD_{550}.

   Note: A higher incubation temperature may lead to reduced folding efficiency of the foreign protein in the bacterial periplasm as well as early onset of cell lysis after induction.

3. Expression is induced at $OD_{550} = 0.5$ (after correction with a LB blank value) by adding 200 µl of 2 mg/ml anhydrotetracycline in DMF. The optimal induction period varies between 2.5 and 3 h under these conditions and may depend on toxic effects on the cell caused by the recombinant antibody fragment. The best time for harvest is when the growth curve reaches a plateau.

4. The culture is quickly transferred to centrifuge tubes (e. g. Sorvall GS3) and centrifuged at 4 400 g (5 000 rpm) for 15 min at 4°C (ensure that tubes and rotor are chilled at 4°C before harvest). After discarding the supernatant the tubes are put on ice and residual culture medium is removed with a pipette.

Subprotocol 2
Preparation of the Periplasmic Extract

▨▨ Materials

- Preparative centrifuge, rotors, tubes (Sorvall or Beckman)

- Bench top centrifuge (Sigma)

- PE buffer
  - 500 mM Sucrose
  - 100 mM Tris
  - 1 mM Na$_2$EDTA
  - adjust to pH 8.0 with HCl

▨▨ Procedure

Materials and solutions should be pre-chilled at 4°C before use.

1. The sedimented bacterial cells from a 2 l culture are carefully resuspended in 20 ml of cold PE buffer, transferred to a 50 ml Falcon tube, and incubated for 30 min on ice.

   Note: Use a 25 ml pipette for repeated pipetting with 10 ml portions of the buffer in order to avoid shear stress. Addition of lysozyme (Sigma) at a final concentration up to 200 µg/ml (from a fresh 10 mg/ml stock solution in PE buffer) may improve the efficiency of the cell fractionation but is normally not necessary.

2. The spheroplasts are sedimented by centrifugation at 5 000 rpm in a Sigma 4K10 bench top centrifuge (4400 g, 15 min, 4°C) and the supernatant is carefully recovered as the periplasmic cell fraction. In order to

clear the supernatant it is transferred to fresh centrifuge tubes (e. g. SS34) and submitted to a second centrifugation step at 15 000 rpm (27 000 g, 15 min, 4°C).

3. The periplasmic protein extract should be dialysed against 2 l of chromatography buffer (CB) overnight at 4°C or directly frozen at - 20°C for storage.

Subprotocol 3
Immobilized Metal Affinity Chromatography (IMAC)

Materials

- Chelating Sepharose Fast Flow (Pharmacia)
- Chromatography station (Pharmacia)

Chromatography buffer I, CB I	alternatively: CB II
1 M NaCl	0.5 M betaine/monohydrate (Fluka)
40 mM NaH_2PO_4	50 mM NaH_2PO_4
adjust to pH 7.5 with NaOH	adjust to pH 7.5 with NaOH

Imidazole buffer I, IB I	alternatively: IB II
1 M NaCl	0.5 M betaine/monohydrate (Fluka)
40 mM NaH_2PO_4	50 mM NaH_2PO_4
300 mM imidazole (Merck)	300 mM imidazole (Merck)
adjust to pH 7.5 with HCl	adjust to pH 7.5 with HCl

Regeneration buffer, R	
1 M NaCl	
50 mM Na_2EDTA	
adjust to pH 8.0 with NaOH	

▨▨ Procedure

All steps are carried out at 4°C. Materials and solutions should be pre-chilled at 4°C before use. Periplasmic extract should be sterile-filtered before application to the column.

1. A chromatography column (diameter: 7 mm) is packed with 2 ml of Chelating Sepharose Fast Flow ("IDA-Sepharose", Pharmacia) and connected to a peristaltic pump, a flow through UV detector, and a fraction collector (e. g. as part of a Gradifrac-station, Pharmacia).

2. The column is first washed with 10 to 20 ml of water at a flow rate of 20 ml/h. Then it is charged with 10 ml of 10 mM $ZnSO_4$ followed by washing with 10 to 20 ml of water. Equilibration is finally achieved with 10 to 20 ml CB (I or II).

   Note: Although not necessary during the purification of Fab fragments the use of betaine instead of NaCl often gives rise to better resolved elution profiles.

3. The dialysed and sterile-filtered periplasmic protein extract is applied to the column. The flow through should be saved for analysis of the purification procedure. The column is then washed with CB until the absorption at 280 nm diminishes to the base line.

4. Elution of the bound proteins is effected by applying an increasing linear concentration gradient of imidazole in CB (0 to 300 mM, cf. below). For this purpose a gradient mixer is filled with 20 ml CB in the first chamber and with 20 ml of imidazole buffer (IB) in the second chamber and connected to the column via the peristaltic pump. In order to achieve better resolution the flow rate is reduced to 10 ml/h and fractions of 1 or 2 ml are collected.

5. The column is regenerated with 20 ml of buffer R and finally washed with water again.

▨▨ Results

A typical elution profile for the purification of a Fab fragment is shown in Figure 3. The A_{280} diminishes quickly when the column is washed after application of the sample. Soon following the start of the imidazole gradient a steep peak arises containing host cell proteins that were weakly bound to the column. The absorption decreases again before a symmetric peak is obtained corresponding to the specifically eluted Fab fragment.

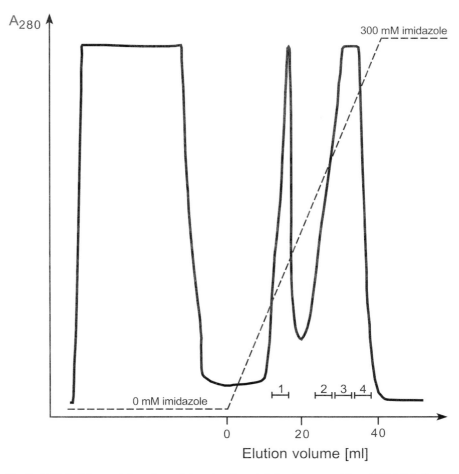

Fig. 3. IMAC of the periplasmic protein extract containing the recombinant IN-1 Fab fragment. The elution profile was monitored via absorption at 280 nm. The increasing concentration of imidazole is indicated as a scattered line.

For a given recombinant Fab fragment this elution behaviour is highly reproducible whereas height and retardation vary when going from one protein to another.

Troubleshooting

- Check pH of buffers (see below) and confirm the proper order of the solutions in the gradient mixer.

- Do not use $ZnCl_2$ instead of $ZnSO_4$ because the solution is unstable and prone to hydrolysis.

- If the peaks for non-specifically bound proteins and for the Ig fragment are not well separated the imidazole gradient can be made less steep.

Subprotocol 4
SDS-Page and Immunoblot Analysis

The purification is analysed by standard discontinuous SDS-PAGE with 15 % (w/v) polyacrylamide gels (0.1 % SDS), followed by staining with Coomassie brilliant-blue. We recommend the buffer system of Fling and Gregerson (1986). Representative samples of the whole cell protein (see below), periplasmic extract, flow-through, and of the elution fractions are reduced with 2-mercaptoethanol prior to gel electrophoresis. Under these conditions the two polypeptide chains of the Fab fragment should appear as separate bands with an approximate size of 25.0 kDa.

Note: The spacing of light and heavy chains in the gel may vary from protein to protein. The separation can often be enhanced by adding 6 M Urea to the gel (Skerra 1994a). In order to investigate whether the Fab fragment contains the light and heavy chains linked by the interchain disulphide bond a non-reduced sample should be analysed as well.

Materials

-   Ni/NTA-AP conjugate (Qiagen)

PBS	PBST
4 mM KH_2PO_4	0.1 % (v/v) Tween 20 in PBS
16 mM Na_2HPO_4	
115 mM NaCl	

AP Buffer	
100 mM Tris	
100 mM NaCl	
5 mM $MgCl_2$	
adjust to pH 8.0 with HCl	

Procedure

Preparation of Whole Cell Protein

1. Prior to cell harvest 1 ml of the culture is transferred in a 1.5 ml reaction tube and the cells are spun down (microfuge, 14 000 rpm, 5 min).

2. The sedimented cells are resuspended in 80 µl 100 mM Tris/HCl pH 8.0; 5 mM $MgCl_2$ with 12.5 u/ml benzonase (Merck). Then 20 µl of loading buffer for SDS-PAGE (7.5 % w/v SDS, 25 % v/v glycerol, 0.25 M Tris/HCl pH 8.0, 12.5 % v/v 2-mercaptoethanol, 0.25 mg/ml bromophenol blue) is added and the lysate is incubated on ice for 1 h.

3. The sample can be stored at -20°C until use.

Western-Blotting

The recombinant protein can either be directly detected using an anti-immunoglobulin antibody or it can be revealed via the His_6 tag with the help of a commercially available reagent (e. g. Ni/NTA-alkaline phosphatase, Qiagen):

All incubation steps are performed under gentle shaking at ambient temperature.

1. After electro-transfer of the proteins from the polyacrylamide gel onto a nitrocellulose membrane (Schleicher & Schuell) using a semi-dry blotting procedure the membrane is placed in a clean dish and washed 3 times for 10 min with 10 ml PBST (Blake et al. 1984).

2. The membrane is incubated with Ni/NTA-AP conjugate at a dilution of 1:500 in 10 ml PBST for 60 min.

3. The membrane is washed twice for 5 min with 10 ml PBST and twice for 5 min with 10 ml PBS.

4. The chromogenic reaction is performed (without shaking) by adding 10 ml of AP buffer with 5 µl NBT (75 mg/ml in 70 % w/v DMF) and 30 µl BCIP (50 mg/ml in DMF) until the bands appear (ca. 15 min). The reaction is stopped by washing with water and air-drying of the membrane.

A typical Coomassie-stained gel and a Western-Blot demonstrating the purification of the IN-1 Fab fragment is shown in Figure 4.

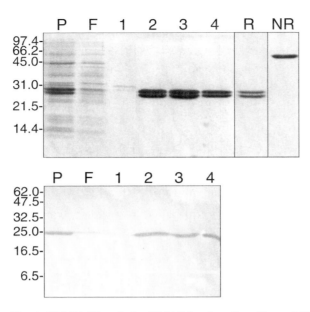

Fig. 4. SDS-PAGE analysis of IMAC fractions from Figure 3. The His$_6$ tag fused to the heavy chain was detected via Ni/NTA-AP conjugate on a Western-Blot (bottom). Molecular sizes are indicated at the left (kDa). Lane P: periplasmic cell extract; lane F: flow-through; lanes 1 to 4: fractions from the IMAC as shown in Figure 3; R: reduced IN-1 Fab (1 µg); NR: non-reduced IN-1 Fab (1 µg).

Subprotocol 5
Detection of a His-Tagged Antibody Fragment in an Enzyme Linked Immunosorbent Assay (ELISA)

▦ Materials

- Microtiter plate, 96 well (Becton Dickson)

- Ni/NTA-AP conjugate (Qiagen)

- ELISA Reader (Molecular Devices)

Procedure

In this standard protocol antigen-binding activity is documented for the His-tagged Fab fragment derived from the anti-lysozyme antibody D1.3, which was purified according to the procedure described above.

All incubation steps are carried out for 60 min at ambient temperature unless otherwise stated.

Residual liquid should be thoroughly removed after each washing or incubation step by draining the plate on a tissue wipe.

1. The wells in a row of a 96 well microtiter plate are each coated with 50 μl 3 mg/ml lysozyme in 50 mM $NaHCO_3$, pH 9.6 overnight at 4°C.

2. The wells are blocked with 200 μl 3 % (w/v) BSA (Sigma), 0.5 % v/v Tween 20 in PBS.

3. The plate is washed three times with PBST.

4. 50 μl of the purified D1.3 Fab fragment is applied in a decreasing concentration series from 0.5 to 0.01 mg/ml. PBST should be used as a blank during this incubation step.

5. The plate is washed three times with PBST.

6. 50 μl of Ni/NTA-alkaline phosphatase conjugate (Qiagen) diluted 1:500 in PBST is applied to each well.

7. The plate is washed twice with PBST and twice with PBS.

8. 100 μl of a solution of 0.5 mg/ml p-nitrophenylphosphate in AP buffer is added to each well.

9. The enzymatic activity is measured at 25°C as the change in absorbance at 405 nm per min (e. g. in a SpectraMAX 250 instrument, Molecular Devices). Alternatively, the end point of absorption can be determined after 15 or 30 min.

References

Bandtlow CE, Schiweck W, Tai HH, Schwab ME, Skerra A (1996) The *Escherichia coli*-derived Fab fragment of the IgM/κ antibody IN-1 recognizes and neutralizes myelin-associated inhibitors of neurite growth. Eur. J. Biochem. 241:468-475

Blake MS, Johnston KH, Russel-Jones GJ, Gotschlich EC (1984) A rapid, sensitive method for detection of alkaline phosphatase-conjugated anti-antibody on Western blots. Anal. Biochem. 136:175-179

Essen LO, Skerra A (1993) Single-step purification of a bacterially expressed antibody Fv fragment by immobilized metal affinity chromatography in the presence of betaine. J. Chromatogr. A 657:55-61

Fling SP, Gregerson DS (1986) Peptide and protein molecular weight determination by electrophoresis using a high-molarity Tris-buffer system without urea. Anal. Biochem. 155:83-88

Hochuli E, Bannwarth W, Döbeli R, Gentz R, Stüber, D (1988) Genetic approach to facilitate purification of recombinant proteins with a novel metal chelate adsorbent. Biotechnology 6:1321-1325

Schiweck W, Skerra A (1995) Fermenter production of an artificial Fab fragment, rationally designed for the antigen cystatin, and its optimized crystallization through constant domain shuffling. Proteins: Struct. Funct. Genet. 23:561-565

Schiweck W, Buxbaum B, Schätzlein C, Neiss HG, Skerra A (1997) Sequence analysis and bacterial production of the anti-c-myc antibody 9E10: the VH domain has an extended CDR-H3 and exhibits unusual solubility. FEBS Lett. 414:33-38

Skerra A, Plückthun A (1988) Assembly of a functional immunoglobulin Fv fragment in *Escherichia coli*. Science 240:1038-1041

Skerra A, Pfitzinger I, Plückthun A (1991) The functional expression of antibody Fv fragments in *Escherichia coli*: Improved vectors and a generally applicable purification technique. Biotechnology 9:273-278

Skerra A (1994a) A general vector, pASK84, for cloning, bacterial production, and single-step purification of antibody Fab fragments. Gene 141:79-84

Skerra A (1994b) Use of the tetracycline promoter for the tightly regulated production of a murine antibody fragment in *Escherichia coli*. Gene 151:131-135

Functional Minipreps of scFv-Antibody Fragments

ANDREAS SCHMIEDL and STEFAN DÜBEL

Introduction

Complete immunoglobulin molecules cannot be produced in *Escherichia coli*, but different strategies have been developed to express antibody fragments like single-chain antibodies (scFvs), disulfide stabilized antibodies (dsFvs), Fab fragments, bispecific antibodies (bsAbs) or various immunotoxins (overview in Breitling and Dübel, 1999). First, antibody fragments can be deposed as cytoplasmic inclusion bodies followed by refolding *in vitro*. This application is usually employed if the functional molecule is expected to kill the producing cell, e. g. for immunotoxins. Due to the general low folding efficiency, an alternative strategy has been developed which better imitates the folding conditions of antibodies in eucaryotic cells. The secretion of the antibody fragments into the periplasmic space of *E. coli* permits their production as soluble and functional proteins with correctly formed intramolecular disulfide bonds. The yields are usually lower than those of inclusion body refolding, however, it is the method of choice for a rapid assessment of antigen binding activity and specificity.

The following procedure describes the rapid functional analysis of scFv-antibody genes by expression from the vector pOPE101 (Genebank accession no. Y14585). It is further applicable to obtain antibody fusion proteins expressed from related vectors like pSTE (Dübel et al., 1995; Genebank accession no. Y14583), pAP (Dübel et al., 1993; Genebank accession no. Y18290), or other vectors providing secretion into the periplasm of *E. coli*.

Andreas Schmiedl, Universität Heidelberg, Institut für Molekulare Genetik,
Im Neuenheimer Feld 230, 69120 Heidelberg, Germany
✉ Stefan Dübel, Universität Heidelberg, Institut für Molekulare Genetik,
Im Neuenheimer Feld 230, 69120 Heidelberg, Germany (*phone* +49-6221-545638;
fax +49-6221-545678; *e-mail* sd@uni-hd.de; *homepage* duebel.uni-hd.de)

In pOPE101-215(Yol) shown as an example (See Figure 1 of the article by Breitling et al. Chapter 3) the genes coding for the heavy and the light chain variable domain of a mouse monoclonal antibody mAb215 directed against the largest subunit (215 kDa) of *Drosophila melanogaster* RNA polymerase II (Krämer et al., 1980; Kontermann et al., 1995) are joined by a flexible linker containing the linear pig brain α-tubulin epitope EE-GEFSEAR (Yol-tag) recognized by the rat monoclonal antibody Yol1/34 (Kilmartin et al., 1982; Breitling and Little, 1986). The variable regions can be exchanged for other variable regions of mouse, rat or human origin using the standard restriction sites *Nco* I/*Hind* III for the heavy chain and *Mlu* I and *Not* I for the light chain variable domain (see Chapter 3 and Chapter 7). The scFv-gene is preceded by a pelB leader sequence (Lei et al., 1987) for secretion of the encoded antibody fragment into the periplasmic space. A short peptide tag of the human c-*myc* proto-oncogene product containing the linear epitope EEKLISEEDL recognized by the mouse monoclonal antibody Myc1-9E10 (Evan et al., 1985) is located at the C-terminus of the V_L domain. Six histidine residues terminate the protein to facilitate its purification by IMAC (Lilius et al., 1991; see also Chapter 17).

Outline

After cloning of an antibody fragment and the identification of productive clones e. g. by colony lifts (see Chapter 3), the correct insertion of the gene fragment should be confirmed by restriction analysis and partial sequencing. The expression of proteins with the expected molecular weight can be analyzed by electrophoresis of total cell lysates and periplasmic extracts on polyacrylamide SDS gels followed by immunoblotting using mAb Myc1-9E10. In addition, the antigen binding activity and specificity of the expressed scFv-antibody fragment has to be confirmed by ELISA or other immunostaining methods (immunoblot, immunofluorescence or FACS). The major steps of all these experiments can be carried out within one day following the procedure described below, which is divided into three sections:

1. expression of the scFv-antibody fragment,

2. extraction from the periplasmic space and

3. functional characterization in ELISA.

Incubation times during the expression of the protein offer enough spare time to prepare plasmid isolations using commercial mini plasmid isolation kits for the analysis of the encoded gene fragment by restriction analysis or sequencing.

Materials

- Bacterial culture equipment   **Equipment**
- Centrifuges
- Gel electrophoresis equipment
- Immunoblot equipment
- ELISA-Reader

Expression   **Reagents**

- Either *E. coli* K12 JM109, TG1 or XL1-blue competent cells
- Bacterial growth medium (LB) containing 100 mM glucose and 100 µg/ml of ampicillin (Biomol, Hamburg, Germany) prepared according to standard protocols (Sambrook et al., 1989)
- 10 mM solution of Isopropyl-ß-D-thiogalactopyranoside (IPTG) (store aliquots frozen in the dark)

Preparation of periplasmic extracts

Spheroblast solution

- 50 mM tris(hydroxymethyl)-aminomethane/HCl (pH 8.0)
- 20% (w/v) sucrose
- 1 mM EDTA (pH 8.0)

Immunoblot and enzyme-linked immunosorbent assay (ELISA)

- 2xSDS-sample buffer
  - 4.0 ml 87% glycerol
  - 3.0 ml 20% SDS
  - 600 µl 2-mercaptoethanol
  - 200 µl 1M tris(hydroxymethyl)-aminomethane/HCl (pH 7.5)
  - ad 10 ml PBS

- PBS (phosphate buffered saline)
  - 137 mM NaCl
  - 3 mM KCl
  - 8 mM Na_2HPO_4
  - 1 mM KH_2PO_4 (pH 7.3)

- PBST: 0.05% (v/v) Tween 20 in PBS

- MPBST: 2% (w/v) skim milk (Oxoid, Basingstoke, UK) in PBST

- 100 mM sodium carbonate/bicarbonate buffer (pH 9.6)

- 100 mM sodium acetate (pH 6.0)

- Protran nitrocellulose transfer membrane (Schleicher und Schuell, Dassel, Germany) 96-well MaxiSorp Immuno Plates (Nunc, Wiesbaden, Germany)

- Mouse monoclonal antibody Myc1-9E10 (Calbiochem, Cambridge, UK) specific for the human c-*myc* proto-oncogene product

- Rat monoclonal antibody Yol1/34 (Harlan Sera-lab, Sussex, UK) recognizing pig brain α-tubulin

- HRP-conjugated antibodies to mouse or rat immunoglobulins (Dianova, Hamburg, Germany)

- Ni-NTA-HRP conjugate (Qiagen, Hilden, Germany)

- TMB Stabilized Substrate for HRP (Promega, Madison, USA)

- 40 mg/ml 3,3',5,5'-tetramethylbenzidine peroxidase substrate (TMB, Sigma-Aldrich, Steinheim, Germany) in DMSO

- 30% H_2O_2 (Merck, Darmstadt, Germany)

Procedure

Expression

1. Prepare an overnight culture of *E. coli* cells transformed with pOPE101 in 5.0 ml of LB medium containing 100 mM glucose and 100 µg/ml of ampicillin.

   Note: If possible, use the original glycerol stock for the inoculation of overnight cultures since clones on agar plates may mutate easily after prolonged storage at 4°C. Glucose should always be present in the bacterial growth medium since it is necessary for the tight suppression of the synthetic promotor of the pOPE vector family and thus for the stability of the insert.

2. Dilute 300 µl of the overnight culture into 6.0 ml (1/20) of LB-medium containing 100 mM glucose and 100 µg/ml of ampicillin and grow at 37°C and 230 rpm to an OD_{600} of 0.6.

   Note: The mainder of the overnight culture may be used for plasmid isolation for the analysis of the encoded gene fragment by restriction analysis or sequencing. Protein expression cannot be induced in bacterial cultures grown to stationary phase.

3. Induce the promotor by the addition of 20 µM IPTG to an aliquot of 3.0 ml of the culture. Use the remaining culture as control.

   Note: Optimal protein secretion from different pOPE vectors in *E. coli* JM109 cells was achieved with 20 µM IPTG. The optimal IPTG concentration, however, varied between different Fv-sequences. Higher IPTG concentrations lead to higher amounts of total protein, but most of the expressed scFv-fragments still carried the leader sequence and formed aggregates (Dübel et al., 1992). We recommend therefore to check concentrations between 20 µM and 100 µM to optimize the conditions for each particular protein.

4. Incubate the induced and the control culture for an additional 3 h of vigorous shaking (280-300 rpm) at 25°C.

   Note: A maximum amount of functional scFv-fragments was achieved at 25°C. Higher temperatures lead to decreased yields of functional proteins. Incubation times longer than three hours did not lead to a significant increase in the amount of secreted protein. However, differences in the solubility of the antibody fragment and its ability to be secreted seemed to be related with its protein sequence. Some antibody constructs even lead to strong growth inhibitions during induction, resulting in problems caused by cytoplasmic contaminants from lysed cells.

5. Incubate the two cultures for 10 min on ice.

6. To prepare a total soluble protein sample for immunoblotting, take out 1.0 ml of each culture and centrifuge at 5000 x g for 5 min. Resuspend the pelleted bacteria in 100 µl of 2xSDS-sample buffer and boil at 95°C for 5 min.

    Note: To confirm the expression of antibody fragments, these cell lysates should be analyzed by SDS-PAGE followed by immunoblotting using e. g. the mouse monoclonal antibody Myc1-9E10 or a Ni-NTA-HRP conjugate (see step. 6 of preparation of periplasmic extracts).

7. Use the remaining culture for the preparation of periplasmic extracts.

Preparation of periplasmic extracts

To extract the maximum amount of scFv-fragments from the periplasma each of the following steps should be carried out on ice and the incubation times should not be exceeded.

1. Harvest the bacteria from 1.5 ml of the cultures by centrifugation at 5000 x g for 10 min at 4°C.

2. Resuspend the pellets in 150 µl (1/10 volume of the initial culture volume) of cold (4°C) spheroblast solution.

    Note: A homogenous suspension of the bacterial pellet is important to extract the maximum amount of scFv-fragments. Use pipetting up and down with a 200 µl pipette for resuspension, if necessary.

3. Incubate the suspensions for 20 min on ice with occasional shaking.

    Note: Incubation times longer than 20 min could decrease the yield.

4. Harvest the bacteria by centrifugation at 6200 x g for 10 min at 4°C and clear the supernatants by centrifugation at 30000 x g for 30 min at 4°C.

5. Carefully collect the supernatants (representing the periplasmic extracts) which contain the functional scFv-fragments.

6. Dilute 25 µl of each extract with 25 µl of 2xSDS-sample buffer and boil at 95°C for 5 min.

Note: To confirm the expression of antibody fragments, total cell lysates of the induced and the control culture as well as their periplasmic extracts should be analyzed by SDS-PAGE (Laemmli, 1970) on 12% polyacrylamide gels followed by immunoblotting (Towbin et al. 1980). ScFv polypeptide chains usually migrate with apparent molecular masses from 29 up to 35 kDa. Two bands corresponding to the complete translation product and the processed Fv-fragment should be detectable in total cell lysates. However, only the slightly smaller processed antibody construct is usually detectable in periplasmic extracts. The expression of antibody fragments from pOPE101 can be analyzed by the mouse monoclonal antibody Myc1-9E10 (diluted 1/1000 in MPBST) or, alternatively, by the rat monoclonal antibody Yol1/34 (diluted 1/1000 in MPBST). Yol 1/34, however, cross-reacts with an *E. coli* protein of about 97 kDa. HRP-conjugated antibodies to mouse or rat immunoglobulins (diluted 1/2000 in MPBST) are employed as second antibodies before TMB Stabilized Substrate for HRP is used for the detection of bound enzymatic activity. Alternatively, a Ni-NTA-HRP conjugate (diluted 1/1000 in MPBST) recognizing the additional His-tag may be used to confirm the expression of the recombinant protein in a one-step staining.

ELISA

Since the stability of newly cloned scFv-fragments is not predictable, it is recommended to start the ELISA as soon as possible after obtaining the periplasmic extracts.

1. Coat 96-well MaxiSorp Immuno ELISA plates with 100 µl/well of a 1 µg/ml solution of antigen and control antigen in 0.1 M sodium carbonate/bicarbonate buffer (pH 9.6) overnight at 4°C.

2. Wash three times with PBST.

3. Block unspecific antibody binding with 400 µl/well of MPBST for at least 2 h at room temperature.

4. Wash three times with PBST.

5. Dilute 100 µl of the periplasmic extract 1/5 in MPBST and incubate with 100 µl per well for another 2 h at room temperature.

6. Wash five times with PBST.

7. Add 100 μl/well of mouse monoclonal antibody Myc1-9E10 (diluted 1/1000 in MPBST) and incubate for 1 h at room temperature.

8. Wash five times with PBST.

9. Add 100 μl/well of a goat anti-mouse IgG-peroxidase conjugate (diluted 1/2000 in MPBST) and incubate for another 1 h at room temperature.

10. Wash five times with PBST and once with PBS.

11. Prepare a solution containing 25 μl of TMB-solution and 8 μl of 30% H_2O_2 in 10 ml of 100 mM sodium acetate (pH 6.0) and add 100 μl per well.

12. Incubate at room temperature for about 10 min.

13. Stop the reaction by adding 50 μl of 1 M H_2SO_4 per well.

14. Read the absorbance of the colored reaction product at 630 nm on an ELISA Reader.

References

Breitling, F. and Dübel, S. (1999) Recombinant Antibodies. John Wiley and Sons, New York.

Breitling, F. and Little, M. (1986) C-terminal regions on the surface of tubulin and microtubules. Epitope locations of Yol1/34, DM1A and DM1B. J. Mol. Biol. 189, 367-370.

Dübel, S., Breitling, F., Klewinghaus, I. and Little, M. (1992) Regulated secretion and purification of recombinant antibodies in E. coli. Cell Biophysics 21, 69-80.

Dübel, S., Breitling, F., Fuchs, P., Klewinghaus, I. and Little, M. (1993) A family of vectors for surface display and production of antibodies. Gene 128, 97-101.

Dübel, S., Breitling, F., Kontermann, R., Schmidt, T., Skerra, A. and Little, M. (1995) Bifunctional and multimeric complexes of streptavidin fused to single chain antibodies (scFv). J. Immunol. Meth. 178, 201-209.

Evan, G. I., Lewis, G. K., Ramsay, G. and Bishop, J. M. (1985) Isolation of Monoclonal Antibodies Specific for Human c-myc Proto-Oncogene Product. Mol. Cell. Biol. 5, 3610-3616.

Kilmartin, J.V., Wright, B. and Milstein, C. (1982) Rat monoclonal antitubulin antibodies derieved by using a new nonsecreting rat cell line. J. Cell. Biol. 93, 576-582.

Kontermann, R.E., Liu, Z., Schulze, R.A., Sommer, K.A., Queitsch, I., Dübel, S., Kipriyanov, S.M., Breitling, F. and Bautz, E.K.F. (1995) Characterization of the epitope recognised by a monoclonal antibody directed against the largest subunit of Drosophila RNA polymerase II. Biol. Chem. Hoppe-Seyler 376, 473-481.

Krämer, A., Haars, R., Kabisch, R., Will, H., Bautz, F.A. and Bautz, E.K.F. (1980) Monoclonal Antibody Directed Against RNA Polymerase II of *Drosophila melanogaster*. Molec. gen. Genet. 180, 193-199.

Laemmli, U.K. (1970) Cleavage of structural proteins during the assembly of the head of the bacteriophage T4. Nature 227, 680-685.

Lei, S.-P., Lin, H.-C., Wang, S.-S., Callaway, J. and Wilcox, G. (1987) Characterization of the *Erwinia carotovora pelB* gene and its product pectate lyase. J. Bacteriol. 169, 4379-4383.

Lilius, G., Person, M., Bülow, L., and Mosbach, K. (1991) Metal affinity precipitation of proteins carrying genetically attached polyhistidine tails. Eur. J. Biochem. 198, 499-504.

Sambrook, J., Fritsch, E.F. and Maniatis, T. (1989) Molecular Cloning. A Laboratory Manual, 2nd ed. Cold Spring Harbor Laboratory Press, Cold Spring Harbor, NY.

Towbin, H., Staehelin, T and Gordon, I. (1980) Electrophoretic transfer of protein from polyacrylamid gels to nitrocellulose sheets: procedures and some applications. Proc. Natl. Acad. Sci. USA 76, 4350-4354.

Expression of scFv in L-Form Bacteria

DIETER MOOSMAYER and JÖRG F. RIPPMANN

Introduction

The production of heterologous protein, including recombinant antibody derivatives like scFv, in the periplasmic space of *E. coli* is accomplished by correct signal sequence processing, formation of disulfide bridges and can therefore lead to soluble functional recombinant product. But the yield of functional, soluble recombinant protein from *E. coli* is often drastically limited by growth inhibition due to a toxic effect of the product and protein aggregation. In accordance with other investigators concerning scFv expression in *E. coli* we found that in the periplasm only a constant threshold concentration of soluble recombinant protein can be reached. Moreover this threshold concentration is dependent on the individual V-regions and can not be significantly influenced by promoter strength and culture conditions (Rippmann et al., 1998).

We recently demonstrated that scFv expression in an L-form strain of Proteus mirabilis (LVI) can overcome the limitations of periplasmic expression in *E. coli* (Rippmann et al., 1998). L-form strains are stable mutants which do not form an outer membrane and murein sacculus, and which have been used for expression and secretion into the culture supernatant of several recombinant protein (Gumpert et al. 1983 and 1996, Gumpert J. and Hoischen C. 1998). Using various scFv which differed significantly upon expression yield and their aggregation tendency in *E. coli* we could demonstrate that with Proteus mirabilis LVI concentrations of

✉ Dieter Moosmayer, University of Stuttgart, Institute of Cell Biology and Immunology, Allmandring 31, 70569 Stuttgart, Germany,
Present address: Schering AG, 13342 Berlin, Germany (*phone* +49-30-46817038; *fax* +49-30-46816707; *e-mail* dieter.moosmayer@schering.de)
Jörg F. Rippmann, University of Stuttgart, Institute of Industrial Genetics, Allmandring 31, 70569 Stuttgart, Germany, Boehringer Ingelheim Pharma KG, 88397 Biberach/Riss, Germany

scFv in the range of 40 to 200 mg/ liter culture supernatant could be reached. With LVI the volume yield scFv was up to 160 fold, and the specific yield up to 32 fold higher than in *E. coli*. Additionally we revealed that secretion of recombinant protein in LVI is an active process coupled to signal processing. LVI has also been used for expression of a bivalent scFv, a miniantibody, but only with a moderate yield (Moosmayer D and Rippmann JF, unpublished data, Kujau et al. 1998).

We constructed a modular scFv-expression vector (pEA40) for LVI which is compatible for exchange of scFv-cassettes (as a *Nco* I/*Not* I fragment with the phagemide vector pSEX and the bacterial expression vectors pOPE developed by Dübel et al. 1992 and 1993. The backbone of pEA40 is based on the vector pACK02scKan containing an ompA signal sequence, kanamycin restistance gene and an IPTG inducible promotor and which has been applied for expression and secretion of recombinant proteins in LVI (Kujau et al. 1998). Cloning of the scFv cassettes into pEA40 is accomplished by standard procedures X.

Materials

Standard culture medium (BHI/Y) for Proteus mirabilis LVI

The medium consists of 3.5% (w/v) Brain-Heart-Infusion medium (Difco Nr. 0502-08-5) in double distilled H_2O with 0.5% (w/v) yeast extract (Difco). BHI medium and yeast extract stock solution are separately prepared and autoclaved.

For selection of the bacteria, harbouring the expression plasmid pEA40, kanamycin with a final concentration of 50 mg/l is applied using stock solution of 50 mg/ml.

Preparation of BHI/Y culture plates

Prepare 200 ml BHI/Y with 2 g agar, autoclave for 20 minutes and cool solution down to 50°C. In individual petri dishes the following solutions are added and carefully mixed:

- 20 µl kanamycin (50 mg/ml stock solution)

- 2.5 ml horse serum (PAA Labor und Forschungsgesellschaft, Linz, Austria, Nr. B15021)

- 0.5 ml sucrose (60% stock solution, sterilized by filtration)

Finally add to each petri dish 30 ml of BHI/Y/agar-solution (50°C) and mix carefully by pipeting the solution up and down. Air bubbles are removed from the petri dish by directing shortly the heat of a gas flame against the surface of the liquid medium.

Procedure

Cultivation and storage of LVI

Liquid cultures are usually prepared in 20 ml BHI/Y by inoculation with 1 ml of stock culture and incubated at 37°C under heavy agitation (200 rpm) for 12 to 15 h. Usually the medium volume is about 1/5 of the flask volume to enable efficient ventilation of the culture. To grow LVI from frozen stock cultures, 100 μl of the thawed culture is used as an inocculum for 20 ml medium and incubated for 1 to 3 days until an OD550 of about 6 is reached.

LVI bacteria can readily be observed with a phase contrast microscope (100 fold magnification) and appear in a logarithmic growth phase as spheroids with a diameter of 1–3 μm which often form threat-like aggregats. In the stationary growth phase the bacteria are mostly solitary with a diameter of 5–10 μm.

Stock cultures of LVI are stored in the standard culture medium at 4°C and passaged once a month. Cryocultures from LVI can be prepared as follows:

An aliquot (5 ml) of an overnight culture is pelleted (10,000 xg, 5 min.) and mixed with 1 ml of the cryomedium (49.8% (v/v) BHI-solution, 30% (v/v) horse serum, 20% (v/v) glycerol, 0.2 % (w/v) glucose, streil filtered and stored at −20°C) subsequently frozen in liquid nitrogen, and stored at −80°C.

Transformation of P. mirabilis LVI

Before transformation LVI are cultured under optimized conditions. For this LVI is passaged twice as a liquid culture in 20 ml BHI/Y medium inocculated with 1 ml stock culture or 1ml of the previous culture and incubated at 37°C for 16 to 24 h.

Transformation procedure

Add the following reagents into a steril reaction tube (1.5 ml) and treat as follows:

1. 20 µl DNA solution containg 10 µg expression plasmid.

   Note: DNA needs to be steril, but the purity of DNA isolated by mini-preparations is sufficient.

2. 160 µl transfection solution (PEG 6000 30 %, (w/v), 0.4 M sucrose, sterile filtrated).

3. 100 µl LVI culture, mix carefully by tapping the reaction tube.

4. incubate the reaction tube with the sample:
   - 10 min on ice
   - 5 min 37°C (mix carefully by tapping)
   - 1 min on ice

5. Add the sample to 1 ml BHI/Y medium containing 1 % (w/v) sucrose and incubate in a 100 ml Erlenmeyer flask for 2.5 to 3 h at 37°C with agitation (200 rpm).

6. 100 µl of the transformed LVI are plated on culture plates (containing kanamycin) and incubated for 2 - 4 days until colonies grow.

Alternative

After step 6.: add 3 ml BHI/Y containing kanamycin (final concentration 50 mg/l) and incubate (37°C) under agitation (200 rpm) for 1-3 days. Plate 100 µl of different dilutions (1:100 - 10,000) of the transformed bacteria on culture plates (containing kanamycin) and incubate (37°C) several days until colonies grow.

Adaption of LVI colonies for growth in liquid culture

1. Cut out an agar block (~ 1 cm^2) with a single LVI colony using a sterile spatula.

2. Turn the agar block upside-down (use sterile tweezers) and move it back and forth on a fresh culture plate to inoculate the culture surface. After incubation (37°C, 1-2 days) of the culture plate a thin layer of vital bacteria grows.

3. From the bacterial layer on the culture plate cut off an agar block (1 cm²) and use it to inoculate 20 ml of BHI/Y/kanamycin. Incubate the flask (37°C) until an OD550 of 6-10 (about 16 h).

4. For generation of a stock culture 1 ml of the liquid culture from 4.) is used to inoculate 20 ml BHI/Y/kanamycin subsequently incubated (37°C, 200 rpm) and stored at 4°C.

Expression of recombinant proteins with LVI

To generate a pre-culture the transformed LVI strain is passaged twice as a liquid culture in BHI/Y/kanamycin at 30°C for 16 h under agitation (200 rpm). For production of recombinant proteins BHI/Y/kanamycin medium (maximum culture volume of 100 ml in a 500 ml Erlenmeyer flask) is inoculated with the pre-culture (1/20 - 1/10 volume of the production culture), induced with IPTG (final concentration 0.5 mM) and incubated at 30°C for at least 14 h with agitation (200 rpm). After cultivation a OD550 of about 6 and a pH of 8 should be reached. A pH below 7 indicates insufficient oxygen supply of the culture.

The recombinant antibody molecules are secreted and accumulate in the culture supernatant and can easily be used for analyses or purification after removal of the bacteria by centrifugation (10000 xg, 10 min).

References

Dübel S, Breitling F, Klewinghaus I, and Little M (1992) Regulated secretion and purification of recombinant antibodies in E. coli. Cell. Biophys. 21: 69-79

Dübel S, Breitling F, Fuchs P, Braunagel M, Klewinghaus I, and Little M (1993) A family of vectors for surface display and production of antibodies. Gene. 128: 97-101

Gumpert J, Cron H, Plapp R, Niersbach H, and Hoischen C (1996) Synthesis and secretion of recombinant penicillin G acylase in bacterial L-forms. J. Basic Microbiol. 36: 89-98

Gumpert, J. and Hoischen, C. (1998). Use of cell wall-less bacteria (L-forms) for efficient expression and secretion of heterologous gene products. Curr. Opin. Biotechnol. 9, 506-509.

Gumpert J and Taubeneck U (1983) Characteristic properties and biological significance of stable protoplast type L-forms. Experientia Suppl. 46: 227-241

Kujau, M.J., Hoischen, C., Riesenberg, D., and Gumpert, J. (1998). Expression and secretion of functional miniantibodies McPC603scFvDhlx in cell-wall-less L-form strains of Proteus mirabilis and Escherichia coli: a comparison of the synthesis capacities of L-form strains with an E. coli producer strain. Appl. Microbiol. Biotechnol. 49, 51-58.

Rippmann JF, Klein M, Hoischen C, Brocks B, Rettig J, Gumpert J, Pfizenmaier K, Mattes R, and Moosmayer D (1998) Procaryotic expression of scFv antibodies: secretion in L-form cells of Proteus mirabilis leads to active product and overcomes limitations of periplasmic expression in Escherichia coli. Applied and Environmental Microbiology 64:4862-4869

Expression and Purification of Fab Fragments

PETER FISCHER

Introduction

Fab or scFv fragments may be easily purified when a special tag is added during the cloning process. However, it may be necessary to produce functional antibody fragments without such artificial enhancements, or systems may be used that do not provide such tags. The most efficient one-step purification is in any case an affinity chromatography. Human Fab fragments may be purified with anti-Fab specific antibodies or the purified antigen coupled to a column. Fabs with a kappa light chain may also be purified with protein L (see Chapter 2). Protein A or G affinity chromatography is well established for purifying complete antibodies or Fc fragments by binding with high affinity to the Fc region. However, it was shown that protein A also binds to the V region of certain antibodies. It was concluded that protein A functions as a superantigen for B cells originating from the V_H3 family (Silverman GJ 1997). Panning of human phage display libraries with protein A resulted in enrichment of V_H3 derived Fab-phages with kappa as well as lambda light chains. However, only 60% of the unselected V_H3 derived Fab-phages could bind to protein A (Sasano M et al. 1993). Therefore, many Fabs with either light chain may also be purified with commercially available, robust protein A affinity columns when the heavy chain is derived from the V_H3 family.

Peter Fischer, Charité Children's Hospital, Humboldt-University, Molecular Biology Laboratory, Ziegelstr. 5-9, 10117 Berlin, Germany (*phone* +49-30-2802-6583; *fax* +49-30-2802-6528; *e-mail* peter.fischer@charite.de)

Generation of Fab displaying phage libraries

The general procedure of generating human phage display libraries is described elsewhere in the book. The following protocols were developed by using the pComb3 and pComb3H cloning vectors (Barbas CFI et al. 1991) as described before (Barbas CFI and Burton DR1993, Barbas CFI and Lerner RA1991, Duchosal MA et al. 1992, Fischer P et al. 1999, Yang YY et al. 1997, Yang YY et al. 1999). With this vector it is necessary to remove the gene III fragment by double restriction and ligation to facilitate the production of soluble Fab fragments (Barbas CFI et al. 1991). Other vector systems may use special suppressor strains instead.

Materials

SB Super Broth Medium

- 60 g Tryptone
- 40 g yeast extract
- 20 g MOPS
- aqua dest. to 2 l
- adjust pH to 7.0
- autoclave

Coupling buffer

- 0.1 M $NaHCO_3$
- 0.5 M NaCl
- adjust pH to 8.3

Elution buffer (anti-Fab CNBr-column)

- 0.05 M glycine
- 0.5 M NaCl
- adjust pH with Tris to 2.7

Elution buffer (anti-Fab Gammabind)

- 0.05 M citric acid

- 0.5 M NaCl

- adjust pH to 2.5

Procedure

Removing the gene III fragment

1. Purify the DNA from the bacterial pellet collected after the last panning step using a midi DNA purification kit (QIAGEN) (Chapter 14).

2. Double-restrict 1 - 5 µg DNA with Nhe1 and Spe1 in buffer M (ROCHE) for 2 h at 37°C. Extract once with phenol/chloroform.

3. Separate the extracted DNA in a 1% low-melting agarose gel containing ethidium bromide for approximately 1 h at 150 V. Cut the appropriate band corresponding to the vector with the heavy and light chain fragment out of the gel and transfer it to a 1.5 or 2 ml reaction tube.

4. Melt the agarose at 60°C.

5. Dilute 7 µl of the molten agarose band with 9 µl prewarmed distilled water and 2 µl of 10x ligation buffer (ROCHE), cool to 37°C.

6. Add 2 µl ligase (ROCHE), mix by pipetting up and down, and incubate 3 h at RT or (better) over night at 18°C.

7. Electroporate up to 10 µl of the ligation mixture directly into 200 µl icecold, electrocompetent E. coli XL1 Blue.

8. After a 20 min incubation in 3 ml SOC medium at 37°C on a shaker at 240 rpm, determine the number of transformed bacteria by plating appropriate dilutions on carbenicillin plates.

9. Dilute the bacteria in 10 ml SB supplemented with 20 mM $MgCl_2$ and a low concentration of carbenicillin (20 µg/ml). Grow them for 1 h as above and then add 90 ml SB/20 mM $MgCl_2$ with a high concentration of carbenicillin (50 µg/ml). Grow them further overnight.

10. The next morning, pellet the bacteria by centrifugation and purify the plasmid DNA.

11. The DNA should be analyzed in a 1% gel to control for removal of the gene III fragment.

Simultaneous production of Fab fragments from single colonies for analytical purpose

1. Pick single colonies from the plates used to determine the transformation efficiency with sterile pipette tips or toothpicks. Inoculate the wells of 48-well plates filled with 300 μl SB containing carbenicillin (50 μg/ml) and 20 mM $MgCl_2$. Leave blank wells to control for cross-over contamination during the following incubation steps (Fischer P et al. 1994). In parallel, plate the colonies on a master petri dish with the inoculating tips.

2. Seal the plates with lids and tape and incubate on a shaker with low rpm at 37°C for 6 h. Add 100 μl SB containing 1 mM IPTG with a multi-stepper pipette to each well.

3. Shake the plate overnight at 30°C.

4. Precipitate the bacteria for 20 min at 2000 rpm using a plate rotor. Resuspend the pellets in 100 μl PBS/200 μM PMSF (protease inhibitor).

5. Freeze (in a -80°C freezer) and thaw the plate (at 37°C on a shaker) for 4 cycles to disrupt the bacteria. Precipitate the bacterial remnants and transfer the supernatants to a blocked low-protein-binding 96-well cell culture plate for short storage or directly to a coated ELISA plate.

Fab-specific ELISA of recombinant (human) Fab fragments

1. Coat microtiter plates with small diameter wells (#3690, COSTAR) with goat-anti-human IgG F(ab')$_2$ specific antibodies (#31122, PIERCE) at 70 ng/well each or PBS for the blanks at 4°C overnight.

2. After 1x washing with PBS containing 0.2% casein, block the plates with 1% casein in PBS for 1 h at 37°C.

Note: Instead of casein, BSA or other blocking agents may be used.

3. After 1x washing with PBS/0.2% casein add 30 - 50 μl/well of the soluble Fabs collected from the 48-well cultures and incubate at 37°C for 1 hour. To quantify the Fab, a serial dilution of a commercial human Fab preparation (#009-000-07, DIANOVA) may be used.

4. After 5x washing with PBS/casein, incubate the plate with peroxidase-labeled anti-(Fab')$_2$ antibody (#31414, PIERCE) diluted in PBS/casein at 40 ng/well at 37°C for 1 h.

5. After 5x washing with PBS, bound Fab are detected by addition of an appropriate staining solution such as TMB (3,3',5,5'-tetramethylbenzidine) substrate.

6. The reaction can be stopped with 1 M H$_3$PO$_4$. This additionally enhances the staining intensity. The absorption may be quantified at 450 nm in an ELISA reader.

Large scale production of Fab fragments

1. Start an over-night culture by inoculating 20 ml SB containing carbenicillin (50 µg/ml) with a positive clone from the master petri dish.

2. Transfer into 0.5 l prewarmed SB with carbenicillin (50 µg/ml) and 20 mM MgCl$_2$, shake at 37°C for approximately 5 h until an OD$_{600}$ of 1.2 is reached.

3. Add 1 ml of a 500 mM IPTG stock and shake over night at 30°C.

4. Pellet the bacteria for 20 min at 1500 xg, resuspend the pellet in 20 ml PBS/PMSF (34 µg/ml).

5. Freeze/thaw or disrupt the bacteria with ultrasound for 2 min at 1 sec intervals of 3 VA in an ice basket.

6. Pellet the debris for 20 min at 1500 xg at 4°C, transfer the supernatant to a new tube and pellet again for 30 min at 27000 xg at 4°C.

7. Sterile-filtrate the clear supernatant and transfer it to the affinity column.

Preparing an anti-Fab CNBr-sepharose affinity chromatography column

1. Resolve 0.5 g CNBr-activated Sepharose 4B (#17-0430-01, PHARMACIA) in 7.5 ml of 1 mM ice-cold HCl.

2. Wash the material with a total of 100 ml HCl in a glass sinter-filter (porosity grade 3) using vacuum.

3. Transfer the gelmaterial into coupling buffer, final volume 2.8 ml.

4. Add 1.7 mg goat-anti-human IgG (Fab')2 (PIERCE #31122H), incubate for 3 h at RT on a rotator.

5. Wash with 1 M ethanolamine, pH 8.0 on the glass filter and incubate the gel material in 5 ml ethanolamine over night at 4°C.

6. Wash 4x with coupling buffer containing 0.1 M NaN$_3$.

7. Resolve in 10 ml PBS and fill into an appropriate column (e.g. HR 5/5, Pharmacia #18-0383-01).

Preparing an anti-Fab protein G-sepharose affinity chromatography column

1. Incubate 2 mg goat-anti-human IgG (Fab')$_2$ (PIERCE #31122H) with 2 ml Gammabind beads (protein G-sepharose, #17-0885-01, PHARMACIA) in a 15 ml centrifuge tube for 1 h at RT on a rotator.

2. Wash the beads 2x by centrifugation at 100 rpm (~ 2 xg), resuspend each time in 10 ml 0.2 M sodium borate, pH 9.0.

3. Crosslink the antibodies with the protein G by adding 200 µl of a 100 mM DMP (dimethyl pimelimidale-2 HCl) stock (#21667, PIERCE) and incubate for 30 min at RT on a rotator.

4. Stop the reaction with 0.2 M ethanolamine, pH 8.0 for 2 h at RT.

5. Precipitate the beads, resuspend them in 2 - 10 ml PBS and fill them into an appropriate column (e.g. HR 5/5, Pharmacia #18-0383-01).

Purification of Fab fragments using anti-Fab affinity chromatography

1. Wash the column with 10 ml of PBS.

2. Load the column at 0.5 ml/min with 2 ml of the Fab supernatant, stop the flow for 10 min.

3. Wash with 10 ml PBS.

4. Elute the bound Fabs with the appropriate elution buffer.

5. Immediately neutralize the collected fractions with 100 µl 1 M Tris, pH 9.0.

6. Equilibrate the column with 10 ml PBS containing 0.02% NaN$_3$ for storage.

7. Analyze the collected fractions for protein content (e.g. BCA assay, Fab-ELISA), purity (SDS-PAGE) and binding activity (ELISA on antigen).

Purification of Fab fragments using protein A affinity chromatography for V_H3 derived immunoglobulins

The procedure is essentially the same as for the anti-Fab affinity chromatography as described above. However, it will only work for Fabs derived from the V_H3 family. Use the following modifications:

1. Use a ready-to-use protein A sepharose column (e.g. #17-5079-01, PHARMACIA).

2. For elution of bound Fabs use 0.1 M sodium citrate, pH 3.0.

Results

Recombinant antibody fragments often are toxic for E. coli. The amount of expressed and purified Fab fragments depends very much on the composition of the antibody, single amino acid exchanges may strongly influence the stability and folding efficiency of the protein (Knappik A and Plückthun A 1995). This may affect the growth rate of E. coli and can control whether the Fabs are excreted into the medium or remain stored in the periplasm. Thus the current protocol should be considered as a basic recommendation (Yang et al. 1997 and 1999). The conditions for bacterial growth (e. g. time, temperature) and affinity purification (choice of column) should be optimized for each individual Fab fragment-producing clone. A typical affinity purification of a Fab fragment in the FPLC is shown in figure 1.

Troubleshooting

- All buffers used for column purification should have the same temperature as the column, preferably RT. Cold temperatures are not recommended for affinity chromatography.

- When purifying small amounts of Fab fragments, it is recommended to elute the bound Fabs in the opposite direction as loaded. This prevents

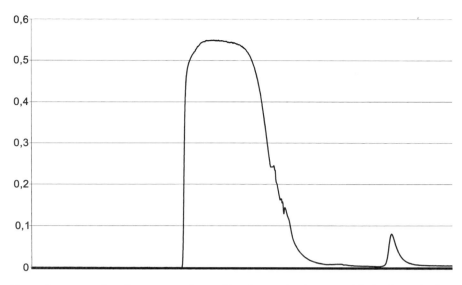

Fig. 1. Representative chromatography profile of an anti-Fab Gammabind column in the FPLC. The purification of a Fab from the bacterial lysate is shown. The large, broad peak represents the loading of the supernatant, the small peak represents eluted Fabs.

Fabs bound at the top of the column from binding again at the bottom during the elution.

- Check and adjust the pH of the Fab supernatant before loading onto the column.

Comments

- To establish the purification procedures, we suggest starting with small column volumes as described and repeating the procedure to recover larger amounts of Fabs. The method can however easily be scaled up by using larger columns and volumes if required.

- The procedures have been optimized for use with a FPLC system (PHARMACIA). Other chromatography systems should be equally suited. If no chromatography system is available, a plastic syringe filled with the column material should work as well, however monitoring and optimized recovery of the peak fraction are then not possible (Yang YY et al. 1997, Yang YY et al. 1999).

- For SDS-PAGE and Western blotting a system requiring only µl amounts of the protein to be analyzed such as the PhastSystem (PHAR-MACIA) is recommended (Scherberich JE et al. 1989).

- For the purification and ELISA of mouse-derived Fabs use appropriate anti-Fab antibodies.

Acknowledgements. I thank Prof. Dr. Gerhard Gaedicke, Dr. Martina Uttenreuther-Fischer, and Ms. Heike Lerch for discussion. Supported by the DFG.

References

Barbas CFI, Burton DR (1993): 1993 Cold Spring Harbor Laboratory Course on: Monoclonal Antibodies from Combinatorial Libraries. The Scripps Research Inst., La Jolla

Barbas CFI, Kang AS, Lerner RA, Benkovic SJ (1991) Assembly of combinatorial antibody libraries on phage surfaces: the gene III site. Proc Natl Acad Sci USA 88: 7978-7982

Barbas CFI, Lerner RA (1991) Combinatorial immunoglobulin libraries on the surface of phage (Phabs): rapid selection of antigen-specific Fabs. Methods Companion Methods Enzymol 2: 119-124

Duchosal MA, Eming SA, Fischer P, Leturcq D, Barbas CFI, McConahey PJ, Caothien RH, Thornton GB, Dixon FJ, Burton DR (1992) Immunization of hu-PBL-SCID mice and the rescue of human monoclonal Fab fragments through combinatorial libraries. Nature 355: 258-262

Fischer P, Jendreyko N, Hoffmann M, Lerch H, Uttenreuther-Fischer MM, Chen PP, Gaedicke G (1999) Platelet reactive IgG antibodies cloned by phage display and panning with IVIG from three patients with autoimmune thrombocytopenia. Br J Haematol 105: 626-640

Fischer P, Leu SJC, Yang YY, Chen PP (1994) Rapid simultaneous screening for DNA integrity and antigen specificity of clones selected by phage display. Biotechniques 16: 828-830

Knappik A, Plückthun A (1995) Engineered turns of a recombinant antibody improve its in vivo folding. Protein Eng 8: 81-89

Sasano M, Burton DR, Silverman GJ (1993) Molecular selection of human antibodies with an unconventional bacterial B cell antigen. J Immunol 151: 5822-5839

Scherberich JE, Fischer P, Bigalke A, Stangl P, Wolf GB, Haimerl M, Schoeppe W (1989) Routine diagnosis with PhastSystem compared to conventional electrophoresis: automated sodium dodecyl sulfate-polyacrylamide gel electrophoresis, silver staining and western blotting of urinary proteins. Electrophoresis 10: 58-62

Silverman GJ (1997) B-cell superantigens. Immunol Today 18: 379-386

Yang YY, Fischer P, Leu SJ, Zhu M, Woods VL, Jr., Chen PP (1999) Possible presence of enhancing antibodies in idiopathic thrombocytopenic purpura. Br J Haematol 104: 69-80

Yang YY, Fischer P, Leu SJC, Olee T, Carson DA, Chen PP (1997) IgG rheumatoid factors isolated by surface displaying phage library technique. Immunogenetics 45: 301-310

Suppliers

Amersham-Pharmacia Biotech Europe GmbH, Munzinger Str. 9, 79111 Freiburg, phone: 0761-49030, FAX: 0761-4903247

Costar *via* Corning Inc., Corning, NY 14831, USA, phone: (800)-492-1110, FAX: (978)-635-2476 or via Multilab, Reichenberger Str. 153, 10999 Berlin, Germany, phone: 030-6112005, FAX: 030-6116028

Dianova *via* Jackson Immuno Research, Klein Fontenay, 20354 Hamburg, phone: 040-450670, FAX: 040-45067390

Roche Molecular Biochemicals, Sandhofer Str. 116, 68305 Mannheim, phone: 0800-759-2226, FAX: 0800-759-8509

Qiagen GmbH, Max-Volmer-Str. 4, 40724 Hilden, phone: 0210-3892230, FAX: 0210-3892233

Purification and Analysis of Antibody Fragments Using Proteins L, A and LA

Morten L. Isaksen and Kevin FitzGerald

Introduction

Since the first demonstration that antibody fragments could be functionally displayed on filamentous phage surfaces (McCafferty et al. 1990) and that the in vivo antibody response could be mimicked in vitro by the affinity selection of large phage antibody repertoires (Marks et al. 1991, Clackson et al. 1991), the selection of antibodies has constituted one of the most widely used applications of phage display. The antibody fragments contained within these libraries generally comprise either single chain variable region fragments (scFv) or Fab fragments and binding proteins with specificity to a very wide range of target structures have now been isolated (see Winter et al. 1994 and Hoogenboom et al. 1998 for reviews). A major motivation for the development of the phage display technique has been the desire to isolate high quality antibodies for therapeutic or diagnostic uses. To this end antibodies with tailor-made properties such as neutralising potency (Thompson et al. 1996, Yang et al. 1995), receptor agonism (Xie et al. 1997), the ability to discriminate between closely related proteins or to react with specific epitopes (Parsons et al. 1996, Osbourn et al. 1998) or cell types (Siegel 1995, de Kruif et al. 1995), have all been successfully isolated using phage display. Further, phage display enables the isolation of antibodies to target antigens hitherto considered 'difficult' by traditional hybridoma-based protocols; examples include self antigens (Griffiths et al. 1993, Cai and Garen 1995), carbohydrate antigens (Griffiths et al. 1993) and toxic substances such as doxorubicin (Vaughan et al. 1996).

✉ Morten L. Isaksen, Affitech AS, Gaustadalleen 21, Oslo, 0371, Norway
(*phone* +47 22 95 88 77; *fax* +47 22 95 83 58; *e-mail* m.isaksen@affitech.com)
Kevin FitzGerald, Actinova Ltd., Babraham Institute, Babraham, Cambridge, CB2 4AT, United Kingdom

A common feature of these fragments is that they lack the Fc portion which is the predominant target for the growing family of bacterial immunoglobulin-binding proteins, of which the best characterised are protein A (Forsgren and Sjoquist 1966) and protein G (Björck and Kronvall 1984). These proteins have been of great value to researchers who work principally with whole immunoglobulins for detection and/or purification purposes. For phage-library derived antibody fragments these functions have been replicated by means of peptide tags appended onto either the amino- or carboxy-terminus of the protein. Most commonly a hexa-histidine tag is used to simplify purification by metal chromatography (Hochuli et al. 1988) and the myc tag is often included to facilitate detection of recombinant antibody fragments using the monoclonal antibody 9E10 (Munro and Pelham 1986). Through use, it has become increasingly apparent that tag-based detection and purification methods have distinct disadvantages. These include the relatively low affinity and/or specificity for the appended tag that is typically exhibited by the detecting antibody, as well as sensitivity of the tags to *E.coli* proteases.

Recently, some new affinity proteins, rProtein L^{TM} and rProtein LA^{TM}, have been shown to overcome the limitations of these affinity tags. Protein L is a surface protein from *Peptostreptococcus magnus* which binds to scFv and Fab fragments containing either human lambda light chains of subclasses I, III and IV or mouse κ chains of subclass I and V (Björck 1988, Åkerström et al. 1994). rProtein L is a recombinant form of the native protein and contains only the 4 immunoglobulin binding domains. The size of each domain is approximately 7.5 kDa, and the 4 domain rProtein L has a molecular weight of 35 kDa (Åkerström and Björck 1989). Protein A binds primarily to the Fc-portion of most IgG antibodies, but can also bind to V_HIII domains present in some Fab and scFv fragments (Sasso et al. 1991). rProtein LA is a recombinant fusion protein consisting of four Ig-binding domains from protein L and four Ig-binding domains of protein A (Svensson et al. 1998). This novel fusion protein combines the antigen-binding properties of both protein L and protein A, resulting in an affinity ligand that binds to a broad range of antibody classes and sub-classes. rProtein L and protein A bind to the light and heavy chains of antibody fragments respectively, without interfering with the antigen binding site. Altogether, these properties make rProtein L, protein A and rProtein LA very useful reagents when working with antibody fragments derived from phage libraries.

This chapter will deal specifically with work involving antibody fragments containing Vκ and/or V_HIII domains. Unlike the affinity tag approach, rProtein L and rProtein LA may also be applied to full length antibodies derived from phage-selected antibody fragments, thus obviating the need to change the detection or purification procedure during process development and subsequent scale up.

rProtein L - HRP™ and rProtein LA - HRP™ as detection reagents in an ELISA

Following rounds of phage selection it is necessary to identify output clones which are specific for the target antigen. This is typically achieved by phage ELISA or ELISA of soluble scFv fragments. We find that rProtein L - HRP and rProtein LA - HRP have a significantly higher sensitivity than anti-tag antibodies or protein A when used as secondary conjugates in these assays. This is most probably due to the multimeric nature of rProtein L and rProtein LA which can greatly improve the avidity of the formed complex. The increased sensitivity afforded by these reagents enables the detection and identification of low affinity clones which may nevertheless be useful due to a very desirable specificity profile or, alternatively, high affinity clones which express poorly.

Rapid estimation of scFv concentration in supernatants

Having identified a panel of antibody fragments that shows activity against a target antigen, it is often necessary to quickly identify those clones that express the highest level of scFv protein, as well as to determine the growth conditions that are optimal for expression. We have found that the incorporation of rProtein L and rProtein A-HRP into a sandwich ELISA provides a convenient method for rapidly screening supernatants to assess approximate fragment concentration. rProtein L is first coated onto a microtiter well to capture the scFv content of bacterial supernatants. Bound protein is then detected using rProtein A-HRP and signals compared to a standard curve to generate a value for the scFv concentration. The technique does not require protein absorbance measurements and therefore, no purification of the fragments is required. The method can also be reversed, using protein A on the plate and rProtein L - HRP conjugate to detect the antibody fragment.

In addition, the technique can be used to detect scFv expressed on phage surfaces simply by replacing the protein A-HRP or rProtein L HRP with a conjugated anti-M13 antibody to detect formed complexes.

rProtein L - HRP™ and rProtein LA - HRP™ as detection reagents on a Western blot

When detecting antigens on a Western blot using a combination of bacterially expressed scFv, rProtein L or rProtein LA and ECL (Amersham Pharmacia Biotech) we have successfully used a purified scFv to detect as little as 1 ng of an immobilised antigen which was considerebly less than that detectable using conventional anti-tag antibodies. The high sensitivity, coupled with low background (especially when using rProtein L) makes this an attractive way to assay for antibody fragment binding to antigen in Western blots.

When protein A is used as the secondary detection reagent, a slightly higher background is sometimes observed. This may be due to residual non-specific binding of protein A to the PVDF membrane or to components of the skimmed milk used in the blocking step that have coated onto the membrane. However, despite this observation, rProtein LA still appears to provide enhanced sensitivity for most Western blot protocols.

Purification of scFv from bacterial supernatants

Purification by immobilised metal affinity chromatography (IMAC) has become the standard method of purifying scFv fragments from *E.coli* periplasmic preparations (Hochuli et al. 1988). Although a well-established method, the technique suffers from the non-specific binding of certain bacterial proteins to the nickel-column. In addition, the technique cannot be applied to the direct purification of heterologous proteins from bacterial supernatants since components in the supernatant appear to strip the nickel from the column and thus interfere with the purification process. In contrast, protein A and L matrices can be routinely used to allow direct capture of expressed antibody fragments from bacterial supernatants. After washing the column, the antibody fragments can be eluted with high purity. Although the binding capacities of protein A agarose and rProtein L agarose appear to be similar, rProtein L seems to result in higher purity preparations. These matrices can be packaged into columns, although batch purification also works well for this procedure.

Materials

- PBS - 10 mM phosphate buffer pH 7.2-7.4 with 150 mM NaCl
- Tween 20 (Sigma)
- MicroTest III 96 well flexible assay plates (Falcon)
- rProtein L (Actigen Ltd., Cambridge, UK)
- rProtein L (Actigen Ltd., Cambridge, UK)
- rProtein L - HRP (Actigen Ltd., Cambridge, UK)
- rProtein LA - HRP(Actigen Ltd., Cambridge, UK)
- TMB substrate (Actigen Ltd., Cambridge, UK)
- Protein A (Sigma)
- Protein A - HRP (Sigma)
- Microtitre plate reader (Molecular Devices)
- NuPage Bis Tris gels (Novex)
- MES PAGE running buffer (Novex)
- PVDF membrane (pre-soaked in methanol before use)
- ECL detection reagents (Amersham)
- X-ray film (Fuji)
- 0.1 M glycine or 0.2 M sodium citrate (pH 3.0)
- 100 mM phosphate buffer pH 8.0

Bacterial cultures were grown, soluble scFv induced and phage prepared according to Marks et al. 1991.

Procedure

rProtein L - HRP™ and rProtein LA - HRP™ as detection reagents in an ELISA

1. Coat a 96 well plate overnight with suitable concentration of the antigen.

2. Incubate the plate for 1 hour in blocking buffer (3% BSA-PBS, 2% TWEEN-PBS or 3% powdered skimmed milk in PBS).

3. Add the test scFv in blocking buffer (3% BSA-PBS, 2% TWEEN-PBS or 3% powdered skimmed milk in PBS) and incubate for one hour to allow binding to the antigen.

4. Wash the plate twice with PBS containing 0.1% Tween and add rProtein L or rProtein LA diluted 1/5000 in blocking buffer (3% BSA-PBS, 2% TWEEN-PBS or 3% powdered skimmed milk in PBS) to detect the bound scFvs.

5. After one hour of incubation, wash the plate three times with PBS 0.1% TWEEN and then once with PBS alone.

6. Add TMB substrate to develop the plate.

7. Stop the reaction using H_2SO_4 and read the absorbance of each well at 650 nm.

8. Determine the OD of each well to compare the sensitivity of the different reagents at equivalent concentrations.

Rapid estimation of scFv concentration in supernatants

1. Coat a 96 well plate overnight using 1 µg/ml of rProtein L in PBS.

2. Incubate the plate for 1 hour in blocking buffer (3% BSA-PBS, 2% TWEEN-PBS or 3% powdered skimmed milk in PBS).

3. Load the scFv to be quantified onto the plate in four fold dilutions in blocking buffer (3% BSA-PBS, 2% TWEEN-PBS or 3% powdered skimmed milk in PBS) and incubate for 1 hour. Set up a series of standards using scFvs of known concentration.

4. Wash the wells 3 times with PBS containing 0.1% TWEEN and add rProtein A - HRP to each well at a concentration of 1/5000 in blocking buffer (3% BSA-PBS, 2% TWEEN-PBS or 3% powdered skimmed milk in PBS).

5. Incubate the plate for one hour and then wash the plate three times with PBS 0.1% TWEEN then once with PBS only.

6. Add TMB substrate to each of the wells to develop the plate.

7. Stop the reaction using H_2SO_4 and read the absorbance of each well at 650nm.

8. Use the values from the standard wells to set up a standard curve.

9. Use the standard curve to determine the scFv values corresponding to the unknowns.

rProtein L - HRP™ and rProtein LA - HRP™ as detection reagents on a Western blot

1. Load a bacterial lysate containing a recombinant protein onto a 4-12% NuPage Bis Tris gel at six concentrations ranging from 800 ng to 25 ng with a 2 fold dilution between each.

2. Run the gel with MES running buffer.

3. Perform a Western blot using a PVDF membrane.

4. Block the membrane using 4% powdered milk in PBS.

5. Incubate for one hour with an scFv specific for the expressed protein in blocking buffer (3% BSA-PBS, 2% TWEEN-PBS or 3% powdered skimmed milk in PBS).

6. Wash the membrane 3 times with PBS containing 0.1% TWEEN.

7. Incubate for one hour using either rProtein L - HRP or rProtein LA - HRP in blocking buffer (3% BSA-PBS, 2% TWEEN-PBS or 3% powdered skimmed milk in PBS).

8. Wash the membrane 3 times with PBS containing 0.1% TWEEN, followed by one wash in PBS only.

9. Detect the bound HRP using the ECL kit (Amersham Pharmacia Biotech, Amersham UK) and expose the blot to Fuji film for one minute.

Purification of scFv from bacterial supernatant

1. Pellet the cells after induction and growth at optimal conditions overnight.

2. Wash the affinity matrix with 5 volumes of PBS.

3. Mix 1 ml of the matrix with the supernatant and incubate on a shaker table for 1 hour.

4. Spin down the matrix at 2000 g.

5. Wash the matrix in 3 volumes of PBS, centrifuge, decant the PBS. Repeat once.

6. Add 1.6 ml of elution buffer (0.1 M glycine, or 0.2 M citrate, pH 3.0). Mix and incubate for 10 minutes. Pellet the matrix and pipette off the elution buffer containing the antibody fragment.

7. Neutralise the eluate by adding 800 µl 0.1M NaPhosphate buffer at pH 8.0.

Comments

The procedures described in this chapter provide an attractive means of detecting and purifying antibody fragments without the need for fused peptide tags. Providing that the antibody fragments bind to protein L, protein A or protein LA, the use of these affinity proteins greatly simplifies both detection and purification of antibody fragments.

Acknowledgements. We thank Lucy Holt and Ian Tomlinson at the MRC Centre for Protein Engineering/MRC Laboratory of Molecular Biology, Cambridge, UK, for their help in establishing the protocols described in this chapter.

References

Åkerström B, Nilson BH, Hoogenboom HR, and Björck L (1994) On the interaction between single chain Fv antibodies and bacterial immunoglobulin-binding proteins. J Immunol Methods 177:151-163

Åkerström B, and Björck L (1989) Protein L: an immunoglobulin light chain-binding bacterial protein. Characterization of binding and physicochemical properties. J Biol Chem 264:19740-19746

Björck L & Kronvall G (1984) Purification and some properties of streptococcal protein G, a novel IgG-binding reagent. J Immunol 133:969-974

Cai X & Garen A (1995) Anti-melanoma antibodies from melanoma patients immunized with genetically modified autologous tumor cells: selection of specific antibodies from single-chain Fv fusion phage libraries. Proc Natl Acad Sci U S A 92:6537-6541

Casey J L, King D J, Chaplin L C, Haines A M, Pedley R B, Mountain A, Yarranton G T & Begent R H (1996) Preparation, characterisation and tumour targeting of cross-linked divalent and trivalent anti-tumour Fab' fragments. Br J Cancer 74:1397-1405

Chester K A, Begent RH, Robson L, Keep P, Pedley R B, Boden J A, Boxer G, Green A, Winter G, Cochet O & Hawkins RE (1994) Phage libraries for generation of clinically useful antibodies. Lancet 343:455-456

Clackson T, Hoogenboom HR, Griffiths AD & Winter G (1991) Making antibody fragments using phage display libraries. Nature 352:624-628

Clark, M. R. (1997). IgG effector mechanisms. Chem Immunol 65, 88-110

de Kruif J, Terstappen L, Boel E & Logtenberg T (1995) Rapid selection of cell sub-population-specific human monoclonal antibodies from a synthetic phage antibody library. Proc Natl Acad Sci U S A 92:3938-3942

Forsgren A & Sjoquist J (1966) "Protein A" from S. aureus. I. Pseudo-immune reaction with human gamma-globulin. J Immunol 97:822-827

Griffiths A D, Malmqvist M, Marks J D, Bye JM, Embleton MJ, McCafferty J, Baier M, Holliger KP, Gorick BD, Hughes-Jones NC, Hoogenboom HR & Winter G (1993) Human anti-self antibodies with high specificity from phage display libraries. EMBO Journal 12:725-734

Hochuli E, Bannwarth W, Döbeli H, Gentz R & Stüber D (1988) Genetic approach to facilitate purification of recombinant proteins with a novel metal chelate adsorbent. Bio/Technology 6:1321-1325

Hoogenboom HR, de Bruine AP, Hufton SE, Hoet RM, Arends JW & Roovers RC (1998) Antibody phage display technology and its applications. Immunotechnology 4:1-20

Junghans RP (1997) Finally! The Brambell receptor (FcRB). Mediator of transmission of immunity and protection from catabolism for IgG. (published erratum appears in Immunol Res (1997)16(2):215 Immunol Res 16:29-57

Marks JD, Hoogenboom HR, Bonnert TP, McCafferty J, Griffiths AD & Winter G (1991) By-passing immunization: Human antibodies from V-gene libraries displayed on phage. J Mol Biol 222:581-597

McCafferty J, Griffiths AD, Winter G & Chiswell DJ (1990) Phage antibodies: filamentous phage displaying antibody variable domains. Nature 348:552-524

Milenic DE, Yokota T, Filpula DR, Finkelman MA, Dodd SW, Wood JF, Whitlow M, Snoy P & Schlom J (1991) Construction, binding properties, metabolism, and tumor targeting of a single-chain Fv derived from the pancarcinoma monoclonal antibody CC49. Cancer Res 51:6363-6371

Munro S & Pelham HR (1986) An Hsp70-like protein in the ER: identity with the 78 kd glucose- regulated protein and immunoglobulin heavy chain binding protein. Cell 46:291-300

Nissim A, Hoogenboom HR, Tomlinson IM, FlynnG, Midgley C, Lane D & Winter G (1994) Antibody fragments from a 'single pot' phage display library as immuno-chemical reagents. Embo J 13:692-698

Parsons HL, Earnshaw JC, Wilton J, Johnson KS, Schueler PA, Mahoney W & McCafferty J (1996) Directing phage selections towards specific epitopes Protein Eng 9:1043-1049

Siegel DL (1995) Isolation of human anti-red blood cell antibodies by repertoire cloning. Ann N Y Acad Sci 764:547-558

Svensson HG, Hoogenboom HR & Sjobring U (1998) Protein LA, a novel hybrid protein with unique single-chain Fv antibody- and Fab-binding properties. Eur J Biochem 258: 890-896

Thompson J, Pope T, Tung J-S, Chan C, Hollis G, Mark G & Johnson K (1996) Affinity maturation of a high-affinity human monoclonal antibody against the third hyper-variable loop of human immunodeficiency virus: use of phage display to improve affinity and broaden strain reactivity. J Mol Biol 256:77-88

Vaughan T, Williams AJ, Pritchard K, Osbourn JK, Pope AR, Earnshaw JC, McCafferty J, Wilton J & Johnson KS (1996) Human antibodies with sub-nanomolar affinities isolated from a large non-immunized phage display library. Nature Biotechnology 14:309-314

Winter G, Griffiths AD, Hawkins RE & Hoogenboom HR (1994) Making antibodies by phage display technology. Annu Rev Immunol 12:433-455

Xie MH, Yuan J, AdamsC & Gurney A (1997) Direct demonstration of MuSK involvement in acetylcholine receptor clustering through identification of agonist ScFv. Nat Biotechnol 15:768-771

Yang WP, Green K, Pinz-Sweeney S, Briones AT, Burton DR & Barbas CF, 3rd (1995) CDR walking mutagenesis for the affinity maturation of a potent human anti-HIV-1 antibody into the picomolar range. J Mol Biol 254:392-403

Yokota T, Milenic DE, Whitlow M & Schlom J (1992) Rapid tumor penetration of a single-chain Fv and comparison with other immunoglobulin forms. Cancer Res 52:3402-3408

Purification and Analysis
of *Strep*-Tagged Antibody-Fragments

Martin Schlapschy and Arne Skerra

Introduction

The development of generic purification techniques for immunoglobulin (Ig) fragments has gained considerable interest, particularly because the corresponding antigens are often too scarce or unstable to prepare a matrix for affinity chromatography. In this respect, the use of a short peptide tag with defined molecular recognition properties has the advantage that it should not interfere with the function of the antibody fragment and, therefore, its removal is not necessary for most *in vitro* applications. The *Strep*-tag constitutes a nine-amino acid peptide with the sequence "Ala-Trp-Arg-His-Pro-Gln-Phe-Gly-Gly", which can easily be fused to scFv, Fv, and Fab fragments. This peptide confers reversible binding activity towards the well known protein-reagent streptavidin. Hence, it enables the purification of a corresponding fusion protein via streptavidin affinity chromatography in one step. Furthermore, the *Strep*-tag can be used for detection on Western blots or in ELISAs using streptavidin-enzyme conjugates.

The *Strep*-tag was originally developed as a generic affinity tag for the rapid isolation of a bacterially produced Fv fragment (Schmidt and Skerra 1993). Its amino acid sequence was selected in a specialized filter sandwich colony screening assay from a plasmid-encoded library of random amino acid sequences. These peptides were displayed at the C-terminus of the V_H domain as part of the recombinant anti-lysozyme D1.3 Fv fragment. The Fv fragment was secreted across the inner membrane of *E. coli*, partially released from the colonies, and captured to an antigen-coated membrane

Martin Schlapschy, Technische Universität München, Lehrstuhl für Biologische Chemie, 85350 Freising-Weihenstephan, Germany

✉ Arne Skerra, Technische Universität München, Lehrstuhl für Biologische Chemie, 85350 Freising-Weihenstephan, Germany (*phone* +49-8161-714351; *fax* +49-8161-714352; *e-mail* skerra@weihenstephan.de)

so that the attached peptide could subsequently be probed for binding activity with streptavidin-alkaline phosphatase conjugate. In repeated rounds of screening the *Strep*-tag was identified as a non-hydrophobic amino acid sequence with considerable affinity towards streptavidin, as it was finally judged according to its practical performance in detection and purification experiments.

The *Strep*-tag has the property of binding to streptavidin in a competitive manner with biotin, the protein's natural ligand. This behaviour permits elution of a bound *Strep*-tag fusion protein from the streptavidin affinity column under very gentle conditions, just in the presence of a diluted solution of biotin or one of its chemical derivatives. Thus, the *Strep*-tag enables the purification of a fully functional, heterodimeric Fv fragment when merely attached to the V_H domain (Schmidt and Skerra 1993), even though this type of antibody fragment is known for the weak association between V_H and V_L. The *Strep*-tag cannot only be used for the purification of Fv fragments (when either fused to V_H or V_L) but also for scFv (Schiweck et al. 1997) and Fab fragments and furthermore for totally different proteins (Schmidt and Skerra 1994). The *Strep*-tag can even be applied for isolating the complex between a recombinant Ig fragment and its cognate antigen (Schmidt and Skerra 1993). This strategy led to the facile purification and successful crystallization of several membrane proteins after the Fv fragments from corresponding monoclonal antibodies had been produced as *Strep*-tag fusion proteins in *E. coli* (Ostermeier et al. 1995; 1997).

In order to establish a standardized purification protocol for *Strep*-tag fusion proteins the heterologous production of a well-defined truncated version of streptavidin turned out to be critical (Schmidt and Skerra 1994). In addition, the *Strep*-tag II with the modified sequence "Asn-Trp-Ser-His-Pro-Gln-Phe-Glu-Lys" was developed as a variant that may not only be fixed to the C-terminus but also to the N-terminus or even amid a polypeptide chain, for example in fusion proteins that are composed of different domains (Schmidt et al. 1996). In its sequence the penultimate Glu residue functionally substitutes the free terminal carboxylate group of the original *Strep*-tag, which participates in a salt bridge when complexed with streptavidin and is thus critical for binding. On the basis of a crystallographic analysis of the complexes between recombinant core streptavidin and each of the peptides a streptavidin mutant was finally engineered with enhanced affinity both for the *Strep*-tag and for the *Strep*-tag II (Voss and Skerra 1997). This mutant, later termed StrepTactin, has the amino acid sequence of residues 44 to 47 changed from "Glu-Ser-Ala-Val" to "Val-Thr-Ala-Arg". It exhibits significantly improved performance, especially in the purification of *Strep*-tag II fusion proteins, when coupled to a chromatographic support.

Generally, all recombinant fusion proteins, carrying either the *Strep*-tag or the *Strep*-tag II, get more tightly adsorbed to the StrepTactin affinity matrix. In this case desthiobiotin is the preferred ligand for elution of a bound protein. Regeneration of the StrepTactin column is performed by applying a buffer containing the organic dye HABA (2'(4-hydroxyazobenzene)benzoic acid), which becomes weakly complexed by the biotin-binding pocket of streptavidin and gives rise to a colour change. Its presence in sufficient excess blocks emerging free binding sites and prevents rebinding of desthiobiotin so that this compound is quickly removed.

Biotin should be absent from the extract that is applied to the affinity column. Its amount in the periplasmic cell fraction of *E. coli* can normally be neglected. Otherwise, especially when working with culture supernatants, biotin should be removed by dialysis or gel filtration. Covalent conjugates between proteins, like *E. coli*'s BCCP, and biotin can be efficiently masked by complexation with avidin from hen egg white, which does not interact with the *Strep*-tag (II).

For the production of a functional Fv or Fab fragment as *Strep*-tag fusion protein the light and heavy chains are co-secreted into the bacterial periplasm, where protein folding and formation of the disulphide bounds can take place (Skerra and Plückthun 1988). In the case of an scFv fragment just one polypeptide needs to be secreted. Selective release of the proteins by periplasmic cell fractionation contributes to the protein purification since cytoplasmic host cell proteins are efficiently removed. Convenient vectors for the cloning and periplasmatic secretion of recombinant antibody fragments carrying the *Strep*-tag were derived from the generic expression plasmid pASK75 (Skerra 1994b), and are available for scFv, Fv, and Fab fragments (Schiweck et al. 1997). Once the two variable genes of an antibody have been cloned via conserved restriction sites (Skerra 1994a) they can easily be transferred from one vector to another, thus enabling their quick production in different formats (Fig. 1).

In the case of the heterodimeric Fv and Fab fragments the *Strep*-tag is best fused to the C-terminus of the heavy chain. In this way the copurification of soluble light chain dimers, which can accompany the production of Ig fragments in the periplasm of *E. coli*, is avoided.

The structural genes for both polypeptide chains, fused with suitable signal peptides, are arranged in a dicistronic operon under transcriptional control of the tightly regulated *tet* promoter/operator (Skerra 1994b). Foreign gene expression can be induced with anhydrotetracycline. As an ex-

pASK90: Fv fragments

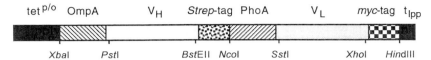

pASK98: scFv fragments

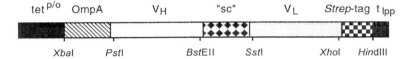

pASK99: Fab fragments

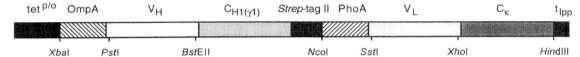

Fig. 1. Expression cassettes for the production of different antibody fragments. tet$^{p/o}$: *tet* promoter/operator; t$_{lpp}$: lipoprotein transcription terminator; OmpA and PhoA: signal sequences; *Strep*-tag and *myc*-tag: affinity tags; V$_H$, V$_L$, C$_{H1}$ and C$_\kappa$ denote the variable and constant domains of heavy and light chain, respectively; *Xba*I, *Pst*I, *Bst*EII, *Nco*I, *Sst*I, *Xho*I, and *Hind*III indicate conserved restriction sites for cloning or exchange of domains; pASK90 is used for the production of Fv fragments. The V$_H$ domain carries the *Strep*-tag at its C-terminus, the V$_L$ domain is equipped with a *myc*-tag (Schiweck et al. 1997). pASK98 is used for the production of scFv fragments. The V$_H$ and V$_L$ domains are linked via a 15 amino acid spacer ("sc"). pASK99 is used for the production of Fab fragments. For description of the vector backbone see chapter "Purification and characterisation of His-tagged antibody fragments".

ample for the production and purification of a recombinant antibody fragment with the *Strep*-tag the D1.3 Fab fragment (Boulot et al. 1990) will here be used.

Outline

An overview of the procedure for production, purification and characterisation of a recombinant Ig fragment carrying the *Strep*-tag is given in Figure 2.

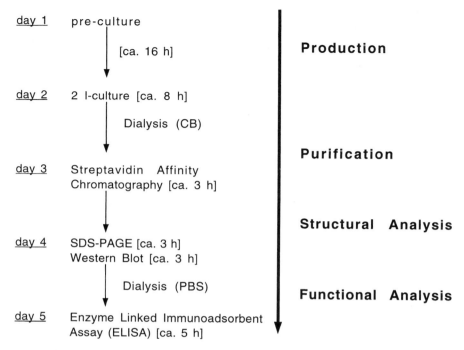

Fig. 2. Short protocol for the production, purification and analysis of Strep-tagged antibody-fragments.

Subprotocol 1
Production of a *Strep*-Tagged Antibody Fragment in the Shaker Flask

▪▪ Materials

- Incubation shaker, 22°C and 37°C (Infors or New Brunswick)

- Preparative centrifuge, rotors, tubes (Sorvall or Beckman)

- Luria-Bertani (LB) Medium
    - 10 g/l Bacto Tryptone (Difco)
    - 5 g/l Bacto Yeast Extract (Difco)
    - 5 g/l NaCl
    - adjust to pH 7.5 with NaOH

Procedure

Materials, media, and solutions should be sterilized by autoclaving or filtration.

1. A fresh single colony of *E. coli* JM83 transformed with pASK99-D1.3 is used for inoculating 50 ml LB medium containing 100 µg/ml ampicillin. The preculture is incubated at 37 or 30°C and 200 rpm overnight.

2. 40 ml of the preculture is added to 2 liter of LB/ampicillin medium in a 5 l Erlenmeyer flask. Cells are incubated at 22°C and 200 rpm and growth should be documented by measuring OD_{550}.

   Note: A higher incubation temperature may lead to reduced folding efficiency for the foreign protein in the bacterial periplasm as well as early onset of cell lysis after induction.

3. Expression is induced at $OD_{550} = 0.5$ (after correction with a LB blank value) by adding 200 µl of 2 mg/ml anhydrotetracycline in DMF. The optimal induction period varies between 2.5 and 3 h under these conditions and may depend on toxic effects on the cell caused by the recombinant antibody fragment. The best time for harvest is when the growth curve just reaches a plateau.

4. The culture is quickly transferred to centrifuge tubes (e. g. Sorvall GS3) and centrifuged at 4400 g (5000 rpm) for 15 min at 4°C (the rotor should be chilled at 4°C before harvest). After discarding the supernatant the tubes are placed on ice and residual culture medium is removed with a pipette.

Subprotocol 2
Preparation of the Periplasmic Extract

Materials

- Preparative centrifuge, rotors, tubes (Sorvall or Beckman)

- Bench top centrifuge (Sigma)

- PE buffer
  - 500 mM Sucrose
  - 100 mM Tris

- – 1 mM Na₂EDTA
- – Adjust to pH 8.0 with HCl
- – Chromatography buffer CB
  - – 100 mM Tris
  - – 150 mM NaCl
  - – 1 mM Na₂EDTA
  - – adjust to pH 8.0 with HCl

Procedure

Materials and solutions should be pre-chilled at 4°C before use.

1. The sedimented bacterial cells from a 2 l culture are carefully resuspended without delay in 20 ml of cold PE buffer, transferred to a 50 ml Falcon tube, and incubated for 30 min on ice.

   Note: Use a 25 ml pipette for repeated pipetting with 10 ml portions of the buffer in order to avoid shear stress. Addition of lysozyme (Sigma) at a final concentration of up to 200 μg/ml (from a fresh 10 mg/ml stock solution in PE buffer) may improve the efficiency of the cell fractionation but is in most cases not necessary.

2. The spheroplasts are sedimented by centrifugation at 5000 rpm in a Sigma 4K10 bench top centrifuge (4400 g, 15 min, 4°C) and the supernatant is carefully recovered as the periplasmic cell fraction. In order to clear the supernatant it is transferred to fresh centrifuge tubes (e. g. SS34) and submitted to a second centrifugation step at 15,000 rpm (27,000 g, 15 min, 4°C).

3. The periplasmic protein extract should be dialysed against 2 l of chromatography buffer (CB) overnight at 4°C or directly frozen at - 20°C for storage.

Subprotocol 3
Streptavidin Affinity Chromatography

Materials

- StrepTactin Sepharose (IBA GmbH, Göttingen, Germany)
- Chromatography station (Pharmacia)
- D-Desthiobiotin (Sigma-Aldrich)
- 2'-(4-Hydroxyazobenzene)benzoic acid, HABA (Sigma-Aldrich)
- Elution Buffer (EB): 2.5 mM Desthiobiotin in CB
- Regeneration Buffer (R): 5 mM HABA in CB

Procedure

All steps are carried out at 4°C. Materials and solutions should be pre-chilled at 4°C before use. The protein solution should be sterile-filtered before application to the column.

1. A chromatography column (diameter: 7 mm) is packed with 2 ml StrepTactin (mutant streptavidin) Sepharose (5 mg/ml; IBA GmbH, Göttingen, Germany) and connected to a peristaltic pump, a flow through UV detector, and a fraction collector (e. g. as part of a Gradifrac station, Pharmacia).

2. The column is equilibrated with 10 to 20 ml of CB at a flow rate of 20 ml/h.

3. The dialysed and sterile-filtered periplasmic protein extract is applied to the column. The flow through should be collected for analysis of the purification procedure. The column is then washed with CB until the absorption at 280 nm diminishes to the base line.

4. Elution of the bound protein is effected by applying the elution buffer containing desthiobiotin (EB). Fractions of 1 or 2 ml are collected. The major fraction of the eluted *Strep*-tag fusion protein usually appears within two 2 ml fractions.

Note: The void volume of the chromatography setup should be considered when not using a fraction collector.

5. The column is regenerated with 20 ml of buffer R containing HABA and finally washed with CB again. Colour change to red indicates quantitative replacement of desthiobiotin and functionality of the StrepTactin matrix due to the reversible complexation of the azo compound. When the column has turned pale again it is ready for the next purification run.

Results

A typical elution profile for the purification of a Fab fragment is shown in Fig. 3. The A_{280} diminishes quickly when the column is washed after application of the sample. StrepTactin has a low tendency for nonspecific binding so that host cell proteins are rapidly removed. Selective elution of the Fab fragment is then effected in the presence of desthiobiotin at a low concentration. Soon after applying the desthiobiotin a steep peak arises, which contains the purified Fab fragment.

Troubleshooting

• In order to avoid the quasi irreversible binding of free biotin or of biotinylated host cell proteins, like BCCP, to the StrepTactin affinity matrix, avidin may be added to the protein extract. This step is important in case of low expression levels, because larger amounts of cell extract can then be applied to the affinity column, or when the *Strep*-tagged Ig fragment is to be purified from a crude lysate or culture supernatant. The total biotin content of an *E. coli* soluble cell extract from a 1 l culture with $OD_{550} = 1.0$ is ca. 1 nmol. 20 μl of a 2 mg/ml stock solution of avidin in CB should be added per 1 ml of the protein extract. After incubation on ice for 30 min an aggregate usually forms, which should be removed by centrifugation. The sample is then directly ready for the StrepTactin affinity chromatography.

• Ordinary streptavidin Sepharose - as compared with StrepTactin - can also be used for the affinity chromatography although recovery of the *Strep*-tagged Ig fragment may be less quantitative. In this case elution is best achieved with a solution of 5 mM diaminobiotin (Sigma) in CB

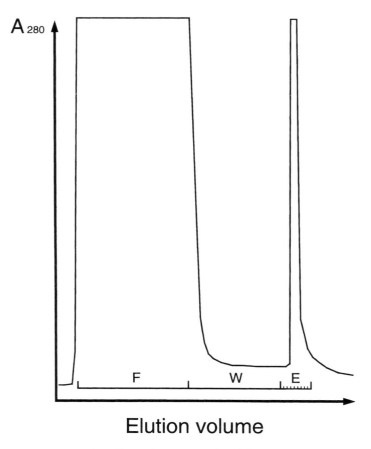

Elution volume

Fig. 3. Streptavidin affinity chromatography of the periplasmic protein extract containing the recombinant D1.3 Fab fragment. The elution profile was monitored via absorption at 280 nm. F: flow through; W: wash; E: elution with 2.5 mM desthiobiotin. Nine fractions were collected as indicated by small ticks.

(Schmidt and Skerra 1994). It should be kept in mind, however, that different commercial preparations of immobilized streptavidin can vary considerably in their ability to complex the *Strep*-tag. (Note that streptavidin is usually sold and tested for the binding of biotin, and not for the binding of a peptide.)

Subprotocol 4
SDS-PAGE and Immunoblot Analysis

The purification is analysed by standard discontinuous SDS-PAGE with 12 % (w/v) polyacrylamide gels (0.1 % SDS), followed by staining with Coomassie brilliant-blue. We recommend the buffer system of Fling and Gregerson (1986). Representative samples of the whole cell protein (see below), periplasmic extract, flow through, and of the elution fractions are reduced with 2-mercaptoethanol prior to gel electrophoresis. Under these conditions the two polypeptide chains of the Fab fragment should appear as separate bands with an approximate size of 25.0 kDa.

Note: The spacing of light and heavy chains may vary from protein to protein. The separation can often be enhanced by adding 6 M urea to the gel (Skerra 1994a). In order to investigate whether the Fab fragment contains the light and heavy chains in stoichiometric composition, linked via the interchain disulphide bond, a non-reduced sample should be analysed as well.

Materials

- Streptavidin-AP conjugate (Amersham)

- Avidin from chicken egg white (Amresco)

- NBT and BCIP (Amresco)

- PBS
    - 4 mM KH_2PO_4
    - 16 mM Na_2HPO_4
    - 115 mM NaCl

- PBST: 0.1% (v/v) Tween 20 in PBS

- AP Buffer
    - 100 mM Tris
    - 100 mM NaCl
    - 5 mM $MgCl_2$
    - adjust to pH 8.8 with HCl

Procedure

Preparation of a Whole Cell *E. coli* Lysate

1. Prior to cell harvest 1 ml of the culture is transferred in a 1.5 ml reaction tube and the cells are spun down (microfuge, 14 000 rpm, 5 min).

2. The sedimented cells are resuspended in 80 μl 100 mM Tris/HCl, pH 8.0; 5 mM MgCl$_2$ with 12.5 u/ml benzonase (Merck) for the degradation of chromosomal DNA. Then 20 μl of loading buffer for SDS-PAGE (7.5 % w/v SDS, 25 % v/v glycerol, 0.25 M Tris/HCl pH 8.0, 12.5 % v/v 2-mercaptoethanol, 0.25 mg/ml bromophenol blue) is added and the lysate is incubated on ice for 1 h whereby the viscosity becomes greatly reduced.

3. The sample can be stored at -20°C and should be heated to 95°C for 5 min prior to SDS PAGE.

Western-Blotting

The recombinant protein can either be detected on the blot using an anti-immunoglobulin antibody or it can be revealed via the *Strep*-tag with the help of a commercially available conjugate (e. g. streptavidin-alkaline phosphatase conjugate, Amersham).

All incubation steps are performed under gentle shaking at ambient temperature.

1. After electro-transfer of the proteins from the polyacrylamide gel onto a nitrocellulose membrane (Schleicher & Schuell) by means of a semi-dry blotting apparatus the membrane is placed in a clean dish and washed 3 times for 10 min with 10 ml PBST (Blake et al. 1984).

2. Prior to detection of the Fab fragment the membrane is incubated for 10 min in 10 ml PBST containing 2 μg/ml egg white avidin. In this way endogenous protein-bound biotin groups are specifically masked.

3. 20 μl of the streptavidin-AP conjugate are then directly added (at a dilution of 1:500) and incubation is continued for up to 60 min.

4. The membrane is washed twice for 5 min with 10 ml PBST and twice for 5 min with 10 ml PBS.

5. The chromogenic reaction is performed (without shaking) by adding 10 ml of AP buffer with 5 µl nitroblue tetrazolium (NBT; 75 mg/ml in 70 % w/v DMF) and 30 µl 5-bromo-4-chloro-3-indolyl-phosphate (BCIP; 50 mg/ml in DMF) until the bands become visible (ca. 15 min). The reaction is stopped by washing with water and air-drying of the membrane.

Results

A typical Coomassie-stained gel and a Western-Blot demonstrating the purification of the D1.3 Fab fragment is shown in Fig. 4.

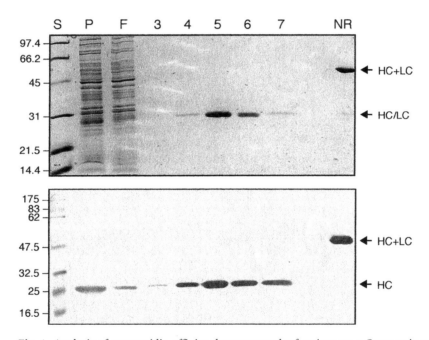

Fig. 4. Analysis of streptavidin affinity chromatography fractions on a Coomassie-stained 12 % SDS-PAGE (top) and corresponding Western Blot (bottom). Lane S: low molecular weight standard; lane P: periplasmic extract; lane F: flow through; lanes 3-7: fractions 3-7 from the streptavidin affinity chromatography as shown in Figure 3 (reduced); lane NR: non-reduced D1.3 Fab fragment from fraction 6. The heavy (HC) and light chains (LC) of the D1.3 Fab fragment have the same mobility in a 12 % SDS-PAGE. The *Strep*-tag fused to the heavy chain was detected on the Western-Blot via streptavidin-AP conjugate.

Subprotocol 5
Detection of a *Strep*-Tagged Antibody Fragment in an Enzyme Linked Immunosorbent Assay (ELISA)

In this standard protocol antigen-binding activity is tested for the *Strep*-tagged recombinant Fab fragment derived from the anti-lysozyme antibody D1.3.

Materials

- Microtiter plate, 96 well (Becton Dickson)
- Streptavidin-AP conjugate (Amersham)
- ELISA Reader (Molecular Devices)
- pNPP (Amresco)

Procedure

All incubation steps are performed at ambient temperature for 60 min unless otherwise stated. Residual liquid should be thoroughly removed after each step by pushing and draining the plate on a tissue wipe.

1. The wells in a row of a 96 well microtiter plate are each coated with 100 µl 3 mg/ml lysozyme in 50 mM NaHCO$_3$, pH 9.6, overnight at 4°C.

2. The wells are blocked with 200 µl 3 % (w/v) BSA (Sigma), 0.5 % v/v Tween 20 in PBS.

3. The plate is washed three times with PBST.

4. 50 µl of the purified D1.3 Fab fragment is applied in a decreasing concentration series from 0.5 to 0.01 mg/ml. PBST should be used as blank during this incubation step.

5. The plate is washed three times with PBST.

6. 50 µl of streptavidin-alkaline phosphatase conjugate (Amersham) diluted 1:500 in PBST is applied to each well.

7. The plate is washed twice with PBST and twice with PBS.

8. 100 µl of a solution of 0.5 mg/ml p-nitrophenylphosphate (pNPP) in AP buffer is added to each well.

9. The enzymatic activity is measured at 25°C as the change in absorbance at 405 nm per min (e. g. in a SpectraMAX 250 instrument, Molecular Devices). Alternatively, the end point of absorption can be determined after 15 or 30 min.

References

Blake MS, Johnston KH, Russel-Jones GJ, Gotschlich EC (1984) A rapid, sensitive method for detection of alkaline phosphatase-conjugated anti-antibody on Western blots. Anal Biochem 136: 175-179

Boulot G, Eisele JL, Bentley GA, Bhat TN, Ward ES, Winter G and Poljak RJ (1990) Crystallization and preliminary X-ray diffraction study of the bacterially expressed Fv from the monoclonal anti-lysozyme antibody D1.3 and of its complex with the antigen, lysozyme. J Mol Biol 213: 617-619

Fling SP, Gregerson DS (1986) Peptide and protein molecular weight determination by electrophoresis using a high-molarity Tris-buffer system without urea. Anal Biochem 155: 83-88

Ostermeier C, Iwata S, Ludwig B and Michel H (1995) Fv fragment-mediated crystallization of the membrane protein bacterial cytochrome c oxidase. Nature Struct Biol 2: 842-846

Ostermeier C, Harrenga A, Ermler U and Michel H (1997) Structure at a 2.7 A resolution of the *Paracoccus denitrificans* two-subunit cytochrome c oxidase complexed with an antibody Fv fragment. Proc Natl Acad Sci USA 94: 10547-10553

Schiweck W, Buxbaum B, Schätzlein C, Neiss HG, Skerra A (1997) Sequence analysis and bacterial production of the anti-c-myc antibody 9E10: the VH domain has an extended CDR-H3 and exhibits unusual solubility. FEBS Lett 414: 33-38

Schmidt TGM and Skerra A (1993) The random peptide library-assisted engineering of a C-terminal affinity peptide, useful for the detection and purification of a functional Ig Fv fragment. Protein Eng 6: 109-122

Schmidt TGM and Skerra A (1994) One-step affinity purification of bacterially produced proteins by means of the "Strep tag" and immobilized recombinant core streptavidin. J Chromatogr A 676: 337-345

Schmidt TGM, Koepke J, Frank R, and Skerra A (1996) Molecular interaction between the *Strep*-tag affinity peptide and ist cognate target, streptavidin. J Mol Biol 255: 753-766

Skerra A (1993) Bacterial expression of immunglobulin fragments. Curr Opin Immunol 5: 256-262

Skerra A (1994a) A general vector, pASK84, for cloning, bacterial production, and single-step purification of antibody Fab fragments. Gene 141: 79-84

Skerra A (1994b) Use of the tetracycline promotor for the tightly regulated production of a murine antibody fragment in *Escherichia coli*. Gene 151: 131-135

Skerra A and Plückthun A (1988) Assembly of a functional immunoglobulin Fv fragment in *Escherichia coli*. Science 240: 1038-1041

Voss S and Skerra A (1997) Mutagenesis of a flexible loop in streptavidin leads to higher affinity for the *Strep*-tag II peptide and improved performance in recombinant protein purification. Protein Eng 10: 975-982

Improving Expression of scFv Fragments by Coexpression of Periplasmic Chaperones

HENDRICK BOTHMANN and ANDREAS PLÜCKTHUN

Introduction

The periplasmic expression in *E. coli* has become the standard technology for preparing functional antibody fragments in a rapid way (Skerra and Plückthun 1988, Plückthun et al. 1996). The consequences of choosing Fab or scFv fragments, the properties of suitable expression vectors or the influence of the *E. coli* strain have been extensively summarized elsewhere (Plückthun et al. 1996). Even when paying attention to all these components and experimental conditions, it has become clear that the yield of recombinant antibody fragments is variable and these variations are a direct consequence of their primary sequences (Wörn and Plückthun 2001). The periplasmic folding is the yield limiting step and is most strongly influenced by the sequence.

Consequently, antibody expression can be greatly improved by altering the sequence. This is possible without losing affinity, and two principal approaches have been found to be successful, a "rational" approach and a "directed evolution" approach. The rational approach gives the choice between aligning the antibody sequence to well-expressing ones (Knappik and Plückthun 1995, Wörn and Plückthun 2001) together with exchanging exposed hydrophobic residues (Nieba et al. 1997) and the possibility of grafting the CDRs directly to a very stable and well expressing framework (Jung and Plückthun 1997, Wörn and Plückthun 2001, Willuda et al. 1999). Alternatively, the protein can be subjected to a directed evolution approach (Proba et al. 1998, Jung and Plückthun 1999). While a picture of important residues and interactions for folding is slowly emerging (Wörn and Plückthun 2001), these approaches are still

Hendrick Bothmann, Universität Zürich, Biochemisches Institut, Winterthurerstr. 190, 8057 Zürich, Switzerland; Present address: Genion Forschungsgesellschaft mbH, Schnackenburgallee 114, 22525 Hamburg, Germany

✉ Andreas Plückthun, Universität Zürich, Biochemisches Institut, Winterthurerstrasse 190, 8057 Zürich, Switzerland (*phone* +41-1-6355570; *fax* +41-1-6355712; *e-mail* PLUECKTHUN@BIOCFEBS.UNIZH.CH)

time consuming and worthwhile only if the antibody will be used frequently and in important applications or if the initial yield is extremely poor.

A faster method, which can of course be used in conjunction with sequence engineering, is to coexpress periplasmic factors which may improve the yield of folded antibody. Up to now, two periplasmic factors have been discovered by using phage display for selection for the improved expression of a poorly folded antibody fragment on filamentous phage (Bothmann and Plückthun 1998, 2000). The first factor identified, Skp, is a basic periplasmic protein of unknown function, which has been implicated in outer membrane protein biogenesis (de Cock et al. 1999, Schäfer et al. 1999). The second factor, FkpA, is a periplasmic peptidyl-prolyl *cis/trans* isomerase, which also acts as a chaperone even on antibodies not containing *cis*-prolines (Bothmann and Plückthun 2000, Ramm and Plückthun 2000). Both factors were found to increase the yield of periplasmic antibody fragments. Which one is to be preferred, however, appears to be molecule-specific among antibodies. Neither additivity nor synergy was observed between the factors, the more effective one of the two appears to set the expression yield.

We have never found a negative influence of the coexpression of either Skp or FkpA up to now. However, in some "well-behaving" antibodies, the effect of coexpression is very small, while in other cases it can increase the yield up to 10-fold. Therefore, it seems to be a good strategy to have them both routinely present on antibody expression plasmids or work with a second compatible plasmid. While the latter strategy might seem more versatile, working with single vectors is generally more convenient, as the loss of a second replicon is a potential source of irreproducibility.

Undoubtedly, many variations of the method of coexpressing Skp and/or FkpA are possible (Bothmann and Plückthun 1998, 2000, Hayhurst and Harris, 1999), and not all have been experimentally tested up till now. The procedure we discuss uses the natural promoters of Skp and FkpA, but other strategies may be possible as well (Bothmann and Plückthun 2000).

Here, we describe the vectors pHB100s, pHB110 and pHB610, which are compatible with the pAK series for phage display (Chapter 2) and miniantibody production (Chapter 43), and as an alternative, the PCR procedure for cloning the genes into another vector. We give the procedure for a small scale experiment to test the effect of the chaperone coexpression and that of a large scale expression for purification. In contrast to the production of poorly folding antibodies in the absence of the chaperone, the expression in the presence of the chaperone can be carried out for longer times, as it leads to less cell lysis, which in turn increases the reachable cell density, again increasing the yield.

Materials

- French Press (Aminco) (Rochester, NY, USA) with 4 ml cell and 40 ml cell

- Optima TLX Ultracentrifuge (Beckmann Instruments) with TLA-100.3 rotor

- DNeasy Tissue Kit (50) (QIAGEN or equivalent)

- Appropriate expression system to produce his6-tagged antibody fragments, such as the pAK system (Chapter 2, Krebber et al. 1997)

- Automated LC-System: BioCAD workstation with dual channel variable wavelength UV/visible detector, semi-preparative flow cell (Perkin Elmer), fraction collector Advantec SF-2120 (Toyo Roshi International) or equivalent system

- POROS20 MC/M 4.6 mm/100 mm (metal chelate)

- POROS20 HQ/M 4.6 mm/100 mm (anion exchange)

- POROS20 HS/M 4.6 mm/100 mm (cation exchange)

- 200 mM imidazole adjusted to pH 7 with acetic acid

- 3 M NaCl stock solution

- distilled water

Solubilization buffer

- 2 M urea

- 1 mM EDTA

- 10 mM glycylglycine (pH 7.5)

PBS (PBST)

- 8 mM Na_2HPO_4

- 1.8 mM KH_2PO_4

- 3 mM KCl

- 137 mM NaCl (pH 7.4)

- For PBST add Tween 20 to 0.05% final concentration

100 mM MHA-buffer (stock solution)

- 33 mM Mes
- 33 mM Hepes
- 33 mM Na-acetate (adjust to pH 7.5 with NaOH)

Hepes extraction buffer

- 20 mM Hepes
- 0.5 M NaCl (adjust to pH 7.0 with NaOH)

▓ Procedure

Cloning of scFv fragments from pAK/pJB vectors in pHB100s/pHB610 vectors

1. Excise the the antibody fragment expression cassette with *XbaI*/*Hind*III (Fig. 4, Chapter 2).

   Note: The vector system and primer are described in detail in Chapter 2

2. Prepare the vector fragment by removing the *tet*-cassette from pHB100s, pHB610 or pJB33 with *XbaI*/*Hind*III (Fig. 1).

3. Ligate the antibody fragment expression cassette into the appropriate vector (pHB100s, pHB610 or equivalent).

———————————————————————————————————————▶

Fig. 1. Vectors and cloning strategies. All vectors contain a chloramphenicol resistance cassette (*camR*) and a tetracycline resistance "stuffer" cassette (*tet*, composed of *tetA* and *tetR*; 2101 bp) (Krebber et al. 1997), which will be replaced by the antibody fragment. This stuffer, shown here only schematically, contains the genes for *tetA* and *tetR*, and it does not make a fusion protein with upstream or downstream elements in the vector. For details, see Chapter 2. The vectors pHB110, pHB100s and pHB610 allow phage display, if an scFv cassette without stop codon is introduced, as detailed in Chapter 2, or periplasmic expression, if a stop codon is present at the end of the scFc gene, and the g3p gene can be replaced by a tag sequence. Because of its strong ribosome binding site vector pJB33 allows strong periplasmic expression and the tag allows IMAC purification of the antibody fragment (see Chapter 2). (a) Vector containing the *skp* cassette with flanking genes, as it was enriched during panning (Bothmann and Plückthun 1998), allowing isolation and subsequent recloning of *skp* after digestion with *Not*I, *Spe*I and *Sal*I/*Xho*I. (b) Vector containing *skp* without the truncated *lpxD*. (c) Vector containing *fkpA*. (d) High yield expression vector (see Chapter 2). *lpxD*: the first 65 aa of UDP-3-O-[hydroxymyristoyl]-glucosamine-N-acyltransferase, *yeat*: the last 49 aa of Yeat (unknown function).

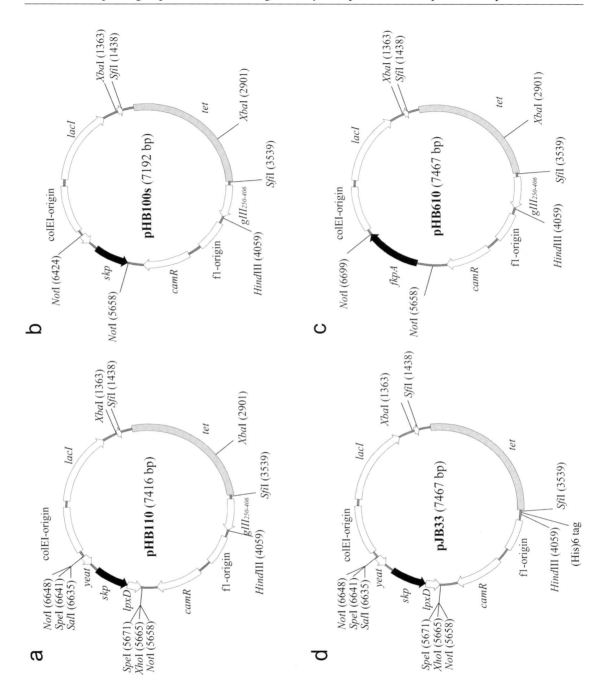

Cloning *skp/fkpA* in other expression vectors

Notes
- *skp* and *fkpA* can be cut out from vectors pHB100s and pB610 by *Not*I (Fig. 1).

- *skp* can also cut out from vector pHB110 with *Spe*I and *Xho*I/*Sal*I (Fig. 1)

- *skp* and *fkpA* can be amplified from genomic template DNA or from the above vectors pHB100s, pHB610 or equivalent and inserted in any expression vector.

For neither *skp* nor *fkpA* has the exact position of the promotor been experimentally determined and we had to rely on predictions. Therefore, we recommend the following PCR-primers for amplifying the genes from the vectors pHB110 or pHB610 or from genomic *E. coli* DNA.

fkpAseq: 5' NNNNNXXXXXXTTATTTTTTAGCAGAATCTGCGGC 3'
fkpArev: 5' NNNNNXXXXXXGATTCACCTCTTTTGTCGAATGG 3'
skpseq: 5' NNNNNXXXXXXGATCCAAGCAATATCCGTATG 3'
skprev: 5' NNNNNXXXXXXTTATTTAACCTGTTTCAGTACGTC 3'

- XXXXXX stands for the restriction site used for subcloning.

- NNNNN stands for additional bases flanking the restriction sites, which are necessary for efficient cleavage (see, e. g. the New England Biolabs catalog for more information).

4. Isolate genomic DNA from *E. coli* as described in the manual of the DNeasy Tissue kit or equivalent.

5. Perform PCR with primers for *skp* or *fkpA* as according to standard protocols.

6. Digest the PCR-product with the appropriate restriction enzymes and clone it into the expression vector. Another strategy is to subclone the PCR-product first in a vector such as pCRscript (Stratagene) and use this construct as starting point for further subcloning.

Small scale expression

7. Inoculate 50 ml LB medium, containing the appropriate antibiotic, with a single colony of *E. coli* JM83 or SB536 (see Chapter 43, harboring a plasmid encoding the respective scFv fragment.

8. Grow the culture at 24°C or 37°C to an OD_{550} of 0.5 and induce with 1 mM IPTG (final concentration).

9. Harvest the cells after 1 h induction at 37°C or 3 h induction at 24°C.

10. Resuspend the cells carefully in 2 ml PBS.

11. Measure OD_{550}.

12. Lyse the cells by two passages through the French Press (20,000 psi).

13. Centrifuge 1 ml of the whole cell extract in an Eppendorf tube for 30 min at 50,000 rpm in a TLA-100.3 rotor at 4°C.

14. Transfer the supernatant carefully to a fresh Eppendorf tube (soluble fraction).

15. Dissolve the pellet in 1 ml solubilization buffer (shaking overnight at 4°C, if necessary) (insoluble fraction).

16. Normalize all fractions to the same OD_{550} of the original culture.

17. For crude extract ELISA, use normalized soluble fraction, for Western blots normalized soluble and insoluble fractions.

18. Coat ELISA plates with antigen ON at 4°C using standard protocols (see, e. g. Thorpe and Kerr 1994).

19. Mix a defined amount of normalized soluble fraction with 2% skimmed milk in PBST in the presence and absence of soluble antigen and apply to the blocked ELISA plate.

20. Perform detection as e. g. described in Thorpe and Kerr 1994.

21. For Western blots load defined amounts of soluble and insoluble fractions on a reducing SDS-PAGE.

22. Perform standard Western blot, according to the protocols described in Sambrook et al. 1989.

Large scale expression

The single-chain Fv fragment with 6-histidine tag can by purified by rapid two-column chromatography. This protocol is given for 5-10 g wet weight of *E. coli* cells, corresponding to about 1 l of baffled shake-flask culture.

23. Resuspend the cell pellet in 30 ml Hepes extraction buffer.

24. Add DNase to a final concentration of 10 μg/ml.

25. Disrupt the cells in a French Press (20,000 psi, 4°C in a cold room). At least two passages are required for optimal lysis of the cells.

26. Centrifugation of the suspension (Sorvall SS-34, 48 200 g, 4°C, >30 min).

27. Filter the supernatant (0.22 μm, use filter with low protein binding properties).

28. Load the filtrate on an Ni-IDA POROS column (1.66 ml) (preloaded with 3 ml 0.1 M NiCl$_2$), pre-equilibrated with 20 mM MHA-buffer, 0.15 mM NaCl, pH 7.0. The flow rate should be 5 column volumes (CV) per minute (Fig. 2).

    Note: The whole procedure with both columns takes only 30 minutes.

29. Wash column with 20 mM MHA-buffer, containing 0.15 mM NaCl until the baseline is reached.

30. Wash column with 10 CV 20 mM MHA-buffer, containing 30 mM NaCl, pH 7.0.

31. Wash column with 10 CV 20 mM MHA-buffer, containing 1 M NaCl, pH 7.0.

32. Wash column with 30 mM imidazole, 0.15 mM NaCl, pH 7.0 for 10 CV.

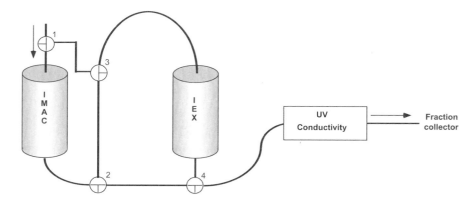

Fig. 2. Tubing diagram for rapid two-column purification of antibody fragments. At the beginning of the chromatography, all flow is through the IMAC column (valve positions as shown). Upon antibody elution, the flow is redirected to the ion-exchange column (IEX), by turning valves 2 and 4. The adsorbed protein is then eluted with a new gradient, by turning valves 1 and 3.

33. Elute either using an imidazole gradient from 30 to 150 mM imidazole (pH 7.0) (**no salt**) (10 CV) or a step elution with 150 mM imidazole (pH 7.0) (**no salt**) (6 CV) (Fig. 2).

34. Load the IDA-elution directly on a second column, without collecting the samples, by using the BIO-CAD workstation or equivalent. This column can be either a cation-exchange or an anion-exchange.

   Note: For the tubing diagram, see Fig. 2. The pH for the second washing step and the elution depends on the pI of the antibody fragment and on the type of the second column (i. e. if the antibody has a pI of 8.5, the pH should be adjusted to 7.0 and the sample should applied to a cation exchange column.

35. Wash the column with 20 mM MHA-buffer, containing 30 mM NaCl, pH 7.0 until the baseline is reached.

36. Elute the cation-exchange column with a salt gradient from 30 to 750 mM NaCl (15 CV). Collect the samples in 1 ml fractions and analyze each of them by SDS-PAGE.

Comments

- Coexpressing Skp together with an antibody fragment sometimes results in a prolonged lag phase and slower doubling phase of the cells. Nevertheless, when the OD_{550} reaches 0.8, these cells recover, showing a higher doubling rate and, finally, higher yield of recombinant protein.

- Most of the *E. coli* host proteins co-purified in IMAC have a pI less then 6.5, therefore they will bind to an anion-exchange column. A salt gradient for separation usually works very well, in conjunction with a running buffer of pH 7.0.

- Imidazole as a storage buffer and as a sample component in SDS-PAGE is not desirable, because it will slowly catalyze the hydrolysis of acid labile bonds. Therefore, the 2-step method might even be useful for those antibody-fragments which are already pure enough after the IMAC step. Alternatively, the IMAC eluate can be dialyzed against a physiological buffer such as PBS immediately after purification.

Acknowledgements. We wish to thank Stefan Ewert for helpful contributions to the methods presented here.

References

Bothmann H, Plückthun A (1998) Selection for a periplasmic factor improving phage display and functional periplasmic expression. Nature Biotechnol 16:376-380

Bothmann H, Plückthun A (2000) The periplasmic *E. coli* peptidylprolyl cis, trans-isomerase FkpA (I): increased functional expression of antibody fragments with and without *cis*-prolines. J Biol Chem 275:17100-17105

de Cock H, Schäfer U, Poteger M, Demel R, Müller M, Tommassen J (1999) Affinity of the periplasmic chaperone Skp of Escherichia coli for phospholipids, lipopolysaccharides and non-native outer membrane proteins. Eur J Biochem 259:96-103

Hayhurst A, Harris WJ (1999) *Escherichia coli* Skp chaperone coexpression improves solubility and phage display of single-chain antibody fragments

Jung S, Plückthun A (1997) Improving *in vivo* folding and stability of a single-chain Fv antibody fragment by loop grafting. Prot Eng 10:959-966

Jung S, Honegger A, Plückthun A (1999) Selection for improved protein stability by phage display. J Mol Biol 294:163-180

Knappik A, Plückthun A (1995) Engineered turns of a recombinant antibody improve its *in vivo* folding. Prot Eng 8:81-89

Krebber A, Bornhauser S, Burmester J, Honegger A, Willuda J, Bosshard HR, Plückthun A (1997) Reliable cloning of functional antibody domains from hybridomas and spleen cell repertoires employing a reengineered phage display system. J Immunol Meth 201:35-55

Missiakas D, Betton JM, Raina S (1996) New components of protein folding in extracytoplasmic compartments of *Escherichia coli* SurA, FkpA and Skp/OmpH. Mol Microbiol 21:871-884

Nieba L, Honegger A, Krebber C, Plückthun A (1997) Disrupting the hydrophobic patches at the antibody variable/constant domain interface: improved *in vivo* folding and physical characterization of an engineered scFv fragment. Prot Eng 10:435-444

Plückthun A, Krebber A, Krebber C, Horn U, Knüpfer U, Wenderoth R, Nieba L, Proba K, Riesenberg D (1996) Producing antibodies in *Escherichia coli*: From PCR to fermentation. In: McCafferty J, Hoogenboom H, Chiswell D (eds) Antibody Engineering. A practical approach, Oxford University Press, Oxford, pp 203-252

Proba K, Wörn A, Honegger A, Plückthun A (1998) Antibody scFv fragments without disulfide bonds made by molecular evolution. J Mol Biol 275:245-253

Ramm K, Plückthun A (2000) The periplasmic *E. coli* peptidyl-prolyl isomerase FkpA (II): isomerase-independent chaperone activity *in vitro*. J Biol Chem 275:17106-17113

Sambrook J, Fritsch EF, Maniatis T (1989) Molecular Cloning: a laboratory manual. Cold Spring Harbor Laboratory Press, New York

Schäfer U, Beck K, Müller M (1999) Skp, a molecular chaperone of Gram-negative bacteria, is required for the formation of soluble periplasmic intermediates or outer membrane proteins. J Biol Chem 274:24567-24574

Skerra A, Plückthun A (1988) Assembly of a functional immunoglobulin F_V fragment in *Escherichia coli*. Science 263:14315-14322

Thorpe SJ, Kerr MA (1994) Common immunological techniques: ELISA, blotting, immunohistochemistry and immunocytochemistry. In: Kerr MA, Thorpe R (eds) Immunochemistry. Labfax, BIOS Scientific Publishers Limited, Oxford, pp 175-209

Willuda J, Honegger A, Waibel R, Schubiger PA, Stahel R, Zangenmeister-Wittke U, Plückthun A (1999) High thermal stability is essential for tumor targeting of antibody fragments: Engineering of a humanized anti-epithelial glycoprotein-2 (epithelial cell adhesion molecule) single-chain Fv fragment. Cancer Research 59:5758-5767

Wörn A, Plückthun A (2001) Stability engineering of antibody single-chain Fv fragments. J Mol Biol, 305: 989–1010

Expression and Purification of Antibody Fragments in Eukaryotic Cells

Expression of scFv Antibody Fragments in the Yeast *Pichia pastoris*

HERMANN GRAM and RÜDIGER RIDDER

Introduction

With the advent of antibody display on phage, antigen binders are now being increasingly isolated as scFv fragments or Fab fragments from hybridomas (see Chapter 2 and Chapter 3) or from phage libraries (see Chapters 5 - 13). Though scFv fragments and Fab fragments have in principle the same binding properties towards antigen, scFv fragments are more easily manipulated for expression in microorganisms. The main advantage is that scFv antibody fragments can be expressed from a single transcription/translation unit, which obviates the need to achieve approximately equal expression levels for both chains of the heterodimeric Fab fragment. Antibody fragments are typically produced in any chosen host organism by secretion across a cell membrane. Bacterial expression systems relying on secretion into the periplasmic space give typical yields of 0.1 - 2 mg/l in shake flask cultures for many scFv fragments, making it difficult to produce 10-100 mg of scFv needed for many applications. When searching for alternative expression systems, we felt that a microbial host system allowing for secretion of scFv fragments would be best suited for fast and economic high level expression of these molecules, and set out therefore to explore the methylotrophic yeast *Pichia pastoris* for expression of scFv fragments.

Pichia pastoris has an excellent secretory capacity and was shown to be a suitable host organism for high-level expression of heterologous proteins (Laroche et al., 1994; White et al., 1994; Petersen et al., 1995; Sun et al., 1995) including scFv fragments (Ridder et al., 1995; Luo et al.,

✉ Hermann Gram, Novartis Pharma AG, Arthritis and Bone Metabolism, Bldg. 386-927, 4002 Basel, Switzerland (*phone* +41-61-324-4376; *fax* +41-61-324-9457; *e-mail* Hermann.Gram@pharma.novartis.com)

Rüdiger Ridder, MTM Laboratories AG, Im Neuenheimer Feld 519, 69120 Heidelberg, Germany

1995; Luo et al., 1996; Eldin et al., 1997). Expression levels of $\gg 50$ mg/l even at the laboratory scale have been reported for many proteins. Typically, secreted proteins are soluble and show post-translational modifications similar to those of higher eukaryotes (Cregg et al., 1993). The well established and commercially available expression plasmids for *Pichia pastoris* rely on chromosomal integration and allow for specific targeting of the alcohol oxidase 1 gene, a highly transcribed and regulatable gene. The available expression vectors allow the selection of transformants by auxotroph or antibiotic markers, and offer the option to select for multicopy integrations. *P. pastoris* combines the advantage of an eukaryotic secretion machinery, which should in general terms be best suited for the expression of antibody fragments, with fast growth in a non-complex growth medium. These features make this yeast an attractive host for scFv expression, in particular as the high expression level of heterologous proteins make the production of considerable amounts of material feasible at a small scale in shake flasks. Production in shake flasks does not require specialized equipment and can be performed with the standard equipment of a molecular biology laboratory. If larger amounts of product are needed, recombinant *P. pastoris* strains can be fermented to high cell density yielding product titers in the gram per liter range. We found *Pichia pastoris* from the ease of handling and production capability indeed well suited for the expression of scFv fragments, and observed accumulation of a scFv fragment to > 100 mg/l in the culture supernatant.

Due to their small size and lack of constant part domains against which generic secondary antibodies are commercially available, scFv fragments need an epitope tag for detection in most of the typical applications. Several tags, such as peptides, the murine C_κ domain, or dimerization motifs were sucessfully used for scFv antibody fragments expressed in *Pichia pastoris* (Gram et al., 1998). These epitope tags greatly facilitate purification of the scFv from *Pichia pastoris* culture supernatant. As the principles of tagging and engineering are adressed in various chapters of this book, the emphasis of this chapter will be on the use of the *P. pastoris* expression system.

Outline

Described is here the amplification of cDNA encoding the scFv from phagemids or plasmid vectors, the subcloning into an expression vector for *Pichia pastoris*, transformation, selection of colonies and analysis of expression. Though in recent years the *Pichia* expression plasmids were further developed to allow for antibiotic selection, we describe here the selection of transformants for the auxotroph His4 marker gene. Random integration of the expression cassette into the *Pichia pastoris* genome as well as homologous integration into the alcohol oxidase 1 gene (AOX1) can be selected in the host strain GS115. Both types of integrations can lead to high level expression, but random integrations may give more variable results, depending on the site of integration. Although it is not necessary in principle to select for integrations into the AOX1 locus and accordingly, for the Mut^S phenotype, we feel that this step gives a better control over the expression system: (i) Transcription of the AOX1 locus is tightly regulated by methanol. AOX1 integrations usually result in a reliable and high level expression upon induction with little variance amongst individual clones. (ii) The Mut^S genotypic yeasts, generated by disrupting the AOX1 gene, cannot efficiently utilize methanol as a carbon source. Therefore, methanol used for induction of the promoter is not quickly metabolized as in Mut$^+$ strains obviating the need to re-feed the cultures during the production phase.

We describe below the cloning and expression of a rabbit scFv fragment. The same procedure can be followed using scFv fragments derived from any other species, however, the primer sequences used for amplification have to be adjusted accordingly. We recommend here using the pre-pro leader sequence of the α-mating factor present in the vector pPIC9 (Fig. 2) or one of its derivatives for secretion. Vectors containing the Pho-1 leader sequence may work in some instances, but not in others (own observation, Eldin et al., 1997). The general scheme for expressing scFv fragments in *Pichia pastoris* is given in Fig. 1.

Materials

– PCR thermocycler	Equipment
– Temperature-controlled shaker unit	
– Electroporator (e.g. BioRad GenePulser) with 0.2 cm cuvettes	
– Lederberg stamp	
– *Pichia pastoris* strain GS115 (*his4*)	Reagents
– *Pichia pastoris* control strain GS115/albumin	
– *Escherichia coli* XL-1 blue or comparable strain	

Amplification of coding frame for scFv from display or expression vector by PCR, introduction of suitable restriction sites

Ligation into *Pichia pastoris* expression vector, transformation into E. coli and selection of recombinant plasmids

Tranformation of *Pichia pastoris* GS115 by elektroporation and selection of transformants (His$^+$ phenotype)

Identification of integration events into AOX1 locus (Mut^s phenotype)

Small-scale expression cultures and analysis of expression by Western Blot or ELISA

Fig. 1. Flow Chart

– Expression vector pPIC9

– Oligonucleotide primers: The primers described here were used for the expression of a rabbit scFv. Sequences for the tags and coding frames may be adjusted according to the planned construct.

5′ primer, addition of *Eco*RI site (bold) upstream of scFv coding frame (double underlined)	5′-CGA**GAATTC**GAGCTCGATCTGACCCAGACT-3′
3′ primer, addition of *Not*I site (bold), His$_6$- tag (underlined), and STOP codon (italics) downstream of scFv coding frame (double underlined)	5′ CGA**GCGGCCGC***TTA*GTGGTGATGGTGATGGTG TGGGCTAGCGACTGA 3′

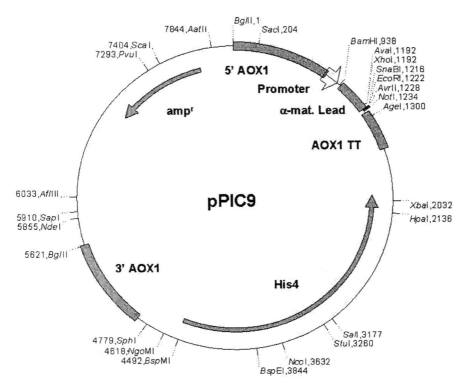

Fig. 2. Plasmid map of the expression vector pPIC9. ScFv coding frames are inserted into the multiple cloning site under the control of the AOX1 promoter. AOX TT denotes the transcriptional terminator downstream of the scFv coding frame. The 5' and 3' portions of the AOX1 gene serve as homology regions for recombination of the linearized vector into the AOX1 locus. The His4 gene serves as selectable marker in *his4 Pichia pastoris* strains.

– Pwo polymerase supplied with reaction buffer (Roche Biochemicals)

– PCR reaction buffer (supplied with the enzyme)

– Bovine serum albumin (BSA)

– YPD Medium : 1% yeast extract, 2% peptone, 2% glucose

– MD Agar: 2% agar, 1.34% yeast nitrogen base, 1% glucose, 1.6 μM Biotin

– MM Agar: 2% agar, 1.34% yeast nitrogen base, 0.5% methanol, 1.6 μM Biotin

- RDB Agar: 2% agar, 2% dextrose, 1.34% yeast nitrogen base, 1 M sorbitol, 1.6 µM Biotin, L-leucine, L-lysine, L-glutamic acid, L-isoleucine, and L-methionine, 0.005% each

- BMGY: 2% peptone, 1.34% yeast nitrogen base, 1% yeast extract, 1% glycerol, 1.6 µM Biotin, 100 mM K_2HPO_4, pH 6.0

- BMMY : 2% peptone, 1.34% yeast nitrogen base, 1% yeast extract, 0.5% methanol, 1.6 (M Biotin, 100 mM K_2HPO_4, pH 6.0

- Anti-tag (monoclonal) antibody

- Anti-murine HRP-conjugated antibody or HRP-streptavidin conjugate

- BM Blue solution (Roche)

- Phosphate-buffered saline (PBS): 137 mM NaCl, 2.7 mM KCl, 4.3 mM Na_2HPO_4, 1.4 mM KH_2PO_4 , pH 7.2

- Phosphate-buffered saline/Tween (PBS-T): PBS+0.05% Tween 20

- Skimmed milk powder

- PCR purification kit

- Gel extraction kit

- DNA Miniprep kit

Procedure

PCR amplification of the single chain antibody and cloning into the expression vector

1. Design the upstream primer such that the scFv coding frame is being fused in frame to the α-mating factor leader sequence and use appropriate restriction site. Include into the downstream primer codons for an appropriate peptide tag followed by a stop codon.

   Note: A hexa-His tag can be used both for purification and detection as specific antibodies are commercially available now. Alternatively, a peptide tag used for detection could be followed by a stretch of hexa-histidine for purification.

2. Assemble the following reaction: 10 ng of template DNA, oligonucleotide primers, 1 mM each, 0.2 mM of dATP, dTTP, dCTP and dGTP each, 0.5 -1 units *Pwo* polymerase in 1 x reaction buffer in a final volume of 100 µl.

3. Perform 25-30 cycles of 30 sec at 95°C, 30 sec at 55°C, and 30 sec at 72°C.

4. Remove enzyme, salts, and oligonucleotides from the resulting PCR product using a PCR purification kit according to the manufacturer's instructions.

5. Digest the entire PCR product with the restriction enzymes chosen for cloning into the expression vector pPIC9.

   Note: Suitable enzymes are *Eco*RI and *Not*I.

6. Purify the digested PCR fragment by gel electrophoresis using one of the commercially available extraction kits.

7. Ligate the purified PCR fragment into appropriately digested pPIC9 vector.

8. Transform *E. coli* XL-1 blue and select recombinant bacteria by analysing miniprep DNA or by PCR analysis.

9. Prepare about 50 μg of recombinant plasmid DNA using a commercial DNA preparation kit.

10. Verify the DNA sequence of the cloned PCR product and the fusion to the vector by DNA sequencing.

    Note: This step usually eliminates rarely occurring mutants or out-of-frame fusions which may cost a substantial effort in trouble shooting when not detected at this stage.

11. Digest 20 μg of the recombinant expression vector with *Bgl*II to allow for site-specific insertion into the AOX 1 gene.

12. Inactivate the restriction enzyme by phenol extraction of the digested DNA. Ethanol precipitate the DNA and dissolve at a concentration of 1 μg/μl in destilled and sterile water.

    Note: It is not necessary to purify the expression cassette from the vector DNA in this step.

Transformation of *Pichia pastoris* by electroporation

1. Inoculate 5 ml of YPD medium with *P. pastoris* strain GS115 in a 50 ml shake flask and incubate with shaking at 30°C overnight.

2. Inoculate 500 ml of fresh YPD medium in a 2 L flask with 0.5 ml of the overnight culture. Grow for 16-28 h with vigorous shaking until an OD_{600} of 1.5 is reached.

3. Pellet the cells in a cooled centrifuge at 1500 x g for 5 min at 4°C and resuspend the cells in 500 ml of ice-cold, sterile distilled water.

   Note: Care must be taken that the water is close to 0°C. Chill the water in an ice bucket well in advance.

4. Pellet the cells again as in step 3 and carefully resuspend the pellet in 250 ml of ice-cold water.

   Note: Handle the pellet with care, as it gets less compact during the entire procedure.

5. Centrifuge the cells again as in step 4, and resuspend the cell in 20 ml of ice-cold 1 M sorbitol solution.

6. Pellet the cells again as above and finally resuspend in 1 ml ice-cold 1 M sorbitol.

7. Transfer 80 μl of the electrocompetent cells to a 1.5 ml reaction vial on ice and mix with 5 -10 μg of linearized plasmid DNA dissolved in water. Transfer the mixture to a pre-chilled 0.2 cm electroporation cuvette and leave on ice for 5 min.

8. Insert the cuvette into the electroporation device and deliver a pulse at a field strength of 7500 V/cm.

   Note: The GenePulser device was used at the following settings: Voltage, 1500 V; Capacitance 25 μF, resistance 200 Ohm. A successful pulse was indicated by a pulse length of about 4.5 ms. Other electroporation devices may require a different setting.

9. Add 1 ml of ice-cold 1 M sorbitol to the cuvette immediately after the pulse.

10. Plate between 50 and 200 μl of the entire mixture on several RDB plates for selection of His$^+$ transformants.

11. Incubate the inverted plates at 30°C until colonies appear.

    Note: It usually takes 4-6 days until colonies are large enough to be transferred.

Screening for Mut^S phenotype

1. Streak single colonies on a 10 cm MD plate using a 52-square grid, and incubate for 2 days at 30°C.

2. Replica plate the colonies onto MD and MM plates using a velvet Lederberg stamp.

3. Incubate the plates at 30°C and inspect after 16-30 h for growth. Mut^S clones, occurring at a frequency of 5-20%, can easily be identified by their much slower growth on MM plates.

   Note: The identification of Mut^S clones may be omitted, and randomly picked colonies could be used for expression analysis. However, it is strongly recommended, especially for the beginner, to perform expression analysis with Mut^S clones first. There should be no great clone-to-clone variation, and the consumption of methanol in Mut$^+$ clones during induction does not pose a potential problem.

Expression of scFv fragments

1. Grow 4-6 Mut^S clones in 10 ml BMGY medium at 30°C in individual 100 ml Erlenmeyer flasks with glycerol as the sole carbon source. It will take about 2 days until an OD_{660} of 20-40 is reached.

2. Dilute cells 1:10 into 25 ml of BMGY medium containing 1% casamino acids.

3. Incubate at 30°C for an additional 6-8 h with shaking.

4. Pellet cells by spinning the cultures at 1500-3000 x g for 5 min at ambient temperature.

5. Resuspend cells in 25 ml BMMY medium containing 0.5% methanol and transfer suspension into a 300 ml Erlenmeyer flask for optimal aeration.

   Note: If the described setting is used, baffled flasks for aeration are not necessary. For initial screening, 10 ml of the cell supsension in 100 ml Erlenmeyer flask may be used.

6. Incubate at 30°C with vigorous shaking and draw samples of 1 ml after 48 and 72 h.

Note: We did not find it necessary to add additional methanol to the culture during induction. However, if increased evaporation could occur, or cells are not of Mut^S phenotype, it may be advisable to supplement the medium with 0.5% methanol every 24 h.

7. Spin samples for 5 min in a table top microfuge at maximum speed to pellet cells and debris.

8. Remove supernatant and keep frozen at -80°C until analysis.

Analysis of supernatant for secreted scFv by SDS-PAGE or ELISA

1. Analyze 10-20 µl of supernatant by SDS-PAGE. Always include a non-induced control or non-recombinant control for comparison. Visualize proteins by Coomassie brilliant blue staining.

   Note: To verify the identity of the secreted scFv, it is advisable to perform a Western blot using antibodies directed against the peptide tag of the scFv.

2. For ELISA, coat a microtiter ELISA plate with antigen by incubation for 16h at 4°C with 100 µl of diluted antigen solution in PBS per well.

   Note: If coating properties of the antigen are unknown, it may be advisable to test various concentrations between 0.1 and 10 µg/ml with available polyclonal or monoclonal antibodies.

3. Wash off excess antigen with PBS.

4. Fill each well with 200 µl of PBS/1% skimmed milk and incubate for 1 h at ambient temperature in order to prevent unspecific binding of antibodies or immunochemical reagents.

5. During incubation in step 4 prepare serial dilutions of the *Pichia pastoris* supernatants ranging from 1:10 to 1:5000 in PBS-T/1% BSA in separate reaction tubes or a polypropylen plate.

6. Flick the blocking solution from step 4 out of the ELISA plate and wash 3 times with PBS-T.

7. Add to each well 100 µl of the diluted antibody and incubate for 1- 2 h at ambient temperature.

8. Remove the scFv antibody and wash the plate as described for step 6.

9. Add monoclonal or polyclonal antibody directed against the tag according to the manufacturer's specification diluted in PBS/1% BSA and incubate for 1 h at ambient temperature.

   Note: Preferably, biotinylated or antibodies directly labelled with horseradish peroxidase (HRP) should be used here.

10. Remove the secondary antibody and wash the plate as described for step 6.

11. Add 100 µl HRP-labelled antibody or HRP-labelled streptavidin diluted in PBS/1% BSA and incubate for 30 min at ambient temperature.

    Note: This step is omitted when HRP-labelled anti-tag antibody is used.

12. Wash plate 3 times with PBS-T.

13. Detect HRP activity by adding 50 µl of BM blue substrate to the well. Allow blue colour to develop for 10 to 30 min.

14. Stop the reaction by adding 50 µl of 1M sulfuric acid.

15. Determine absorbance at 450 nm using the reference wavelength of 650 nm.

Results

ScFv antibodies can in general be expressed to a high level and secreted into the culture medium by *Pichia pastoris*. This was not only reported by us (Ridder et al., 1995), but also by other groups (Luo et al., 1996; Eldin et al., 1997). Usually, binding activity can be detected by ELISA (Fig.3), and a hexa-His tag facilitates rapid purification of the scFv fragments.

Troubleshooting

Expression of scFv may not be readily detectable in Coomassie blue-stained gels.

To control the conditions used for induction of protein expression, use the control strain GS115/albumin expressing human serum albumin (HSA) at a high level in parallel to the scFv clones. In a successful experiment, the HSA should be clearly visible in Coomassie blue-stained gels. If the expression level of the scFv is lower than 20 µg/ml, the protein samples

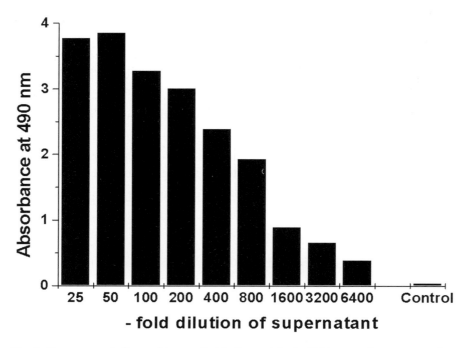

Fig. 3. Measurement of recombinant scFv binding activity in Pichia pastoris supernatant by ELISA. Supernatant from a small-scale expression culture was taken 48 h after induction of gene expression. Samples of serially diluted supernatant were assayed for binding activity to recombinant human LIF in ELISA. Detection of binding of scFv to ELISA plates coated with human recombinant LIF was via a monoclonal antibody directed against the epitope tag and alkaline phosphatase-labelled secondary antibody. Supernatant from a non-recombinant *Pichia pastoris* strain served as a control.

may be concentrated by spin dialysis or precipitation with TCA. In any case, it is recommmended to express the scFv antibody with a tag and perform a Western blot followed by detection with a suitable antibody.

References

Cregg JM, Vedvick TS, Raschke WC. (1993) Recent advances in the expression of foreign genes in *Pichia pastoris*. Biotechnology, 11:905-910.

Eldin P, Pauza ME, Hieda Y, et al. (1997) High-level secretion of two antibody single chain Fv fragments by *Pichia pastoris*. J. Immunol. Methods, 201:67-75.

Gram H, Schmitz R, Ridder R. (1998) Secretion of scFv antibody fragments. Methods Mol. Biol. 103:179-192.

Laroche Y, Storme V, De Meutter J, Messens J, Lauwereys M. (1994) High-level secretion and very efficient isotopic labeling of tick anticoagulant peptide (TAP) expressed in the methylotrophic yeast, *Pichia pastoris*. Biotechnology, 12:1119-1124.

Luo D, Mah N, Krantz M, et al. (1995) Vl-linker-Vh orientation-dependent expression of single chain Fv-containing an engineered disulfide-stabilized bond in the framework regions. J. Biochem. (Tokyo), 118:825-831.

Luo D, Mah N, Wishart D, Zhang Y, Jacobs F, Martin L. (1996) Construction and expression of bi-functional proteins of single-chain Fv with effector domains. J. Biochem. (Tokyo), 120:229-232.

Petersen CE, Ha CE, Mandel M, Bhagavan NV. (1995) Expression of a human serum albumin variant with high affinity for thyroxine. Biochem. Biophys. Res. Commun. 214:1121-1129.

Ridder R, Schmitz R, Legay F, Gram H. (1995) Generation of rabbit monoclonal antibody fragments from a combinatorial phage display library and their production in the yeast *Pichia pastoris*. Biotechnology, 13:255-260.

Sun J, Coughlin P, Salem HH, Bird P. (1995) Production and characterization of recombinant human proteinase inhibitor 6 expressed in *Pichia pastoris*. Biochim. Biophys. Acta:1252, 28-34.

White CE, Kempi NM, Komives EA. (1994) Expression of highly disulfide-bonded proteins in *Pichia pastoris*. Structure. 2:1003-1005.

Suppliers

Invitrogen Inc., Carlsbad, CA

Roche Diagnostics GmbH, Mannheim, Germany

Bio-Rad Laboratories, Hercules, CA

Abbreviations

BSA	Bovine serum albumin
ELISA	Enzyme-linked immunosorbent assay
HRP	Horseradish peroxidase
LIF	Leukemia inhibitory factor
PBS	Phosphate-buffered saline
PCR	Polymerase chain reaction
SDS-PAGE	Sodiumdodecylsulfate-polyacrylamide gel electrophoresis
TCA	trichloric acid

Production of Recombinant Human IgG Antibodies in the Baculovirus Expression System

Mifang Liang and Stefan Dübel

Introduction

As a well-established method, phage display has been widely used to select scFv or Fab antibody fragments specific for many different antigens (see Chapters 9-16 of this book). For some purposes, it is required to reassemble the variable regions of the selected antibodies with constant regions to generate complete recombinant human antibodies. Expression of entire IgG molecules, however, is not feasible in *E. coli*, thus an eukaryotic IgG expression vector system is needed.

The baculovirus expression system has been used for the expression of a wide variety of heterologous genes. Since its first description for Ig expression (Hasemann and Capra, 1990), it has been continuously improved to become one of the most powerful and convenient eukaryotic expression systems available for the production of recombinant chimeric, humanized or human antibodies (Nesbit et al., 1992; Poul et al., 1995; Liang et al., 1997) Using baculovirus systems for antibody production has several advantages over either prokaryotic or transfection-based eukaryotic expression systems. First, the baculovirus expression system provides appropriate signal peptide cleavage, folding and disulfide bond formation. The glycosylation is different from mammalian cells, but although the addition of high mannose oligosaccharides has been reported, it does not appear to affect the immunoglobulin expression. Second, compared to other eukaryotic expression systems, baculovirus expression allows the production of IgGs with yields significantly larger than those typically obtained from transfected eukaryotic cells. About 10 mg secreted human IgG antibodies per liter can be obtained from the supernatant of infected SF9 cells cul-

✉ Mifang Liang, Chinese Academy of Preventive Medicines, Institute of Virology, 100 Ying Xin Jie, Xuan Wu Qu, Beijing, 100052, China (*e-mail* liangmf@public3.bta.net.cn.)
Stefan Dübel, University of Heidelberg, Institute of Molecular Genetics, Im Neuheimer Feld 230, 69120 Heidelberg, Germany

tured in flasks (Liang et al., 1997). Third, it is not necessary to store transfected productive cells in liquid nitrogen or repeatedly passage them, since the recombinant baculovirus stock can be easily stored for weeks at 4°C or for years in normal deep freezers and used to infect fresh cells for the next production batch.

The most common baculovirus strain used in the baculovirus expression system is *Autographa californica* nuclear polyhedrosis virus (AcNPV), which has a double stranded circular DNA genome of 131kb (Miller, 1996). Infectious AcNPV particles enter susceptible insect cells by an endocytosis or fusion mechanism, and the viral DNA is uncoated in the nucleus where the viral DNA is transcribed, replicated and packaged into nucleocapsid protein. During the early phase of viral infection, the infectious viral particles are released extracellularly by budding from infected cell membranes. Late in infection (>48 hours), viral particles are found in occlusion bodies, crystal-like structures mainly composed of a 29 kd protein named polyhedrin, and accumulated within the nucleus of the infected cells. The later phase is marked by a dramatic increase in the transcription of the polyhedrin and p10 genes (Weyer, 1990). The temporarily regulated hyperexpression of the late genes, as polyhedrin and p10, which is nonesssential for baculovirus production, made baculoviruses perfect vectors for the expression of foreign genes. Current commercial baculovirus transfection kits provide linearized modified AcNPV baculovirus DNA which contain a lethal deletion with Bsu361, resulting in removal of an essential part of the open reading frame ORF1629. The antibody gene is supplied on a transfer vector plasmid under the control of polyhedrin and/or p10 promoters and flanked by regions homologous to AcNPV. When the linearized modified AcNPV baculovirus DNA is cotransfected with the shuttle vector plasmid into insect cells, recombination takes place between the homologous regions (ORF603 and ORF1629) of viral DNA and plasmid, resulting in recombinant Baculoviruses carrying and expressing the foreign genes. If the expressed protein is preceded by a signal sequence, the insect cells secrete it into the medium in large amounts. In addition, the recombinant Baculoviruses will be released into the supernatant, from where they can be recovered and stored until they will be used to infect fresh insect cells for a new batch of antibody production.

To achieve expression of complete human immunoglobulins, we have constructed a set of "cassette" baculovirus expression vectors which were designed for the convenient insertion of heavy and light chain genes of Fab or scFv antibodies selected from phage display libraries. The IgG expression vector system (Fig. 1) is based on the backbone vector pAcUW51 (Pharmingen Cat# 21205P), which contains a SV40 transcription terminator, the f1 origin of phage DNA replication, an Amplicillin resistance gene for selection in *E. coli* and the two very late baculovirus expression promoters of polyhedrin and p10, located in back-to-back orientation. The IgG expression elements were cloned separately into pAcUW51 via the BglII and BamHI sites. The heavy chain gene is under the control of the polyhedrin promoter, is preceded by the authentic IgG signal sequence from IgG1 subgloup VHIII and contains the mutant in-frame cloning sites XhoI and NheI for inserting the VH gene of the scFv (Fig. 1A) or XhoI and SpeI for inserting the Fab Fd gene (Fig. 1B). The entire constant region genes of human IgG1 or the Fc region gene of human IgG1, respectively, are located downstream of signal sequence and cloning sites. In opposite orientation to the heavy chain operon, the light chain elements start with the authentic signal sequence of human lambda chain or kappa chain, in the scFv cassette vectors followed by the constant region of lambda or kappa chains (Fig. 1A). The cloning sites for the insertion of lambda or kappa VL genes are SacI and HindIII; or SacI and EcoRV for the insertion of lambda or kappa VL Fab genes. The original HindII site in the vector pACUW51 was removed. Thus, either VL or Fab-L and VH or Fab-Fd genes can be cloned into one of above vectors. This design of the vectors results in the secretion of correctly processed and assembled immunoglobulins from recombinant baculovirus infected insect cells, without apparent differences between the expression level of heavy and light chains.

Outline

The key steps to the successful expression of recombinant IgG antibodies are:

1. The phage library derived Fab or scFv genes must be cloned in frame into the baculovirus expression vectors, the cloning sites of your Fab or scFv gene fragments must match the correct amino acid positions which are calculated from the first start codon of IgG heavy or light chain genes mRNA.

2. The Fab genes derived from pComb3 phagemid vector system (Barbas et al., 1991) can be directly cloned into our expression vectors pAc-L(K)-Fc. To clone scFv single chain genes derived from panning using the pSEX phagemid vector system (Welschof et al., 1997) or others, the primers may need to be designed according to the gene sequences.

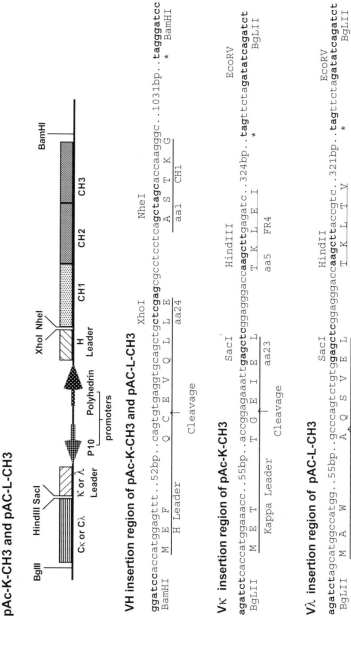

Fig. 1. The antibody coding regions of the baculovirus vectors for human IgG expression used in this protocol. **A,** vectors for the cloning of V region DNA obtained from scFv phage display systems; **B,** vectors for the cloning of Fab region DNA obtained from Fab phage display systems. The vector backbone in all cases (outside (Bgl II / BamHI) is pAcUW51. Abbreviations: aa, amino acid number; C(x), constant human immunoglobulin regions; H, κ or λ leader, sequence coding for the respective human immunoglobuline signal peptides.

B

pAc-K-Fc or pAC-L-Fc

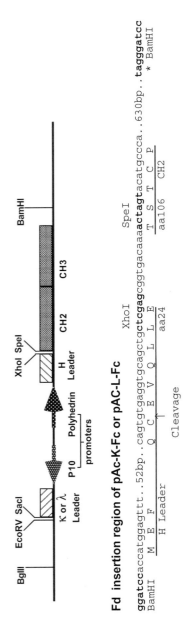

Fd insertion region of pAc-K-Fc or pAC-L-Fc

```
                                                              XhoI                              SpeI
ggatccaccatggagtt..52bp..cagtgtgagtgcagctgctcgagcggtgacaaaactagtacatgccca..630bp..tagggatcc
BamHI  M  E  F     Q  C  E  V  Q  L  L  E           T  S  T  C  P              *  BamHI
       H Leader                  aa24                aa106     CH2
       Cleavage
```

Kappa chain insertion region of pAc-K-Fc

```
                                     SacI          EcoRV
agatctcaccatggagaacc..52bp..accggagaaattgagctcacgcaggatatcagatct
BglII   M  E  T     T  G  E  I  E  L                    BglII
        Kappa Leader          aa23
        Cleavage
```

Lambda chain insertion region of pAC-L-Fc

```
                                        SacI         EcoRV
agatctcaccatggccatgg..55bp..gcccagtctgtggagctcacgcgcaggatatcagatct
BglII   M  A  W     A  Q  S  V  E  L                      BglII
        Lambda Leader          aa23
        Cleavage
```

Fig. 1. Continuous.

3. The sequences of the inserts have to be checked to confirm that no mutation happened in the ORF of complete IgG heavy or light chain genes. The consecutive steps for the baculovirus expression of IgG antibodies can be summarised as follows:

4. Choose appropriate expression vector (depending on the antibody format).

5. Clone Fab or scFv genes into one of expression vectors pAc-l(k)-CH3 or pAc-l(k)-Fc.

6. Propagate and purify the vector DNA containing Fab or scFv genes from *E. coli*.

7. Co-transfer the purified plasmid DNA with linearized AcNPV baculovirus DNA into SF9 insect cells.

8. Verify the production of functional human IgG antibodies in SF9 cells by immunofluorescent assay (IFA) and ELISA.

9. Amplify and plaque purify the recombinant virus, and prepare the recombinant virus stock for storage and reinfection.

10. Use recombinant virus to reinfect insect cells for the production of recombinant IgG immunoglobulin molecules.

11. Affinity purify the expressed IgG from the supernatants of infected cells and characterize recombinant antibodies by SDS-PAGE, Western-Blot, IFA, ELISA etc.

Materials

– 27°C cell culture incubator   **Equipment**

– sterile workbench and basic cell culture equipment

– Immunofluorescence, gel electrophoresis and ELISA equipment

– Cloning and expression vectors: pAc-H(k) -CH3 , pAc-H(k)-Fc, pAc-   **Reagents**
  H(l)-CH3 , pAc-H(l)-Fc

– Insect cell lines: SF9 cells and H5 cells (Pharmingen or Invitrogen)

– *Escherichia coli* DH5α cells (GIBCO-BRL), or XLI- Blue cells (Stratagene)

- Restriction endonucleases: Xho1, SpeI, NheI, HindIII, SacI, EcoRV (New England BioLabs or Life Technology)

- PCR reagents: ELONGase Enzyme Mix and buffers (Life Technology), 10 mM dNTP

- Insect cells culture media: TC100 or TNF-FH, IPL41, Sf900-II SFM (Serum Free Medium) (Life Technology or Pharmingen)

- QIAGEN Spin Miniprep kit (Cat# 27104) , QIAGEN-tip 100 Midiprep kit (Cat# 12143 or 12144), QIAquick Gel Extraction Kit (cat # 28704)

- BaculoGold DNA (PharMingen Cat #21227M), Transfection buffer A and B set (PharMingen cat# 21483A)

- Low temperature melting agarose (Agarplaque Agarose, PharMingen Cat# 21403A)

- Protein-free insect medium (PharMingen Cat# 21228M)

- T25, T75 cell culture flasks, spinner flasks, 60mm tissue culture plates, 6-wells tissue culture plates, 96-wells tissue culture plates (Greiner)

- 96-well polyvinylchloride ELISA plates (Nunc, Wiesbaden, Germany)

- Skimmed milk powder (Oxoid, Basingstoke, UK)

- TMB (3,3',5,5'-Tetramethylbenzidine) substrate solution (Sigma Aldrich, Steinheim, Germany)

- 10 or 12 wells glass slides for microscopy

- PBS (140 mM NaCl, 20 mM phosphate, pH7.4)

- FITC or HRP or non conjugated anti-human IgG , Fc specific, Fab specific antibodies (Sigma, Deisenhofen, Germany)

- Protein G-sepharose affinity column (Pharmacia Biotech) and suitable pump

Procedure

Selecting an appropriate expression vector for IgG expression

As we described in the instruction, the vector system was specially designed for cloning of Fab or scFv genes obtained from a phage display library.

- If your antibody genes were derived from scFV single chain antibody expression system, use vectors pAc-KL-CH3. If your light chain gene is kappa, the pAc-K-CH3 vector is recommended. If your light gene is lambda, the vector pAc-L-CH3 is recommended.

- If your antibody genes were derived from Fab fragments antibody expression system, use vectors pAc-KL-Fc. If your light chain gene is kappa, the vector pAC-K-Fc is recommended. If your light gene is lambda, the vector pAC-L-Fc is recommended.

PCR Amplification of the variable region of light and heavy chain genes

Primer design for the PCR amplification of VH and VL genes
The vectors pAC-K(L)-Fc were specially designed for cloning from the pComb3 phage display system. If the cloning sites of your antibody genes do not match our baculo vector system, you have to design primers according to the sequences of your antibody genes. The key issue for your cloning strategies is that the insert gene must be in frame of the vector that you are going to use. As Fig 1 indicated, the heavy chain 5' cloning site XhoI must start at 72bp, at amino acid (aa) position 24 since ATG of the IgG1 gene amplified from mRNA, with CTC=Leu; the 3' cloning site NheI must start at first amino acid of constant region, with GCT=Ala; SpeI must start at 318 bp, aa106 of constant region, with ACT=Thr; the kappa or lambda 5' cloning site SacI must start at 69bp, aa23 (ATG of kappa or lambda gene from mRNA); the 3'cloning site HindIII must start at 18bp, aa6 (first bp of FR4), with AAG=Lys. Table 1 outlines the reference primers for cloning VH and VL genes into our vector system.

2. PCR amplification of ScFV or Fab antibody genes

DNA (plasmid)	1 μl (0.1 μg)
Forward Primer	1 μl (60 pm)
Reverse Primer	1 μl (60 pm)
E-longase (GIBCO-BRL)	2 μl
Buffer A	10 μl
Buffer B	10 μl
H_2O to	100 μl

Table 1. Primers for cloning human antibody genes into baculovirus IgG expression vectors

---

Human IgG variable chain 5' primers

VH3a	5'-GAG GTG CAG CTC GAG GAG TCT GGG-3'*
VH1f	5'-CAG GTG CAG CTG CTC GAG TCT GGG-3'*
VH3f	5'-GAG GTG CAG CTG CTC GAG TCT GGG-3'*
VH4f	5'-CAG GTG CAG CTG CTC GAG TCG GG-3'*
VH6a	5'-CAG GTA CAG CTG CTC GAG TCA GG-3'

Human kappa variable chain 5' primers*

VK1a	5'-GAC ATC GAG CTC ACC CAG TCT CCA-3'
VK3a	5'-GAA ATT GAG CTC ACG CAG TCT CCA-3'
VK2a	5'-GAT ATT GAG CTC ACT CAG TCT CCA-3'

Human Lambda variable chain 5' primers

VL1	5'-AAT TTT GAG GAG CTC CAG CCC CAC-3'
VL2	5'-TCT GCC GAG CTC CAG CCT GCC TCC GTG-3'*
VL3	5'-TCT GTG GAG CTC CAG CCG CCC TCA GTG-3'*
VL4	5'-TCT GAA GAG CTC CAG GAC CCT GTT GTG TCT GTG-3'*
VL5	5'-CAG TCT GTG GAG CTC CAG CCG CCC-3'
VL6	5'-5'-CAG ACT GAG GAG CTC CAG GAG CCC-3'*

Human IgG (Fd) heavy chain 3' primers (SpeI) *

CG1z	5'-GCA TGT ACT AGT TTT GTC ACA AGA TTT GGG-3'

Human IgG variable chain 3' primer

NheI-HR	5'- TGG GCC CTT GGT GCT AGC TGA GGA GAC GGT GACC-3'

Human lambda variable chain 3' primer

HindII-LR	5'- GAC GGT AAG CTT GGT CCC TCC-3'

Human Kappa variable chain 3' primer

HindIII-KR1	5'-CAG TTC GTT TGA TTT CAA GCT TGG TCCC-3'
HindIII-KR2	5'-CAG TTC GTC TGA TCT CAA GCT TGG TCCC-3'

Human Lambda chain 3' primer

EcoR-CL	5'-CCG GAT ATC TAG AAC TAT GAA CAT TCT GTA GG-3'

Human Kappa chain 3' primer

EcoRV-CK	5'- CCG GAT ATC TAG AAC TAA CAC TCT CCC CTG TTGA-3'

---

* Primers identical to the pComb 3 system primers (Kang et al., 1991)

PCR reaction: 94°C 2 minutes, followed by 25 cycles with 94°C 50", 54°C 50", 72°C 1 min, followed by a final incubation at 72°C for 10 min, cool down to 4°C .

3. Purify the PCR products via agarose gel electrophoresis using a QIA-quick Gel Extraction Kit.

Cloning into a baculovirus IgG expression vector

1. To clone the light chain gene into the IgG expression vector, double digest the gel purified PCR products of your light gene (or your phagemid DNA) and the vector pAc-k-CH3 or pAc-L-CH3 with SacI and HindIII (For scFv genes) or the vector pAc-k-Fc or pAc-L-Fc with SacI and EcoRV (for Fab genes). Ligate the insert DNA with the vector DNA at concentration molar ratio of 1:4. Ligate with 1 U ligase (Life Technology) at 16°C overnight. Following the ligation reaction, transform the ligated plasmid DNA into competent cells of *E.coli* XLI-Blue or DH5α, plate the bacteria on LB plates containing 100ug/ml ampicillin, incubate at 37°C overnight. Pick colonies for DNA miniprep and restriction endonuclease analysis to select the clones with insert.

2. To clone the heavy chain gene into the IgG expression vector that contains the whole light chain gene, double digest gel purified VH or Fd PCR products (or your phagemid DNA) and the vector DNA containing your light chain insert with XhoI and NheI (for the VH gene) or XhoI and SpeI (for Fd gene). Ligate the insert DNA with the vector DNA at concentration molar ratio of 1:4. The subsequent steps are identical to those of step 1.

3. Prepare the plasmid DNA by using the QIAGEN miniprep kit and sequence the inserts of the new constructs obtained from step 2 or 3. Use following primers for sequencing:
   - For VH and Fd:
     Forward Primer: 5'-TCCACCATGGAGTTTGGGCTGAGC-3'
     Reverse Primer: 5'-TCCCATGTGACCTCAGGGGTCCGGGAGAT-3'
   - For VL or VL-CL:
     Forward Primer: 5'-CCGGGACCTTTAATTCAACCCAACAC-3
     Reverse Primer: 5'-GGCAGACATGGCCTGCCCGGTTAT-3'

4. Propagate and purify the plasmid DNA for co-transfection. The quality of the plasmid DNA is critical for successful co-transfections. We re-

commend using the QIAGEN-tip100 Midiprep Kit. The IgG expression vectors are high copy number vectors, usually, 40 ml of a 10-12hours culture of bacteria should yield a total of 100 µg DNA per column (tip-100) with a concentration of at least 0.5 µg/ul.

Notes
- Generally, we suggest cloning the light chain first, followed by the heavy chain. When you start to clone your Fab or scFv antibody genes into our IgG expression vectors, however, you may need to check the variable region sequences for internal restriction sites and decide upon cloning strategy.

- For cloning the light chain gene of Fab antibody, make sure your light chain gene fragment contains a stop codon. If not, you need to reamplify your gene with the respective primer (see table 1). If your gene comes from a pComb3 phage display system, you can digest your gene from plasmid DNA with XbaI first, make it blunt, then digest it with SacI.

- It is very important to sequence your insert genes completely to make sure that no open reading frame mutation occurred after the insertion of your genes.

- When you are preparing the high quality plasmid DNA for co-transfection, the bacteria should not be growing too much into the stationary phase, i.e., the O.D $_{600\,nm}$ value of the bacterial culture should not exceed 1.5.

Preparation of insect cells

Insect cell lines
Several insect cell lines are established to be highly susceptible to AcNPV virus infection. The two most frequently used insect cell lines for virus infection and protein expression are Sf9 and H5 cells. Sf9 cells are most frequently used for co-tranfection, recombinant virus propagation and protein expression, H5 cells are generally used for high expression of foreign proteins.

Thawing frozen insect cells
1. Remove a vial from liquid nitrogen and thaw it rapidly with gentle agitation in 37°C waterbath.

2. Transfer the 1 ml of cell suspension into 7 ml of complete TC-100 medium (other insect cell culture medium, such as Grace's, TNM-FH or IPL41 are suitable as well) in a 15 ml centrifugation tube, centrifuge at 230xg , 4°C for 5-8 min.

3. Carefully discard the supernatant, resuspend the cells in 5 ml (for $1X10^6$ cells/ml) or 15 ml (for $1X10^7$ cells/ml) of complete TC-100 medium, transfer the cell suspension into a T25 ($25cm^2$) or T75 ($75\ cm^2$) flask.

4. Put the flask in a 27°C incubator and allow the cells to attach to the bottom of flask. Change the medium after 12 hours or 24 hours. Viability of the recovered cells should be more than 70%. Check the cells daily until a confluent monolayer has formed. Once a confluent monolayer has formed, the cells can be divided (see next step).

1. Sf9 cells can be cultured either in flasks or in spinner flask (Suspension culture) at 27 °C. To maintain a healthy culture, the cells should be passaged 1:3-4 when they reach confluent monolayer in T25 or T75 flask, or reach $1x10^6$ cells/ml in suspension culture. Complete insect cell culture medium with serum can be used to subculture cells in both culture systems, the serum free medium SF-900-II (GIBCO-BRL), however, can only be used to subculture cells in suspension culture.

Culturing and passaging insect cells

2. To passage cells from regular flasks, remove all but 3 ml or 9 ml of medium from the T25 or T75 flask with confluent monolayer cells, respectively. Dislodge the cells by gently blowing down medium from a 10 ml pipette while sweeping the tip of the pipette across the monolayer from side to side. Start at the bottom and work your way to the top of the flask.

3. Transfer the 3 ml (or 9 ml) of cells suspension into 3 T25 (or T75) flasks containing 4 ml (or 12 ml) of fresh complete medium. Rock gently to humidify the entire growth surface and distribute cells evenly.

4. Incubate at 27°C until the cells are confluent, then passage as described above.

1. Cells must be frozen in their logarithmic growth phase at 98% viability.

Freezing insect cells

2. Remove cells from flasks and transfer into 15 ml (or 50 ml) centrifugation tubes, pellet down the cells by centrifugation at 230 xg for 10 minutes, carefully remove all of the supernatant.

3. Gently resuspend cells in certain volume of cells storage solution containing 10% dimethylsulfoxide (DMSO) and 90% fetal bovine serum, at final cell density of $1x10^7$ cells/ml.

4. Aliquot the cells into cryogenic tubes and place tubes into a slowly cooling container, store the container at -70°C overnight. The tubes are then very quickly transferred to liquid nitrogen for long term storage.

Notes
- Sf9 cells can be grown as monolayer or suspension cultures and transitioned repeatedly between either type of culture without significant changes in cell viability or growth rate. After 30-35 passages, if the cell viability or growth rate tends to decrease over time, we recommend that you recover new Sf9 cells from your storage.

- Sf9 cells are sensitive to centrifugal forces and any other forces. Do not exceed 250 xg in centrifugation. When dislodging cells from flasks, avoid creating aerosols that can promote cell damage.

- Always allow your cell culture medium to acquire room temperature before use.

Generation of recombinant Baculoviruses by co-transfection

1. Prepare at least 10 µg of highly purified plasmid DNA using the QIAGEN tip-100 midiprep kit. Resuspend the DNA in pure water with a final concentration of at least 0.5 µg /ul.

2. Seed $2x10^6$ Sf9 cells into 60 mm tissue culture plates and incubate the plates at 27°C for 20-30 minutes. The cells should be attached and form a flat and even surface in the plate. If cells do not attach well after 30 minutes, they are not suitable to be used for transfection.

3. To co-transfect your plasmid DNA with linerized baculovirus DNA, combine 0.5 µg (5 µl) BaculoGold DNA (Pharmingen, Cat# 21100D) and 2-5 µg (5-10 µl) plasmid DNA in a microcentrifuge tube, mix well by pipetting up and down, incubate the mixture at room temperature for 5 min, then add 1 ml of transfection Buffer B (Pharmingen Cat# 21483A) to the mixture.

4. Remove the medium from the cells attached on the co-transfection plate and replace with 1 ml of transfection Buffer A (Pharmingen, Cat# 21100D). Make sure that all cells are covered to prevent their dehydration.

5. Slowly add 1 ml of above mixture of Transfection buffer B plus DNA drop by drop to the cotransfection plate. After every three drops, gently

rock the plate to mix the drops with medium. After the addition of transfection buffer B plusDNA, the medium in the plate should appear slightly cloudy due to the calcium phosphate/DNA precipitate.

6. Incubate the cotransfection plate at 27°C for 4 hours, remove the medium from the plate, wash the plate with 4 ml of fresh TC-100 complete medium (or the insect cell culture complete medium) by gently rocking the plate. Remove the wash medium, add 3.5 ml of fresh medium to the plate and incubate at 27°C for 4-5 days.

7. After 4-5 days, check the plate for signs of infection (larger cells, enlarged nuclei - for comparison, keep a nontransfected culture in parallel. Please note that the appearance of cytopathic effects in recombinant virus infected Sf9 cells is delayed in comparison to wildtype virus infected cells, and does not contain crystal like inclusion bodies). Harvest the supernatant medium, it is the original recombinant virus stock, designated P0 (passage zero). This stock can be stored at 4°C for at least half a year. The transfected Sf9 cells can be used to prepare cell slides to check for antibody expression by immunofluorecence staining (see below).

• The density of the Sf9 cells in co-transfection plate should not be over 60-70%, since in 24-48 hours after co-transfection, uninfected cells still will divide and can be infected.   **Notes**

• Always keep transfection buffer B on ice before adding to the mixture of Baculovirus DNA and plasmid DNA in order to maintain high transfection efficiency.

Fast detection of recombinant IgG expression in SF9 cells after transfection

1. Remove medium from recombinant baculovirus infected SF9 cells attached on 60 mm co-transfection plates or T25 flask, gently wash the cells twice with PBS.   **Immunofluorescent assay**

2. Add 3-3.5 ml PBS, gently dislodge cells from monolayer by pipetting, drop the cell suspension to multiple well slides. Air dry at room temperature.

3. Put the slides into a slide fixing glass container and cover the entire slide with acetone for 10 min at room temperature for fixation. Remove the slides from the container. Air dry.

4. Add 25-30 µl of FITC-conjugated anti-human IgG, Fc or Fab antibodies per well (concentration according to manufacturer), incubate at 37°C for 30 minutes.

5. Wash the slide 3 times with PBS, and observe by regular or confocal immunofluorescent microscopy.

Notes
- Infected insect cells must be washed with PBS before fixing the cells to multiple slides since cells cultured in medium containing serum are diffucult to fix to the slides.

- The observation time can be prolonged by embedding the stained cells in 50% glycerol in PBS (or commercial embedding media, e.g. Moviol).

Detection by sandwich ELISA

1. Dilute first antibody (unconjugated anti-human IgG Fc or Fab antibody) in 0.1M NaHCO$_3$, pH 8.6 coating buffer to final concentration about 5- 10 µg/ml. Coat the unconjugated antibodies to the wells of a 96 well polyvinylchloride ELISA plate by adding 100 µl of the diluted antibody solution per well. Incubate at 4°C overnight.

2. Wash the wells three times with PBS.

3. Dilute the supernatant medium from recombinant virus infected SF9 cells 1:1 with 3% skimmed milk powder in PBS. Add 100 µl of the dilution to each well, and incubate at 37°C for 1 hour.

4. Wash the wells three times with PBS.

5. Add 100 µl of diluted HRP-conjugated anti-human IgG antibody. The amount of conjugated second antibody depends on the recommendation of the manufacturer. Incubate the plate at 37°C for 1 hour.

6. Wash the wells 5 times with PBS.

7. Develop the color by adding TMB substrate solution and detect the O. D value at 490 mm.

Purification and titration of recombinant baculoviruses by plaque assay

The plaque assay can be used to purify recombinant virus in order to increase virus titer and to determine recombinant virus titer in plaque-forming units per milliliter (pfu/ml).

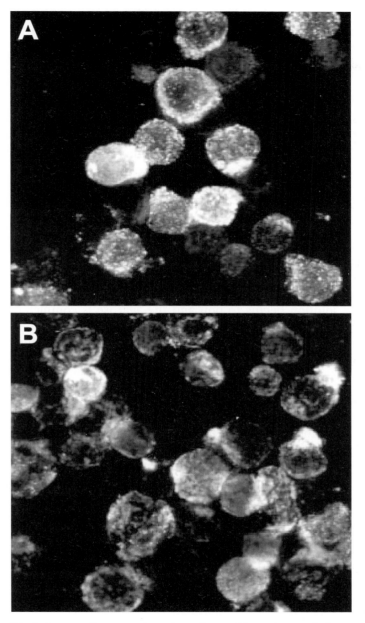

Fig. 2. Immunoflorescent detection of recombinant human IgG expressed in SF9 cells. **A,** detection by FITC conjugated anti-human IgG Fc antibodies. **B,** detection by FITC conjugated anti-human IgG Fab antibodies.

1. Use a six-well cell culture plate. Remove the medium from a T75 flask with a confluent monolayer of SF9 cells. Add 20 ml of fresh TC-100 complete medium, dislodge the cells by gently pipetting. Add 3 ml of the SF9 cell suspension to each well (about 2×10^6 cells) of the six-well cell culture plate. Allow the cells to attach to the plate at 27°C for 1 hour.

2. Prepare a dilution series (from 10^{-2} to 10^{-7}) of the recombinant virus stock with SF-900 II medium. Carefully remove the medium from the six well plate, add 200 μl of the respective virus dilution per well. Make sure that the entire bottom surface of the well is covered with the solution. Incubate at 27°C in a humid atmosphere for 1 hour (e.g. by putting the six well pate into a larger box containing a moistured tissue) to allow infection of the cells by the virus particles.

3. During the above incubation, prepare a 2% agarose solution of cell culture quality low melting point agarose (Agarplaque Agarose, Pharmingen, cat# 21403A or Seakem ME, FMC Corp) in protein-free medium (Pharmingen Cat#21228M#), heat the solution in a microwave oven until the solution begins to boil and the agarose is completely dissolved (about 30 seconds to 1 minute). Cool to 40°C by putting into a waterbath 15 minutes before use. Let the Sf-900-II SFM medium (Life Technology) warm up to room temperature (takes approx. 20 minutes).

4. Mix equal volumes of agarose solution and Sf-900-II SFM medium. The final solution therefore contains approx. 1% agarose. Apply a first agarose layer onto the plate by carefully adding 3 ml of the mixed argarose solution per well to the side of the titled plate. Allow the agarose to harden to a levelled surface. Incubate the plate in a humid environment at 27°C for 4-5 days.

5. On day 5 to 6, prepare a second layer of agarose similar to the first layer of agarose described above but with the addition of 2% Neutral Red Solution (GIBCO-BRL). Incubate at 27°C for 12-24 hours.

6. Plaques can be observed on the next day after adding second layer containing Neutral Red solution. Plaques can be visualized by putting the plate against natural light or white illuminating light. Count the plaque number. Virus titers can be determined using the following formula:

pfu/ml = Plaque number x highest dilution x 5.

7. *(The next steps are required for recombinant virus purification only)* Remove all medium from one T75 flask containg more than 90% confluent SF9 cells, add 30 ml of fresh medium, dislodge the cells, seed 1 ml of the cell suspension into each well of 24 -well cell culture plate, let cells attach at 27 °C for 30 min.

8. Remove all medium, replace with fresh medium (culture medium or SF-900-II serum free medium). Mark plaques on the back side of the agar plates from step 5 with a color pen. Pick up the plaques by using a sterile Pasteur pipette and bulb. Carefully penetrate the agarose containing a recombinant virus plaque, and transfer the agarose plug into a well of the 24 well plate with medium, pick at least 5 plaques of each transfection. Incubate the plate at 27°C in a humid environment until visible infection appears (5-7 days).

9. Harvest the supernatant, store at 4 C. These clonally purified stocks are designated P1. Check the cells in each well to make sure the P1 stocks from positive plaques do not contain polyhedra.

• The quality of the SF9 cells used for plaque assay and the cell density of confluent cell monolayers are very important factors for efficient plaque formation. The viability of the cells should be more than 98% and the density of the monolayer cells should be around 70%. Use of cell densities higher or lower than 70% will be not result in efficient plaque formation.

• To obtain high titer stocks and stable recombinant virus, plaque purification is always recommended.

Preparation and amplification of high-titer viral stock

1. Seed a T25 flask with $2x10^6$ log phase SF9 cells in 5 ml of complete TC-100 medium. Allow the cells to attach at 27°C for one hour, then change the medium (serum medium or serum free medium are ok).

2. Add 50 µl of P1 stock from the plaque purification to the T25 flask, incubate at 27°C for 5-7 days or until the cells are completely lysed. Transfer the infected medium into a sterile 15 ml centrifuge tube, centrifuge for 1000 xg for 10 minutes at 4°C, harvest the supernatant and store at 4°C as P2 stock.

3. Infect T75 flasks with each 200-300 µl of P2 recombinant virus stock, incubate at 27°C for 5-6 days, harvest the supernatant as step 2. Aliquot the virus stock, store part of them at -70°C for long term storage, keep the rest at 4°C as reserve stock. Baculovirus will be stable at 4°C at least half year, usually even more than one year.

4. *(Optional step instead of step3 for the generation of larger amounts of virus)* Infect a 250 ml suspension culture of log phase SF9 cells with 2.5 ml of P2 recombinant virus stock, put the flask in an incubator shaker at 27°C, with 80 rpm shaking for 7-10 days. Transfer the cell suspension to a centrifuge tube, centrifuge at 4°C, 1000 xg for 15 minutes, collect the supernatant, store at -70°C or 4°C as described in step 3.

5. Determine the virus titer of your stocks by plaque assay (see above). The virus titer of recombinant baculovirus virus stock after amplification usually reaches $1x10^8$/ml. The use of this high titer recombinant virus is optimal for high level expression of IgG antibodies.

Note   Always check the IgG expression for every amplifying step by the immunofluorescent assay as described above. PCR may also be used to monitor the virus replication.

Expression of recombinant antibodies in insect cells

1. Seed T75 culture flasks with about $3x10^7$ SF9 cells or H5 cells per flask.

2. For optimal antibody production, infect cells with a multiplicity of infection (MOI) of 10^{-3} (see notes) of high titer recombinant baculovirus. Usually, for virus stocks with a titer around $0.5 x10^7$ to $1x10^8$ or higher, add 2 ml of recombinant virus into one T75 flask containing > 90% confluent SF9 cells or 100% confluent H5 cells.

3. Incubate the cells at 27°C for 4-5 days. Examine the cells for viability and cytopathic effects. The cells should be enlarged in size and large nuclei appear visible inside of cell. The viability of total cells should be around 30-40% at the time of harvest. Higher or lower viabilities will result in lower yields of antibodies.

4. Harvest the culture supernatant. Recombinant human IgG antibody is produced in SF9 or H5 cells as a secreted protein. After centrifugation at 1000 xg for 15 minutes, keep the supernatant at 4°C for further purification and characterization.

5. Check the cells in the flask by the immunofluorescent assay described above. Fig. 2 shows a typical immunofluorescence of recombinant human IgG in SF9 cells stained with FITC conjugated anti human IgG Fab (A) or Fc (B) antibodies.

- To obtain optimal protein production, the cells should be infected at a particular multiplicity of infection (MOI = 10^{-3}). Therefore, it is essential to know the concentration of the recombinant virus to be inoculated. MOI can be calculated using the following formula:

$$\text{MOI (pfu/cell)} = \frac{\text{Virus titer (pfu/ml) X ml of inoculum}}{\text{Number of total cells to be infected}}$$

Accordingly, the volume of viral inoculum can be calculated from the following formula:

$$\text{ml of inoculum} = \frac{\text{MOI X number of total cells to be infected}}{\text{Virus titer (pfu/ml)}}$$

- Both SF9 and H5 insect cells can be used to express recombinant human antibodies. Although H5 cells are particularly well suited for expression of secreted recombinant protein, IgG expression with our vector system in H5 cells did not differ significantly from that observed using SF9 cells. The secreted IgG protein can be expressed at levels that approach 10-20 mg per liter.

Notes

Affinity purification of recombinant human IgG

1. Harvest 100- 500 ml of the supernatants of recombinant virus infected Sf9 or H5 cells from flask or suspension culture 4-5 days after virus infection. Clear the supernatants by centrifugation at 1000 xg at 4°C, and filter through 0.45μm filters. Store it at 4° C. Warm to room temperature before loading (step 3).

2. Equilibrate a 1 ml (or 5 ml) prepackaged protein G-sepharose affinity column (Pharmacia Biotech) to room temperature before use. Equilibrate the column with at least three volumes of start buffer (20 mM sodium phosphate, pH 7.0).

3. Apply at least 100 ml (or 500 ml) of above filtered supernatants onto 1 ml (or 5 ml, respectively) affinity column by using a syringe or by slow pumping.

4. Wash the column with 5 column volumes of start buffer.

5. Elute with 3 column volumes of elution buffer (0.1 M glycine-HCL, pH 2.7). Collect 1 ml fractions. To each fraction, add about 25-30 µl of 1 M Tris-HCL, pH 9.0, to adjust the fractions to pH 7-8.

6. Estimate the IgG concentration by determining the absorption at 260 nm and 280 nm in a photometer using the formula:
    IgG concentration mg/ml = (1.55 x A280) - (0.76 x A260).
    The eluted IgG can be concentrated by using centricon-10 protein concentrators (Amicon). Usually, the peak concentration of the eluted IgG fractions reaches or exceeds 200 µg/ml. Therefore, the concentration step may not be necessary for the subsequent experiments.

7. Check the purity of IgG protein by SDS-PAGE under reducing and non-reducing conditions.
    A typical result is shown in Fig 3.

Fig. 3. SDS-PAGE and immunoblot analysis of recombinant human IgG produced by SF9 cells. A,B, coomassie blue stain, C, Immunoblot. Antibodies purified from the supernatants of recombinant baculovirus infected SF9 cells were analysed by SDS-PAGE using reducing (A) or nonreducing (B) conditions. Lane A-1, size Marker, lane A-2, IgG from vector pAc-L-CH3; lane A-3, IgG from vector pAc-L-Fc. Lane B-1, IgG from vector pAc-L-CH3, lane B-2, IgG from vector pAc-L-Fc. Lane B-3, Control (Human IgG franction obtained from Sigma, Deisenhofen, Germany). C, Immunoblot staining with HRP conjugated anti-human IgG (Lane C-1, IgG from vector pAc-L-CH3, lane C-2, IgG from vector pAc-L-Fc).

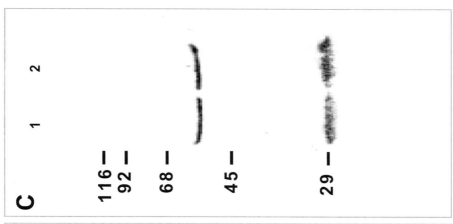

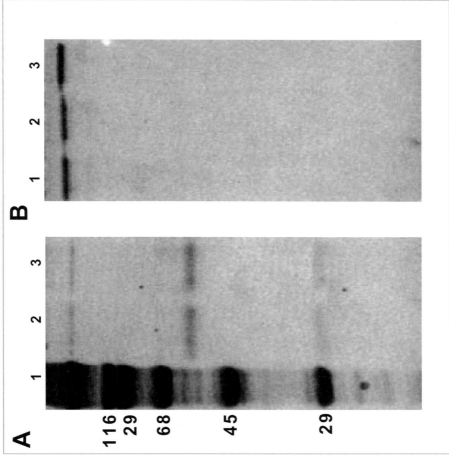

References

Barbas III CF and Lerner RA.(1991) Combinatorial Immunoglobulin Libraries on the surface of phage (Phabs): Rapid selection of antigen -specific Fabs

Hasemann CA; Capra JD (1990) High level production of a functional immunoglobulin heterodimer in a baculovirus expression system.Proc Natl Acad Sci U S A. 87: 3942-3946

Kang AS, Burton DR and Lerner R. (1991) Combinatorial immunoglobulin libraries. Methods: A Companion to Methods in Enzymology, 2: 106:110.

Liang MF, Guttieri Mary, Lunkvist A and Schmaljohn CS. (1997) Baculovirus expression of a human G2-Specific, neutralizing IgG moclonal antibody to Puumala virus. Virology 235:252-260

Lois K. Miller. (1996) Insect Virus. Fields Virology, Third Edition, Lippincott-Raven Publisher, p533

Nesbit M., Fu ZF., McDonald-Smith J., Steplewski Z and Curtis P (1992) Production of a functional monoclonal antibody recgnizing human colorectal carcinoma cells from a baculovirus expression system. J.Immuno.Methods 151:201-208

Poul MA, Cerutti M., Chaabihl H., Devauchelle G., Kaczorek M and Lefranc MP (1995) Design of cassette baculovirus vectors for the production of therapeutic antibodies in sect cells Immunotechnology 1:189-196.

Welschof, M., Terness, P., Kipriyanov, S., Stanescu, D., Breitling, F., Dorsam, H., Dübel, S., Little, M. and Opelz, G. (1997). The antigen binding domain of a human IgG-anti-F(ab')2 autoantibody. Proc.Natl.Acad.Sci. USA 94, 1902-1907

Weyer G., Knight S and Possee RD (1990) Analysis of very late gene expression by Autographa californica nuclear polyhedrosis virus and the further development of multiple expression fectors. J. Gen. Virology 71, 1525-1534

Expression of Antibodies in Mammalian Cells

ANDREW BRADBURY

Introduction

When expressing the minimum binding region of antibodies, comprising
VH and VL domains, in either the scFv or Fab formats, the use of bacteria
is usually sufficient for most research purposes. Expression levels are relatively high (0.1-10 mg/litre) and the use of tags, such as the his tag (which
permits purification by immobilised metal affinity chromatography), or
one recognised by monoclonal antibodies, such as Myc1-9E10 (which re-
cognises the myc tag), allow purification and detection of the antibodies.
In some cases, though, bacterial expression is not appropriate. This is particularly true when complete antibodies are required for therapeutic, diagnostic or immunological applications. Sensitive biological screening assays such as viral or cytokine neutralization are often required to identify
the best of a panel of selected phage antibodies. Such tests may be compromised by contaminants such as LPS in bacterially produced antibodies.
Furthermore, complete antibodies have longer serum half lives, and provide effector functions encoded by the Fc portion. Such full length antibodies are difficult to produce in bacteria, usually resulting in low yields of
poorly folded, unglycosylated and aggregated product.

The sources of V regions which are to be expressed in mammalian cells
are nowadays most likely to be phage antibody libraries, either large naive
libraries, or smaller ones created from immunised sources or by mutation
of appropriate lead antibodies, although humanised antibodies derived
from traditionally obtained hybridomas are also important. Assuming
that one is starting from either scFv or Fab fragments, a number of factors
need to be taken into account when making the transition from bacterial

Andrew Bradbury, Los Alamos National Laboratory, Biosciences Division, MS-M888,
Los Alamos, NM, 87545, USA (*phone* +1-505-665-0281/0287; *fax* +1-505-665-3024 /
667-2891; *e-mail* amb@telomere.lanl.gov), and SISSA, Via Beirut 2-4, Trieste, 34014,
Italy

recombinant V region manipulation to mammalian complete antibody expression. These include:

1. The antibody constant region to be used

2. The promoter to be used

3. Transient or stable expression

4. Mammalian selectable markers

5. Testing antibody production

Recently an integrated vector system (IVS) has been created which permits direct cloning of either V genes, or Fab fragments, directly from selected phage antibody clones (Persic et al. 1997, see Fig. 1). This vector system exploits restriction sites which are rare in immunoglobulin genes to clone either V genes or Fab fragments into either a heavy chain or a light chain expression vector. The promoter used is the EF-1α promoter, a strong constitutive promoter which permits expression in all cell types. The constant regions, which are in the genomic format, are flanked by rare restriction sites which permit the exchange of the constant region and so the expression of antibodies with any constant region. Similarly, the selectable markers and promoter may be removed and exchanged using the appropriate restriction sites. Two vectors were made, one to express heavy chains and one to express light chains (Fig. 1).

The constant region to be used depends very much upon the application. Antibodies to be used in intravenous therapy, will tend to have the IgG isotypes, while for those prospected for oral use (for example to treat dental caries) IgA is likely to be better. Positive controls in diagnostic tests can also be made from recombinant antibodies, and in this case, the isotype of the antibody used will depend upon the isotype of the serum antibodies to be detected. The IgM isotype is likely to be useful when high multivalency is required. cDNA or genomic constant region constructs can be used, although it is generally felt that at least one intron should be present to maintain high expression levels. In the integrated vector system described above, genomic constructs are used. These have the disadvantage that the plasmids are somewhat larger, but the advantage that restriction sites in each of the introns permit the creation of variations of normal antibodies such as immunoadhesins (Byrn et al. 1990).

A

BACKBONES

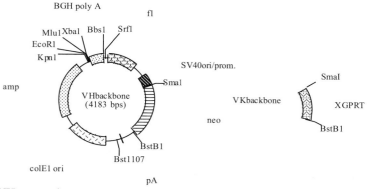

B VH expression cassette

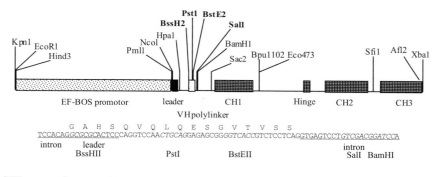

VK expression cassette

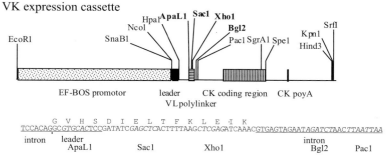

Fig. 1. VHExpress and VKExpress expression cassette maps. *(A)* Maps of the VH and VK backbone plasmids used to create the final constructs. The different functional cassettes are indicated together with the restriction sites flanking them. *(B)* The structure and sequence of the VHExpress and VKExpress polylinkers. Indicated are also the introns and the C terminal four amino acids of the leader sequence which forms part of the V region exon. V region cloning sites are indicated in bold.

Promoter to be used

For standard expression of full length antibodies, a strong promoter should be used. The choice here is between a general promoter which is active in all cells, such as the CMV promoter, or the EF1α promoter (Mizushima and Nagata 1990), an inducible one such as that based on the recombinant tetracycline system (Gossen and Bujard 1992), or cell type specific promoters. Ig promoters may be useful when full length antibodies are to be expressed in myeloma cells, while general promoters allow the same construct to be used in many different cell types.

Transient or stable expression

Expression in transient cell systems [such as COS, see (Trill et al. 1995) for a discussion] has the advantage that enough full length antibody (up to 4 μg/ml) can be produced two to three days after transfection for most analytical purposes. To permit transient expression in COS cells (or other cells expressing the large T antigen), the SV40 origin of replication is required. This is found in the IVS described above. This allows many constructs to be tested in a true mammalian cell context, allowing the selection of the best antibody to be then expressed in stable cell lines. The disadvantage, however, is that it is not feasible to repeat multiple transient transfections when larger quantities of antibody are required.

For this the establishment of stable cell lines is more appropriate. The protocol given below is for transfection using calcium phosphate with reagents made in the laboratory, however, the number of kits provided by many different manufacturers has proliferated, and it is reccommended that these are used instead, since they tend to be extremely reliable. However, no single kit can be guaranteed for all cell lines, so if one kit does not work another should be tried. The yields obtained with these (and similar vectors) range from 200-500 ng per million cells per day with stable clones. This allows the accumulation of up to about 5 μg/ml of antibody, levels similar to those obtained after three days of a good transient transfection in COS cells.

Mammalian selectable markers

When establishing cell lines, it is preferable to have a selectable marker on each plasmid which is to guide expression. In the case of antibody expression where one vector to express the heavy chain and another to express the light chain are used, two different selectable markers should be present, although this can be avoided by relying on the fact that when transfecting mammalian cells with two plasmids, most cells will take up and express both plasmids. In the IVS, the *neo* gene is used to select the heavy chain plasmid and XGPRT to select the light chain plasmid. Both can be selected for simultaneously, by using G418and mycophenolic acid.

The inclusion of amplifiable markers such as glutamine synthetase (Bebbington et al. 1992) or dihydrofolate reductase (Page and Sydenham, 1991) permits the amplification of the stably integrated plasmid by appropriate selective pressure and subsequently an increase in antibody expression levels. Similar effects have been obtained by providing a sub-optimal Kozak sequence to the selectable marker. Under these conditions, cells can be selected if the plasmid has integrated into a region of the genome which is very transcriptionally active. This results in fewer transformants, but they tend to express at higher levels.

Testing antibody production

Once a cell line has been transfected, either stably or transiently, with the immunoglobulin vectors of interest, expression of full length immunoglobulin needs to be assessed. The easiest way of doing this is by ELISA, by coupling the recognised antigen, and using appropriate second layer (e.g. anti-human IgG) anti-sera. One can also obtain a qualitative assessment of the degree of antibody production at the individual cell level by doing immunofluorescence using anti-heavy and anti-light antibodies labelled with different fluors (e.g. fluorescein and rhodamine). This can be especially important when transient cell transfection does not yield antibody, since the problem is often related to inefficient cell transfection.

Subprotocol 1
Transient Transfection in COS Cells With DEAE-Dextran

Efficient transient transfection is absolutely dependent upon the quality of the cells and the quality of the DNA. Cells which have been allowed to overgrow may be irreversibly damaged (from a transfection point of view) however healthy they look afterwards. We suggest that when transfection efficiency decreases one should return to (healthy) frozen stocks.

Figures in parentheses in the protocol indicate the volumes and compositions which should be used when performing parallel transfections in order to use the supernatant for assays. Step 10 uses much less DMEM in order to concentrate the antibody, and includes HEPES to prevent acidification of the medium.

▪▪ Procedure

1. Plate the cells at semiconfluence in a 90 mm (60 mm) Petri dish the day before the experiment.

2. Add 10 µg plasmid DNA to 2.7 ml DMEM.

3. Add 0.3 ml 5 mg/ml DEAE-dextran (Pharmacia, mol. wt. 500,000).

   Note: DEAE-dextran should be autoclaved and stored at -20°C in small aliquots.

4. Wash the cells twice in DMEM.

5. Add DMEM/DNA/DEAE-dextran to the dish (final concentration = 0,5 mg/ml).

6. Incubate 30 min at 37°C.

7. Add 15 ml 100 mm chloroquine in DMEM/10% FCS.

   Note: Chloroquine (Sigma) should be filter sterilised and kept at -20°C in small aliquots.

8. Incubate 3 hrs at 37°C in CO_2 incubator.

9. Wash with DMEM.

10. Feed with DMEM/10% FCS (2.5ml DMEM/10%FCS containing 20 mM Hepes pH 7.2).

11. Harvest 2 (3) days later.

Subprotocol 2
Indirect Immunofluorescence Detection of the Two Chains

Procedure

1. Plate COS cells on polylysine-coated glass coverslips which have been placed at the bottom of the petri dish and transfect them following Subprotocol 1.

2. Wash three times by dipping into a beaker containing PBS.

3. Fix the cells with 3.7% paraformaldhehyde in PBS for 10 min at room temperature.

4. Wash once with PBS.

5. Permeabilize the cells with 0.1 M Tris-Cl pH 7.6/ 0.2% Triton X100 for 4 min at room temperature.

6. Wash three times by dipping into a beaker containing PBS.

7. Incubate with the appropriate dilution of anti-human isotype primary antibody (1-2 hours at room temperature) and subsequently with the fluorescein-labelled secondary antibody (1-2 hours at room temperature), with washes in PBS between incubations. Incubations should be done in a moist chamber (an airtight box with a piece of damp tissue paper at the bottom) to prevent drying out of the cells. Biotin avidin based systems can also be used.

Subprotocol 3
Stable Transfection of Cells by Calcium Phosphate

Materials

- TE
  - 1 mM Tris (pH 7.5)
  - 0.1 mM EDTA
- 2.5 M $CaCl_2$

- 2X HeBS (pH 7.1)
  Different batches of apparently identical HeBS can have different transfection efficiencies. For this reason, a number of batches are made in parallel and tested for transfection efficiency. The batch(es) which give high transfection efficiencies are then used for further experiments. Once good HeBS is found, it can be kept at RT.
  - 280 mM NaCl
  - 0.5 mM Na_2HPO_4
  - 1 mM NaH_2PO_4
  - 50 mM Hepes

- G418 in medium (50 mg/ml) without serum, add 1N NaOH to neutral pH; store in aliquots at -20°C for few months. Use 500 µg/ml final. This is for the selection of Neo resistance.

Procedure

1. Seed 10^6 cells per 100 mm Petri dish one day before in 7 ml medium.

2. Dissolve DNA at 40 µg/ml in TE (solution 1), using 0.5 ml per plate. Use Falcon 2059 or 2063 tubes.

3. Add 50 µl 2.5 M $CaCl_2$ (250 mM final) per plate, dropwise, mixing between drops; DNA may precipitate.

4. Add DNA/$CaCl_2$ solution dropwise to equal volume of sterile 2xHeBS (X+X/2 ml) (0.55 ml). This is best done by gently dripping the DNA/$CaCl_2$ mixture down the outside of a pipette attached to a standard electric pipetter. By blowing air through the pipette, the DNA/$CaCl_2$ solution is mixed gently with the HeBS. The precipitate should be floccular or filamentous and should form immediately. Should be left to mature over 45 min at room temperature.

5. After 45 min, gently mix by pipetting, add 1 ml to each plate (to 7 ml of medium). Then rock gently.

6. After one day at 37°C, pour off the medium and start selection. (DMEM+20% FCS; never overgrow the cells). This can be for one or both of the plasmids (VH and VL) described above. In general, clones containing one of the two plasmids will usually contain both. The concentration of selectable marker to be used should be determined empirically for each cell line, and is the lowest concentration which will kill the untransfected cell line. In general, G418 (to select for

the neo gene) is used at 300-1000 μg/ml final concentration, while my-cophenolic acid (to select for the XGPRT gene) is used at 10 μg/ml. Cells can be more susceptible to selectable markers when two markers are used simultaneously. This is why the use of one marker is often preferred. pCO_2 is generally used at 5%.

Individual clones, when they contain about 50 cells, can be gently picked up with a p200 plastic pipette tip and transferred to 96 well plates. This is best done by direct view under the tissue culture micro-scope and requires a very steady hand.

Troubleshooting

In general, when transfections do not work, the cells are to blame. This is usually because the cells have overgrown, even at a previous passage. The best solution is to return to an original aliquot of cells and start again.

Comments

Different assays can be done to check the binding activity of the cloned antibody, depending on the availibility and nature of the antigen. In some cases (e.g. neurotrophins), biological assays can be done. In these cases, controls for the effect of the (COS cell conditioned) serum should be done since we have found cases of a biological activity which overcomes the effect of the antibody (even when added exogenously). ELISA assays are relatively straightforward if the antigen is available in sufficient quan-tities, since there is enough specific antibody in supernatant collected from the transiently transfected COS cells. If the antibody recognizes the denatured antigen, Western blot analysis can also be used. Generally whichever assay was used to characterise the selected antibody can also be used to characterise the COS cell supernatant.

Transient transfections are usually done in parallel, half for immuno-fluorescence, and half for supernatant harvesting. This allows one to be sure that the supernatant used for assays is derived from cultures which have positive cells present. The presence of singly positive cells as well as double positive cells serves as a good internal control.

If the percentage of positive cells is lower than 1%, it is not worth pro-ceeding with specific assays. Poor transfection efficiencies are usually due to poor DNA quality or unhealthy COS cells (overgrowth, mycoplasma contamination or late passages prior to transfection). When 1-10% of cells are positive one can expect up to 50 ng/ml immunoglobulin in the supernatant three days after the transfection, which is sufficient for most tests of specificity.

References

Bebbington C. R., Thomson G. R. S., King D., Abrams D. and Yarranton G. T. (1992) High level of a recombinant antibody from myeloma cells using a glutamine synthetase gene as an amplifiable selectable marker. BioTech 10: 169-175

Byrn R. A., Mordenti J., et al. (1990) Biological properties of a CD4 immunoadhesin. Nature 344: 667-70

Gossen M. and Bujard H. (1992) Tight control of gene expression in mammalian cells by tetracycline-responsive promoters. Proc. Natl. Acad. Sci. U.S.A 89: 5547-5551

Mizushima S. and Nagata S. (1990) pEF-BOS, a powerful mammalian expression vector. Nucl. Acids Res. 18: 5322

Page M. J. and Sydenham M. A. (1991) High level expression of the humanized monoclonal antibody Campath-1H in Chinese hamster ovary cells. Bio/Technology 9: 64-68

Persic L., Roberts A., Wilton J., Cattaneo A., Bradbury A. and Hoogenboom H. (1997) An integrated vector system for the eukaryotic expression of antibodies or their fragments after selection from phage display libraries. Gene 187: 9-18

Trill J. J., Shatzman A. R. and Ganguly S. (1995) Production of monoclonal antibodies in COS and CHO cells. Curr. Opin. Biotech. 6: 553-560

Expression of Antibody Fragments in Plant Cells

UDO CONRAD and ULRIKE FIEDLER

Introduction

Stable expression of active antibodies and antibody fragments has been achieved in plant cells. It has been shown for complete antibodies, Fab fragments, scFv's and single VH domains (for review see Conrad and Fiedler 1998). Production of antibodies in transgenic plants and plant cells is generally of interest if high quantitites of a specific antibody fragment are needed, i.e. therapeutic antibodies in human medicine. Other fields of application are monitoring of diseases and purification of specific products. Plant organs and especially storage organs like seeds and tubers are traditionally used as human and animal food. Therefore, a direct application by feeding the plant material containing the therapeutic antibody fragments could force the development of new passive vaccination strategies in human and veterinary medicine. A general advantage of plant expression systems are the relatively low production costs. Specific media and sterile conditions are avoided if the plants are grown in soil. Furthermore, human and animal pathogens such as viruses and bacteria can be excluded per se. A main advantage is the possibility to produce in specific plant storage organs that are normally used in nutrition. Growing and harvesting technologies for such basic plant products have already been developed and have only to be adapted for the development of new plantibodies for passive immunization.

Antibody fragments have been successfully accumulated in plant leaves as documented by several papers (for review see Conrad and Fiedler 1998). Leaf-accumulated antibodies can serve as a source for purification and are used in different studies and applications. The seed-specific ex-

✉ Udo Conrad, IPK Gatersleben, Corrensstrasse 3, 06466 Gatersleben, Germany (*phone* +49-039482-5253; *fax* +49-039482-5382; *e-mail* conradu@ipk-gatersleben.de)
Ulrike Fiedler, TU Dresden, Klinik und Poliklinik für Urologie, Fetscherstrasse 74, 01307 Dresden, Germany

pression of antibodies is the method of choice if a cheap and valuable storage of crude material is necessary. Transgenic seeds accumulating active antibody fragments can be stored for years at room temperature without any detectable loss of antibody amount and activity (Fiedler and Conrad 1995). Seeds contain much more protein than leaves and are therefore well suited as "protein biofactories". Potato tubers have also been used successfully to produce recombinant antibodies. Here high accumulation has been measured but storage should occur at $4°C$ (Artsaenko et al. 1998). To optimize the level of accumulation immunoglobulins were directed into different compartments of the plant cell: the cytosol, the apoplastic space and the endoplasmic reticulum. Targeting into the endoplasmic reticulum (ER) was performed by using a signal sequence in front of the antibody fragments. Signal sequences of animal and plant origin have been successfully used (Table 1). Without any further signals the antibody fragments are transported to the apoplastic space. Retention of the recombinant antibody fragments in the ER can be achieved by fusion of a KDEL coding sequence to the 3' end of the immunoglobulin sequences. Avoidance of a signal sequence at the N-terminus of the antibody fragment should force the cytosolic accumulation. In most cases cytosolic expression of antibody fragment proteins was rather low. The accumulation level could be improved by fusion of KDEL to the C-terminus (Schouten et al. 1997) or by constructing a fusion protein with an N-terminal thioredoxin part (Askari 1999). Generally the highest accumulation of antibody fragments in transgenic plants has been measured after retention in the ER (for review see Conrad and Fiedler 1998). This seems to be the method of choice for the production of high amounts of antibody fragments for different purposes. Depending on the promoter the combination of seed-specific expression and retention in the ER results in rather high accumulation rates of transgenic antibody fragments (Fiedler et al. 1997). Accumulation in other compartments such as apoplast or cytosol lead to new concepts in plant defense (Fecker et al. 1996, Zimmermann et al. 1998). Furthermore, the compartment-specific accumulation of specific antibody fragments against physiological factors was used to construct specifically immunomodulated plants (for review see Conrad and Fiedler 1998).

Outline

The production of antibody fragments in plants will be described for an scFv expressed in transgenic tobacco plants. The following procedure has to be carried out. The scFv is generated from hybridoma (see Chapters 2-4)

Table 1. Expression levels of scFv antibody fragments in transgenic plants. DNA constructs, used signal sequences, targeted cell compartments, the transformed plant species as well as the obtained expression levels are shown. CaMV: Cauliflower Mosaic Virus, ER: endoplasmic reticulum, KDEL: ER retentions signal sequence, LeB4: Legumine B4, PGIP: polygalacturonase inhibiting protein, PR1a: tobacco pathogenesis related protein SS: signal sequence, TSP: total soluble protein, TSWV: tomato spotted wilt tospo virus, USP: unknown seed protein.

scFv directed against	construct	signal sequence	compartment/ organ	plant species	expression level in % of TSP
phytochrome (Owen et al. 1992)	CaMV 35S-scFv	none	cytosol/ leaf	N. tabacum	0.06-0.1
artichoke mottled crinkle virus (Tavladoraki et al. 1993)	CaMV 35S-scFv	none	cytosol/ leaf	N. benthamiana	0.1
phytochrome (Firek et al. 1993)	CaMV 35S-SS-scFv	PR1a	secretory pathway/ leaf	N. tabacum	0.5
abcisic acid (Artsaenko et al. 1995)	CaMV 35S-SS-sFv-KDEL	LeB4	ER/ leaf	N. tabacum	0.05-4.8
cutinase (Schouten et al. 1996)	CaMV 35S-SS-scFv-KDEL	murine κ light chain	ER/ leaf	N. tabacum	1
beet necrotic yellow vein virus coat protein (Fecker et al. 1996)	CaMV 35S-SS-scFv-KDEL	phytohem-agglutinin	secretory pathway/leaf	*N. bethamiana*	0.1
human creatine kinase MM (Bruyns et al. 1996)	CaMV 35S-SS-scFv CaMV 35S-scFv	2S2 storage protein	secretory pathway/ leaf cytosol/ leaf	Nicotiana tabacum	0.01
oxazolone (Fiedler et al. 1997)	CaMV 35S-SS-scFv-KDEL	LeB4	ER/ leaf	N. tabacum	4
oxazolone (Fiedler et al. 1997)	USP - SS-scFv-KDEL	LeB4	ER/ seed	N. tabacum	2.6
β-1,4-endoglucanase (Schouten et al. 1997)	TR2'-1-scFv-KDEL	none	ER/ leaf	Solanum tuberosum	0.3
oxazolone (Artsaenko et al. 1998)	CaMV 35S-SS-scFv-KDEL	LeB4	ER/ tuber	Solanum tuberosum	2.0
Tobacco Mosaic Virus (Zimmermann et al. 1998)	?? -SS- scFv	murine VL IG	secretory pathway/ leaf	N. tabacum	1-2000 ng/g tissue
dihydroflavonol 4-reductase (De Jaeger et al. 1999)	CaMV 35S-SS-scFv-KDEL	2S2 storage protein	ER/ leaf and petals	Petunia hybrida	3
TSWV (Franconi et al. 1999)	CaMV 35S-SS-scFv	PGIP	secretory pathway	N. benthamiana	n. d.

or isolated from a combinatorial antibody library (see Chapters 5-16). The constructed scFv is characterized concerning functionality and affinity by bacterial expression (see Chapter 17-23). Afterwards the scFv gene is cloned into a plant expression cassette. Depending on the final application different promoters and signal sequences for ER entry or ER retention (KDEL sequence) can be included in the expression cassette (Figure 1, Table 1). For the overall plant expression the CaMV promoter (Franck et al. 1980) is most widely used. For seed specific-expression in transgenic tobacco the LeB4 promoter (Bäumlein et al. 1987) or the USP promoter (Bäumlein et al. 1991) from *Vicia faba* can be used. Detection of scFv's in plants and their purification is more simple if a peptide tag sequence such as the c-myc tag (Munroe and Pelham 1987) is fused at the 3' end of the scFv gene. Figure 1 shows a general scheme of possible plant expression cassettes.

Beside direct gene transfer the most convenient method for the production of transgenic tobacco plants is the agrobacterium mediated gene

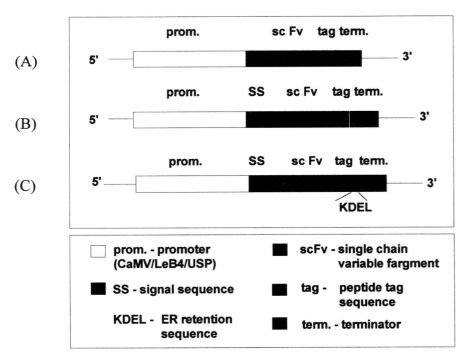

Fig. 1. Schematic presentation of cassettes used for expression of antibody fragments in plants. *(A)* cytoplasmic expression, *(B)* apoplastic expression, *(C)* endoplasmic reticulum expression.

transfer (Zambryski et al. 1983). Therefore, the designed plant expression cassette has to be cloned into a shuttle vector. The most suitable plant binary vectors are pBin 19 (Frisch et al. 1995) and pGSGLUC1 (Saito et al. 1990). These vectors contain a single plant-expressed kanamycin resistance gene (NPT II) allowing the selection of transformed plants by kanamycin. Furthermore, these vectors have normally two origins of replication which allow replication in *E. coli* and *Agro-bacterium tumefaciens*. After transfer of plasmids into *A. tumefaciens* plant leaves are infected with recombinant agrobacteria (leaf disk transformation) and transgenic plants are regenerated. Stable transformed tobacco plants can be regenerated by kanamycin selection. These plants can be tested by Southern or PCR analyses for the presence of the transgene in the plant genome. Transgenic plant material (leaves or seeds) can be investigated for scFv expression and activity by western blot analyses or ELISA. For detection of the scFv fragments antibodies recognizing the tagged peptide (c-myc) or the scFv itself can be used. ScFv purification from crude plant extracts can be performed by chromatography. The entire protocol for the expression of antibody fragments in plant cells is outlined in Fig. 2.

Materials

- electroporation device (e.g. Gene-Pulser, BioRad Richmond CA, USA)

- electrophoresis and transfer unit (e.g. Transfor, Hoefer San Francisco CA, USA)

- laminar flow cabinet (e.g. HERAsafe, Heraeus Instruments GmbH, Hanau, Germany)

CBY-medium

- 1 g/l yeast extract

- 5 g/l tryptone

- 5 g/l sucrose

- 20 g/l agar

- Adjust the pH to 7.2, sterilize by autoclaving and add filter sterilized $MgSO_4$ to a final concentration of 2 mM.

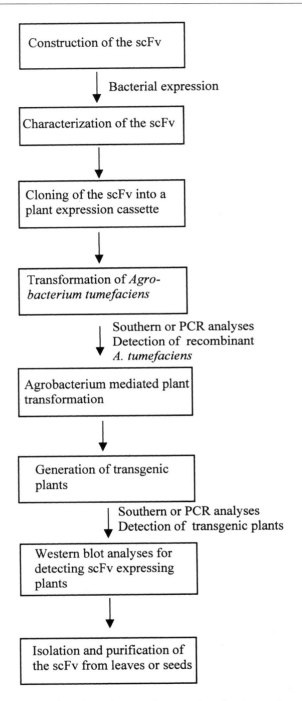

Fig. 2. Flow chart showing the procedure for the production of recombinant antibodies in transgenic plants.

CPY-medium

- 1 g/l yeast extract
- 5 g/l tryptone
- 5 g/l sucrose
- Adjust the pH to 7.2, sterilize by autoclaving and add filter sterilized $MgSO_4$ to a final concentration of 2 mM.

DNA lysis buffer

- 10 mM Tris-HCl (pH 8.0)
- 10 mM EDTA
- 0.1 M NaCl

MS-medium

- 4.49 g/l Murashige and Skoog medium basal salt mixture (Duchefa)
- 1 ml/l vitamin stock solution (100 g/l myo-inositol, 2 g/l glycine, 0.5 g/l nicotinic acid, 0.5 g/l pyridoxine-HCl, 0.1 g/l thiamine-HCl)
- 30 g/l sucrose
- Adjust the pH to 5.8 with 1 M KOH and sterilize for 15 min.

PBS buffer

- 0.008 M Na_2HPO_4
- 0.002 M KH_2PO_4
- 0.15 M NaCl
- Adjust the pH to 7.2

SDS gel loading buffer

- 1 ml 10% glycerol
- 1.4 ml separating buffer (1.5 M Tris-HCl pH 8.8, 10% SDS)
- 0.5 ml β-mercaptoethanol
- 5.1 ml water
- 2 ml 10% SDS
- 0.001% bromphenol blue

Seed protein extraction buffer

- 50 mM Tris-HCl (pH 7.7)
- 200 mM NaCl
- 5 mM EDTA
- 0.1% Tween-20

SOC-medium

- 5 g/l yeast extract
- 20 g/l tryptone
- 0.58 g/l NaCl
- 0.186 g/l KCl
- Sterilize by autoclaving and add 1 ml/l filter sterilized 2 M glucose afterwards.

TE-buffer

- 10 mM Tris-HCl (pH 8.0)
- 1 mM EDTA

Procedure

Agrobacterium tumefaciens transformation

Competent *A. tumefaciens* cells are transformed with the shuttle plasmid containing the plant expression cassette with the scFv gene.

1. Inoculate 10 ml CPY-medium containing the appropriate antibiotics with a single *A. tumefaciens* colony from a plate.
2. Incubate by shaking at 28°C, 150-200 rpm for 2 days.
3. 0.5 ml of the culture are used to inoculate 100 ml CPY-medium containing the appropriate antibiotics.
4. Cells are harvested by centrifugation at 6,000 rpm, 4°C for 15 min.
5. Resuspend the pellet in 100 ml sterile, distilled water and centrifuge as described before.

6. Wash the cells twice more with 50 ml water and resuspend in 500 µl 10% glycerol.

7. Aliquots of 45 µl are frozen in liquid nitrogen and stored at -80°C.

8. 100-1000 ng of desalted shuttle plasmid DNA are added and mixed with an aliquote of the competent cells on ice.

9. Electroporation is carried out in an ice cold electroporation cuvette at 25 µF, 2.5 V and 200 using the Gene-Pulser.

10. Add 1 ml SOC- medium to the cuvette, transfer into a sterile micro-centrifuge tube, incubate by shaking gently at 28°C for 2-4 hours and plate onto selective CBY-medium plates.

Isolation of total DNA from *A. tumefaciens*

After electroporation different *A. tumefaciens* clones are investigated for the presence of the transgene. Therefore, total DNA is isolated according to Armitage et al. 1988 and used for Southern blot or PCR analyses.

1. Inoculate 5 ml CPY-medium containing selective antibiotics with a single colony and incubate for 2 days shaking at 28°C.

2. Isolate cells of 1.5 ml culture by centrifugation at 12,000 rpm for 5 min.

3. Resuspend the cells in 300 µl TE-buffer and add 100 µl 5% Sarcosyl and 150 µl pronase (5 mg/ml).

4. Incubate at 37°C for 1 hour.

5. Add 500 µl phenol/chloroform and pass the mixture 5 times through a pipette tip. After centrifugation at 12,000 rpm for 5 min the obtained supernatant is transferred to a new tube and extracted 3 more times with 500 µl phenol/chloroform.

6. Isolate the upper aqueous phase by centrifugation, transfer it to a new tube and add 0.05 volumes of 5 M NaCl and 3 volumes of ethanol.

7. Incubate the tubes at -20°C for 2 hours and collect the DNA by centrifugation at 12,000 rpm for 10 min. After washing the DNA with 70% ethanol the DNA pellet is air dried and resuspended in 50 µl water.

Plant transformation and regeneration of transgenic tobacco plants

Tobacco plant tissue culture is performed on MS-medium at 22°C, 50% humidity, a light regime of 16 h light and 8 h darkness and a light intensity of 500 μmol m^{-2}s^{-1}. In the green house tobacco plants grow with a 16 h light and 8 h darkness regime.

1. Tobacco leaves from *Nicotiana tabacum* (Samsun NN) are washed with tap water and are surface sterilized by immersion in 30% Domestos for 15 min. Work is carried out in the laminar flow cabinet.

2. Wash the leaves several times in sterile water until no detergence is left.

3. Remove the midrip and cut the leaf into 1 cm squares. Place the leaf discs on MS-medium plates containing 0.1 mg/l NAA (α-naphthaleneacetic acid) and 1.0 mg/l BAP (6-benzyl-amino-purine). Seal the plates and incubate for two days in dim light at 25°C .

4. An *A. tumefaciens* culture grown for 2 days is diluted 1:50 in liquid MS-medium in a petri dish. The leaf explants are dipped briefly in the bacterial suspension, put back on the MS-medium plates and are cocultivated for two further days with the recombinant agrobacteria. After this the discs are tranferred onto plates with MS-medium containig cefotaxime (250 mg/l), kanamycin (100 mg/l), NAA (0.1 mg/l) and BAP (1.0 mg/l).

5. After 4 weeks shoots are cut off, put into MS-medium containing 100 mg/l kanamycin and cefotaxime and are incubated in the light at 25°C.

6. After 2-3 weeks roots occure and plantlets are transferred to soil. First 1-2 weeks plants acclimatize in a propagator with a plastic cover before they grow in the green house.

Analyses for presence of the transgene

To determine whether the transformed tobacco plants are transgenic for the scFv genomic DNA is isolated from tobacco leafs. The isolated genomic DNA can be used for PCR or Southern blot analyses (Sambrook et al. 1989). For Southern blot analyses the DNA is digested with a suitable restriction enzyme, DNA fragments are separated by agarose gel electrophoresis, denatured, transferred to a nitrocellulose filter and immobilized. The DNA attached to the filter is then hybridized to ^{32}P-labelled DNA fragment containing the scFv gene. Plants giving a positive signal of the expected size in autoradiography or PCR are chosen for further experiments.

Isolation of genomic DNA from transformed tobacco plants

1. Grind 200-300 mg of a young tobacco leaf in liquid nitrogen to a fine powder.

2. Collect the frozen powder into a 15 ml Falcon tube containing 20 µl proteinase K (25 mg/ml).

3. Add 3 ml DNA lysis buffer and 0.8 ml 10% SDS and shake well.

4. Incubate at 56°C for at least 2 hour.

5. Cool to room temperature, add 1 ml saturated NaCl and shake for 15 seconds.

6. Centrifuge for 15 minutes at room temperature at 4,000 rpm.

7. Pipette supernatant into a beaker containing 10 ml cold ethanol.

8. Precipitated DNA can be collected by stirring with a glass rod.

9. Put the DNA into a 1.5 ml tube and wash with 70 % ethanol at room temperature.

10. Dry the DNA pellet for 10 minutes at room temperature and resuspend in 200 µl RNase in TE-buffer and leave in the fridge overnight.

11. After adding 400 µl TE-buffer to the dissolved DNA proteins are removed by extraction once with 500 µl phenol, once with 500 µl of a 1:1 mixture of penol/chloroform and once with 500 µl chloroform. Always mix the contents of the tubes until an emulsion forms and centrifuge for about 5 minutes at full speed in an Eppendorf centrifuge at room temperature. The upper, aqueous phase is used for the next extraction.

12. DNA is precipitated by adding 0.1 volume 3 M sodium acetate, pH 5.2 and 1 ml ethanol.

13. After leaving for a few minutes on ice spin down and wash the DNA with 70 % ethanol.

14. The air dried DNA pellet is resuspended in 100-200 µl TE-buffer.

Characterization of plant expressed scFv's

Expression of scFv fragments in the transgenic plant can be investigated by western blot analyses or ELISA (Sambrook et al. 1989). After purification plant expressed scFv's can be further characterized concerning their affinity constants (Chapter 38 and Chapter 39) or used for first practicability studies. For all these purposes extracts of total soluble leaf or seed proteins are used. Whereas for western blot analyses denatured proteins can be used (extraction with SDS gel loading buffer) ELISA and purification procedures need functionally active antibody fragments. For western blot analyses the proteins are separated by SDS-PAGE electrophoresis and transferred onto nitrocellulose (Sambrook et al. 1989). ScFv tagged to c-myc peptide can be detected by incubation with a monoclonal antibody recognizing this tag (9E10 developed in mice), followed by incubation with a secondary antibody (anti-mouse conjugated to horseradish peroxidase). The ECL detection system can be used for developing the enzymatic reaction.

Preparation of protein extracts from leaves for western blot analyses

1. Homogenize a piece of a leaf (100 mg) with an electric drill and add 200 µl SDS gel loading buffer.

2. Boil the homogenate for 10 minutes.

3. Centrifuge at full speed at room temperature for 15 minutes.

4. Use the supernatant for determination of protein content by the Bio-Rad Coomassie assay.

5. For western blot subject 40 µg of total soluble protein.

Preparation of leaf protein extracts for ELISA measurements or purification

1. Grind leaves (200-300 mg) from transgenic plants expresssing scFv with liquid nitrogen to a fine powder.

2. Resuspend powder with 400 µl PBS buffer/1% Triton as extraction buffer.

3. Centrifuge for 10 minutes at 4°C, full speed. The supernatant contains soluble proteins, including scFv and can be applied to the column.

Preparation of protein extracts from seeds

1. Grind 50 - 150 mg ripe seeds of transgenic and control plants with liquid nitrogen and acid washed sand.

2. Add 0.5-0.8 ml of seed protein extraction buffer to the frozen powder.

3. Transfer the homogenate to an Eppendorf tube and centrifuge for 10 minutes at 4°C, 12,000 rpm.

4. Collect the supernatant and determine the protein concentration by the Bio-Rad Coomassie assay.

5. For western blot subject 40 μg of total soluble protein of each plant for SDS-PAGE.

Purification of scFv fragments with c-myc-tag fusion by affinity chromatography

The technique used for affinity purification depends on the C- or N-terminal tag which is used. For protein fusions with the penta-histidine tag the expressed scFv can be purified by metal affinity purification. Purifications of scFv tagged with c-myc peptide can be done by coupling the antibody 9E10 recognizing the c-myc peptide tag to activated agarose. Affi-Gel 10 (BIO-RAD) is an activated immunoaffinity support that offers rapid, high efficiency coupling for all ligands with a primary amino group, including proteins. Affi-Gel 10 support is a N-hydroxysuccinimide ester of a derivatized crosslinked agarose gel bead support. It couples to ligands spontaneously in aqueous or non-aqueous solution.

Preparation of a 9E10 column

1. Mix 22.3 ml 0.1 M MOPS pH 7.5 and 2.5 ml ascites containing 9E10 antibodies. Concentrate to 2.5 ml by Centriprep (Amicon) concentration.

2. Mix the concentrated solution again with 22.3 ml 0.1 M MOPS and concentrate to a final volume of 1.5 ml. There should be 25-50 mg protein in the concentrated solution.

3. Shake the Affi-gel 10 agarose vial to get a uniform suspension.

4. 1 ml of the slurry are washed 3 times with 3 beed volumes of ice cold millipore water. Centrifuge at low speed to separate agarose and water. Care should be taken, not to allow the gel bed to get dry.

5. Transfer the moist gel cake to a test tube and add cold ligand solution. Agitate sufficiently to make a uniform suspension.

6. Continue gentle agatitation of the gel slurry overnight at 4°C.

7. After coupling any remaining active esters are blocked by adding 0.1 ml 1 M monoethanolamin pH 8.0. Allow 1 hour for completion of the blocking reaction at 4°C.

8. Transfer the 9E10-agarose slurry to a column and wash extensively with at least 10 beed volumes of PBS buffer.

9. Proteins or other solutes are not bound, or are weakly bound by non-specific interactions, and must be washed off prior to elution. Elute with 2 ml glycin-NaCl pH 2.2. Immediately wash the column with 10 beed volumes PBS buffer until the pH of the effluate is around 7. The column is now equilibrated.

Purification of scFv tagged with myc peptide on 9E10-Affi-gel 10 column

1. Applicate the plant extract to the column. Allow the antigen to bind for 10 minutes at 4°C.

2. Repeat binding by running the effluate eight times through the column. Following steps can be carried out at room temperature.

3. Wash the column 3 times with 10 beed volumes PBS.

4. The scFv is eluted with 3 ml 0.1 M glycin-HCl pH 2.2, 0.02 M NaCl. It is important to remove the eluted antigen from the eluant as quickly as possible to minimize the chance of denaturation. Therefore, eight times 400 µl fractions are collected in 100 µl 1 M Tris-HCl, pH 8.0. Mix immediately after collection. Normally the eluted antigen can be detected in fraction 3-5.

5. Regenerate the column by washing with 1 ml eluant to remove maintained proteins and neutralize immediatly with 0.1 M Tris-HCl in PBS buffer pH 8.0.

6. Equilibrate the column with 2 times 10 beed volumes PBS buffer.

7. Columns can be stored at 4°C in PBS buffer containing 0.2 % sodium azide.

References

Armitage, P., Walden, R. and Draper, J. (1988): Vectors for the transformation of plant cells using *Agrobacterium*. In: Plant Genetic Transformation and Gene Expression. Draper, J., Scott, R., Armitage, P., Walden, R. (Eds.), The Alden Press, Oxford.

Askari, B. (1999): Expression and characterisation of thioredoxin-single-chain-Fv fusion proteins in bacteria and plants. Dissertation an der Martin-Luther-Universität Halle-Wittenberg.

Artsaenko, O., Kettig, B., Fiedler, U., Conrad, U. and Düring, K. (1998): Potato tubers as a biofactory for recombinant antibodies. Molecular Breeding 4, 313-319.

Artsaenko, O., Peisker, M., zur Nieden, U., Fiedler, U., Weiler, E.W., Müntz, K. and Conrad, U. (1995): Expression of a single-chain Fv antibody against abscisic acid creates a wilty phenotype in transgenic tobacco. Plant J. 8, 745-750.

Bäumlein, H., Boerjan, W., Nagy, I., Bassüner, R., van Montagu, M., Inze, D. and Wobus, U. (1991): A novel seed protein gene from *Vicia faba* is developmentally regulated in transgenic tobacco and *Arabidopsis* plants. Mol. Gen. Genet. 225, 459-467.

Bäumlein, H., Müller, A.J., Schiemann, J., Helbing, D., Manteuffel, R. and Wobus, U. (1987): A legumin B gene of *Vicia faba* is expressed in developing seeds of transgenic tobacco. Biol. Zentralbl. 106, 569-575.

Bruyns, A.M., de Jaeger, G., de Neve, M., de Wilde, C., van Montagu, M. and Depicker, A. (1996): Bacterial and plant-produced scFv proteins have similar antigen-binding properties. FEBS Letters 386, 5-10.

Conrad, U. and Fiedler, U. (1998): Compartment-specific accumulation of recombinant immunoglobulins in plant cells: an essential tool for antibody production and immunomodulation of physiological functions and pathogen activity. Plant Mol. Biol. 38, 101-109.

De Jaeger, G., Buys, E., Eeckhout, D., De Wilde, C., Jacobs, A., Kapila, J., Angemon, G., van Montagu, M., Gerats, T. and Depicker, A. (1999): High level accumulation of single-chain variable fragments in the cytosol of transgenic *Petunia hybrida*. Eur. J. Biochem. 259, 426-434.

Fecker, L., Kaufmann, N., Commandeur, U., Commandeur, J., Koenig, R. and Burgemeister, W. (1996): Expression of single-chain antibody fragments (scFv) specific for beet necrotic yellow vein virus coat protein or 25 kDa protein in *Escherichia coli* and *Nicotiana benthamiana*. Plant Mol. Biol. 32, 979-986.

Fiedler, U. and Conrad, U. (1995): High-level production and long-term storage of engineered antibodies in transgenic tobacco seeds. Bio/Technology 13, 1090-1093.

Fiedler, U., Phillips, J., Artsaenko, O. and Conrad, U. (1997): Optimization of scFv antibody production in transgenic plants. Immunotechnology 3, 205-216.

Firek, S., Draper, J., Owen, M.R.L., Gandecha, A., Cockburn, B. and Whitelam, G.C. (1993): Secretion of a functional single-chain Fv protein in transgenic tobacco plants and cell suspension cultures. Plant Mol. Biol. 23, 861-870.

Franck, A., Guilley, H., Jonard, G., Richards, K. and Hirth, L. (1980): Nucleotide sequence of the Cauliflower Mosaic Virus DNA. Cell 21, 285-294.

Franconi, R., Roggero, P., Pirazzi P., Arias, F.J., Desiderio, A., Bitti, O., Pashkoulov D., Mattei B., Bracci L., Masenga V., Milne, R.G. and Benvenuto, E. (1999): Functional expression in bacteria and plants of an scFv antibody fragment against tospoviruses. Immunotechnology 4, 189-201.

Frisch, D.A., Harris-Haller, L.W., Yokubaitis, N.T., Thomas, T.L., Hardin, D.H. and Hall, T.C. (1995): Complete sequence of binary vector Bin 19. Plant Mol. Biol. 27, 405-409.

Munroe, S. and Pelham, H. (1987): A C-terminal signal prevents secretion of luminal ER proteins. Cell 48, 899-907.

Owen, M.R.L., Gandecha, A., Cockburn, B., Whitelam, G. (1992): Synthesis of a functional anti-phytochrome single-chain Fv protein in transgenic tobacco. Bio/Technology 10, 790-794.

Saito, K., Kaneko, H., Yamazaki, M., Yoshida, M., and Murakoshi, I. (1990): Stable transfer and expression of chimeric genes in licorice (*Glycyrrhiza uralensis*) using an Ri plasmid binary vector. Plant Cell Reports 8, 718-721.

Sambrook, J., Maniatis, T. and Fritsch, E. F. (1989): Molecular Cloning: A laboratory manual. Cold Spring Harbor. Cold Spring Harbour Laboratory Press, New York

Schouten, A., Roosien, J., de Boer, J.M., Wilmink, A., Rosso, M.N., Bosch, D., Stiekema, W.J., Gommers, F.J., Bakker, J. and Schots, A. (1997): Improving scFv antibody expression levels in the plant cytosol. FEBS Letters 415, 235-241.

Schouten, A., Roosien, J., van Engelen, F.A., de Jong, G.A.M., Borst-Vrenssen, A.W.M., Zilverentant, J.F., Bosch, D., Stiekema, W.J., Gommer, F.J., Schots, A. and Bakker, J. (1996): The C-terminal KDEL sequence increases the expression level of a single-chain antibody designed to be targeted to both the cytosol and the secretory pathway in transgenic tobacco. Plant Mol. Biol. 30, 781-793.

Tavladoraki, P., Benvenuto, E., Trinca, S., Martinis, D.D., Cattaneo, A. and Galeffi, P. (1993): Transgenic plants expressing a functional single-chain Fv antibody are specifically protected from virus attack. Nature 366, 469-472.

Zambryski, P., Joos, H., Gentello, J., Leemans, J., Van Montagu, M. and Schell, J. (1983): Ti-plasmid vector for introduction of DNA into plant cells without altering their normal regeneration capacity. EMBO J. 2, 2143-2150.

Zimmermann, S., Schillberg, S., Yu-Cai Liao and Fischer, R. (1998): Intracellular expression of TMV-specific single- chain Fv fragments leads to improved virus resisitance in *Nicotiana tabacum*. Molecular Breeding 4, 369-379.

Determination of Affinities

Affinity Measurements by Band Shift and Competition ELISA

MAJA A. BUMKE and DARIO NERI

Introduction

The affinity constant (see Neri, Montigiani et al. 1996 for a simple introduction) is a useful parameter for the characterisation of antibody-antigen interactions. This constant can be seen as a "thermometre" which allows us to answer the following question: at certain conditions of concentration, to which extent does antibody A bind to its antigen B?

Competition ELISA and band-shift assays are two simple and inexpensive methods which allow the determination of affinity constants of antibodies in solution, with negligible *avidity* effects (see Neri, Montigiani et al. 1996 for a simple introduction).

For the discussion of the techniques presented in this Chapter, it is worthwhile to introduce the following equations and concepts.

Let us consider a monomeric antibody A (e.g., an antibody fragment) binding in solution to a monomeric antigen B, according to the chemical equation:

$$A + B \Leftrightarrow AB$$

At equilibrium, the dissociation constant Kd (the reciprocal of the affinity constant Ka) can be defined as:

$$Kd = [A][B] / [AB]$$

[A], [B] and [AB] are the concentrations of the free antibody, free antigen and complex, repectively, at equilibrium.

Maja A. Bumke, ETH Zürich, Institut für Pharmazeutische Wissenschaften, Winterthurerstrasse 190, 8057 Zürich, Switzerland

✉ Dario Neri, ETH Zürich, Institut für Pharmazeutische Wissenschaften, Winterthurerstrasse 190, 8057 Zürich, Switzerland (*phone* +41-1-635-6082; *fax* +41-1-635-6886; *e-mail* neri@pharma.anbi.ethz.ch)

Let us consider the case in which the total amount of antibody used in the reaction is much lower than the dissociation constant (i.e., $[A]_{total} \ll$ Kd). At semisaturation (i.e., when [A]=[AB]), it follows that:
Kd = [A][B] / [AB] = [B] $\approx$ [B] $_{total}$

In other words, in the conditions outlined above, the dissociation constant Kd is approximately equal to the total concentration of antigen at which semisaturation of the antibody is achieved. A corollary of this equation, is that at concentrations of antigen > Kd most of the antibody will be in complex with the antigen, whereas at concentrations of antigen < Kd most of the antibody will be free.

In the implementation of the methods presented here, we will use the experimental conditions outlined above ($[A]_{total} \ll$ Kd), and titrate the low amount of antibody with increasing concentrations of antigen. Due to the symmetry of the dissociation constant equation, of course, the experiment could be performed by titrating low concentrations of antigen with increasing concentrations of antibody. Both band-shift and competition ELISA methodologies may allow the determination of the fraction of antibody bound to the antigen in different experimental conditions. Since the antibody is used at low concentrations ($[A]_{total} \ll$ Kd), sensitive detection methods are used, in order to monitor the fraction of antibody bound (let us remember that at semisaturation Kd $\approx$ [B] $_{total}$).

In band-shift assays, antibodies are labelled (either radioactively or fluorescently), and then incubated (in parallel reactions) with increasing concentrations of antigen. When a binding equilibrium is established, the portions of antibody free or bound to the antigen are separated by polyacrylamide electrophoresis in non-denaturing conditions (Figure 1). The antibody bands in the gel are detected either by autoradiography (Neri, Petrul et al. 1996) or by fluorescence imaging (Neri, Prospero et al. 1996). Normally, only antibody-antigen complexes of a sufficient kinetic stability (typically $k_{off} < 10^{-3}$ s^{-1}) do not dissociate during electrophoretic separation and can be detected in the gel autoradiography (Neri, Petrul et al. 1996; Montigiani, Neri et al. 1996). This fact hinders the applicability of the method for the low affinity, fast dissociating antibody-antigen complexes.

In competition ELISA assays (Friguet, Chaffotte et al. 1985) antibodies are incubated with increasing concentrations of antigen in parallel reactions performed in aqueous solution. When a binding equilibrium is established, the fraction of unbound antibody can be captured on an antigen-coated microtitre plate and can be detected by sensitive conventional ELISA procedures (using chromogenic or fluorogenic substrates; Figure 2). Care has to be taken that during the capture step the concentrations

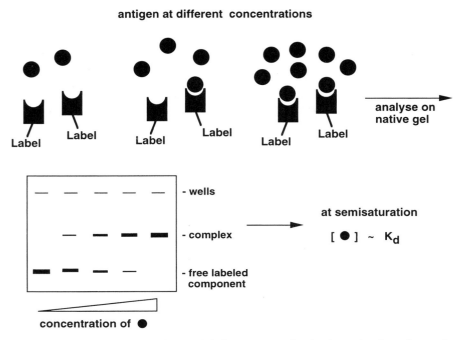

Fig. 1. Schematic illustration of a band-shift experiment for the determination of an antibody-antigen dissociation constant. If the experiment is performed at antibody concentrations $\ll$ Kd, the concentration of antigen at which the antibody semisaturation is observed, is approximately equal to the dissociation constant Kd.

of free and bound antibody are not significantly altered (for example, by a preferential accumulation of antibody on the antigen coated plate by virtue of avidity effects). This can be controlled experimentally by:

- keeping capture times short

- after the capture step on the antigen-coated microtitre plate, transfer the content of each well to a second antigen-coated microtitre plate. The two plates should give comparable ELISA results if the amount of capture antibody is small (e.g., <10% of the free antibody available in the reaction mixture)

The method does not require the use of purified antibody preparations, since in practice, the antibody is affinity-purified during the ELISA capture step. Indeed, antibody-containing crude protein mixtures (such as ammonium sulfate pellets or bacterial supernatants) can be used for the assay. However, in view of the requirements on the stability of the antibody-antigen complex, competition ELISA assays, similar to band-shift assays, are useful in practice only for antibodies with intermediate-high affinity.

Band-shift assays can also be used for the measurement of low kinetic dissociation constants ($k_{off} < 10^{-3}$ s^{-1}; Montigiani, Neri et al. 1996). Here it is worthwhile to remind ourselves that, for a bimolecular reaction, the dissociation constant Kd depends on the kinetic dissociation and association constants according to the following relation:

Kd = k_{off} / k_{on}

The intuitive meaning of k_{off} is easy to grasp when an antibody-antigen complex is forced to dissociate irreversibly (for example by dilution to concentrations below Kd). In these conditions, k_{off} is related to the half-life of the complex ($t_{1/2}$) by the relation:

k_{off} = (ln 2) / ($t_{1/2}$) = 0.692 / ($t_{1/2}$)

In band-shift assays for the determination of k_{off}, an aqueous solution of the antibody-antigen complex, in which one of the two binding partners is labelled, is incubated for different times with a large molar excess of the unlabelled binding partner (Figure 3). The resulting reaction mixtures are

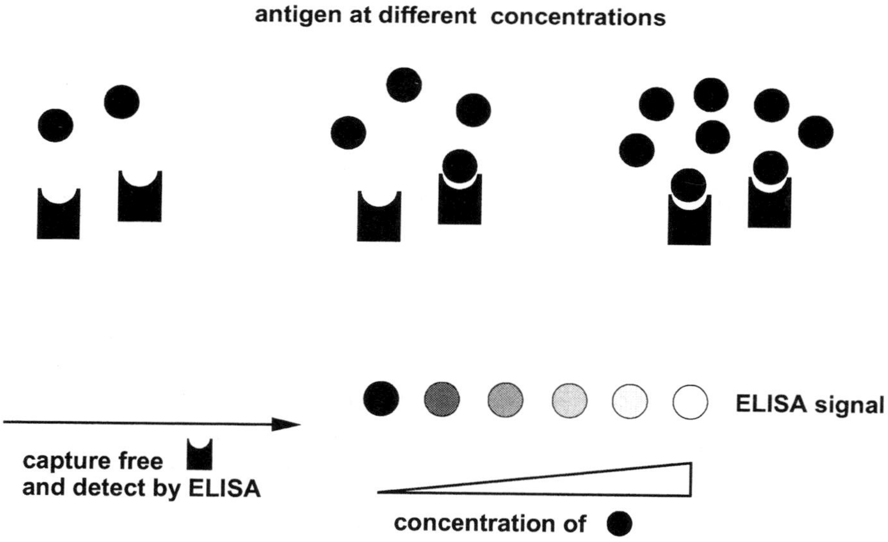

Fig. 2. Schematic illustration of a competition ELISA experiment for the determination of an antibody-antigen dissociation constant. If the experiment is performed at antibody concentrations << Kd, the concentration of antigen at which the antibody semisaturation is observed (i.e., the ELISA signal is half-maximal), is approximately equal to the dissociation constant Kd.

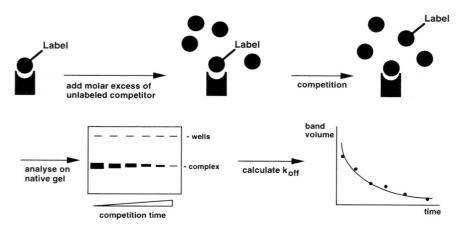

Fig. 3. Schematic illustration of the experimental scheme for the measurement of the kinetic dissociation constant of an antibody fragment binding with moderate-high affinity to a florescently labelled antigen. The time course of the competition reaction of the labelled complex with a molar excess of unlabelled antigen is monitored by polyacrylamide gel electrophoresis in native conditions. The intensity of the labelled bands on the gel follows an exponential profile, which can be fitted to obtain the kinetic dissociation constant k_{off}. Alternatively, the competition time at which the initial band volume is halved corresponds to $t_{1/2}$; this value can be used for the calculation of k_{off} according to the formula $k_{off} = 0.692 / (t_{1/2})$.

analysed by band-shift analysis on a non-denaturing polyacrylamide gel. One should observe an exponential decay of the intensity of the labelled complex band, as the kinetic competition takes place. The time at which the band volume becomes half of the initial volume corresponds to $t_{1/2}$ and this value can be used for the calculation of k_{off} according to the formula reported above.

The experimental protocols reported below are written for recombinant antibody fragments. We made this decision based on the fact that such fragments are easily obtained using antibody phage technology (Winter, Griffiths et al. 1994) and may represent the reagents of choice for *in vitro* (Nissim, Hoogenboom et al. 1994) and *in vivo* applications (Viti, Tarli et al. 1999). However, with small variations, the protocols can also be used with other antibody formats.

Materials

- Buffer A: 50 mM Tris, pH 7.4, 200 mM NaCl, 12 mM $MgCl_2$
- PBS: 8 g NaCl, 0.2 g KCl, 1.44 g Na_2HPO_4 and 0.24 g KH_2PO_4 in 1 l distilled H_2O, pH adjusted to 7.4
- gel mix: 0.4 g sucrose and 0.5 mg bromophenol blue in 10 ml PBS
- 3% MPBS: 3% skimmed dry milk in PBS
- 10% MPBS: 10% skimmed dry milk in PBS

Subprotocol 1
Band-Shift Assay

The procedure described in this chapter for the determination of affinity constants and kinetic dissociation constants by band-shift assay refers to an ideal antibody fragment (e.g., a scFv or an Fab fragment) binding to a well-behaved protein antigen (pure, of well-defined oligomeric state, migrating as a single band in non-denaturing gel electrophoresis.

As mentioned in the introduction, either the antibody or the antigen has to be labelled in these assays. In order to illustrate different labelling methods, we will use radiolabelled recombinant antibodies in the protocol for the determination of Kd, and fluorescently labelled antigen in the protocol for the determination of k_{off}.

Recombinant antibodies can conveniently be radiolabelled site-specifically and without loss of immunoreactivity using genetically introduced phosphorylation sites (Neri, Petrul et al. 1996). This procedure is described in this Chapter. However, other radiolabelling methods can be used (e.g., radioiodination using Iodogen tubes), provided that immunoreactivity is preserved and that a sufficiently high specific activity is achieved.

The fluorescent antigen labelling method presented in this Chapter features the use of N-hydroxysuccinimido ester derivatives of a commercially available cyanine dye (Cy5-NHS; Amersham Pharmacia Biotech). These reagents react with primary amino groups in the antigen molecule, and may impair the antigen's ability to react with the antibody after labelling.

Procedure

Phosphorylation of recombinant antibodies

1. Dissolve in buffer A the purified antibody carrying an engineered phosphorylation site (e.g., the sequence DDDSDDDD; Neri, Petrul et al. 1996) at 1 mg/ml concentration (alternatively, exchange buffer by dialysis).

2. To 100 μl antibody solution, add 500 units casein kinase II (4 μl; New England Biolabs) and 1.6 mCi of $[\gamma\text{-}^{32}P]$ATP (12 μl, Cat. No. 35020 from ICN Biomedicals).

3. Incubate at room temperature for 1 hour.

4. Remove unreacted ATP using a disposable PD-10 column (Amersham Pharmacia Biotech), preblocked with bovine serum albumin dissolved in PBS (2 mg/ml) and extensively washed with PBS. This gel filtration procedure is performed by loading the 116 μl reaction mixture onto the column, waiting until the column is dry, adding 2.4 ml PBS (without collecting the liquid), waiting until the column is dry, and finally adding 2 ml of PBS, collecting the liquid from the column. This fraction of radiolabelled antibody can be used in band-shift assays.

Labelling the antigen with a red fluorophore

1. Dissolve in PBS the purified antigen at 1 mg/ml concentration (alternatively, exchange buffer by dialysis).

2. Add 1 ml to a dry tube containing an adequate amount of Cy5-NHS for the labelling of 1 mg protein (Cat. No. PA25001 - monoreactive Cy5 dye pack from Amersham Pharmacia Biotech).

3. Incubate 30 min. at room temperature.

4. Remove unreacted Cy5 dye using a disposable PD-10 column (Amersham Pharmacia Biotech), extensively pre-washed with PBS (> 25 ml). This gel filtration procedure is performed by loading the 1 ml reaction mixture onto the column, waiting until the column is dry, adding 1.5 ml PBS (without collecting the liquid), waiting until the column is dry, and finally adding 2.5 ml of PBS, collecting the liquid from the column. A clear separation between incorporated and unincorporated dye

should be visible. The collected blue-coloured fraction of fluorescently labelled antigen can be used in band-shift assays.

Native gel electrophoresis

1. Protein samples (4 μl) are diluted with 2 μl of gel mix immediately before the electrophoretic separation.

2. Apply protein samples onto a Gradient 8-25 Phast Gel (Amersham Pharmacia Biotech; a different gel percentage may be required, depending on the mobility of the proteins which have to be measured), equipped with native buffer strips in a Phast System separation module (Amersham Pharmacia Biotech) and run the gel according to the manufacturer's instructions. Alternatively, vertical non-denaturing polyacrylamide gel electrophoresis can be performed, for example as described in Montigiani, Neri et al. 1996.

3. Remove the gel from the Phast System and image with a suitable imaging technique (autoradiography with a PhosphorImager in the case of radioactive band-shifts; fluorescence imaging in the case of fluorescent band-shifts; Neri, Prospero et al. 1996).

Determination of the concentration of radiolabelled antibody

1. In parallel tubes, incubate a ≈ 1 μM radiolabelled antibody solution in PBS with increasing concentrations of antigen (e.g., ranging between 0.1 μM and 10 μM). At these high concentrations ($>$ Kd), the antibody should rapidly and quantitatively bind to the available antigen.

2. After few minutes of incubation, run the reaction mixtures on a non-denaturing gel as described in the previous section.

3. Image the gel by autoradiography (preferably on a PhosphorImager). The concentration of antigen at which a "band shift" is observed (Figure 1) is equal to the concentration of radiolabelled antibody used in the assay (N.B.: it is NOT equal to the Kd, since we are working in a concentration range $>$ Kd).

Band-shift assay for the determination of the dissociation constant Kd

In these experiments, it is convenient to work at antibody concentrations lower than the dissociation constant. This range of concentrations is typically not known *a priori*, and is determined by performing band-shift experiments at low antibody concentrations (e.g., 0.1- 1 nM). The intensity of the bands is integrated with appropriate methods and fitted to the equation: Kd = [A][B]/[AB] (this procedure relies on the assumption that band intensity is proportional to the concentrations at the moment the sample is loaded on the gel). This fitting may yield an acceptable Kd value, or may indicate in which concentration range one should work to obtain more accurate results.

Another important issue to consider is how long should antibody and antigen be allowed to react before analysing the reaction mixture on a native gel. In pseudo-first order conditions (e.g., when the antibody concentration is kept constant and $<< Kd$) this reaction time is a function of the kinetic association constant k_{on} and of the antigen concentration. The concentrations of antibody free and in complex will exponentially tend to a limit value, and one should wait at least $t > 1 / (k_{on} \times [Antigen]_{total})$ before considering the reaction close to completion. Whenever the k_{on} of the antibody is not known, it is important to measure affinity constants by band-shift with two different incubation times. If the reaction has reached equilibrium, the affinity values in the two experiments should not differ significantly.

The protocol listed below is for a labelled antibody with Kd = 3 nM and $k_{on} = 10^6$ s^{-1}M^{-1}.

1. In parallel tubes, incubate a 0.3 nM radiolabelled antibody solution with increasing concentrations of antigen (e.g., ranging between 0.3 nM and 30 nM) in PBS.

2. After 30 minutes incubation at room temperature, run the reaction mixtures on a non-denaturing gel as described in the previous section.

3. Image the gel by autoradiography (preferably on a PhosphorImager). The concentration of antigen at which a "band shift" is observed (Figure 1) is equal to the dissociation constant Kd. Alternatively, the Kd value can be obtained by fitting the intensity of the bands to the equation: Kd = [A][B]/[AB].

4. To be sure that the equilibrium is reached when the samples are applied onto the gel, repeat the procedure, with 2 hours incubation time in step 2.

Band-shift assay for the determination of the kinetic dissociation constant k_{off}

A prerequisite of this technique is that the antibody-antigen complex can be detected in a non-denaturing gel (if possible as a single band). If the antigen is fluorescently labelled, it is essential to be able to resolve the band of the complex from the band of the free antigen.

It is also worth mentioning that only negatively charged proteins (or protein complexes) migrate towards the anode in a standard non-denaturing polyacrylamide gel with Tris/Glycine buffer, pH 8.3 (see above). If one wishes to study positively charged proteins (or protein complexes), one needs to work with reverse polarity electrodes and possibly with buffer systems different from Tris/Glycine (see manufacturer's instructions of the Phast System, Amersham Pharmacia Biotech).

The following protocol is written for the negatively charged complex of a recombinant antibody with a fluorescently labelled antigen. Although the reaction conditions are given here in the micromolar range, much lower concentrations can also be used if a high-sensitivity luminescence analyser is available (see for example Neri, Prospero et al. 1996).

1. In separate reaction tubes, incubate a 1 μM solution of fluorescently labelled antigen with a 1 μM solution of antibody in PBS for few minutes at room temperature.

2. At different times (e.g., at time 0 h, 24 h, 48 h, 60 h, 66 h, 69 h, 71 h, 71 h 30 min, 71 h 40 min; the choice of time intervals depends on the k_{off} value) add unlabelled antigen to one of the tubes, reaching a final antigen concentration of 20 μM. By this procedure, in the different tubes the antibody-(labelled antigen) complex is allowed to compete with a molar excess of unlabelled antigen for different amounts of time.

3. At the end of the competition reactions (e.g., at 71 h 41 min), run the reaction mixtures on a non-denaturing gel as described in a previous section.

4. Image the gel with a suitable fluorescence imager, making sure that even the strongest bands fall in the linear detection range of the imager (Neri, Prospero et al. 1996).

5. Integrate the fluorescence intensity of the bands.

6. Plot band volume versus time (Figure 3). An exponential fit of the band volume decaying with time will yield the characteristic constant k_{off}.

Subprotocol 2
Competition ELISA

The ELISA protocols for detection of the antibody binding to an antigen-coated microtitre plate are standard laboratory techniques and will not be described here. We will just mention that most recombinant antibody fragments are typically detected using monoclonal antibodies directed against a peptidic tag engineered at the C-terminal extremity of the recombinant antibody. In certain experimental conditions, such peptidic tags may undergo proteolytic cleavage, thereby lowering the sensitivity of antibody detection. Reagents that bind to the antibody molecule without impairing antigen binding (e.g., protein A or protein L) may therefore be preferable. Alternatively, the experimental scheme described below can be performed in a similar fashion, using radiolabelled antibodies and radioacitve detection of antibody-mediated antigen binding.

The concentration of purified antibody preparations is typically determined spectrophotometrically (1 mg/ml antibody solution absorbs 1.4 absorption units at 280 nm). If necessary (for example when using supernatants), the concentration of active antibody can be detected with a straightforward ELISA adaptation of the protocol mentioned above for the determination of antibody concentration by band-shift assay.

■ ■ Procedure

1. Coat with antigen (in identical fashion) an appropriate number of wells of two microtitre plates. Preblock the wells with 3% MPBS for 2 hours at room temperature, then wash with PBS.

2. In parallel tubes, incubate an antibody solution (at concentrations below Kd, e.g. 0.5 nM) with increasing concentrations of antigen (e.g., ranging between 0.1 nM and 1 μM) in PBS [total volume of each reaction: 100 μl].

3. After 30 minutes incubation at room temperature, apply 90 μl of the reaction mixtures to the wells of the first antigen-coated microtitre plate (perform the experiment in duplicate or triplicate), containing 30 μl 10% MPBS.

4. Incubate the reaction mixture on the antigen-coated plate for a suitably short time (e.g., 10 min.).

5. After incubation, transfer the reaction mixtures to the second antigen-coated microtitre plate. The ELISA assay using this second plate will now be performed exactly as for the first microtitre plate. The purpose of the second ELISA assay is to check that only a small fraction of the free antibody is captured on the first microtitre plate and, therefore, no readjustment of the equilibrium occurred during the first capture step.

6. Wash extensively the first ELISA plate and perform the remaining steps of an ELISA procedure, aimed at the determination of the antibody binding to the coated antigen.

7. Develop the ELISA with a suitable chromogenic, fluorogenic or chemiluminescent substrate, and measure the individual wells with an appropriate ELISA plate reader. The highest ELISA signal should be observed at low concentrations of antigen. No ELISA signal should be observed at high concentrations of antigen. The concentration of antigen at which the half-maximal ELISA signal is detected corresponds to the dissociation constant Kd. Alternatively, the Kd value can be obtained by fitting the ELISA signal of the individual wells to the equation: $Kd = [A][B]/[AB]$ (Friguet, Chaffotte et al. 1985).

References

Friguet B, Chaffotte AF, Djavadi-Ohaniance L, Goldberg ME (1985) Measurements of the true affinity constant in solution of antigen- antibody complexes by enzyme-linked immunosorbent assay. J Immunol Methods 77: 305-319

Montigiani S, Neri G, Neri P, Neri D (1996) Alanine substitutions in calmodulin-binding peptides result in unexpected affinity enhancement. J Mol Biol 258: 6-13

Neri D, Montigiani S, Kirkham PM (1996) Biophysical methods for the determination of antibody-antigen affinities. Trends Biotechnol 14: 465-470

Neri D, Petrul H, Winter G, Light Y, Marais R, Britton KE, Creighton AM (1996) Radioactive labeling of recombinant antibody fragments by phosphorylation using human casein kinase II and [gamma-32P]-ATP. Nat Biotechnol 14: 485-490

Neri D, Prospero T, Petrul H, Winter G, Browne M, Vanderpant L (1996) Multipurpose high sensitivity luminescence analyzer (LUANA): use in gel electrophoresis. Biotechniques 20: 708-713

Nissim A, Hoogenboom HR, Tomlinson IM, Flynn G, Midgley C, Lane D, Winter G (1994) Antibody fragments from a 'single pot' phage display library as immuno-chemical reagents. Embo J 13: 692-698

Viti F, Tarli L, Giovannoni L, Zardi L, Neri D (1999) Increased binding affinity and valence of recombinant antibody fragments lead to improved targeting of tumoral angiogenesis. Cancer Res 59: 347-352

Winter G, Griffiths AD, Hawkins RE, Hoogenboom HR (1994) Making antibodies by phage display technology. Annu Rev Immunol 12: 433-455

Affinity Measurements of Antibody Fragments on Phage by Quartz Crystal Microbalance (QCM)

ELKE PROHASKA, CONRAD KÖSSLINGER, ARNE HENGERER,
JOCHEN DECKER, SABINE HAUCK, and STEFAN DÜBEL

Introduction

The screening of large combinatorial phage display gene libraries usually leads to a set of diverse recombinant antibody fragment clones specifically binding to the antigen which have to be characterised further and ranked for their usefulness. Beside specificity, the key parameter to be determined is the affinity to the antigen. The dissociation rate is of particular significance for diagnostic and therapeutic applications. For the affinity determination, the most convenient method would be one which directly utilises phage bound antibody fragments from the culture supernatants of isolated *E. coli* clones. However, affinity determinations of phage particles carrying recombinant antibodies are not easy with established methods like equilibrium ELISA or surface plasmon resonance due to a lack of sensitivity and unspecific effects caused by the large size and filamentous shape of the phage particles. QCM (Quartz Crystal Microbalance), however, provides a method to directly monitor association and dissociation of phage antibodies to their antigen, thus allowing to determine an apparent binding constant. The QCM is an acoustic sensor based on a piezoelectric crystal which can be used for the measurement of specific inter-

✉ Elke Prohaska, Universität Regensburg, Institut für Medizinische Mikrobiologie und Hygiene, Franz-Josef-Strauß Allee 11, 93053 Regensburg, Germany (*phone* +49-941-944-6451; *fax* +49-941-944-6402; *e-mail* Elke.Prohaska@klinik.uni-regensburg.de)
Conrad Kösslinger, BioTul, Gollierstr. 70b, 80339 München, Germany
Arne Hengerer, Siemens AG, Medizintechnik, Abt. GT, Henkestr. 127, 91052 Erlangen, Germany
Jochen Decker, Universität Regensburg, Institut für Medizinische Mikrobiologie und Hygiene, Franz-Josef-Strauß Allee 11, 93053 Regensburg, Germany
Sabine Hauck, Fraunhofer-Institut für Mikroelektronische Schaltungen und Systeme, Hansastr. 27d, 80686 München, Germany
Stefan Dübel, Universität Heidelberg, Molekulare Genetik, Im Neuenheimer Feld 230, 69120 Heidelberg, Germany

actions between immobilized molecules and analytes in solution. Binding of a soluble analyte causes a frequency shift in the resonance frequency which can be recorded by a frequency counter with high resolution. The method, known for four decades (Sauerbrey, 1959), had recently been adapted for the measurement of binding kinetics of phage presented proteins (Hengerer et al., 1999a, Decker et al., 2000) and recombinant antibody fragments fused to filamentous phage gene product III (Hengerer et al., 1999b). It applies an antigen monolayer immobilised on a flat gold surface which is not influenced by mass transport effects at the liquid solid interface. It should be noted that the method does not allow the measurement of true affinities in the given protocol, since the fraction of total phage carrying functional scFv fragments might not be 100% and a fraction of pIII surface expressed scFv antibodies may be shed by proteolysis. Further, in some phage display systems, soluble scFv fragments may be present in the phage supernatants due to the existence of an amber codon between scFv and pIII, or more than one fragment may be presented per phage. Thus, for the determination of true affinities, an accurate determination of true molarities of scFv fragments would be required. Nevertheless, for the ranking of a set of antibody clones from a screening experiment, the apparent affinities as determined by the presented method are sufficient. The minimal mass of the ligand to be analysed has been calculated to be around 5000 Da.

Materials

Equipment
- *AFFco 2000*: Quartz crystal microbalance with flow-injection analysis system and peristaltic pump, 20 MHz quartz crystals with gold electrodes (Fig. 1), all supplied by Fraunhofer Institute of Microelectronic Circuits and Systems, Munich, Germany (Kößlinger et al., 1992).
- 500 µl syringe, Hamilton
- Software: Origin 5.0, Microcal Software Inc.

Reagents
- Phosphate-buffered-saline (PBS)
- Dithiobissuccinimidyl-propionate (DSP), Sigma
- Dimethylacetamide (DMA), Fluka
- Bovine serum albumine (BSA), Sigma or Biomol

– Regeneration buffers: Standard buffer, 100 mM Glycin/HCl pH 2.5; 100 mM NaCl; or commercially available borate buffer, glycine buffer, urea, sodium thiocyanate, guanidine chloride, guanidine thiocyanate (all by Merck)

Optionally

– Streptavidin, Sigma
– Biotinylation Kit, Boehringer
– Pinpoint vector, Promega

Procedure

Binding of molecules in solution to targets immobilized on the quartz crystal of QCM causes a resonance frequency change of the quartz crystal. The time course of this frequency change can be evaluated to determine kinetic and equilibrium constants of the binding reaction.

Affinity measurements of antibody phage by QCM, therefore, consist of three parts: Immobilization of the target molecule, measurement of phage adsorption and data evaluation.

We immobilize the antigen of interest by a chemical linker (DSP) (Uttenthaler et al., 1998a, Uttenthaler et al., 1998 b, Hengerer et al., 1999a) or via biotin-streptavidin interaction (Hengerer et al., 1999b, Decker et al., 2000) onto the gold surface of the quartz crystal. Immobilization via DSP is an easy and fast method to achieve a stable protein layer on the quartz crystal. However, the antigens are immobilized in random orientation. In contrast, site-specifically biotinylated antigens fixed to immobilized streptavidin build up an ordered quasi-monolayered structure. This will result in well interpretable binding curves as each coated molecule is equally accessible to the phage.

Site-specific in vivo biotinylation can be achieved by cloning the gene of interest into a Pinpoint vector downstream to a biotinylation signal. As a consequence, the expressed antigen molecules are modified with exactly one biotin within the respective signal sequence (Schatz, 1993). These biotinylated antigens will bind unidirectional to a streptavidin layer.

Of course any antigen can be biotinylated chemically as well using commercially available biotinylation kits.

Direct immobilization via DSP

All solutes and buffers are applied to the quartz crystal by a micropipette (10 µl or 100 µl) before inserting it into the flow-injection analysis system.

After cleaning the surface of the crystal with 100 µl acetone the quartz is incubated with 0.4% DSP in water free DMA for 15 minutes at room temperature. The quartz is then rinsed with 100 µl PBS and incubated with 10 µl of antigen solution at 4°C for at least 12 h. Finally, the quartz crystal is rinsed with PBS (100 µl) and kept at 4°C in a humid chamber until used.

Immobilization of biotinylated antigen via streptavidin

After activating the quartz crystal with DSP and rinsing (see above), it is incubated with 10 µl streptavidin in PBS (1 mg/ml) at 4°C for at least 12 h. Subsequently, it is rinsed with 100 µl PBS and the biotinylated antigen is applied (1 hour, room temperature). At last, the quartz is washed with 100 µl PBS and kept at 4°C .

Note: For an accurate determination of dissociation kinetics care has to be taken that rebinding of dissociated phage cannot take place. It is therefore important to work with low surface concentrations of the antigen on the gold electrode. For the immobilization procedure protein concentrations should be in the range of 10 to 100 µg/ml.

Measurement

Apparative parameters
The size of the sample loop depends on the association kinetics (our standard: 250 µl). Slow kinetics may require larger sample loops to increase the incubation time for the association reaction. Flow rates are typically 10 to 30 µl/min. The base line should be stable, without a drift. Injection of running buffer must not cause a signal.

Note: Small constant drifts (<2Hz/min) can be tolerated especially if working with fast kinetics. They can be corrected mathematically by multiplication of the whole curve with the slope of the drift. Larger drifts indicate a malfunction of the fluid system. In this case the whole system should be checked for leakage and air bubbles and cleaned with water (optionally with urea).

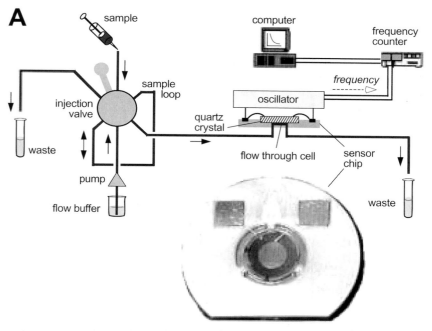

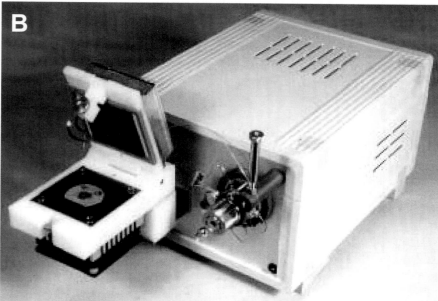

Fig. 1. The QCM sensor system. **A,** experimental setup, **B,** the *AFFco 2000* biosensor machine.

We are routinely working with PBS as running buffer. In principle, however, any other buffer can be used as well. Phage should be diluted in the same buffer to minimize signals arising from differences in viscosity.

As differences in viscosity between running buffer and sample cannot be avoided completely (resuspended phage themselves increase the viscosity of the buffer), it is important to separate such signals from the binding and dissociation kinetics. By injection of a non-binding sample (e.g. a suspension of helper phage in comparable concentration) the time required for the frequency change resulting from viscosity change can be assessed. This time depends on dispersion and on the slowness of the detector. Dispersion causes an inconstant concentration of phage. For these reasons this time should be regarded as the dead time of the experiment and should be excluded when determining the time range for the fitting procedure of rate constants. An example of the response of the QCM to an extreme change of viscosity is shown in Fig. 2.

Measurement   Before starting the measurements, we inject BSA (0.025% in PBS) to block uncovered gold sites to which unspecific binding might occur. The injection of a sample of helper phage might serve to determine the dead time of the experiment (see above).

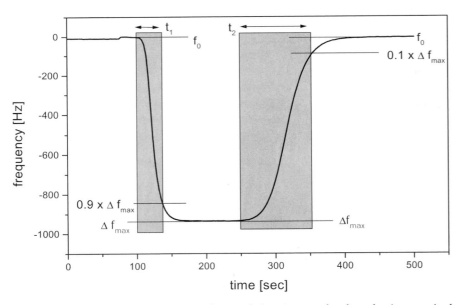

Fig. 2. Response of QCM to an extreme change of viscosity. t_1 and t_2 show the time required for 90% of signal change.

After these controls the actual measurements of the phage under investigation can be started. Various concentrations of phage should be injected. Binding and dissociation processes are recorded (Fig. 3).

Note: The range of applicable concentrations depends on the binding constant and on the sensitivity of the QCM. Suspensions of 10^{11} to 10^{13} phage/ml may give reasonable signals. One should start with low concentrations and then inject higher concentrations. Although it is not necessary to observe slow binding or dissociation reactions up to their very ends, it can be advantageous to record at least one dissociation curve over night.

Various buffers can be used to regenerate the protein layer. We successfully applied acidic and basic buffers (glycine buffer, borate buffer) and chaotropic reagents (urea, sodium thiocyanate, guanidine chloride, guanidine thiocyanate). These buffers might partly damage the receptor layer causing thereby a decreased signal amplitude. However, the rate constants are not affected and therefore the dissociation constant can be calculated nevertheless.

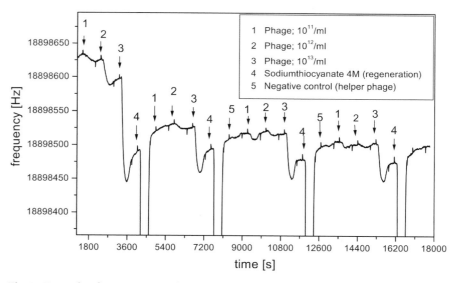

Fig. 3. Example of a sensorgram showing 17 consecutive measurements on the same chip (for details about the samples see Hengerer at al., 1999b). Three different concentrations of antibody phage were tested. Helper phage served as negative control. Each set of controls and samples was followed by regeneration of the receptor layer with 4 M sodium thiocyanate. In spite of incomplete regeneration of the surface, the kinetic parameters obtained from consecutive experiments were reproducible.

Note: A standard buffer containing 100 mM Glycin/HCl pH 2.5; 100 mM NaCl has been proven to be sufficient in many cases. However, it is best to establish the optimal regeneration buffer conditions for each antigen/antibody pair. To do this, start with mild conditions and proceed to more extreme ones. For example, several injections of acidic buffers of pH 4, pH 3.5, pH 3 etc. can be tried. When working with chaotropic buffers, the range from relatively low concentrations (1 M, 1.5 M, 2 M) to saturation should be tested. Optimally, these buffers do not denature the proteins, but merely inhibit their interaction. In the cases where the antigen structure is known to be a hapten or a short peptide epitope, more rigid conditions may be applied from the start.

Determination of dissociation constant

Dissociation constants are determined by non-linear regression of the data by applying Origin software. The rate constant of dissociation k_{dis} is assessed using equation (1):

$$F_{(t)} = F_0 * e^{k_{dis}*(t-t_0)} \tag{1}$$

The kinetics of association is fitted to equation (2):

$$F_{(t)} = \frac{k_{ass} * C * \Delta F_{max}}{k_{ass} * C + k_{dis}} * (e^{(k_{ass}*C+k_{dis})*(t-t_0)} - 1) + F_0 \tag{2}$$

The dissociation constant is the quotient of k_{dis} and k_{ass} (3):

$$K_D = \frac{k_{dis}}{k_{ass}} \tag{3}$$

$F_{(t)}$: frequency [Hz], k_{ass}: rate constant of association [$s^{-1}M^{-1}$], k_{dis}: rate constant of dissociation [s^{-1}], C: concentration [M], ΔF_{max}: maximal frequency shift for completely occupied surface [Hz], t: time [s], t_0: start time for fitting procedure [s], F_0: frequency [Hz] at t_0.

First, the dissociation reaction should be analyzed, as equation 1 depends only on k_{dis}. k_{dis} can then be inserted into equation 2 to determine k_{ass} by non-linear regression analysis. K_D can be calculated by using equation 3.

Note: When choosing the fitting range it is important to exclude the dead time of the experiment in which the signal of viscosity change overlaps phage binding (see above). Kinetics from different concentrations of phage should be measured and should result in the same values for rate constants of association and dissociation. Phage concentrations can be determined by determination of pfu/cfu or by quantitative ELISA using monoclonal antibodies to p8 major phage protein. Please be aware that the resulting apparent affinity does not necessarily represent the absolute affinity of the scFv fragment (see introduction for details).

An example of determination of association and dissociation is presented below (Fig. 4):

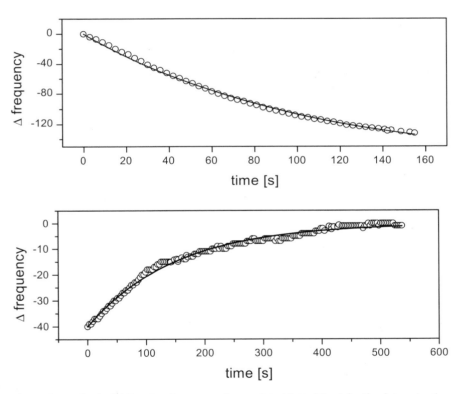

Fig. 4. Curve fits (solid lines) to frequency change data (dotted line) for the determination of k_{ass} and k_{dis} based on one of the injections shown in Fig. 3 (using 10^{13} antibody phage, for details see Hengerer at al., 1999b). A: Frequency change in Hz due to specific adsorption of phage to the respective antigen. B: Frequency change in Hz due to desorption of binding phage from the receptor layer. By nonlinear regression, the following values were derived: $\Delta F_{max} = 569 \pm 41$ Hz, $k_{ass} = 1.7\ (\pm\ 0.1) \times 10^{5}\mathrm{M}^{-1}\mathrm{s}^{-1}$, $k_{dis} = 6.6\ (\pm\ 0.05) \times 10^{-3}\mathrm{s}^{-1}$.

References

Kößlinger, C., Drost, S., Aberl, F., Wolf, H., Koch, S. and Woias, P. (1992) A Quartz Crystal Biosensor for Measurement in Liquids, Biosensors & Bioelectronics 7, 397.

Decker, J., Weinberger, K., Prohaska, E., Hauck, S., Kößlinger, C., Wolf, H. and Hengerer, A. (2000) Characterization of a human pancreatic secretory trypsin inhibitor mutant binding to Legionella pneumophila by a quartz crystal microbalance. J. Immunol. Meth. 233, 159.

Hengerer, A., Kösslinger, C., Decker, J., Hauck, S., Queitsch, I., Wolf, H. and Dübel, S. (1999a) Determination of phage antibody affinities to antigen by a microbalance sensor system. Biotechniques 26, 956.

Hengerer, A., Decker, J., Prohaska, E., Hauck, S., Kösslinger, C. and Wolf, H. (1999b) Quartz crystal microbalance (QCM) as a device for the screening of phage libraries. Biosens Bioelectron 14, 139.

Uttenthaler, E., Kösslinger, C. and Drost, S. (1998a) Quartz crystal biosensor for detection of the african swine fewer disease. Analytica Chimica Acta 362, 91.

Uttenthaler, E., Kösslinger, C. and Drost, S. (1998b) Characterization of immobilization methods for African Swine Fever virus protein and antibodies with a piezoelectric immunosensor. Biosens Bioelectron submitted.

Sauerbrey, G.Z. (1959) The use of oscillators for weighting thin layers and for microweighting. Z. Phys. 155, 209-212.

Schatz, P.J. (1993) Use of peptide libraries to map the substrate specificity of a peptide-modifying enzyme: a 13 residue consensus peptide specifies biotinylation in Escherichia coli. Biotechnology (N Y) 11, 1138.

Sequence and Structure
Analysis and Modelling

Sequence Analysis of Human Antibody Genes

OLGA IGNATOVICH and IAN M. TOMLINSON

Introduction

The antibody molecule, or immunoglobulin (Ig), is composed of two identical heavy (H) and two identical light (L) polypeptide chains, with the light chains being of either kappa (κ) or lambda (λ) type. Each heavy chain consists of a variable domain (V_H) and three or four constant domains (C_H); each light chain consists of a variable domain (V_L: V_κ or V_λ) and a single constant domain (C_L: C_κ or C_λ). The immunoglobulin molecule is bifunctional: the variable domains are involved in binding to the antigen, whereas the constant domains mediate effector functions (interaction of the antibody with other components of the immune system, such as phagocytic cells and/or complement). The antigen binding site is formed by the association of six polypeptide loops, three from the variable domain of the heavy chain (H1, H2 and H3) and three from the variable domain of the light chain (L1, L2 and L3) (see Chothia et al., 1989). The diversity in this site enables antibodies to recognise a wealth of antigenic structures.

In humans, V_H and V_L repertoires are created by the combinatorial rearrangement of a relatively small number of gene segments (Table 1): variable (V_H), diversity (D) and joining (J_H) segments for the V_H domain and variable (V_L: V_κ or V_λ) and joining (J_L: J_κ or J_λ) segments for the V_L (V_κ or V_λ) domain (Tonegawa, 1983; 1993). In addition to enabling different germline V, (D) and J segments to come together (combinatorial diversity), the mechanism of recombination itself generates significant diversity at the V_H-D-J_H and V_L-J_L junctions due to imprecise joining and

✉ Olga Ignatovich, MRC Centre for Protein Engineering and Laboratory of Molecular Biology, Hills Road, Cambridge, CB2 2QH, UK (*phone* +44-1223-402155; *fax* +44-1223-402140; *e-mail* oi1@mrc-lmb.cam.ac.uk)
Ian M. Tomlinson, MRC Centre for Protein Engineering and Laboratory of Molecular Biology, Hills Road, Cambridge, CB2 2QH, UK (*phone* +44-1223-402103; *fax* +44-1223-402140; *e-mail* imt@mrc-lmb.cam.ac.uk)

incorporation of untemplated nucleotides (junctional diversity). Together with the pairing of different heavy and light chains, this creates a highly diverse primary antibody repertoire. Exposure to antigen stimulates additional processes, including somatic hypermutation and receptor editing that further diversify the Ig repertoire (Berek & Milstein, 1988; Radic & Zouali, 1996).

Since all human V, D and J segments and constant region genes have been completely mapped and sequenced and the functional antibody repertoire defined (Table 1), it is now possible to determine the germline segments and somatic mutations that constitute any rearranged V_H, V_κ or V_λ gene. This chapter deals with the amplification of rearranged human V genes, their sequencing and their analysis using a database of germline V gene segments.

Table 1. Number of functional variable and constant region genes

	Light chain		Heavy chain
	κ	λ	H
Variable segments (V)	40	31	51
Diversity segments (D)	–	–	25
Joining segments (J)	5	4	6
Constant genes (C)	1	4	9

Subprotocol 1
Isolation of B Lymphocytes and mRNA Preparation

Materials

- Ficoll-Paque (Pharmacia Biotech, Uppsala, Sweden)
- RNAzol B (Cinna/Biotecx Laboratories, Houston, TX)
- Oligo(dT)-cellulose columns (Pharmacia Biotech, Uppsala, Sweden)
- Phosphate-buffered saline (PBS) (Sambrook et al., 1989)
- Diethyl pyrocarbonate (DEPC)-treated water (Sambrook et al., 1989)

Procedure

1. Collect 9 ml of human peripheral blood into 1 ml of 3.8% trisodium citrate (anticoagulant). Recover B lymphocytes using a Ficoll-Paque gradient (Pharmacia Biotech, Uppsala, Sweden) according to the manufacturer's instructions.

2. Isolate total RNA using the RNAzol B RNA isolation reagent (Cinna/Biotecx Laboratories, Houston, TX). Briefly, wash cells twice with PBS and resuspend in 2 ml RNAzol B. Extract RNA by adding 200 μl chloroform, vortex and keep on ice for 5 minutes. Centrifuge at 13,000g (4°C), collect the aqueous phase containing RNA and precipitate with propan-2-ol. Wash RNA pellet with 75% ethanol, air-dry and resuspend in 10 μl of DEPC-treated water.

3. Recover polyadenylated mRNA by passing isolated total RNA through oligo(dT)-cellulose columns (Pharmacia Biotech, Uppsala, Sweden) according to the manufacturer's instructions.

Subprotocol 2
Amplification and Cloning of Rearranged V Genes

The approach described here is based on the method of Gubler and Hoffman (1983) and allows the construction of unbiased cDNA libraries of rearranged V genes. The procedure involves preparation of double-stranded cDNA using external C region primers (Table 2a). Linkers (Table 2c) are then attached to both ends of the cDNA and PCR amplification (following the removal of the linker from the 3' end) is performed using Linker1 as the 5' (back) primer and a nested C region oligonucleotide (Table 2b) as the 3' (forward) primer. Since none of the primers at any step are based in the V region, the procedure does not introduce a bias towards any particular V gene population. Also note that the C region primers (Table 2a, b and d) anneal to all functional C genes: in the case of the C_H genes, the CH.G primers are specific for all γ heavy chains and the CH.M primers are specific for the μ heavy chain.

Table 2. Primers for amplification and sequencing of the rearranged V genes

a. Primers for the first strand cDNA synthesis[a]

Name	Location	5' Sequence 3'
CL290	C_λ region	GAGGCGCGCCGGGCCACTGTCTTCTCC
CK	C_κ region	GAGGCGCGCCAGCAGGCACACAACAGAGG
CH.G	C_H region	GAGGCGCGCCCACGCTGCTGAGGGAGTAG
CH.M	C_H region	GAGGCGCGCCGTTGGGGTGCTGGACTTTG

b. Nested C region primers[b]

Name	Location	5' Sequence 3'
CL234	C_λ region	TGTGGGACTTCCACTG
CK	C_κ region	GGCGGGAAGATGAAGACAG
CH.G	C_H region	GCCTGAGTTCCACGACACC
CH.M	C_H region	GGAGTCGGGAAGGAAGTC

c. Linkers[b]

Name	5' Sequence 3'
Linker1	GCGGTGACCCGGGAGATCT
Linker 2	AGATC

d. Primers for sequencing

Name	Location	5' Sequence 3'
CL.SEQ	C_λ region	AGTGTGGCCTTGTTGGCTTG
CK.SEQ	C_κ region	GGCGGGAAGATGAAGACAG
CH.G.SEQ	C_H region	GTAGTCCTTGACCAGGCAGCCCAG
CH.M.SEQ	C_H region	GGAGACGAGGGGGAAAAGGG

[a] All primers contain an *Asc*I restriction site

[b] Restriction sites have been omitted and are to be added according to the vectors used for cloning.

▪▪ Materials

– RiboClone cDNA Synthesis System M-MLV(H⁻) (Promega, Madison, WI)

– QIAquick PCR Purification kit (Qiagen, Hilden, Germany)

– QIAquick Nucleotide Removal kit (Qiagen, Hilden, Germany)

- QIAquick Gel Extraction kit (Qiagen, Hilden, Germany)

- Gene Pulser system (Bio-Rad, Philadelphia, PA)

- Thermal cycler (Biometra, Göttingen, Germany or PE Applied Biosystems, Foster City, CA)

- Electro-competent *Esherichia coli* TG1 (Gibson, 1984), (can be bought from Stratagene, La Jolla, CA)

- ØX174 RF DNA/*Hae*III digest (HT Biotechnology, Cambridge, UK)

- Restriction enzymes, T4 DNA ligase, M13mp18 RF DNA (New England Biolabs, Beverly, MA)

- 10X Taq polymerase buffer (500 mM KCl, 100 mM Tris-HCl, pH 8.8, 15 mM $MgCl_2$, 1% TritonX-100)

- SuperTaq polymerase (5U/ µl) (HT Biotechnology, Cambridge, UK)

- dNTPs (5 mM each) (Pharmacia Biotech, Uppsala, Sweden)

- Mineral oil (Sigma, St. Louis, MO)

- TE buffer, TBE buffer, TYE plates, H-top with X-Gal and IPTG (Sambrook et al., 1989)

Procedure

1. Synthesise cDNA with the RiboClone cDNA Synthesis System M-MLV(H⁻) (Promega, Madison, WI) according to the manufacturer's instructions. Use external C region primers (Table 2a) to synthesise the first cDNA strand. Extract synthesised cDNA with phenol/chloroform, precipitate with ethanol and wash with 75% ethanol. Air-dry cDNA pellet and dissolve in TE buffer (Sambrook et al., 1989).

2. Anneal Linker1 and Linker2 (Table 2c) (300 pmol of each) by heating to 94°C for 2 min and by cooling slowly to room temperature. Ligate linkers to both ends of the cDNA overnight at 16°C. Note that restriction sites need to be added to the linkers according to the vector used for cloning. Restriction sites should be chosen so as not to cut within the V genes (see Analysis of Antibody Sequences below).

3. Purify ligation products using the QIAquick Nucleotide Removal kit (Qiagen, Hilden, Germany) according to the manufacturer's instructions. Cut overnight with *Asc*I to remove the linker from the 3' end.

Note that no functional V, D and J segments or C region genes contain *Asc*I restriction sites and, therefore, all will remain intact during this procedure. Isolate DNA from digestion mixture using the QIAquick Nucleotide Removal kit (Qiagen, Hilden, Germany).

4. Amplify rearranged V genes using Linker1 (Table 2c) as a 5' (back) primer and a corresponding nested C region 3' (forward) primer (Table 2b). Note that restriction sites need to be added to the nested C region primers according to the vector used for cloning. Restriction sites should be chosen so as not to cut within the V genes (see Analysis of Antibody Sequences below). Set up 50 µl PCR reactions as follows:

10X Taq polymerase buffer	5.0 µl
dNTP mix (5 mM each dNTP)	2.0 µl
back primer (10 µM)	2.5 µl
forward primer (10 µM)	2.5 µl
SuperTaq polymerase (5 U/µl)	0.5 µl

Use cDNA from step 3 as a template and add Millipore (Bedford, MA) filtered water to 50 µl. Perform a negative control in parallel with no cDNA added to check for possible contamination. Overlay the reaction mixture with mineral oil and transfer to thermal cycler. Hold at 94°C for 5 min, then perform 30 cycles of amplification: 94°C for 1 min (denaturation), 55°C for 1 min (annealing) and 72°C for 1 min (extension); final extension at 65°C for 5 min. Hold at 4°C.

5. After cycling, run 5 µl of the PCR mixture on a 1.5% agarose gel in 1X TBE buffer alongside molecular weight markers (ØX174 RF DNA/*Hae*III digest) (HT Biotechnology, Cambridge, UK).

6. Purify remaining 45 µl using QIAquick PCR Purification kit (Qiagen, Hilden, Germany) according to the manufacturer's protocol.

7. Cut overnight using appropriate restriction enzymes and gel purify DNA using QIAquick Gel Extraction kit (Qiagen, Hilden, Germany).

8. Ligate overnight into precut and purified M13mp18 RF DNA (New England Biolabs, Beverly, MA).

9. Clean ligated products by extracting twice with phenol/chloroform followed by precipitation with ethanol.

10. Transform electro-competent *E. coli* TG1 (Gibson, 1984) bacteria using Gene Pulser system (Bio-Rad, Philadelphia, PA).

11. Add transformed bacteria to 3 ml of molten H-top media containing X-Gal and IPTG (Sambrook et al., 1989) and pour onto TYE plates (Sambrook et al., 1989). Grow overnight at 37°C and toothpick colourless plaques either directly for sequencing or onto TYE plates (Sambrook et al., 1989) and grow as colonies for 30 hrs.

Subprotocol 3
Templates for DNA Sequencing

Both single- and double-stranded DNA can be used as a template for DNA sequencing. Although double-stranded DNA tends to give shorter sequencing reads than single-stranded DNA, it is normally more than adequate for sequencing antibody V genes which are 300-400 nucleotides in length.

■ ■ Materials

- QIAquick PCR Purification kit (Qiagen, Hilden, Germany)

- QIAprep Spin Miniprep kit (Qiagen, Hilden, Germany)

- Thermal cycler (Biometra, Göttingen, Germany or PE Applied Biosystems, Foster City, CA)

- 10X Taq polymerase buffer (500 mM KCl, 100 mM Tris-HCl, pH 8.8, 15 mM $MgCl_2$, 1% TritonX-100)

- M13 specific primers (New England Biolabs, Beverly, MA)

- dNTPs (5 mM each) (Pharmacia Biotech, Uppsala, Sweden)

- SuperTaq polymerase (5 U/μl) (HT Biotechnology, Cambridge, UK)

- Mineral oil (Sigma, St. Louis, MO)

The crucial factor for successful sequencing is the purity of the template DNA. High-purity preparations of double stranded vector DNA can be achieved using the QIAprep Spin Miniprep Kit (Qiagen, Hilden, Germany), according to the manufacturer's instructions. Best results are achieved when using 200-500ng of the purified double-stranded DNA per sequencing reaction.

Nevertheless, PCR products are the most popular choice of sequencing templates today, as their preparation is very quick and straightforward. The procedure for preparation of the PCR templates is outlined below.

▪▪ Procedure

1. Set up 50 µl PCR reactions as described above. Use M13 specific primers (New England Biolabs, Beverly, MA) to amplify the cloned V gene inserts.

2. Toothpick bacterial colonies or phage plaques (see above) into the reaction mixture, overlay with a drop of mineral oil and perform 25 cycles of PCR amplification as described above.

3. After cycling, run 5 µl of the PCR mixture on a 1.5% agarose gel in 1X TBE buffer alongside with molecular weight markers (ØX174 RF DNA/ HaeIII digest) (HT Biotechnology, Cambridge, UK).

4. Purify remaining 45 µl using QIAquick PCR Purification kit (Qiagen, Beverly, MA) according to the recommended protocol.

5. Use 30-90ng of the purified PCR product per sequencing reaction.

Subprotocol 4
DNA Sequencing of Rearranged V Genes

The DNA sequencing method described here is based on the Sanger dideoxy chain termination technique (Sanger et al., 1977). In our laboratory we use a 373A Automated Sequencer with an ABI 373 BigDye Filter Wheel (PE Applied Biosystems, Foster City, CA) which allows the use of the new dichlororhodamine (dRhodamine) dye-labelled terminators (PE Applied Biosystems, Foster City, CA). dRhodamine-based chemistries are an improvement over conventional rhodamine dyes. They are better spectrally resolved and produce data with much more uniform peaks that result in reduced background noise, cleaner signals, improved basecalling accuracy and longer read lengths. Using dye-labelled terminators also means a single-tube reaction per template which reduces liquid handling and enables increased sample throughput.

Materials

- One "notched" and one "plain" glass sequencing plate, pair of gel spacers, gel former and sharkstooth comb (PE Applied Biosystems, Foster City, CA)

- ABI PRISM BigDye Terminator Cycle Sequencing kit (PE Applied Biosystems, Foster City, CA)

- 373A Automated Sequencer with an ABI 373 BigDye Filter Wheel (PE Applied Biosystems, Foster City, CA) and an Apple Macintosh computer

- Gelmix: 250 g urea, 75 ml 40% acrylamide/bis-acrylamide (19:1 ratio) (Severn Biotech, Kidderminster, UK), add Millipore (Bedford, MA) filtered water to approximately 450 ml, stir at 50°C until dissolved, adjust to 450 ml, deionise by adding 55 g Amberlite MB1 resin (Fisher Scientific, Loughborough, UK) and stirring gently for 30 minutes, filter through a 0.45 µm filter and degas for 5 minutes, add 50 ml 10X TBE, mix, and store at 4°C for no longer than a month

- Thermal cycler (Biometra, Göttingen, Germany or PE Applied Biosystems, Foster City, CA)

- Mineral oil (Sigma, St. Louis, MO)

- TBE buffer: 108 g Tris base, 55 g boric acid, 8.3 g EDTA, Millipore-filtered water (Bedford, MA) to 1 litre, filter through 0.45 µm filter, pH should be 8.0-8.5

- deionised formamide/EDTA: 100 ml formamide, 10 g Amberlite resin MB1 (Fisher Scientific, Loughborough, UK), stir gently for 30 minutes, filter through a 0.45 µm filter and store at 4°C. Take 200 µl deionised formamide and add 40 µl 50 mM EDTA, pH 8.0 immediately before resuspending sequencing reactions

- Sequencing buffer: 200 mM Tris, pH 9, 5 mM $MgCl_2$

Procedure

Set up sequencing reactions and purify extension products according to the following protocol: **Set up**

1. For each reaction, add the following reagents to a 0.5 ml microcentrifuge tube or a well of the 96-well thermostable microtitre plate:

BigDye Terminator Ready reaction mix kit (PE Applied Biosystems, Foster City, CA)	4 µl
Sequencing buffer	4 µl
Primer (Table 2d)	3.2 pmol
Template DNA (see above)	
Millipore-filtered water (Bedford, MA) to the final volume of 20 µl	

2. Mix and spin briefly. Add one drop of mineral oil to each tube/well. Place the tubes/plate into the thermal cycler and perform the following programme: 25 cycles of 96°C for 30 sec, 50°C for 15 sec and 60°C for 4 min. Hold at 4°C.

3. For each sequencing reaction, prepare a 0.5 ml microcentrifuge tube containing 2 µl of 3M sodium acetate, pH 5.2 and 50 µl of 95% ethanol.

4. Take the entire 20 µl of the reaction mix from under the oil and transfer into the tube containing the sodium acetate/ethanol mixture.

5. Vortex the tubes and place on ice for at least 15 min to precipitate the extension products.

6. Spin the tubes in a microcentrifuge for 15-20 min at 13,000g.

7. Aspirate ethanol and leave pellets to air-dry. If samples are to be stored, store them as dried pellets at -20°C.

Sequencing gel Prepare and run sequencing gel as follows:

1. Wash glass plates and spacers in deionised water and wipe dry with window wiper or Kimwipes. Do not use organic solvents!

2. Assemble glass plates and spacers and clamp plate assembly with two bulldog clamps on each side.

3. To a 50 ml aliquot of the gelmix add 100 µl 25% ammonium persulphate and 22.5 µl TEMED (BDH, Poole, UK). Mix, take up the gelmix into a 50 ml disposable syringe and inject (horizontally) from the centre of the plate assembly. Allow mix to enter plate assembly by capillary action. Avoid air bubbles (tapping on the plates at problem areas will prevent bubble formation).

4. Insert gel former and clamp the plate assembly with three more bull-dog clamps across the top. Allow the gel to polymerise for at least 1.5 hours. Gels can be kept overnight by wrapping gel former and top of plate assembly in a Kimwipe moistened with 1X TBE and covering with Saran wrap.

5. When the gel is set, remove gel former, rinse the plate assemby and wipe dry as before. Put the lower buffer chamber into the ABI 373A and put in the plate assembly.

6. Switch on the machine and the computer. Launch the ABI Prism 373XL Collection software. Create a new Sequence Run file, set up Run Mode (Full Scan for 24/36 lanes and XL Scan for 48/64 lanes), Collect time and Sample Sheet. Select a "Plate Check" option and check that the plates are clean (blue, red, yellow and green lines should all be straight). Reclean plates if necessary. Set the PMT voltage so that the blue line is between 800 and 1200.

7. Insert the appropriate sharkstooth comb with the teeth just penetrating the gel surface.

8. Assemble the upper buffer chamber and fill chambers with 1X TBE buffer.

9. Dissolve air-dried sequencing samples (see above) in 2-6 µl of de-ionised dormamide/EDTA (2 for 48, 4 for 36 and 6 for 24 lanes), denature at 90°C for 2 min and place on ice.

10. Flush sample slots with 1X TBE buffer and load every other sample.

11. Select "Prerun" and prerun the gel for 5 min. Abort prerun. Flush sample slots as before and load remaining samples.

12. Select "Run".

13. After the run, check the gel image for correct lane tracking, retrack and reanalyse data if necessary.

Results

Analysis of antibody sequences

Analysis of a rearranged V gene is most conveniently performed by alignment of its nucleotide sequence to a database of germline gene segments.

Over the past 10 years, we have compiled a comprehensive directory of all human germline V, D and J segments, V BASE, which incorporates our own data and all published sequences, including those in the current releases of the Genbank and EMBL data libraries (see http://www.mrc-cpe.cam.ac.uk/imt-doc/public/intro.html). In addition to software for the analysis of rearranged genes (developed in collaboration with Werner Müller and colleagues from the University of Köln, Germany) V BASE includes alignments of the germline sequences and tables that list the number of functional segments, the numbers of segments cut by different restriction enzymes and PCR primers for amplifying rearranged V genes. V BASE can also be downloaded in FASTA format for use on your own computer.

Having determined the closest germline V, D and J segment counterparts of your rearranged gene by alignment to V BASE, any nucleotide differences must be checked against the original chromatogram file before being confirmed as new polymorphisms, somatic mutations or PCR errors (in the case of cloned genes). In our experience a single nucleotide change is as likely to be due to a miscalled base rather than a genuine difference between the rearranged and germline sequences. The software package, Sequence Navigator (PE Applied Biosystems, Foster City, CA) can be used both to compare the imported chromatogram file of the rearranged

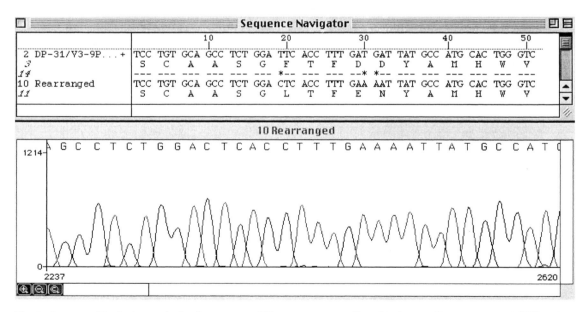

Fig. 1. Sequence Navigator analysis of a rearranged V gene sequence aligned to its germline counterpart. Differences between the two sequences are denoted by stars.

sequence with the germline V, (D) and J segment counterparts copied from V BASE (Figure 1) and also to produce putative protein translations of the rearranged and germline genes.

References

Berek C, Milstein C (1988) The dynamic nature of the antibody repertoire. Immunol. Rev. 105: 5-26

Chothia C, Lesk AM, Tramontano A, Levitt M, Smith-Gil SJ, Air G, Sheriff S, Padlan EA, Davies D, Tulip WR, et al. (1989) Conformations of immunoglobulin hypervariable regions. Nature 342: 877-883

Gibson TJ (1984) PhD thesis. University of Cambridge

Gubbler U, Hoffman BJ (1983) A simple and very efficient method for generating cDNA libraries. Gene 25: 263-269

Sambrook J, Fritsch EF, Maniatis T (1989) Molecular cloning- a laboratory manual. Cold Spring Harbor Laboratory

Radic MZ, Zoulali M (1996) Receptor editing, immune diversification, and self tolerance. Immunity 5: 505-511

Sanger F, Nicklen S, Coulson AR (1977) DNA sequencing with chain-terminating inhibitors. Proc. Natl. Acad. Sci. USA 74: 5463-5467

Tonegawa S (1983) Somatic generation of antibody diversity. Nature 302:575-581

Tonegawa S (1993) Somatic generation of immune diversity. Scand. J. Immunol. 38: 305-317

Protein Sequence and Structure Analysis of Antibody Variable Domains

ANDREW C.R. MARTIN

Introduction

The protocols described here provide methods for computational analysis of antibody sequence and structure. With the availability of the World Wide Web, many online analysis tools have been made available and URLs for these are cited throughout the text. The author has provided a Web site at http://how.to/AnalyseAntibodies/ containing links to all the tools described here.

Brief review of early work on sequence variability and antibody structure

Porter [1] first proposed the four chain model for antibodies consisting of two light chains and two heavy chains, the latter linked by disulphide bonds. The structure of antibodies has been reviewed in detail by a number of authors [2, 3, 4] while the structural basis of antibody/antigen interactions has also been reviewed extensively [5, 6, 7, 8]. The major points will be briefly covered here.

Edelman and Gall [9] analysed sequences of IgG chains and identified homologous regions which they proposed related to domains of specific function [10]. Wu and Kabat [11] examined the sequences of the region proposed to form the variable domain and identified 'hypervariable' segments within that domain which, they proposed, formed the actual antigen combining site. These were termed 'complementarity determining regions' (CDRs) and it was suggested that the CDRs are supported on a framework formed by the rest of the variable domain.

Andrew C.R. Martin, University of Reading, School of Animal and Microbial Sciences, P. O. Box 228, Whiteknights, Reading, RG6 6AJ, UK (*phone* +44-118-987-5123 x 7022; *fax* +44-118-931-0180; *e-mail* a.c.r.martin@reading.ac.uk)

IgG is the best studied of the immunoglobulin classes and electron micrographs revealed the 'Y' shape [12]. The structure of antibodies is divided into variable regions able to bind to a virtually infinite range of substrates and constant regions able to perform a given set of common functions for all antibodies within a class (IgG, IgM, etc.).

Light chains consist of V_L and C_L domains, while heavy chains consist of V_H, C_H1, C_H2 and C_H3 domains (IgM has an additional C_H4 domain). Various fragments are generated by proteolytic digestion or artificially:

Fv	V_H/V_L dimer
Fab	a single arm of the 'Y', consisting of a $V_H,C_H1/V_L,C_L$ dimer (from Papain cleavage)
F(ab')$_2$	the two Fab arms of the 'Y' joined by the disulphide(s) between the heavy chains (from Pepsin cleavage)
Fc	The stem consisting of $C_H2,C_H3/C_H2,C_H3$ (from Papain cleavage)

The first x-ray crystal structure of a Fab fragment was solved in 1973 [13] and showed that the hypervariable regions corresponded approximately to structural loops. The anti-lysozyme antibody D1.3 was the first antibody crystal structure to be solved complexed with antigen [14] confirming the role of the CDRs in binding antigen.

Variable and constant domains all consist of two twisted antiparallel β-sheets which form a β-sandwich. Constant regions have three and four stranded sheets while variable regions have a further two short strands forming sheets of four and five strands. The two sheets are linked by a conserved disulphide bond and are inclined by approximately 30° to one another [15] varying by up to 18° in variable domains and up to 10° in constant domains [16].

As well as this variability in V_L /V_H packing angle, the 'elbow angle' describes the flexibility between the V_L/V_H and C_L/C_H1 pseudo-diads. The angle between the arms of the 'Y' is variable as a result of a flexible hinge region (deleted in IgM). The role of flexibility in antigen binding is reviewed by Huber and Bennet [17].

Linking sequence and structure

Once a structure has been solved for a protein in a homologous family only one simple ingredient is needed in order to link the sequences of other homologous family members to that structure: a standardized numbering scheme. In this way, one always know that, for example, residue number 35 is at the start of a β-strand. Insertions in the sequence relative to that standard numbering scheme are given numbers such as 27A, while deletions are accomodated by simply skipping numbers.

Ideally, such schemes are designed in the light of both large amounts of sequence information and multiple structures. Insertion sites (i.e. residue 27A, etc.) are placed only in loop regions (or form β-bulges) and have structural meaning such that topologically equivalent residues in these loops get the same label.

The Kabat numbering scheme

The Kabat numbering scheme is the most widely adopted standard for numbering the residues in an antibody in a consistent manner. However the scheme does have problems.

The numbering scheme was developed solely from somewhat limited sequence data. Unfortunately, the position at which insertions are placed in CDR-L1 and CDR-H1 does not match the structural insertion position. Thus topologically equivalent residues in these loops do not get the same number.

The second problem is that the numbering adopts a rigid specification. For example in the potentially very long CDR-H3, insertions are numbered between residue H100 and H101 with letters up to K (i.e. H100, H100A ... H100K, H101). If there are more residues than that, there is no standard way of numbering them. Such situations occur at other positions too. The raw Kabat data files in these cases simply state what the additional insertions are and where they occur, but do not assign numbers to them.

The numbering throughout the chains is shown in Table 1.

The Chothia numbering scheme

The Chothia numbering scheme is identical to the Kabat scheme with the exception of CDR-L1 and CDR-H1, where the insertions are placed at the structurally correct positions. This means that topologically equivalent residues in these loops do get the same label.

There are two disadvantages. First, the Kabat scheme is so widely used that some confusion can arise. Second, Chothia et al. erroneously changed their numbering scheme in their 1989 Nature paper [18] such that insertions in CDR-L1 are placed after residue L31 rather than L30. A visual

Table 1. Kabat numbering scheme

Light chain

```
  0    1    2    3    4    5    6    7    8    9   10   11   12   13   14   15   16   17   18   19
 20   21   22   23   24   25   26   27
27A  27B  27C  27D  27E  27F
                                        28   29   30   31   32   33   34   35   36   37   38   39
 40   41   42   43   44   45   46   47   48   49   50   51   52   53   54   55   56   57   58   59
 60   61   62   63   64   65   66   67   68   69   70   71   72   73   74   75   76   77   78   79
 80   81   82   83   84   85   86   87   88   89   90   91   92   93   94   95
95A  95B  95C  95D  95E  95F
                                                                           96   97   98   99
100  101  102  103  104  105  106
106A
                        107  108  109
```

Heavy chain

```
  0    1    2    3    4    5    6    7    8    9   10   11   12   13   14   15   16   17   18   19
 20   21   22   23   24   25   26   27   28   29   30   31   32   33   34   35
35A  35B
                                                                           36   37   38   39
 40   41   42   43   44   45   46   47   48   49   50   51   52
52A  52B  52C
                                                            53   54   55   56   57   58   59
 60   61   62   63   64   65   66   67   68   69   70   71   72   73   74   75   76   77   78   79
 80   81   82
82A  82B  82C
                 83   84   85   86   87   88   89   90   91   92   93   94   95   96   97   98   99
100
100A 100B 100C 100D 100E 100F 100G 100H 100I 100J 100K
     101  102  103  104  105  106  107  108  109  110  111  112  113
```

examination of the conformations of CDR-L1 loops shows that L30 is the correct position. In a more recent paper on CDR conformation [19], Chothia's group returns to using residue L30 as the insertion site in CDR-L1.

The structurally correct version of the Chothia numbering (as used before 1989 and since 1997) throughout the chains is shown in Table 2.

CDR definitions

Table 3 illustrates the main definitions of the CDRs which are commonly in use:

- The Kabat definition is based on sequence variability and is the most commonly used.

- The Chothia definition is based on the location of the structural loop regions.

Table 2. Chothia numbering scheme

Light chain

0	1	2	3	4	5	6	7	8	9	10	11	12	13	14	15	16	17	18	19
20	21	22	23	24	25	26	27	28	29	30									
30A	30B	30C	30D	30E	30F														
											31	32	33	34	35	36	37	38	39
40	41	42	43	44	45	46	47	48	49	50	51	52	53	54	55	56	57	58	59
60	61	62	63	64	65	66	67	68	69	70	71	72	73	74	75	76	77	78	79
80	81	82	83	84	85	86	87	88	89	90	91	92	93	94	95				
95A	95B	95C	95D	95E	95F														
																96	97	98	99
100	101	102	103	104	105	106													
106A																			
							107	108	109										

Heavy chain

0	1	2	3	4	5	6	7	8	9	10	11	12	13	14	15	16	17	18	19
20	21	22	23	24	25	26	27	28	29	30	31								
31A	31B																		
												32	33	34	35	36	37	38	39
40	41	42	43	44	45	46	47	48	49	50	51	52							
52A	52B	52C																	
													53	54	55	56	57	58	59
60	61	62	63	64	65	66	67	68	69	70	71	72	73	74	75	76	77	78	79
80	81	82																	
82A	82B	82C																	
			83	84	85	86	87	88	89	90	91	92	93	94	95	96	97	98	99
100																			
100A	100B	100C	100D	100E	100F	100G	100H	100I	100J	100K									
	101	102	103	104	105	106	107	108	109	110	111	112	113						

- The AbM definition is a compromise between Kabat and Chothia definitions based on that used by Martin et al. [20] used by Oxford Molecular's AbM antibody modelling software.

- The contact definition has been recently introduced by MacCallum et al. [21] and is based on an analysis of which residues are involved in antigen binding in the available antibody:antigen complex crystal structures. This definition is likely to be the most useful for people wishing to perform mutagenesis to modify the affinity of an antibody since these are residues which are most likely to take part in interactions with antigen.

Note that the end of the Chothia CDR-H1 loop, when numbered using the Kabat numbering convention, varies between H32 and H34 depending on the length of the loop as illustrated in Figure 1.

Table 3. Different definitions of the CDRs

Loop	Kabat	AbM	Chothia	Contact
L1	L24 - L34	L24 - L34	L24 - L34	L30 - L36
L2	L50 - L56	L50 - L56	L50 - L56	L46 - L55
L3	L89 - L97	L89 - L97	L89 - L97	L89 - L97
H1 (Kabat numbering)	H31 - H35B	H26 - H35B	H26 - H32..34	H30 - H35B
H1 (Chothia numbering)	H31 - H35	H26 - H35	H26 - H32	H30 - H35
H2	H50 - H65	H50 - H58	H52 - H56	H47 - H58
H3	H95 - H102	H95 - H102	H95 - H102	H93 - H101

[Note that a recent paper from Chothia's group has used the AbM definition for CDR-H2]

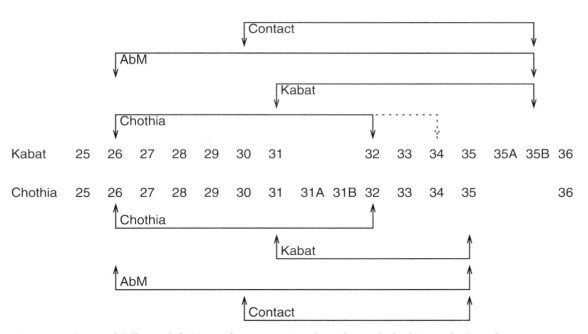

Fig. 1. Boundaries of different definitions of CDR-H1 using the Kabat and Chothia numbering schemes.

Procedure

The following subprotocols describe a number of aspects of sequence/ structure analysis.

Accessing Kabat sequence data

Kabat data may be accessed in a number of ways described below:

Download the raw data

The raw Kabat sequence data may be downloaded for local analysis from either:

- ftp://ncbi.nlm.nih.gov/repository/kabat/
- ftp://ftp.ebi.ac.uk/pub/databases/kabat/

The most up-to-date raw data is in the fixlen subdirectory (or in FASTA format in the fasta_format subdirectory). This site also provides analysed data in PostScript format, but this is not kept up to date so regularly.

EMail query access

The Kabat data may be queried at Kabat and Wu's site by EMail using the SeqhuntII server [22] at seqhunt2@immuno.bme.nwu.edu. Sending mail to this address containing only the word 'help' in the message will return a help file. This document should be obtained for full details of using the SeqhuntII server. In brief, the server allows all sequence classes and annotations to be searched and returned and allows a novel antibody sequence to be compared against the database. Conditions for the search may be combined using and/or/not and nucleotide and amino acid patterns may be specified allowing mismatches.

As well as returning the sequences themselves, the server can be used to calculate the variability of the group of sequences hit by your sequence and residue distribution plots may be calculated.

A query has two parts separated by the word 'Begin'. In the first part, one specifies overall parameters for the search such as the format of the output, number of hits to return, etc. In the second part, one can specify restrictions on the search and patterns to match. Numerous examples are given in the help document. After the restriction comes a pattern which is either a nucleotide or amino acid sequence (fragment) to be matched while allowing a specified number of mismatches.

Access via web site

The Kabat database web site set up by Johnson and Wu (http://immuno. bme.nwu.edu) provides a number of facilities for antibody sequence analysis and for searching the Kabat data.

The standard Kabat numbering for a new sequence may be obtained at http://imuno.bme.nwu.edu/align.html by automated alignment with consensus sequences. Currently only nucleotide sequences of Ig κ or λ chains is supported, though the Web page promises that other types of sequence will be supported in future.

Variability within a sequence type (e.g. κ chains, T-cell receptor, etc) can be calculated at http://imuno.bme.nwu.edu/variability.html . You may restrict the sequences within a group by specifying patterns to match.

The Seqhunt service (http://imuno.bme.nwu.edu/seqhunt.html) allows you to search through a set of fields for a value which you enter. Alternatively, you may search all fields. This is a straight text search of a sequence pattern match. Available fields are: Identification Number, Sequence Type, Sequence Name, Species, Authors, Journals, Antigen and Notes.

KabatMan

KabatMan [23] is a specialised database for the analysis of Kabat antibody sequence data. It may be queried using a language similar to SQL ('structured query language' – a standard for querying relational databases) or via a point and click interface at http://www.bioinf.org.uk/abs/simkab.html.

While the software provided by Johnson, Wu and Kabat at http://imuno.bme.nwu.edu is largely aimed at finding all the information about a single antibody, KabatMan is more suited to global analysis of the antibody data. It also provides a more direct link with structural information by allowing searches to specify individual amino acids or the contents of one of the six CDRs. For example, one could find all the antibodies which bind to DNA, but do not contain arginine in CDR-H3 using the query:

SELECT	name
WHERE	antigen INC 'dna'
	h3 <>' ' AND
	h3 INC 'R' NOT AND

The 'SELECT' statement specifies which results are to be returned (in this case the name of the antibody). The 'WHERE' statement first specifies that the antigen should be DNA, then specifies that the sequence of CDR-H3 should not be blank (i.e. unknown) and finally specifies that CDR-H3 should not include the letter 'R' (i.e. arginine). More examples are given in the paper which describes KabatMan and in the on-line help.

When using KabatMan to find details of a specific antibody, it is important not to overspecify the antibody. For example if you want to find details of the mouse anti-lysozyme antibody HYHEL-5, then the name, HYHEL-5 uniquely identifies the antibody - there is no point in specifying the antigen, class or animal source.

SUBIM   Deret et al. [24] have written a program, SUBIM, for the analysis of Kabat sequence data. Many of the data access functions of this program are also available in KabatMan (described above). The program may be downloaded from ftp://ftp.lmcp.jussieu.fr/pub/subim/ and must be compiled and installed locally. It has no form of graphical user interface.

Accessing IMGT sequence data

Because so many immune-related sequences are being deposited with the major sequence databanks (EMBL, Genbank and DDBJ), they are starting to skew the data these contain and EMBL have begun to separate the immune sequences into a separate databank known as IMGT [25]. The data in IMGT are updated rather more frequently than the Kabat data (which is hand-verified by Johnson and Wu).

The chief advantage of IMGT is the volume of data it contains and the rate at which this is updated. There are however, a number of disadvantages compared with the Kabat data. First, they have introduced a new numbering scheme such that insertion codes (like 27A) are not needed for any normal antibodies (see http://imgt.cnusc.fr:8104/textes/IMGTindex/numbering.html). This, of course, is incompatible with the numbering schemes used by Kabat or Chothia. Second, the raw data (ftp://ftp.ebi.ac.uk/pub/databases/imgt/ligm/) which is stored in a format similar to the SwissProt databank, contains only DNA sequence (no protein translation), does not contain a mapping to the numbering scheme and does not specify the antigen in a standardised easily parsable form (description records may contain specificity subrecords, though on other occasions the information may just be in free text, or may be missing).

IMGT provides a number of services via the web:

- http://www.ebi.ac.uk/imgt/ EBI site which provides access to the main site and to a Sequence Retrieval Service (SRS) interface

- http://imgt.cnusc.fr:8104 Main IMGT server

- http://imgt.cnusc.fr:8104/dnaplot A facility for aligning a DNA sequence against the sequence library and finding the most similar sequences

- http://imgt.cnusc.fr:8104/informatics A Java programming interface for accessing IMGT from computer programs written in Java

The main IMGT server (http://imgt.cnusc.fr:8104) provides a hierarchical interface to the data allowing one to home in on a particular sequence. This is not suited to global analysis of the data. In future, IMGT data will be integrated into KabatMan.

Family assignment

The sequence family can be assigned through Johnson and Wu's page at http://immuno.bme.nwu.edu/famgroup.html. The nucleotide sequence for the variable region must be entered and is compared with the variable region of reference sequences for each family. No attempt is made to find out where the variable region begins if it is a fragment, although Johnson, Kabat and Wu's SeqhuntII programs (http://immuno.bme.nwu.edu/seqhunt.html) may be used to do this. It is then necessary to add leading spaces to the sequence to align the fragment correctly. Sequences must be at least 225 bases in length. Ideally one should have the entire sequence of the variable region.

Assigning subgroups

Deret's SUBIM program [24] which may be downloaded from ftp://ftp.lmcp.jussieu.fr/pub/subim/ allows the assignment of the variability subgroup of novel human sequences by comparison of the N-terminal 15 residues with consensus sequences determined by Kabat et al. [26].

Sophie Deret has kindly made this part of the SUBIM program available to the author who has made this accessible via a Web page (http://www.bioinf.org.uk/abs/hsubgroup.html). In addition, the functionality has been incorporated into KabatMan to allow human subgroup assignment for sequences in the Kabat data.

Kabat and Wu's page (http://immuno.bme.nwu.edu/subgroup.html) also supports sub-group assignment in a more extensive manor allowing input of both amino acids and nucleotides and mouse/human heavy/κ/λ. Note that unlike the Deret method, it is necessary to know whether the input sequence is heavy/κ/λ.

Identifying the CDRs

This protocol describes how to identify the CDRs (Kabat definition) by examining the sequence. Of course there are always (minor) exceptions to these rules, so the word 'always' should be interpreted with care! For example, CDR-L2 is always 7 residues, but antibody NEW (Protein Databank code: 7FAB, http://www.rcsb.org) has a deletion in this region. This also means that the position of the start of CDR-L3 is no longer 33 residues after the end of CDR-L2.

CDR-L1

- **Start** Approx residue 24
- **Residue before** is always C
- **Residue after** is always W. Typically WYQ, but also, WLQ, WFQ, WYL
- **Length** 10 to 17 residues

CDR-L2

- **Start** Always 16 residues after the end of CDR-L1
- **Residues before** generally IY, but also, VY, IK, IF
- **Length** always 7 residues

CDR-L3

- **Start** Always 33 residues after end of CDR-L2
- **Residue before** is always C
- **Residues after** always FGXG
- **Length** 7 to 11 residues

CDR-H1

- **Start** Approximately residue 31 (always 9 after a C) (Chothia/AbM definition starts 5 residues earlier)
- **Residues before** always CXXXXXXXX
- **Residues after** always W. Typically WV, but also WI, WA
- **Length** 5 to 7 residues (Kabat definition); 7 to 9 residues (Chothia definition); 10 to 12 residues (AbM definition)

CDR-H2

- **Start** Always 15 residues after the end of Kabat/AbM definition of CDR-H1

- **Residues before** typically LEWIG, but a number of variations

- **Residues after** K[RL]IVFT[AT]SIA (where residues in square brackets are alternatives at that position)

- **Length** Kabat definition 16 to 19 residues (AbM definition and most recent Chothia definition ends 7 residues earlier; earlier Chothia definition starts 2 residues later and ends 9 earlier)

CDR-H3

- **Start** Always 33 residues after end of CDR-H2 (always 3 after a C)

- **Residues before** always CXX (typically CAR)

- **Residues after** always WGXG

- **Length** 4 to 24 residues

Screening new antibody sequences

Given a new antibody sequence, one is likely to wish to assign families and subgroups using the tools described above. An additional facility is available at http://www.bioinf.org.uk/abs/seqtest.html to identify unusual features in the sequence.

It is simply necessary to enter the amino acid sequence of your Fv fragment (one or both chains). Optionally you may include the whole Fab fragment, but only the Fv portion will be tested.

The tool aligns the provided sequence with a standard sequence in order to assign standard Kabat numbering and then uses the KabatMan database to identify unusual amino acids (i.e. those occurring in less than 1% of the data in the database). This allows the identification of potential cloning artifacts and sequencing errors. If unusual features are verified as being correct, then these residues are likely to be critical to the specificity of the antibody. The method is described in detail at http://www.bioinf.org.uk/abs/seqmethod.html.

The results need to be examined carefully. A typical sequence has 1-2 'unusual' residues. Very unusual sequence features and loops longer than anything observed in the current Kabat database may cause the alignment stage to fail causing errors in the Kabat numbering. Errors can also occur at the C-terminus of the chains. These problems can lead to residues being flagged as 'unusual' when they are not.

Thus, if more than 2 or 3 unusual residues are seen (especially outside the CDRs and if insertions/deletions are observed), the first step is to ensure that the alignment and assignment of Kabat numbering is correct (check the features described in the section "Identifying the CDRs"). If all is judged correct with the alignment, the clone should be checked. If it is confirmed that no cloning errors have occurred, then it is likely that these are key features of the antibody.

Assigning canonicals

The analysis by Chothia and co-workers introduced the concept of canonical conformations for the CDRs. It was proposed that these conformations were influenced by just a small number of residues either in the CDRs or in the framework regions which pack with them. This allows a direct prediction of three-dimensional conformation from sequence.

Chothia and co-workers have published around 10 papers describing this analysis, but unfortunately do not provide a summary of the required amino acids to assign canonical classes. Table 4 attempts to summarise the data from their publications together with results from Martin and Thornton [27]. Chothia numbering of the sequences is used throughout.

Modelling antibodies

There are various approaches to modelling antibodies, but it is widely accepted that methods based on Chothia's analysis of CDR canonical conformations provide the best results where they can be applied.

Any modelling procedure involves the following steps:

Build the framework Antibody crystal structures from the Protein Databank (PDB, http://www.rcsb.org) are searched to identify the most similar light and heavy chains. These are identified separately. If the best match for the light chain is La (paired with Ha) and the best match for the heavy chain is Hb (paired with Lb) then the structure of La is least squared fitted to Lb and chains La and Hb are retained, deleting Ha and Lb. In this way, the V_L/V_H packing angle is inherited from Lb/Hb. To inherit the packing angle from La/Ha, then the fitting is performed on the heavy chains. The choice is relatively

Table 4. Key residues which define the Chothia canonical classes. For CDR-H1, Chothia et al. suggest that residue H27 is also a key residue, but Martin & Thornton did not find this residue influences the conformation. Similarly, for CDR-H2, Chothia et al. identify residue H52A as a key residue in determining the conformation of Class 2 and 3, but Martin & Thornton found that this residue does not influence the conformation.

CDR-L1

Class	1	2	3	4	5	6	5λ	6λ	7λ
Length	10	11	17	16	15	12	13	14	14
L2	I	I	I	VIL	I	N			
L25	A	A	S	S	A	A	G	G	S
L29	VIL	VIL	VIL	L	V	V			
L30	–						I	I	V
L30D				G					
L33	ML	VIL	L	L	L	L	V	V	A
L71	YF	YF	YF	F	F	Y	A	A	A

CDR-L2

Class	1
Length	7
L34	N

CDR-L3

Class	1	2	3	4	5	4λ	5λ
Length	9	9	8	7	10	9	11
L90	QNH	Q	Q	Q	Q		
L94		P					DNG
L95	P	L	P				
L97	T	T		–	T	IV	VG

CDR-H1

Class	1	2	3
Length	10	11	12
H24	TAVGS	VF	VFG
H26	G	G	G
H29	IFLVS	IL	VIL
H34	IVMWTL	WC	WV
H94	RKGSHNTA	HR	HR

CDR-H2

Class	1	2	3	4
Length	9	10	10	12
H54			SGND	KS
H55	GD	GS	GS	Y
H71	RHVI	VALT	R	R

arbitrary and it may be worth constructing two models. Structural fitting is best performed using a program such as ProFit (http://www.bioinf.org.uk/software/).

The sidechains of the framework are then replaced using automated processes available in molecular graphics programs, or software such as CONGEN [28]. Sidechains are generally built using the 'Maximum Overlap Protocol' where the atom positions are inherited from the parent wherever possible and where not possible, conformations are taken from a rotamer library.

Build the CDRs　The methods described above are used to identify canonical classes. This is generally possible for 4 or 5 of the 6 CDRs. CDR-H3 is too variable to be classified at present (it may become possible once sufficient structures are available and a number of authors have begun to suggest limited sets of rules). Antibody crystal structures from the PDB are then searched to find CDRs of the correct canonical classes with maximum sequence identity to the sequence to be modelled. These loops are then attached to the framework either manually using molecular graphics, or by using a least squares fitting program, such as ProFit, to fit the 3 residues on either side of the loop itself (i.e. within the framework region).

CDRs which cannot be built using canonicals may be built by conformational search using CONGEN, by searching the PDB for loops of the same length and with similar distance between the attachment points to the framework, or by combined methods such as CAMAL [20]. Such loops can never be built with a very high degree of confidence in their accuracy.

Sidechains of the CDRs are then built as described above.

Refinement and assessment　Finally the model may be refined by energy minimisation using a package such as GROMOS, AMBER, CHARMM or DISCOVER and structural quality should be assessed using a program such as ProCheck (http://www.biochem.ucl.ac.uk/~roman/procheck/procheck.html) or WhatCheck (http://swift.embl-heidelberg.de/whatcheck). Both programs may be accessed via an online server at http://biotech.embl-heidelberg.de:8400.

Automated methods　The above description details the stages which are necessary in a manual modelling protocol. As an alternative, a number of automated procedures are available. Two of these are general automatic modelling programs and may be used to generate models in a quick and simple manner. However, they do not take advantage of the special properties of antibodies. The third and fourth are programs specially designed for the automated modelling of antibodies.

- **MODELLER** is a general purpose automated protein modelling program (http://guitar.rockefeller.edu/modeller/modeller.html). As such, it is able to produce reasonably reliable models of structures given just a sequence or a sequence aligned to one or more templates from the Protein Databank. However, since it is not designed specifically for antibody modelling, it does not make use of Chothia's canonical analysis and will not model the CDRs as accurately. The software must be downloaded and run locally on a Unix type computer system.

- **SwissModel** is another general purpose automated protein modelling program which is accessible over the Web and does not need to be installed and run locally (http://www.expasy.ch/swissmod/SWISS-MODEL.html). The quality of models is generally not as high as those created by MODELLER and the same caveats about not being antibody-specific apply.

- **AbM** is a commercial program available from Oxford Molecular (http://www.oxmol.co.uk) being antibody specific, it automates the manual procedure described above taking account of canonical structures and using the CAMAL modelling method described by Martin et al. [20] for modelling those loops which cannot be built using canonicals.

- **WAM** is a method based on AbM available on line (http://antibody.bath.ac.uk/). The first model is free, but further runs incur a charge.

Analysis tools applied to humanisation

Sequence/structure analysis programs such as KabatMan can be applied to problems such as humanisation as well as being used for general analysis. Recent work by Saldanha, Martin and Leger [29] used KabatMan to identify a residue which restored the binding of a humanized antibody.

In brief, the humanisation protocol was as follows. Mouse CDRs were grafted onto human frameworks with highest sequence identity. Residues in the framework were then considered for back-mutation to restore full binding. First, key residues (identified by Chothia's canonical analysis) were identified in the framework and back-mutated to those seen in the original mouse antibody. KabatMan was then used to identify residues in the human framework which are particularly unusual in mouse frameworks, even though they may be remote from the combining site. Nine such positions were identified and these were examined on a computer model. Seven of these were conservative and one was a surface residue. However, position 9 in the light chain was unique to the human kappa IV subclass and only seen in one of 1848 mouse kappa sequences. Back-mutation of this residue to that seen in the mouse sequence completely restored binding.

References

[1] Porter, R. R. The hydrolysis of rabbit γ-globulin and antibodies with crystalline papain. Biochem. J. 73:119-127, 1959.

[2] Alzari, P. M., Lascombe, M.-B., and Poljak, R. J. Three-dimensional structure of antibodies. Annu. Rev. Immunol. 6:555-580, 1988.

[3] Padlan, E. A. Anatomy of the antibody molecule. Mol. Immunol. 31:169-217, 1994.

[4] Searle, S. J., Pedersen, J. T., Henry, A. H., Webster, D. M., and Rees, A. R. Antibody structure and function. In: "Antibody Engineering", (Borreback, C. A. K., ed.). Oxford University Press, 1994:3-51.

[5] Padlan, E. A. The structural basis for the specificity of antibody-antigen reactions and structural mechanisms for the diversification of antigen-binding specificities. Quant. Rev. Biophys. 10:35-65, 1977.

[6] Mariuzza, R. A., Phillips, S. E. V., and Poljak, R. J. The structural basis of antigen-antibody recognition. Annu. Rev. Biophys. Bioeng. pages 139-159, 1987.

[7] Davies, D. R., Padlan, E. A., and Sheriff, S. Antibody-antigen complexes. Annu. Rev. Biochem. 59:439-473, 1990.

[8] Wilson, I. A. and Stanfield, R. L. Antibody-antigen interactions. Curr. Opin. Struct. Biol. 3:113-118, 1993.

[9] Edelman, G. M. and Gall, W. E. The antibody problem. Annu. Rev. Biochem. 38:415-466, 1969.

[10] Edelman, G. M. The covalent structure of a human γ-immunoglobulin: XI. functional implications. Biochemistry 9:3197-3205, 1970.

[11] Wu, T. T. and Kabat, E. A. An analysis of the sequences of the variable regions of Bence Jones proteins and myeloma light chains and their implications for antibody complementarity. J. Exp. Med. 132:211-250, 1970.

[12] Valentine, R. C. and Green, N. M. Electron microscopy of an antibody-hapten complex. J. Mol. Biol. 27:615-617, 1967.

[13] Poljak, R. J., Amzel, L. M., Avey, H. P., Chen, B. L., Phizackerley, R. P., and Saul, F. The three-dimensional structure of the Fab' fragment of a human immunoglobulin at 2.8 Å resolution. Proc. Natl. Acad. Sci. USA 70:3305-3310, 1973.

[14] Amit, A. G., Mariuzza, R. A., Phillips, S. E. V., and Poljak, R. J. Three-dimensional structure of an antigen-antibody complex at 6Å resolution. Nature (London) 313:156-158, 1985.

[15] Chothia, C. and Janin, J. Relative orientation of close-packed β-pleated sheets in proteins. Proc. Natl. Acad. Sci. USA 78:4146-4150, 1981.

[16] Lesk, A. M. and Chothia, C. Evolution of proteins formed by β-sheets: II. The core of the immunoglobulin domains. J. Mol. Biol. 160:325-342, 1982.

[17] Huber, R. and Bennett, W. S. Antibody-antigen flexibility. Nature (London) 326:334-335, 1987.

[18] Chothia, C., Lesk, A. M., Tramontano, A., Levitt, M., Smith-Gill, S. J., Air, G., Sheriff, S., Padlan, E. A., Davies, D., Tulip, W. R., Colman, P. M., Spinelli, S., Alzari, P. M., and Poljak, R. J. Conformations of immunoglobulin hypervariable regions. Nature (London) 342:877-883, 1989.

[19] Al-Lazikani, B., Lesk, A. M., and Chothia, C. Standard conformations for the canonical structures of immunoglobulins. J. Mol. Biol. 273:927-948, 1997.

[20] Martin, A. C. R., Cheetham, J. C., and Rees, A. R. Modelling antibody hypervariable loops: A combined algorithm. Proc. Natl. Acad. Sci. USA 86:9268-9272, 1989.

[21] MacCallum, R. M., Martin, A. C. R., and Thornton, J. M. Antibody-antigen interactions: Contact analysis and binding site topography. J. Mol. Biol. 262:732-745, 1996.

[22] Johnson, G., Wu, T.-T., and Kabat, E. A. SEQHUNT. A program to screen aligned nucleotide and amino acid sequences. Methods Mol. Biol. 51:1-15, 1995.

[23] Martin, A. C. R. Accessing the Kabat antibody sequence database by computer. Proteins: Struct., Funct., Genet. 25:130-133, 1996.

[24] Deret, S., Maissiat, C., Aucouturier, P., and Chomilier, J. SUBIM: A program for analysing the Kabat database and determining the variability subgroup of a new immunoglobulin sequence. Comput. Appl. Biosci. 11:435-439, 1995.

[25] Giudicelli, V., Chaume, D., Bodmer, J., Muller, W., Busin, C., Marsh, S., Bontrop, R., Marc, L., Malik, A., and LeFranc, M.-P. IMGT, the international ImMunoGeneTics database. Nuc. Ac. Res. 25:206-211, 1997.

[26] Kabat, E. A., Wu, T. T., Perry, H. M., Gottesman, K. S., and Foeller, C. "Sequences of Proteins of Immunological Interest". U.S. Department of Health and Human Services, National Institutes for Health, Bethesda, MD, Fifth edition, 1991.

[27] Martin, A. C. R. and Thornton, J. M. Structural families in homologous proteins: Automatic classification, modelling and application to antibodies. J. Mol. Biol. 263:800-815, 1996.

[28] Bruccoleri, R. E. and Karplus, M. Prediction of the folding of short polypeptide segments by uniform conformational sampling. Biopolymers 26:137-168, 1987.

[29] Saldanha, J. W., Martin, A. C. R. and Léger, O. B. P. A single backmutation in the human κIV framework of a previously unsuccessfully humanized antibody restores the binding activity and increases the secretion in cos cells. Molecular Immunology 36:709-719, 1999.

Epitope Mapping

Epitope Mapping With Synthetic Peptides Prepared by SPOT Synthesis

ULRICH REINEKE, ACHIM KRAMER, and JENS SCHNEIDER-MERGENER

Introduction

Information about the epitopes of proteins recognized by antibodies or antibody fragments is important for their use as biological or diagnostic tools as well as for understanding molecular recognition events. Several techniques for mapping epitopes have been described, among them X-ray crystallography of antigen-antibody fragment complexes, a method leading to the most detailed results but usually being obtained with expensive and time consuming procedures (Davies 1996). For many purposes, information about the antigen-antibody interaction at the amino acid level, but not necessarily at the atomic level is needed. Protein digestion and chemical or enzymatic protein modifications in protein footprinting experiments, site-directed mutagenesis, binding assays with naturally occurring antigen variants such as species or tissue specific proteins or viral escape mutants and the screening of biological protein and peptide libraries are only some methods to map epitopes at the amino acid level (Morris 1996). Regarding the time required for the experiments and resolution of the data obtained, the use of synthetic peptides derived from the antigen sequence is a recommended strategy (Reineke 1999a).

Scans of overlapping peptides, so-called peptide scans, derived from the amino acid sequence of the antigens (Geysen 1987) are a widely used tool for epitope mapping. The entire antigen sequence is synthesized as linear

Peptide scan

✉ Ulrich Reineke, Jerini AG, Rudower Chaussee 29, 12489 Berlin, Germany
(*phone* +49-30-6392-6359; *fax* +49-30-6392-6395; *e-mail* reineke@jerini.de)
Achim Kramer, Humboldt Universität zu Berlin, Institut für Medizinische Immunologie, Universitätsklinikum Charité, Schumannstraße 20-21, 10098 Berlin, Germany
Jens Schneider-Mergener, Jerini AG, Rudower Chaussee 29, 12489 Berlin, Germany, Humboldt Universität zu Berlin, Institut für Medizinische Immunologie, Universitätsklinikum Charité, Schumannstraße 20-21, 10098 Berlin, Germany

8- to 15-mer peptides that are subsequently tested for binding of the antibody. For the preparation of the peptides, SPOT synthesis is an easy and very flexible technique for simultaneous parallel chemical synthesis on membrane supports and has been described in detail (Frank 1996, Kramer 1998). The peptides are covalently bound to a Whatman 50 cellulose support (Whatman, Maidstone, England) by the C-terminus and have a free N-terminus. These solid phase-bound peptides are used for binding studies directly on the membrane (Fig. 3).

Linear and discontinuous epitopes

Two different types of epitopes have to be considered (Fig.1):

- In linear (continuous) binding sites the key amino acids which mediate the contacts to the antibody are located within one part of the primary structure usually not exceeding 15 amino acids in lenght. Peptides covering these sequences have affinities to the antibody which are within the range shown by the entire protein antigen.

- In discontinuous (conformational) binding sites the key residues are distributed over two or more binding regions which are separated in the primary structure. Upon folding, these binding regions are brought together on the protein surface to form a composite epitope. Even if the complete epitope forms a high affinity interaction, peptides covering only one binding region, as synthesized in a scan of overlapping peptides, have very low affinities which often cannot be measured in normal ELISA or Biacore experiments.

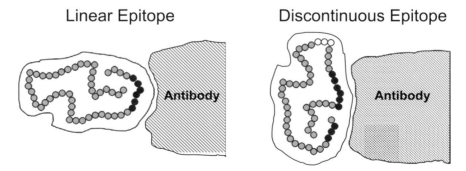

Linear Epitope   Discontinuous Epitope

Fig. 1. In linear (continuous) epitopes the key residues for binding are located within one stretch of the primary structure of the antigen (left) whereas discontinuous (conformational) epitopes are composed of two or more binding regions which are separated in the primary but adjacent in the tertiary structure (right).

In many linear epitopes and single binding regions of discontinuous epitopes the amino acids which are in contact with the antibody (key residues) are separated by residues which are buried in the antigen structure, thus not interacting with the binding partner. To identify the critical residues and side chains, substitutional analyses of the epitopes are carried out substituting every position - but only one at a time - by all other L-amino acids (Fig. 2). This means all single site substitution analogs are synthesized and tested for binding. Key residues may not be exchanged by any other, or only by physicochemically similar amino acids (e. g. leucine/isoleucine) without loss of binding.

Substitutional analysis

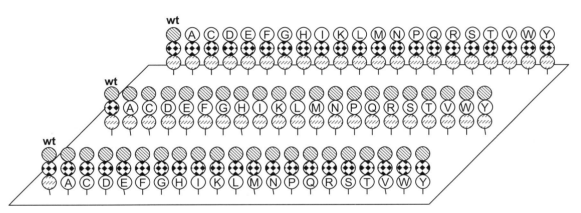

Fig. 2. Substitutional analysis for the identification of key residues for binding: Each residue of the wild-type peptide is substituted (rows) by all other L-amino acids. The sequences corresponding to the left column are identical and represent the wild-type peptide. Other spots are single substitution analogs.

▦ Outline

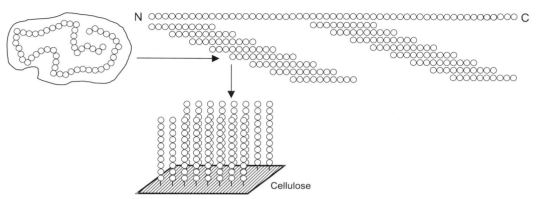

N ○○○ C

Cellulose

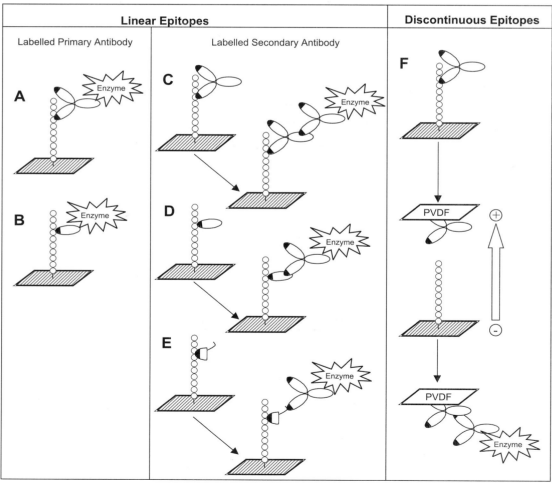

Linear Epitopes	Discontinuous Epitopes
Labelled Primary Antibody	Labelled Secondary Antibody	

A   Enzyme

B   Enzyme

C   Enzyme

D   Enzyme

E   Enzyme

F   PVDF ⊕ ⊖ PVDF   Enzyme

Materials

- Semi-dry blotter

- Standard X-ray film and film cassette, developing machine

- or a chemiluminescence imaging system e. g. the LumiImager™ (Roche Diagnostics, Mannheim, Germany)

- Ethanol

- Methanol

- T-TBS: TBS containing 0.05% Tween 20

- Detection reagent: just before developing prepare the detection reagent ("BM Chemiluminescence Western Blotting Reagents" of Boehringer Mannheim) by mixing the two liquids 1 : 100

- Supersensitive detection reagent: SuperSignal (Pierce, Rockford, IL, USA)

- Blotting paper (0.8 mm thick), GB 003 (Schleicher & Schuell, Dassel, Germany)

- Polyvinylene difluoride (PVDF) membrane, Immobilon-P (Millipore, Eschborn, Germany)

TBS: Tris buffered saline	T-TBS	Blocking buffer
50 mM TRIS	TBS pH 8.0	1 eq. blocking reagent (CRB, Northwitch, England)
137 mM NaCl	0.05 % Tween 20	9 eq. T-TBS
2.7 mM KCl		10 % (w/v) sucrose
adjust pH to 8.0 with HCl		

◄

Fig. 3. Upper part: The entire sequence of a protein antigen is synthesized as overlapping linear peptides covalently bound to a continuous cellulose membrane by the SPOT synthesis technique.
Lower part: Binding assays are carried out directly on the membrane. Linear epitopes are detected either with labelled primary antibodies (A) and antibody fragments (B) or with labelled secondary antibodies directed against these antibodies (C and D) or affinity purification tags (E). For mapping discontinuous epitops the primary antibody to be immobilized on a polyvinylene difluoride (PVDF) membrane by electrotransfer prior to decetion due to the very low affinities of separated epitope binding regions to the antibody.

Cathode buffer	Anode buffer I	Anode buffer II
25 mM TRIS	30 mM TRIS	300 mM TRIS
40 mM 6-aminohexane acid	20% methanol	20% methanol
20% methanol		

Regeneration buffer I	Regeneration buffer IIA	Regeneration buffer IIB
62.5 mM TRIS	8 M urea	400 ml water
2% SDS	1% SDS	500 ml ethanol
Adjust pH to 6.7 with HCl	0.1 % 2-mercapto-ethanol	100 ml acetic acid
Add 70 µl 2-mercaptoethanol per 10 ml SDS buffer (= 100 mM)		

Procedure

Mapping of linear epitopes

The screening strategy depends on whether the antibody of interest is available in an enzyme-labelled form or if a second enzyme-labelled antibody has to be used (Fig. 3). In this section we describe the procedure for horseradish peroxidase-conjugated antibodies, which is the enzyme of choice in our lab, but other enzymes, such as alkaline phosphatase may also be used (Kramer, 1998).

Note: Results with directly labelled antibodies are often much better, especially for low-affinity binding antibodies, and may even be essential for the mapping of discontinuous epitopes. Conjugation of antibodies with peroxidase is easy to perform and has been described in detail (Wilson and Nakane, 1978). The following procedure is practical for the detection of both peroxidase-conjugated primary antibodies and labelled second antibodies.

Note: Do not use sodium azide as a preservative for buffers with peroxidase as it is an inhibitor of the enzyme. The presence of azide will greatly reduce or eliminate the signal.

1. Rinse the membrane with a small volume of ethanol for 1 min. This is done to avoid the precipitation of hydrophobic peptides during the following TBS washing procedure.

2. Wash the membrane three times with an appropriate volume of TBS for 10 min. The volume depends on the membrane and the vessel size. The membrane should be sufficiently covered by the solution.

3. Block the membrane with the same volume of blocking buffer overnight at room temperature with shaking.

   Note: Do not overlay two or more peptide membranes in one vessel.

4. Wash the membrane once with the same volume of T-TBS for 10 min.

5. Incubate the membrane with the same volume of 0.1-1 µg/ml antibody solution (in blocking buffer) for 3 h. If the primary antibody is peroxidase-conjugated go to step 8, if not proceed with the following steps:

6. Wash the membrane three times with the same volume of T-TBS for 10 min.

7. Incubate the membrane with an appropriate volume of peroxidase-labelled second antibody solution (antibody, directed against the primary antibody) for 2 h with agitation. Dilute standard anti-sera 1 : 10,000 in blocking buffer.

8. Wash the membrane three times with the same volume of T-TBS for 10 min.

9. The following steps should be carried out in a dark-room unless a LumiImager^{TM} or another chemiluminescence imager is available: incubate the membrane with 75 µl/cm^2 of detection reagent for about 2 min. Make sure that each part of the membrane is covered by detection solution. Wash the reagent over the membrane repeatedly with a pipette or shake the vessel gently.

10. Insert the membrane antibody side up into a film cassette or the LumiImager^{TM}. Cover the membrane with a transparent plastic film.

11. If needed, switch off the light, place a sheet of film onto the membrane and close the cassette. Expose for 60 s (in the case of working with a LumiImager^{TM} expose 30 s).

12. Immediately do another exposure for a suitable time (up to 1 h) estimated from the signal intensity of the first trial.

Make sure that the detection molecules themselves (enzymes and secondary antibodies) do not produce signals on the membranes.

Control experiments

- Enzyme-labelled primary antibody: The membrane has to be incubated according to the protocol above with horseradish peroxidase (or alkaline phosphatase) in step 5. It has often been observed that alkaline phosphatase alone binds to peptide spots depending on their sequences whereas horseradish peroxidase in almost all cases does not interact with the peptide.

- Enzyme-labelled second antibody: The membrane has to be incubated according to the above protocol omitting steps 5 and 6. If recombinant antibody fragments are detected using antibodies directed against the purification tag (Fig. 3E), binding of the detection antibodies to the peptides on the membrane is often observed. These peptides contain or mimic the key residues for binding of the anti-tag antibody.

- In order to show that the antibody binds the peptides at the paratope the whole procedure, either with enzyme-labelled first or second antibody, has to be performed with an antibody of the same subclass but different specificity. Sequences binding this control antibody as well most likely interact with the constant regions.

We recommend doing the control experiments prior to incubation with the primary antibody under investigation since in some cases the membranes cannot be regenerated completely (see below).

Mapping of discontinuous epitopes

For the mapping of discontinuous epitopes peptide-antibody interaction with low affinities have to be detected. Detection with directly labelled primary antibodies (see above) or with a labelled second antibody after immobilization of the primary antibody on a PVDF membrane (see below) (Rüdiger 1997) is important since direct detection of molecules binding to the peptides with low affinities is extremely difficult.

Note: Detection with directly labelled primary antibodies usally result in more precise images in comparison to the electrotransfer technique.

This is due to the shifting of the binding equilibrium towards the non-complexed antibody during incubation with the detection antibody. In this section a procedure is described in which the pattern of the peptide-bound antibody on the peptide membrane is transferred and immobilized onto a PVDF membrane. For a better performance the electrotransfer should be carried out in a fractionated manner. Subsequently, the antibody is detected on the PVDF membranes with an enzyme-labelled second antibody.

Note: For the following procedure an ether-linked oligo-ethyleneglycol modified cellulose peptide membrane (Ast 1999) is recommended, since these membranes have peptide-cellulose linkages which are stable at basic pH as found in the blotting buffers. If membranes with ester-linked peptides are used, the peptides themselves can be transferred onto the PVDF membranes and the control experiments as described for linear epitopes have to be performed.

1. Cut out six blotting papers per peptide membrane and blotting step. The blot sandwich should be a few millimeters larger than the peptide membrane.

   Preparations

2. Cut out a PVDF membrane with the same size as the blotting paper.

3. Rinse the PVDF membrane shortly with methanol.

4. Equilibrate the PVDF membrane in anode I buffer for at least 10 min.

5. Equilibrate two blotting papers in cathode buffer, two in anode buffer I and the other anode buffer II.

6. Rinse the peptide membrane with a small volume of ethanol for 1 min. This is done to avoid the precipitation of hydrophobic peptides during the following TBS washing procedure.

   Membrane incubation

   Note: Do not overlay two or more peptide membranes in one vessel.

7. Wash the membrane three times with an appropriate volume of TBS for 10 min. The volume depends on the membrane and the vessel size. The membrane should be sufficiently covered by the solution.

8. Block the membrane with the same volume of blocking buffer for 3 h at room temperature with shaking.

9. Wash the membrane with the same volume of T-TBS for 10 min.

10. Incubate the membrane with the same volume of 1-5 µg/ml antibody solution in blocking buffer overnight at room temperature.

11. Wash the membrane three times for 1 min with the same volume of T-TBS. Turn the membrane over repeatedly during this procedure.

Blotting   12. Prepare the blot sandwich as shown in Fig. 4. Depending on the orientation of the electrodes of the semi-dry blotter the whole blot-sandwich can be inverted. Avoid air bubbles between the layers and place the sandwich in the center of the electrodes to obtain a homogeneous field. Remove excess buffer with a paper towel before laying on the upper electrode.

13. Blot at $1.0 \ mA/cm^2$ for 30 min. The voltage can change slightly during the process.

14. Remove the first blotting sandwich and prepare the second with new blotting paper and a new PVDF membrane. Make sure that the peptide membrane remains wet during this process. Blot again at $1.0 \ mA/cm^2$ for 30 min.

15. Repeat step 14 and blot for the third time for 1 h.

16. Collect the PVDF membranes in T-TBS. Wash the membranes in T-TBS twice for 10 min.

17. Block the PVDF membranes with blocking buffer for 3 h.

18. Continue with steps 7 to 12 from the procedure for Mapping of linear epitopes.

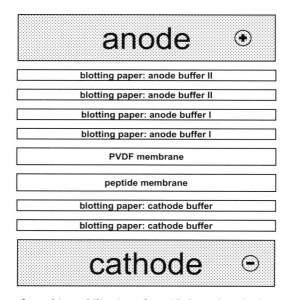

Fig. 4. Electrotransfer and immobilization of peptide-bound antibody onto a PVDF membrane.

Regeneration of peptide membranes

Two different regeneration protocols can be carried out. Start with protocol I and proceed with protocol II if it was not successful.

Note: Usually the peptide membranes can be used several times, but in a few cases regeneration fails due to strong binding of mature or denatured antibodies to the peptides or the cellulose.

1. Wash the membrane three times with water for 10 min.

2. Wash the membrane three times for 10 min with regeneration buffer I at 50°C.

   Note: Temperatures above 50°C can harm the membrane and/or the peptides.

3. Wash three times for 10 min with 10 x T-TBS at room temperature.

4. Wash three times for 10 min with T-TBS at room temperature.

5. If the membrane was incubated with a directly labelled antibody check the success of the regeneration by rinsing the membrane in substrate solution and then exposing it at least as long as in the original exposure. If spots are still detected repeat regeneration protocol I or go to regeneration protocol II.

6. If the membrane was incubated with a primary antibody in combination with an enzyme-labelled second antibody check the regeneration as described in step 5 of regeneration protocol I. If no signals are observed, then no second antibody remains on the membrane. Re-incubate the membrane with the second antibody and substrate solution and make an exposure at least as long as in the original exposure to show that the primary antibody is also completely removed. If spots are still detected repeat regeneration protocol I or go to regeneration protocol II.

7. If the indirect detection method for discontinuous epitopes was used check the regeneration by blotting the membrane at 1.0 mA/cm^2 for 2 h and proceed with step 17 of the procedure for mapping discontinuous epitopes. If spots are still detected repeat regeneration protocol I or go to regeneration protocol II.

Regeneration protocol I

Regeneration protocol II

1. Wash the membrane twice with water for 10 min.

2. Incubate the membrane three times with regeneration buffer IIA for 10 min.

3. Incubate the membrane three times with regeneration buffer IIB for 10 min.

4. Wash the membrane with water for 10 min.

5. Wash the membrane three times for 10 min with T-TBS.

6. If the membrane was incubated with a directly labelled antibody check the success of the regeneration by rinsing the membrane in substrate solution and exposing it at least as long as in the original exposure. If spots are still detected repeat regeneration protocol II.

7. If the membrane was incubated with a primary antibody in combination with an enzyme-labelled second antibody check the regeneration as described in step 5 of the regeneration protocol I. If no signals are observed, then no second antibody remains on the membrane. Re-incubate the membrane with the second antibody and substrate solution and expose at least as long as in the original exposure to show that the primary antibody is also completely removed. If spots are still detected repeat regeneration protocol II.

8. If the indirect detection method for discontinuous epitopes was used check the regeneration by blotting the membrane at $1.0 \, mA/cm^2$ for 2 h and proceed with step 17 of the procedure for mapping discontinuous epitopes. If spots are still detected repeat regeneration protocol II.

Storage of peptide membranes

- New membranes should be stored at -20°C until use.

- Incubated membranes which will be used again after only a few days should be washed three times with T-TBS for 10 min and kept with a small volume of T-TBS in a petri dish at 4°C. Drying out of the membrane sometimes leads to poor results in subsequent experiments.

- Incubated membranes which will be stored for a longer period should be regenerated, washed with methanol twice, air dried and kept at −20°C.

Results

Two epitopes of monoclonal antibodies raised against human interleu-
kin-10 (hIL-10) were mapped. Mab CB/RS/3 recognizes a linear (Fig. 5,
left) (Reineke 1999a) and mab CB/RS/1 a discontinuous epitope (Fig. 5,
right) (Reineke 1999b). Scans of overlapping hIL-10-derived 15-mer
peptides shifted by three (CB/RS/3) or one (CB/RS/1) amino acid(s)

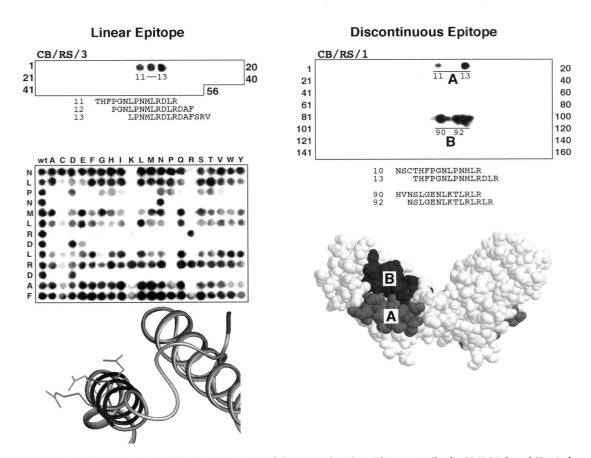

Fig. 5. Left: Characterization of the linear epitope of the monoclonal anti-hIL-10 antibody CB/RS/3 by a hIL-10-de-
rived peptide scan of overlapping peptides (15-mers, shifted by three amino acids) (above) and a substitutional
analysis (middle). Key residues for antibody binding are displayed with their side chains in the tertiary structure
of hIL-10 (below).
Right: Mapping of the discontinuous epitope of the monoclonal anti-hIL-10 antibody CB/RS/1 by a hIL-10-derived
peptide scan of overlapping peptides (15-mers, shifted by one amino acid) (above). The two binding regions are
highlighted on the surface of the three-dimensional hIL-10 structure.

were prepared by SPOT synthesis and tested for antibody binding following the protocols described above. CB/RS/3 binds to three consecutive peptides (11 to 13, upper left). This linear epitope was characterized by a substitutional analysis (middle left) thus identifying the key residues for binding, which are shown with their side chains in the ribbon presentation of the respective part of the three-dimensional hIL-10 structure (lower left). In contrast, mab CB/RS/1 binds to a discontinuous epitope composed of two binding regions A and B that are separated by 62 amino acids (upper right). These binding regions form a connected epitope shown in the space filling hIL-10 structure (lower right).

Troubleshooting

Mapping of linear epitopes

- No signal
  - Increase the antibody concentrations.
  - Prolong the incubation time with primary antibody to overnight at 4°C.
  - Perform a simultaneous incubation of first and second antibody.
  - Shorten the washing times. Use washing buffer without Tween 20.
  - Shorten the blocking time to 3 h.
  - Use a supersensitive chemiluminescence substrate (e. g. Super-Signal, Pierce).
  - Check the antibodies and enzymes in an alternative system.

- High background
  - Increase the detergent concentration in washing buffer.
  - Increase the washing times and/or the washing volumes.

- Clear spots on dark background
  In this case the primary and/or second antibody concentrations may be too high. On the spots a high amount of antibody conjugate results in all the substrate being used up before the X-ray film can be exposed on the membrane or the imaging system can be started, resulting in clear spots.
  - Wash extensively with T-TBS and re-detect or, if the problem persists,
  - Regenerate the membrane and incubate with lower concentrations of proteins.

- "Ring-spots"
  If spots have a white center and a dark ring a peptide membrane with a lower peptide density in the range of 20 to 40 nmol/cm^2 is recommended (standard membranes have 150 to 200 nmol/cm^2) (Kramer 1999).

Mapping of discontinuous epitopes

- High background
  Usually the background of the first blot is higher in comparison to the following blots. If the background of the last blot is still too high prolong the washing times before the electrotransfer.

- No signals
  - Increase the antibody concentrations.
  - Prolong the incubation time with primary antibody to overnight at 4°C.
  - Shorten the washing time before elctrotransfer.
  - Use blocking buffer without Tween 20 or no blocking buffer.
  - Wash with TBS (no Tween 20).
  - Perform only one blotting step for 2 h.
  - Increase the current.
  - Check the antibodies and enzymes in an alternative system.

- The transfer is only sufficient if the isoelectric point of the antibody is below 8.5. In very few cases higher values can occur depending on the amino acid composition of the paratope. This can be checked experimentally or if the antibody sequence is known from prediction programs.

- If by the second blot there is no background, it is sufficient to do only two fractionated blotting steps. The spot pattern can change between subsequent blots. This is probably due to different peptide-antibody affinities.

- Correlation between signals and spot number (sequence)
  - A long exposure results in a background that includes the edge of the membrane. The spot position relative to the edge gives the spot number and the sequence.

- The rows and columns of the peptide membrane can be marked with a pencil. This is also visible on the chemiluminescence exposure of the PVDF membrane.
- The rows and columns of the peptide membrane can be marked on the overlayed PVDF membrane when preparing the blot sandwich.

Applications

- The techniques described are primarily used for the mapping of monoclonal antibody epitopes. Analogous procedures can be applied for the characterization of polyclonal antibodies or patients' sera (Valle 1999). In these cases it cannot be decided if the antibody binding peptides are derived from linear epitopes or from different binding regions of discontinuous epitopes.

- Although antibody epitope mapping was the first application of the SPOT synthesis technique, it has also been used extensively for characterization of protein-protein interactions in general such as cytokine/receptor, chaperone/substrate and protein-domain/ligand contacts (Reineke 1999a).

- Another approach for the identification of antibody binding peptides is the use of synthetic combinatorial peptide libraries. Peptides identified by this procedure either resemble the epitope sequence of the natural protein antigen or have a novel sequence mimicking a natural epitope called a mimotope (Geysen 1986, Kramer 1997).

Acknowledgements. We are grateful to Robert Sabat for the cooperation in the mab CB/RS/1 and CB/RS/3 mapping.

References

Ast T, Heine N, Germeroth L, Schneider-Mergener J, Wenschuh H (1999) Efficient assembly of peptomers on continuous surfaces. Tetrahedron Lett 40:4317-4318

Davies DR, Cohen GH (1996) Interactions of protein antigens with antibodies. Proc Natl Acad Sci USA 93:7-12

Frank R, Overwin H (1996) Spot Synthesis: Epitope analysis with arrays of synthetic peptides prepared on cellulose membranes. Meth Mol Biol 66:149-169

Geysen HM, Rodda SJ, Mason TJ, Tribbick G, Schofs PG (1987) Strategies for epitope analysis using peptide synthesis. J Immunol Methods 102:259-274

Geysen HM, Rodda SJ, Mason TJ (1988) A priori delineation of a peptide which mimics a discontinuous antigenic determinant. Mol Immunol 23:709-715

Kramer A, Keitel T, Winkler K, Stöcklein W, Höhne W, Schneider-Mergener J (1997) Molecular basis for the binding promiscuity of an anti-p24 (HIV-1) monoclonal antibody. Cell 91:799-809

Kramer A, Schneider-Mergener J (1998) Synthesis and application of peptide libraries bound to continuous cellulose membranes. Meth Mol Biol 87:25-39

Kramer A, Reineke U, Dong L, Hoffmann B, Hoffmüller U, Winkler D, Volkmer-Engert R, Schneider-Mergener J (1999) Spot synthesis: observations and optimizations. J Peptide Res 54:319-357

Morris GE (1996) Choosing a method for epitope Mapping. In: Morris GE (ed) Epitope mapping protocols. Humana Press Inc., Totowa, New Jersey, pp 1-9

Reineke U, Kramer A, Schneider-Mergener J (1999a) Antigen sequence- and library-based mapping of linear and discontinuous protein-protein interaction sites by spot synthesis. Curr Top Microbiol Immunol 243:23-36

Reineke U, Sabat R, Misselwitz R, Welfle H, Volk HD, Schneider-Mergener J (1999b) A synthetic mimic of a discontinuous binding site on interleukin-10. Nat Biotechnol 17:271-275

Rüdiger S, Germeroth L, Schneider-Mergener J, Bukau B (1997) Substrate specificity of the DnaK chaperone determined by screening of cellulose-bound peptide libraries. EMBO J 16:1501-1507

Valle M, Munoz M, Kremer L, Valpuesta JM, Martinez-AC, Carrascosa JL, Albar JP (1999) Selection of antibody probes to correlate protein sequence domains with their structural distribution. Prot Sci 8:883-889

Wilson MB, Nakane, PK (1978) Recent developments in the periodate method of conjugating horseradish peroxidase (HRPO) to antibodies. In: Knapp W, Hlubar K and Wick G (Eds. Immunofluorescence Related Staining Techniques. Elsevier, Amsterdam, Netherlands, pp. 215-224

Suppliers

Custom peptide synthesis of cellulose bound protein-derived peptide scans and combinatorial peptide libraries:
JERINI AG
Ost-West-Zentrum Berlin-Adlershof
Rudower Chaussee 29
12489 Berlin
Germany
Phone: +49-30-6392-6392
Fax: +49-30-6392 6395
e-mail: biotools@jerini.de
web: www.jerini.com

Epitope Mapping with Random Peptide Libraries

VOLKER BÖTTGER

Introduction

The display of random peptides on the surface of filamentous bacteriophage is an ingenious methodology developed in the late 1980s by George Smith & colleagues to define antibody epitopes. Synthetic oligonucleotides of defined length but completely randomized sequence are cloned as fusions to gene III or VIII of M13 related filamentous bacteriophage generating phage libraries of enormous molecular diversity. These libraries of hundreds of millions of phage clones each displaying a unique peptide sequence on the phage surface are ideal and cost-effective tools to select novel ligands for target molecules of interest. Over the years, this method has been mainly used to define epitopes of monoclonal and polyclonal antibodies but has also been employed to find ligands for receptors, substrates for enzymes and to obtain new insights into protein-protein interactions.

This chapter will deal with methods used to select phagotopes for antibodies but they can also be used to investigate other ligate-ligand interactions.

In a process known as affinity selection or biopanning the antibody of interest is incubated with a sample of a phage peptide library. Subsequently, antibody-phage complexes are captured and recovered phage are amplified and again selected for binding to the antibody. Several rounds of biopanning lead to an enrichment of a few specific antibody-binding phagotopes out of hundreds of millions of irrelevant library phage. The sequence of binding peptides is deduced by sequencing the nucleotide inserts of individual phage clones.

Volker Böttger, Wilex AG, Grillparzerstr. 10 B, 81675 München, Germany
(*phone* +49-89-457-47378; *fax* +49-89-457-47389; *e-mail* volker.boettger@wilex.de)

When several unique but clearly related peptide sequences are obtained a phage consensus motif can be defined showing only amino acids present in almost all phage sequences in identical positions (see: Data analysis). The consensus motif can then be used to scan the sequence of the original antigen to map the antibody epitope.

The success of epitope mapping depends first and foremost on the nature of the original antibody epitope. If an antibody binds to a linear stretch of amino acids, i.e. a linear or sequential epitope, it is very likely that it will select phagotopes with sequence similarity to the epitope, unless the phage displayed peptides are shorter in length than the linear epitope. Unfortunately, many antibodies recognize conformational or discontinuous epitopes on the natural antigen. Here the epitope is assembled by protein folding bringing amino acids from different parts of the protein into spacial vicinity to each other. In many cases these antibodies fail to select epitope phage from phage display libraries since the short linear peptides can not adopt the conformation of the discontinous epitope. Under these circumstances libraries should be tried where cysteine flanked random peptides are displayed as peptide loops under nonreducing conditions. This constrains a three-dimensional structure of the peptide and might facilitate antibody binding. However, very often unique mimotopes are being selected from these libraries, i.e. sequences with no obvious sequence similarity to each other and to the epitope.

Materials

Sources for libraries, phage vectors and bacteria

Protocols and procedures explained in this chapter are based on the assumption that a pre-made phage peptide library is at hand. These days there are academic as well as commercial sources to get hold of a library sample. A great deal of work which the author has done over the years to characterize antibody epitopes was made possible with material generously provided by G. Smith at the University of Missouri, Columbia. Many experimental details described in this chapter are modifications of Smith's original methods. You can contact Prof. Smith to ask for a free sample of random peptide libraries (6 and 15mers), *E. coli* host strains and even phage vectors to make your "personal libraries". He also maintains a WEB page where you can find a comprehensive list of downloadable documents about constructing and handling phage peptide libraries.

http://www.biosci.missouri.edu/smithgp/index.html
gpsmith@biosci.mbp.missouri.edu

Therefore, protocols about amplification and maintenance of these libraries have not been included in this chapter.

As commercial source for pre-made phage display peptide libraries and a kit to construct your own library NEW ENGLAND BioLabs should be mentioned.

http://www.neb.com/neb/frame_cat.html

The results described in **Data analysis** were obtained by probing antibody LL001 with a sample of a 12mer phage library, kindly provided by W. Dower (Affymax Research Institute, CA.).

Procedure

Affinity selection of antibody-binding phage (biopanning)

Biopanning is carried out by incubating an aliquot of a phage display library with a certain amount of a monoclonal antibody. This interaction can take place in solution (with subsequent affinity capture of the phage-antibody complexes onto a matrix) or on a solid phase precoated with the respective antibody. The following three protocols describe alternative ways for phage selection.

Capture of antibody-phage complexes via biotinylated anti-mouse immunoglobulin on streptavidin-coated plates

1. Mix 50 µl of phage library containing 1×10^{11} phage particles with 5 µl monoclonal antibody (0.1 mg/ml final concentration). To save precious library material phage can be probed with several different antibodies at the same time in this first round. Incubate overnight at 4°C.

2. Coat a polystyrene petri dish (Falcon 3001) with 1 ml streptavidin at 10 µg/ml in 0.1 M NaHCO$_3$ overnight at 4°C in a humidified box.

3. Start a 2 ml *E. coli* K91 culture in LB broth from a single colony (picked from an LB plate) or a glycerol stock and grow overnight at 37°C and 200 rpm. Next day inoculate 10 ml LB with 10 to 100 µl of the overnight culture and grow to late log phase (OD$_{600nm} \approx 1$).

4. Add biotinylated goat anti-mouse IgG (Sigma B-7264) to the phage-antibody mixture at an amount equimolar to the monoclonal antibody/ies and incubate for 3 hours at room temperature.

5. Remove streptavidin solution from the petri dish and fill to the brim with 0.1 M NaHCO₃, supplemented with 5 mg/ml BSA (blocking solution) for 1 to 2 hours at room temperature.

6. Discard solution and wash dish 3 times with TBS, 0.5% Tween 20 (wash solution).

7. Dilute phage-antibody mix with 1 ml TBS, 0.5% Tween 20 and transfer to blocked and washed streptavidin plate.

8. Incubate for 30 to 60 minutes at room temperature on a belly dancer.

9. Remove unbound material in 10 to 20 wash cycles, each time adding 2.5 ml wash solution, swirling it round and pouring it off immediately.

10. Add 400 µl elution buffer (0.1 M HCl, adjusted to pH 2.2 with glycine and supplemented with 1 mg/ml BSA) and incubate 20 minutes with occasional shaking to elute bound phage.

11. Transfer eluted phage to an 1.5 ml Eppendorf tube and neutralize with 24 µl 2 M Tris-base.

12. Use 400 µl to infect 2 ml log-phase K91 bacteria and keep remaining phage for titration.

13. Incubate 15 minutes at 37°C without shaking followed by another 15 minutes with shaking at 200 rpm.

14. Transfer infected bacteria to a 50 ml Falcon tube (Polypropylene) with 10 ml 2xYT medium, supplemented with 20 µg/ml tetracycline (Smith's fd-tet library phage transfer tetracycline resistance to infected bacteria).

15. Grow suspension for 24 hours at 37°C and 250 rpm.

16. Spin suspension in a bench top centrifuge at 3400xg and 4°C for 30 minutes.

17. Transfer supernatant to a new 50 ml Falcon tube and keep 100 µl for titration.

18. For phage purification/precipitation add 2 ml of PEG/NaCl solution (add 100 g polyethylene glycol 8000 and 117 g NaCl to 475 ml water and autoclave) to the 10 ml supernatant, mix by inverting tube several times and incubate on ice for at least 2 hours.

19. Spin as described in step 16.

20. Discard supernatant and resuspend phage pellet in 1 ml aqua dest.

21. Transfer to 1.5 ml Eppendorf tube and spin at 10000xg and 4°C for 5 to 10 minutes to remove remaining bacteria and cellular debris.

22. Transfer phage supernatant to fresh 1.5 ml tube and add 200 µl PEG/NaCl for a second precipitation, mix and leave on ice for 1 hour.

23. Spin 10 minutes at 10000xg, 4°C and pour off supernatant.

24. Respin briefly and remove remaining supernatant with pipette.

25. Resuspend phage pellet in 1 ml TBS and filter through 0.2 µm syringe filter (optional).

26. Determine phage concentration (titration) as described below (Phage titration) and store phage at 4°C. This is the phage library stock after the first round of biopanning and is used as starting material for further rounds of phage selection essentially carried out as described above. In order to select phage clones with higher affinity to the monoclonal antibody, the concentration of the antibody can be decreased to 1 and 0.1 µg/ml final concentration in the second and third round of biopanning. The amount of the biotinylated secondary antibody would have to be adjusted accordingly to give roughly an equimolar concentration to the monoclonal antibody. If several antibodies had been used in a single pot for phage selection in the first round they should now be probed separately with the newly prepared phage stock.

Separation of antibody-phage complexes with magnetic beads

1. Mix 200 µl hybridoma supernatant of a monoclonal antibody (any other antibody source will do as well as long as the binding capacity of Dynabeads is not exceeded) with a 10 µl sample of a phage peptide library (10^{11} to 10^{12} phage) in a 1.5 ml Eppendorf tube and incubate overnight at 4°C.

2. Wash 100 µl slurry of Dynabeads M-280, coupled with sheep anti-mouse IgG, several times with PBS in a 1.5 ml Eppendorf tube, pelleting the beads in between with a magnetic particle concentrator.

3. Do the final wash in PBS, 2% BSA to block any non-specific protein binding sites on the beads.

4. Resuspend beads in 100 µl PBS, 2% BSA and transfer beads to the antibody-phage mixture.

5. Incubate tube for 2 hours at room temperature on a rotary shaker.

6. Pellet the beads, remove supernatant and wash beads 10 to 20 times with 1 ml PBS, 0.1% Tween 20.

7. Proceed with Step 10 (Capture of antibody-phage complexes ...) to continue.

Selection of phage on solid phase bound monoclonal antibody

1. Coat a polystyrene petri dish (Falcon 3001) or tube (NUNC Maxisorp 4-44202) with purified monoclonal antibody at a concentration of 5 to 20 µg/ml in PBS. The experimentalist should use coating conditions which he knows are suitable for his/her antibody to bind the appropriate antigen.

2. For Maxisorp immunotubes use 4 ml antibody solution, seal tube with parafilm and keep it in the fridge overnight.

3. Remove solution and wash tube 3 times , each time filling the tube to the brim with PBS and pouring it off immediately.

4. Block unoccupied binding sites with PBS, 5% (w/v) low fat milk powder for 1 hour at room temperature.

5. Wash tube 3 times with PBS and add 10^{11} to 10^{12} phage in 4 ml PBS, 0.1% Tween 20, 5% milk powder.

6. Incubate for 1 hour at room temperature on a rotary shaker.

7. Wash 10 to 20 times with PBS, 0.1% Tween 20 as described in Step 3.

8. Add 800 µl elution buffer (see above Step 10, Capture of antibody-phage complexes ...) and incubate 20 minutes at room temperatures with constant shaking.

   Note: As an alternative for elution bound phage can be used directly in the tube (dish) to infect bacteria. For this, 4 ml log phase K91 bacteria can be added to the immunotube and incubated in a 37°C waterbath for 30 to 60 minutes.

9. Transfer to Eppendorf tube and neutralize with 48 µl 2 M Tris-base.

10. Proceed with Step 12 (Capture of antibody-phage complexes ...) to continue.

Phage titration

To determine the concentration of phage after elution or amplification/ purification a phage titration is carried out. It is based on counting bacterial colonies, which had been infected with a dilution series of phage, making them grow on tetracycline selection plates. Therefore, only transforming units, i.e. infectious phage particles, are determined.

1. Prepare LB-agar plates with 20 µg/ml tetracycline, dry plates with open lid for several hours and draw a grid with 1.5x1.5 cm squares (4x4 for a 9 cm petri dish) on the bottom of each plate.

2. Fill 45 µl TBS into each well of a sterile 96 well microtiterplate.

3. Add 5 µl phage sample (eluate, phage supernatant or PEG-purified phage) to first well, mix properly by aspirating and dispensing at least ten times and transfer 5 µl into the next well. Continue until 10 serial dilutions have been made. Change pipette tip every third well to minimize unwanted carry over of phage.

4. Transfer 155 µl log phase K91 to each well, mix carefully with the phage dilution and incubate at 37°C for 30 minutes.

5. Resuspend bacteria in the wells and spot 20 µl of each dilution onto the prepared agar plate and wait until the liquid has completely disappeared.

6. Close the lid and incubate plate face down overnight at 37°C.

7. Count the colonies at dilutions with well separated clones and calculate the phage concentration of the starting material.

Determining the success of biopanning

There are several ways to monitor phage selection.

From the second round of biopanning on, a control selection without antibody can be included to estimate antibody-driven phage selection over background binding of phage. In an ideal scenario, the eluates from the antibody selection should contain significantly more phage than the control eluates. The difference should increase with the number of rounds (see Results, Details of biopanning). This, however, is not always the case. Sometimes only phage with a very low affinity to the antibody are selected. Applying stringent panning conditions can cause most of these phage to be washed off leading to no statistically relevant phage enrichment over the negative control. On the other hand, a very gentle washing procedure can cause a high number of phage to nonspecifically stick to the plastic or proteins possibly obscuring an underlying specific selection. In any case, it is worthwhile continuing with the ELISA analysis described below.

1. Coat the wells of a 96 well ELISA plate (Nunc Immunoplate Maxisorp 439454 or Falcon 3912 MicroTest III flexible Assay Plate) with 50 µl monoclonal antibody (used to select peptide phage) at 1 to 20 µg/ml in PBS. Incubate overnight at 4°C. Use different coating conditions, if they are more suitable for your antibody.

2. Wash plate with water (or suitable buffer) and block with 200 µl PBS, 5% (w/v) low fat milk powder, 0.1% Tween 20 (blocking solution) for 1 hour at room temperature. Remove solution and wash as before.

3. In the meantime dilute phage stocks, prepared after each round of biopanning, 1:1 with blocking buffer.

4. Transfer 50 µl to the blocked and washed antibody plate and incubate for 2 hours at room temperature.

5. Wash plate and add 50 µl of an 1:1000 dilution of HRP-labelled anti-M13 antibody (Pharmacia Biotech, 27-9421-01) for 1 hour at room temperature, wash.

6. Add 50 µl substrate (0.1 mg/ml TMB, 0.03% H_2O_2 in 0.1 M sodium acetate, pH 6.0) for 15 minutes and stop colour development by adding 50 µl of 1 M sulphuric acid.

7. Measure the O.D. at 450 nm.

An emerging positive ELISA signal with increasing rounds of biopanning indicates that peptide phage have been selected by the antibody. Include controls with irrelevant antibodies and phage to prove the specificity of the antibody-phage interaction. If nonspecific binding is an issue, use higher phage dilutions.

8. Use a dilution series of phage from ELISA-positive stock(s) to infect log phase K91 cells and spread infected bacteria on LB-agar plates with 20 µg/ml tetracycline.

9. Grow overnight at 37°C.

10. Transfer 96 well-separated colonies into the wells of a sterile microtiterplate filled with 200 µl 2xYT medium, 20 µg/ml tetracycline.

11. Place plate with lid into a humidified box and grow for 24 hours at 37°C, 250 rpm (a white Boehringer enzyme box with a layer of wet tissue is ideal. Fix the plate position with spacers of styrofoam and put the box on the sticky mat of a bacterial incubator).

12. Spin plates 15 minutes at 1000xg and 4°C.

13. Transfer 100 µl phage supernatant into a new 96 well plate with 100 µl PBS, 5% (w/v) low fat milk powder, 0.1% Tween 20 and use for ELISA on antibody-coated plates, as described before.

14. Seal original plate with parafilm and keep as master plate in fridge; run phage ELISA.

15. Transfer 20 µl of bacterial suspension of ELISA-positive clones from the master plate into 5 ml 2xYT, 20 µg/ml tetracycline and grow overnight at 37°C, 250 rpm.

16. Spin at 3400xg and 4°C for 30 minutes to separate bacteria from phage.

17. Transfer phage containing supernatant into fresh tube and store in fridge as backup.

18. Prepare double stranded replicative form of phage DNA from the bacterial pellets using standard DNA miniprep procedures. Single stranded phage DNA can also be extracted from PEG precipitated phage with Phenol/Chloroform.

19. Use a phage-specific primer to sequence the nucleotide inserts coding for the phage-displayed peptides.

20. Look for sequence similarities between different phage clones and try to align the peptide sequences to define a phage consensus motif.

21. This consensus motif is used to scan the protein sequence of the original antigen to find a similar motif indicative for the epitope location of the tested monoclonal antibody.

Results

Data analysis

The following table features the result of a successful biopanning experiment with a monoclonal antibody.

Sequences of phage peptides selected with the human keratin 14-specific antibody LL001

Synthetic peptide*	D	G	K	V	V	S	T	H	E	Q	V	L	R	T	K	N			
Epitope motifs:																			
Peptide 14							A	N	S	G	V	L	R	I	K	T	E	F	
Peptide 8				S	D	R	F	S	E		V	L	R	L	K	N			
Peptide A				V	F	S	L	P	D		I	L	R	D	K	G			
Peptide C				V	V	K	V	G	S		V	L	R	D	K	S			
Peptide 11							A	G	D		V	L	R	N	K	L	E	P	N
Peptide 13				K	P	N	F	C	E		V	L	R		K	C	V		
Mimotope motifs:																			
Peptide 7	G	W	L	N	P	G													
peptide 10	G	W	L	V	P	H	P	S	S	N	F	G							
Peptide B	G	W	D	L	F	F	G	E	S	H	S	P							
Unique mimotopes:																			
peptide 12	Y	F	Y	G	P	T	W	K	E	R	N	N							
Peptide F	H	K	P	G	F	Y	I	I	R										

* synthetic peptide, representing the C-terminal sequence of human keratin 14, used to raise monoclonal antibody LL001 (Purkis et al.)

Details of biopanning

1st and 2nd round of biopanning were carried out on Streptavidin-coated plates, the 3rd round on antibody-coated plates (see Procedure).

- 1st round:
  50 µl phage library ($1x10^{11}$ phage) + 5 µl antibody (5 µg); eluate: $3.8x10^6$ phage/ml; amplified library after 1st round: $2x10^{12}$ phage/ml.

- 2nd round:
  50 µl phage library after 1st round ($1x10^{11}$ phage) + 5 µl antibody 1/100 (50 ng); eluate with antibody: $4x10^7$ phage/ml; eluate without antibody: $3.8x10^6$ phage/ml; selectivity: 10:1; amplified library after 2nd round: $2x10^{12}$ phage/ml.

- 3rd round:
  50 µl phage library after 2nd round ($1x10^{11}$ phage); dish coated with 10 µg/ml antibody; eluate with antibody: $4.4x10^9$ phage/ml; eluate without antibody: $2.2x10^5$ phage/ml; selectivity: 20000:1; binding of single clones to antibody confirmed by phage ELISA; sequencing of DNA inserts.

Monoclonal antibody LL001 was used in a model biopanning experiment. The epitope location on the native antigen was already known since the antibody was raised against a synthetic peptide, representing the C-terminal sequence of this antigen. The antibody was probed in 3 rounds of biopanning with aliquots of two different phage libraries, displaying 6 and 12 amino acids as pIII fusions on the surface of filamentous phage. The DNA inserts of 20 individual clones were sequenced for each library and the corresponding peptide sequence deduced.

For the 6mer library the antibody had only selected relatives of a single phage clone since all phage were identical in their nucleotide insert sequence. On the other hand, 10 independent phage clones with unique insert sequences were found in the 12mer pool. Comparing the sequences of all 11 peptides with each other three different sequence groups could be defined. In the first group, six 12mer peptides show a clear sequence similarity to each other. The amino acids, which are shared by all or almost all of these peptides in identical positions, are marked. A phage consensus motif can be defined as **V L R X K** which most certainly reflects the core epitope motif of antibody LL001 since the same sequence is present in the synthetic epitope peptide. Sequence spacing has been introduced into peptides A and C to bring further sequence similarities into register with the epitope peptide. These additional amino acids might reflect additional weak interaction points with the antibody. In peptide 13 the unspecified amino acid position X of the phage consensus motif is missing. It seems possible that a loop formation caused by two flanking cysteine residues can compensate for this missing amino acid by bringing the V L R K amino acid side chains back into the right spacing (conformation) necessary for antibody recognition.

Five additional peptides show no obvious sequence similarity with the original antigen / peptide. They are called mimotopes since they mimic the epitope structure as being recognized by the antibody. Whether these sequences represent real structural mimics of the epitope is not certain. They might also fit in a different way into the antibody combining site using other side chains as interaction points.

Anyway, it is interesting to see that 3 of the selected peptides show a considerable degree of sequence homology to each other with a **G W L X P X X X S** mimotope consensus motif.

The two remaining mimotopes (one was found to be only 9 amino acids long) are unique in a way that they do not show any substantial sequence similarities to each other and to the peptides of the other two groups.

What is the value of mimotopes if they can not contribute to epitope definition?

There are clinically relevant antibodies for which the nature of the antigen is unknown:

- mimotopes selected with autoimmune antibodies could be used to suppress autoimmune reactions

- antibodies against pathogenic microorganisms : mimotopes could be used for vaccination against the pathogens

- tumor-reactive antibodies: mimotopes could be employed for the generation of second generation anti-tumor antibodies or used directly for active immunization

Troubleshooting

- Very low phage concentration (>1000 phage/ml) in eluate
    - decrease no of cycles and stringency of washing
    - check pH of elution buffer and neutralized eluate with pH paper; to be on the safe side, infect bacteria directly in the biopanning vessel instead of or after acid elution

- Very high phage concentration ($>10^8$ phage/ml) in eluate of negative control
    - increase no of cycles and stringency of washing
    - Type and material of biopanning vessel and kind of blocking agent can greatly influence non-specific background binding $>$ try different combinations.

- No selective enrichment of phage after several rounds of biopanning
    - structure and size (length) of the natural antibody epitope can't be matched by any of the displayed phage peptides $>$ use phage libraries with longer peptide insertions and/or structurally constrained (e.g. C-C loop) libraries
    - due to a low affinity interaction between the antibody and phage peptide sequences and an unfavorable wash protocol an underlying specific selection can be obscured $>$ it is worth continuing with selection of single phage clones or repeating particular rounds of biopanning by changing selection conditions
    - sometimes a suboptimal biopanning procedure can be improved by switching the biopanning method in subsequent rounds of selection
    - for phage selection on solid phase bound antibody make sure that bound antibody is still able to bind antigen efficiently

- All selected and analyzed phage clones have the same insert sequence
  - there is only one peptide sequence present in the library that the antibody can bind to
  - due to extensive and stringent rounds of biopanning other peptide phage (with lower affinity to the antibody) are lost
  - usually it is good practice to start phage selection with a high concentration of antibody and a non-stringent wash regime to select phage with a broad range of affinities to the antibody. In subsequent rounds the selection conditions can be varied in order to select only high affinity ligands or to maintain a high degree of diversity in sequence and affinity.

References

Smith, G. P., and Scott, J. K. (1993). Libraries of peptides and proteins displayed on filamentous phage. In: Methods in Enzymology 217, pp 228-257 (R. Wu, ed.), Academic Press, San Diego, CA.

Phage Display of Peptides and Proteins. A Laboratory Manual. (B. K. Kay, J. Winter, J. McCafferty, eds.), Academic Press, San Diego, 1996.

Purkis, P. E., Steel, J. B., Mackenzie, I. C., Nathrath, W. B., Leigh, I. M., Lane, E. B. (1990). Antibody markers of basal cells in complex epithelia. J. Cell Sci. 97, 39-50

Epitope Mapping with Gene Fragment Libraries

CHRISTOPH H. WINTER and ROLAND E. KONTERMANN

Introduction

Linear epitopes, recognized by a large number of antibodies, are easily identified by expression of small fragments of the known antigen at the tip of filamentous phage. Immobilized antibodies, incubated with such a library of peptide-displaying phage, bind selectively to the fragments containing the epitope, therefore immobilizing the peptide-phage construct as well. Unspecific non-binding phage can then be washed away. Finally, by eluting the epitope-displaying phage following their propagation in *E. coli*, repetitive rounds of panning are performed to further enrich for specific phage. After approximately three rounds of panning, several single phage clones are amplified and the antigen-encoding inserts are then sequenced. Aligning the insert sequences with the antigen sequence identifies the region in which the epitope lies. The comparison of different fragments usually yields overlapping sequences, each containing the epitope, hence narrowing down the number of amino acids essential for binding. Compared to the previously described random peptide libraries, the gene-fragment phage display offers the major advantage of using the correct antigen of an antibody to find its epitope, thus eliminating guesses and sequence comparisons with unknown or irrelevant proteins. Therefore, the chances of finding one or more sequences binding to the antibody that are unrelated to the antigen (*mimotopes*) are drastically reduced in favor of the correct epitope. Furthermore, the epitope sequence is almost guaranteed to be contained in a library of reasonable size, whereas ran-

Christoph H. Winter, Yale University, School of Medicine, Dept. of Molecular Biophysics and Biochemistry, 333 Cedar Street, P.O. Box 208024, New Haven, CT, 06520-8024, USA. Present address: Kybeidos GmbH, Gaibergstr. 8, 69115 Heidelberg, Germany

✉ Roland E. Kontermann, Philipps-Universität, Institut für Molekularbiologie und Tumorforschung, Emil-Mannkopff-Straße 2, 35033 Marburg, Germany
(*phone* +49-6421-286-6727; *fax* +49-6421-286-8923; *e-mail* rek@imt.uni-marburg.de)

dom fragment libraries often contain only such similar to the epitope - due to the fact that only a small portion of all possible peptide sequences are encoded by a single random library. Gene fragment libraries have been successfully applied to map epitopes of monoclonal antibodies (Petersen et al., 1995; Wang et al., 1995; van Zonneveld et al., 1995; Klinger et al., 1996; Winter, 1997) as well as polyclonal sera (Blüthner et al., 1996; 1999).

Outline

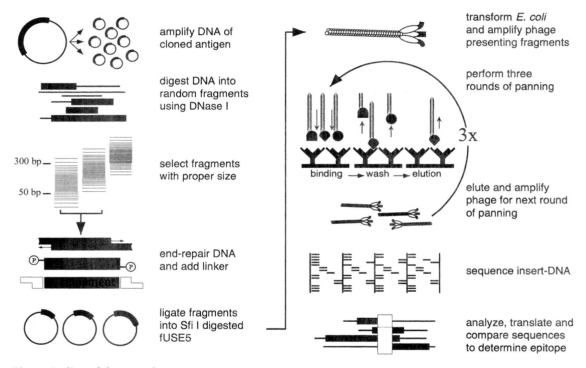

Fig. 1. Outline of the procedure

Materials

Equipment   – Mini-PROTEAN® 3 Electrophoresis System (Bio-Rad)

– Electroporator Gene Pulser II (Bio-Rad)

– Gene Pulser Cuvette #165-2086 (0.2 cm gap) (Bio-Rad)

– Submarine Electrophoresis Systems (Bio-Rad)

- Incubator, non-agitated, 37°C
- Incubator with variable shaking speed, 37°C

Chemicals

- Acrylamide/bisacrylamide 40% (Sigma)
- TEMED (N,N,N',N'-tetramethylethylenediamine) (Sigma)
- APS (Ammonium peroxodisulfate) (Sigma)

Plastic and glassware

- 96 well microtiter plates Maxi-Sorb (Nunc)
- 1.5 ml microcentrifuge tubes (Eppendorf)
- 15 ml plastic culture tubes (Falcon)
- 150 ml flask (Corning)
- 3 l flask (Corning)

Bacteria and phage

- E. coli MC1061 (George P. Smith, Division of Biological Sciences, Tucker Hall, University of Missouri, Columbia, MO 65211-7400; (573) 882-3344 (office), (573) 882-2720 (lab), (573) 882-0123 (fax), gpsmith@biosci.mbp.missouri.edu (e-mail), use fax or e-mail to request strains or libraries. See also: http://www.biosci.missouri.edu/smithgp/).
- E. coli K91 (see above)
- E. coli K91kan (see above)
- E. coli K802 (see above)
- Phage fUSE5 replicative form - dsDNA (see above)

Media and plates

- LB: 10 g bacto-tryptone, 10 g yeast extract, 5 g NaCl in 1 liter H_2O
- SOC: 20 g bacto-tryptone, 5 g yeast extract 0.58 g NaCl, 0.19 g KCl, dissolve in 1 liter H_2O and autoclave. Add 10 ml 2 M $MgCl_2$ (filter-sterilized) to 1 liter medium
- terrific broth media: 12 g bacto-tryptone, 24 g yeast extract, 4 ml glycerol, dissolve in 900 ml H_2O and autoclave. Add 100 ml of autoclaved potassium phosphate buffer (0.17 M KH_2PO_4 (2.31 g/100 ml), 0.72 M K_2HPO_4 (12.54 g/ 100 ml))
- LB/tet: LB medium containing 40 µg/ml tetracycline
- LB/tet/kan: LB medium containing 40 µg/ml tetracycline, 100 µg/ml kanamycin

- LB/tet plates: LB with 1.5% agar, 40 µg/ml tetracycline

- LB/tet/kan plates: LB with 1.5% agar, 40 µg/ml tetracycline, 100 µg/ml kanamycin

- LB/tet/strept plates: LB with 1.5% agar, 40 µg/ml tetracycline, 100 µg/ml streptomycin

Enzymes

- DNase I from bovine pancreas; 171,600 units/ml; usually provided in a concentration range of 50,000-375,000 units/ml (Gibco)

- T4-Polynucleotide kinase (PNK); 10,000 units/ml (New England Biolabs)

- Klenow-Fragment; 5,000 units/ml (New England Biolabs)

- T4 DNA-Ligase, 400,000 units/ml (New England Biolabs)

- *Sfi* I, 20,000 units/ml (New England Biolabs)

Antibiotics

- Tetracycline (20 mg/ml stock): In a 100-ml bottle autoclave 40 ml (50.4 g) glycerol. When cooled down to room temperature, filter-sterilize 40 ml 40 mg/ml tetracycline in H_2O into the bottle and mix thoroughly. Store at -20°C.

- Kanamycin (100 mg/ml stock in H_2O). Store at -20°C.

- Streptomycin (50 mg/ml stock in H_2O). Store at +4°C.

Buffers

- Elution buffer: 0.1 M HCl, pH adjusted to pH 2.2 with glycine, 1 mg/ml BSA

- NAP: 80 mM NaCl, 50 mM $NH_4H_2PO_4$ pH 7.0

- Tris/HCl: 1 M, pH 7.5

- Tris/HCl: 1 M, pH 9.1

- TE: 10 mM Tris/HCl pH 7.6, 1 mM EDTA pH 8.0

- TBS: 50 mM Tris, 150 mM NaCl, adjust pH to 7.5 using HCl. Store at room temperature.

- TBE (5x stock): 60.5 g Tris, 31 g H_3BO_3, 3.7 g $Na_2EDTA \cdot 2H_2O$; adjust volume to 1 l H_2O. Store at room temperature

- Coating buffer: 0.1 M $NaHCO_3$, pH 9.1

- DNase I dilution buffer: 20 mM Na-acetate, 5 mM $CaCl_2$, 0.1 mM PMSF, 50% (v/v) glycerol

- TE-saturated Phenol/Chloroform

- Phenol/chloroform/isoamyl alcohol: Before use, mix 49 parts of a TE-saturated phenol/chloroform solution with one part isoamyl alcohol

- $MnSO_4$, 25 mM

- ATP, 10 mM

- dNTP, 10 mM each

- Tween 20

- Skimmed milk powder

- 20% PEG6000/2.5 M NaCl

- BSA 10 mg/ml

- Proline linkers
  - P-Oligos 1+2: Fus-P1 (5' CTGGCGGTG 3')/(3' TGCGACCGCCAC 5')
  - P-Oligos 3+4: Fus-P2 (5' CACCGCCAGCTG 3')/(3' GTGGCGGTC 5')

- Cysteine linkers
  - C-Oligos 1+2: Fus-C1 (5' CTGGCTGCG 3')/(3' TGCGACCGACGC 5')
  - C-Oligos 3+4: Fus-C2 (5' TGCGGTCCAGCTG 3')/(3' ACGC-CAGGTC 5')

- Sequencing primer: 5' ACTTAAAAGACATACTCC 3' (downstream primer)

Procedure

Preparation of the library

1. Isolate the antigen-encoding DNA by digestion of plasmid DNA or amplify by PCR with appropriate primers. The amount of DNA necessary to create one library is about 50 µg for antigen inserts with average size.

2. Dilute DNase I to a final concentration of about 15 µ/ml with DNase I dilution buffer.

3. Digest approximately 15 µg of DNA containing the cloned antigen gene into random length fragments using 3 different final DNase I concentrations (table 1).

4. Incubate for 10 minutes at 15°C. Stop reaction by adding 8.3 µl 100 mM EDTA, pH 8.0.

5. Use 6 µl of total volume for analysis in 12 % TBE-PAGE minigel.

6. In parallel to the gel run, perform a phenol-chloroform extraction: Add an equal volume of phenol/chloroform/isoamyl alcohol to the digest and vortex for 15-30 seconds. Centrifuge the sample for at least 1 minute at room temperature to separate the phases. Remove the upper aqueous layer into a clean tube. Discard the lower organic phase. Add an equal volume of chloroform, vortex briefly, and centrifuge for 1 minute. Transfer the upper aqueous layer into a clean tube and discard the lower chloroform phase. Repeat the chloroform extraction a second time. Finally, ethanol-precipitate DNA and dissolve each DNA pellet in 20 µl H_2O.

7. Use the gel to determine which digests should be used: The main proportion of digested fragments should range between 45 to 300 bp in length, equivalent to about 15 to 100 amino acids. Combine these samples and end-repair the fragments.

Table 1. Digest of antigen-encoding DNA using different concentrations of DNase I.

Sample:	I	II	III
DNA (approx. 15 µg)	x µl	x µl	x µl
Tris/HCl 1 M, pH 7.5	7.5 µl	7.5 µl	7.5 µl
$MnSO_4$ 25 mM	6 µl	6 µl	6 µl
BSA 1 mg/ml	15 µl	15 µl	15 µl
DNase I dilution	7.5 µl	5 µl	3.75 µl
H_2O	ad 150 µl	ad 150 µl	ad 150 µl

DNA fragment end repair

1. Phosphorylate 5'-ends of DNA using T4 PNK. Incubate for 30 minutes at 37°C (table 2).

2. Use the same tube and reaction mix to complete a fill-in reaction using Klenow Fragment (KF). Incubate for 30 minutes at 37°C (table 3).

3. Phenol-chloroform extract the reaction mix, ethanol-precipitate the DNA and dissolve pellet in 30 µl TE buffer. The approximate concentration of DNA should be 1 µg/µl.

4. Save 1 µg for a test gel.

Table 2. DNA Phosphorylation of 5' ends

DNA fragments	40 µl
PNK buffer	10 µl
ATP 10 mM	10 µl
T4 PNK 10 u/µl	5 µl
H_2O	35 µl
Total volume	100 µl

Table 3. Fill-in reaction

Kinase reaction	100 µl
dNTP (10 mM each)	8.1 µl
KF 5 u/µl	4 µl
Total volume	112.1 µl

Linker ligation of DNA fragments

Note: Two pairs of linkers are suggested: the use of *C-linkers* may result in a peptide which at its base (beginning and ending) contains a disulfide bridge, possibly forming a loop and especially bending and exposing short inserts to a certain degree. On the other hand, the *P-linkers* yield prolines at the beginning of each insert, most likely allowing it to rotate into a more relaxed form. To obtain the benefits of both linkers, it is considered advantageous to generate and screen two libraries: one using C-linkers, the other using P-linkers.

1. Anneal C- and P-oligonucleotides to form the linkers.

2. Determine concentration of dissolved oligos by UV absorption at 260 nm.

3. Mix equimolar amounts of oligos 1+2 to obtain linker 1 and oligos 3+4 for linker 2.

4. The final linker concentration should be about 2 μg/μl.

5. Heat solution to 95°C for 4 minutes.

6. At bench, let solution cool down to room temperature. Store at -20°C.

   Ligate fragments with linkers (ligation (a), table 4 and perform a control ligation (ligation (b), table 5 to prove efficiency of DNA fragment end repair.

1. Incubate ligations for 16 h at 16°C.

2. Save 1 μl of ligation (a) for test gel.

3. Run ligation (a) on preparative 12 % TBE-PAGE minigel.

4. Run saved volume of ligation (a) and control ligation (b) on 12 % TBE - PAGE minigel. The lane containing the linker-ligated DNA (a) should

Table 4. Ligation (a) of DNA fragments with linkers

DNA fragments	5 μl
Linker 1 (= 4 μg linker per μg fragments)	x μl
or Linker 2 (= 4 μg linker per μg fragments)	y μl
Ligase buffer 10x	2.5 μl
Ligase 1 u/μl	2.5 μl
H_2O	ad 25 μl

Table 5. Ligation (b): Control of fill-in and kinase efficiency by self ligation of fragments

DNA fragments	1 μl
Ligase buffer 10x	1 μl
Ligase 1 u/μl	1 μl
H_2O	7 μl
total volume	10 μl

show fragments of smaller size (about 50-300 bp) compared to the self-ligated DNA (b).

5. Cut gel slice of ligation (a) containing fragments ranging in size from 45 bp to 300 bp.

6. Electro-elute fragments into TE buffer.

7. Cut gel slice containing DNA into halves and transfer both parts into a dialysis bag (cut-off 16000 kDa, ID 1 cm) with one side closed.

8. Add 600 µl TBE, close other bag side, and submerge in electrophoresis gel box filled with TBE buffer.

9. Apply 100 V for 50 minutes, then reverse polarity for 15 seconds.

10. Remove buffer from bag into a 1.5 ml microcentrifuge tube and rinse the bag with another 200 µl TBE.

11. Purify DNA by phenol-chloroform extraction, ethanol-precipitate the DNA and dissolve the pellet in 100 µl TE buffer. Save 10 µl for test gel.

Preparation of fUSE5 double-stranded DNA

Non-infective phage vectors with gene-III frameshifts like fUSE5 tend to be overgrown by infective pseudorevertants when propagated in male (infectible) strains like K91 or K91BluKan. Due to the lack of possible reinfection in a female (uninfectible) strain such as K802, the amplification of fUSE5 in this host does not select for pseudorevertants and is therefore highly advised. The original phage stock is usually not used to obtain the single-stranded phage DNA necessary to transform the K802 cells. Instead, phage from a single colony derived from the original stock are produced, and their ssDNA is then extracted.

1. Prepare 20 ml starved male K91 (not K91kan) cells.

2. Grow K91 cells at 37°C in 20 ml LB to mid log phase (OD_{600} ~0.45).

3. Incubate with gentle shaking for an additional 5 min.

4. Centrifuge cells at 4°C and 3,000 g for 10 min.

5. Discard the supernatant and resuspend the cells in 20 ml of 80 mM NaCl.

6. Transfer cells to a sterile 125-ml culture flask, shake gently at 37°C for 45 min.

7. Centrifuge cells at 4°C and 3000 g for 10 min.

8. Gently resuspend cells in 1 ml cold NAP buffer. Store cells at 4°C. Cells remain infectible for 3-5 days.

9. To determine the viability of the original phage stock as well as to obtain single phage colonies, titer 10 μl portions of 10^{-3}, 10^{-4} and 10^{-5} serial dilutions of the fUSE5 stock on 10 μl of the starved K91 cells. Incubate for 10 minutes at room temperature.

10. Add 1 ml LB/0.2 mg/ml tetracycline and incubate at 37°C for 30 minutes.

11. Spread 200 μl of each culture on separate LB/tet agar plates. Incubate overnight at 37°C.

12. A phage titer of down to about 15% of the expected (nominal) titer is perfectly usable. Grow under agitation at 37°C a single colony overnight in 1.7 ml LB medium containing 20 μg/ml tetracycline in a 15 ml glass or plastic culture tube.

13. Centrifuge about 1.5 ml of culture in a 1.5 ml microcentrifuge tube for 1 minute and transfer approximately 1.2 ml of the supernatant into a fresh 1.5 ml tube.

14. Re-centrifuge the culture supernatant again and transfer 1 ml of supernatant to another 1.5 ml tube. Add 150 μl PEG/2.5 M NaCl, mix, and incubate overnight at 4°C.

15. Precipitate phage by centrifugation for 30 minutes at room temperature and re-dissolve pellet in 500 μl *TE*.

16. Phenol-chloroform extract the phage solution, ethanol-precipitate, and re-dissolve the ssDNA in 100 μl *TE*.

17. Dilute 20 μl of the ssDNA with 180 μl *TE*, and use the dilution to transform 200 μl female (uninfectable) K802 cells by the CaCl₂ method. Plate cells on LB/tet agar and incubate at 37°C overnight.

18. Grow an individual colony in 1 liter LB/20 μg/ml tetracycline under agitation at 37°C overnight.

19. Separate the cells from the supernatant by centrifugation (15 minutes, 5,000 g).

20. Use a commercial DNA maxiprep kit to extract the double-stranded, covalently closed circular RF DNA. Determine the DNA concentration and purity by UV absorption.

21. Digest the double-stranded DNA with *Sfi* I at 50°C using approximately 10 units of enzyme per µg of DNA to ensure a good digestion in about 2-4 h incubation. About 3-5 µg digested DNA are necessary per phage library.

22. After digestion, run approximately 100-300 ng DNA on a 1% agarose gel to confirm cleavage. Phenol-chloroform extract the restriction digest, ethanol-precipitate the DNA and resuspend the pellet in an appropriate volume of TE buffer.

23. Run the digested DNA on a preparative 15% TBE-PAGE mini gel to separate the digestion products. λ*Hin*D III is an appropriate marker because of the 9.4 Kbp band which virtually co-migrates at the same height as the 9.2 Kb double-cleaved fUSE5 band.

24. Excise the band from the gel, electro-elute for 2 h at 100 V, and precipitate the DNA as described above. Dissolve the DNA pellet in TE buffer to yield a concentration of about 1 µg/µl.

25. Run approximately 0.5 µg DNA on a gel. Estimate the concentration by comparison with the known amounts of DNA in the marker bands.

   Note: Alternatively, a much more detailed protocol to obtain RF DNA using 'classical' methods like $CsCl_2$ gradient centrifugation etc., as well as a significantly more explicit digestion protocol can be obtained from G.P. Smith (see Materials).

Ligation of fragments into fUSE5

1. To ligate the linked fragments into fUSE5 vector, combine reagents as displayed in table 6 and incubate for 16 h at 16°C.

2. Perform phenol-chloroform extraction of ligation mix, ethanol-precipitate DNA and dissolve pellet in 30 µl H_2O.

Table 6. Ligation into fUSE5

DNA fragments	90 µl
fUSE5 Sfi I digested; 1 µg/µl	3 µl
Ligase buffer 10x	11 µl
Ligase 1 u/µl	3 µl
H_2O	3 µl
Total volume	110 µl

Transformation of *E. coli* MC1061 electro-competent cells

1. Transform 5x 2.5 µl of the purified ligation into 50 µl MC1061 electro-competent cells using a Gene Pulser II at settings 2500 V, 25 µF, 200 W. (The preparation of electro-competent cells is described in Chapter 39). Dilute cells of each transformation in 1 ml SOC/0.2 µg/ml tetracycline and shake for 1 h at 37°C.

2. Combine transformed cells.

3. To estimate transformation efficiency and library size, plate 250 µl of 1:10, 1:100, and 1:1000 dilutions in SOC on LB/tet/strept plates.

4. Plate 20 x 250 µl of transformed cells as well as the three dilutions onto LB/tet/strept agar plates (alternatively, one large 24x24 cm agar plate can be used to plate the cells).

5. Incubate overnight at 37°C.

6. Count colonies on plates of transformation dilution. Usually you can obtain a total of 10^4-10^5 transformed cells or better. This equals the number of primary transformants.

7. Scrape off and resuspend colonies from plates in 200 ml LB/tet/strept. Shake for 1.5 h at 37°C.

8. Cool for 1 h at 4°C. Separate cells from supernatant by centrifugation at 10,000 g for 10 minutes. Store cleared phage-containing supernatant at 4°C.

9. Resuspend cell pellet in 3.2 ml LB/tet. Prepare glycerol stocks by aliquoting 850 µl bacteria solution and adding 150 µl glycerol. Mix well. Freeze glycerol stocks at -80°C.

10. Save 1 ml of supernatant for phage titer determination of the amplified library at 4°C, and precipitate phage from remaining supernatant by adding 0.15 vol. PEG6000/NaCl. Incubate overnight at 4°C.

11. Spin solution for 15 minutes at 13,000 g and dissolve the phage-containing pellet at 37°C in 10 ml TBS.

12. Clear solution once more by centrifugation for 10 minutes at 13,000 g.

13. Precipitate phage from supernatant by adding 0.15 vol. PEG6000/NaCl. Incubate > 3h on ice.

14. Spin phage for 15 minutes at 13,000 g.

15. Resuspend the amplified *gene-fragment phage library* in 120 µl TBS.

Determining phage infectivity

The number of the primary transformants obtained includes colonies producing non-infective phage (e.g. by frame-shift in gIII). To determine the infectivity and therefore the number of 'active' phage of the primary *gene-fragment phage library* the following titration is performed.

1. Plate 200 µl K91kan terrific broth cells (see below) on LB/tet/kan plate. Incubate plate at room temperature until surface is dry.

2. With a toothpick, transfer 25 colonies from the MC1061 transformation efficiency test (primary phage) to the plate.

3. Incubate overnight at 37°C.

4. Count number of grown colonies to calculate infectivity. Normal values are in the range of 50% of grown colonies, giving a library size of approximately one half the number obtained in the preceding protocol hence reflecting the diversity of the *gene-fragment phage library*.

Phage titer determination

To determine the phage titer of the amplified *gene-fragment phage library* finally available for the panning, a serial dilution of the saved supernatant is performed. The number obtained equals the total number of phage in the library.

1. Grow K91kan from an overnight culture in terrific broth/100 µg/ml kanamycin until the OD_{600} of a 1:10 dilution of the then log-phase culture reaches ~ 0.15.

2. Perform a serial dilution (10^{-1} - 10^{-4}) of the phage supernatant saved from transformation of *E. coli* MC1061 electro-competent cells with TBS.

3. To 10 µl phage dilution add 10 µl log-phase K91kan cells.

4. Incubate at room temperature for 15 minutes.

5. Add 1 ml LB/0.2 µg/ml tetracycline.

6. Shake at 37°C for 30 minutes.

7. Plate 1/5 (200 µl) of each dilution on LB/tet/kan plates and incubate overnight at 37°C.

8. Count colonies to calculate transforming units (t.u.). This equals the size of the amplified *gene-fragment phage library* (usually shows a total of around 10^{10} or more phage) contained in the 120 µl TBS in which the library was resuspended.

Panning

1. Per antibody and library, coat one microtiter plate well with approximately 10 µg antibody or antibody fragments (scFv, Fab, etc.) in 100 µl NaHCO₃ (0.1 M, pH 9.1). Incubate overnight at 4°C.

2. Wash wells 3 times with 300 µl TBS/0.5 % Tween 20.

3. Block well using 300 µl 2 % skimmed milk powder in TBS. Incubate at room temperature for 3 h.

4. Wash wells 10 times with 300 µl TBS/0.5 % Tween 20.

5. Dilute phage to a concentration of about 10^{8}–10^{10} t.u./100 µl.

6. To each coated well add 100 µl phage dilution of the corresponding library (C or P) and incubate overnight at 4°C.

7. Save phage dilution from well. The 'depleted' library may be used for another antibody or to repeat a panning to obtain less affine phage.

8. Wash wells twice with 200 µl TBS/0.5 % Tween 20; add 300 µl 2 % skimmed milk powder in TBS and incubate for 10 minutes at room temperature; wash wells again using twice 200 µl TBS/0.5 % Tween 20; add 300 µl 2 % skimmed milk powder in TBS and incubate for 5 minutes at room temperature; wash wells twice with 200 µl TBS/ 0.5 % Tween 20.

9. Elute phage using 100 µl elution buffer per well. Incubate for not longer than 10 minutes at room temperature.

10. Transfer phage solution into 1.5 ml reaction tube; neutralize phage using 12 µl 1 M Tris/HCl pH 9.1.

11. Grow K91kan cells derived from an overnight culture in terrific broth to a log-phase OD_{600} (1:10) of ~ 0.15.

12. Transfer 300 µl of the cells into a 1.5 ml reaction tube.

13. Add neutralized phage solution and incubate at room temperature for 20 minutes.

14. Transfer cells into 20 ml prewarmed (37°C) LB/0.2 µg/ml tetracycline. Shake at 37°C for 1h.

15. Plate 1 ml, 500 µl and 50 µl of solution on LB/tet/kan to determine number of eluted phage. Incubate overnight at 37°C.

16. Increase tetracycline concentration of the remaining 18 ml culture to 20 µg/ml and incubate under agitation overnight at 37°C.

17. Clear the culture by centrifugation for 10 minutes at 5,000 g.

18. Re-centrifuge supernatant to achieve optimal clearance.

19. Save 1 ml of supernatant to determine phage titer of reamplified eluate as in subprotocol Phage titer deterination.

20. Precipitate phage from supernatant by adding 0.15 vol. PEG6000/NaCl. Incubate 3 h or more on ice or overnight at 4°C.

21. Spin phage solution for 15 minutes at 13,000 g and re-dissolve pellet at 37°C in 10 ml TBS.

22. Clear solution once more by centrifugation for 10 minutes at 13,000 g.

23. Precipitate phage again by adding 0.15 vol. PEG6000/NaCl. Incubate > 3h on ice.

24. Spin phage for 15 minutes at 13,000 g.

25. Resuspend pellet in 120 µl TBS yielding phage from 1st panning.

26. Repeat panning using amplified phage. We routinely perform 3 rounds of selection in total.

Determine epitope by sequencing of phage

1. Pick colonies randomly from LB/tet/kan titer plates of panning rounds 2 and 3. Alternatively, an ELISA with amplified phage supernatant of single colonies can be used to determine binding phage. However, we found that several phage containing the epitope did not show a significant ELISA signal. On the other hand, every random-picked phage sequenced in the example shown below contained an epitope-including segment of the antigen.

2. Grow cells in 2 ml LB/tet/kan overnight at 37°C.

3. Separate cells from supernatant by centrifugation at 5,000 g. Transfer supernatant into new 2 ml reaction tube. Resuspend cells in 100 µl 50 % glycerol and freeze at -80°C.

4. Precipitate phage from supernatant by adding 0.15 vol. PEG6000/NaCl. Shake for 5 minutes at room temperature, then incubate for 3 h or more on ice without agitation.

5. Pellet phage by centrifugation at 4°C.

6. Resuspend pellet in 200 µl TBS and perform phenol-chloroform extraction. Use dissolved phage-ssDNA for sequencing. The sequencing primer reads in upstream direction of the pIII (sequence appears antiparallel).

7. Compare sequencing results with the known sequence of the gene to determine the epitope region.

Results

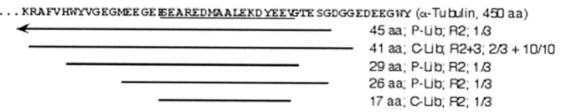

Fig. 2. Result of an epitope mapping using a recombinant human anti-α-tubulin F$_{ab}$ (Winter, 1997): Shown are the 50 C-terminal amino acids of the total 450 residues of *Xenopus laevis* α-tubulin and the aligned sequences obtained from the inserts of amplified phage. Two libraries (C- and P-library) were used to identify the epitope region of the antibody fragment. The lines indicating the insert position are labelled with the length of the insert (xx aa), the kind of library (C or P), the round of panning (R2 or R3) from which the phage was obtained, and the number of phage clones yielding these sequences per number of phage sequenced of this round and library. Further fine-mapping using short overlapping synthetic peptides showed that the epitope of this F$_{ab}$ consists of only 7 amino acids (DMAALEK).

Troubleshooting

- **No overlapping sequences:** When the antibody does not bind denatured protein (e.g. on western blots) it is likely to be directed against a conformational epitope. The affinity to only one half of the epitope might be high enough to select either one or both parts with the phage library separately; however, it is unlikely that you can align the sequences. On the other hand, non-specific binding of phage under high concentrations (either antibody or phage) might lead to non-related insert sequences. This can be true for less affine antibodies as well.

- **Phage show wild type sequence:** Pseudorevertants in the generation of RF phage DNA may 'outgrow' fragment-containing phage. A high concentration might then lead to non-specific binding and amplification.

- **All phage show the same insert:** The affinity and number of this particular amplified phage might compete with other matching sequences. However, they might be available from the first round of panning. When sequencing randomly picked phage from the primary gene library all sequences should be different; otherwise, one phage might have overgrown the library.

Applications

The gene-fragment phage library promises an easy identification of the epitope region on the antigen , especially for antibodies recognizing linear epitopes. However, there are limitations to this technique:

(and limitations)

- The antigen has to be cloned. Cloned homologue proteins from other species may be used if the antibody recognizes them as well, but may yield in poor results if the affinity for the replacement is lower than for the original.

- Since the antigen fragments displayed at the phage tip are usually small compared to the total antigen, the chance of finding a split epitope or conformational epitope - that is, an epitope consisting of two separate portions of one or more peptide chains - is reduced either by inhibition of correct folding or just by completely missing the second part of the antigen in the insert. Partial (and a usually less strong) binding to either part of the epitope on different phage might yield sequences containing the complete epitope, but is less likely to succeed due to the lower affinity of an antibody to the separated epitope parts. This might be especially true for antibodies not binding to the denatured antigen.

- The generated library is only specific for the tested antibody or other antibodies reacting with the same antigen. In contrast, random peptide libraries can be used with antibodies recognizing different antigens.

- The resulting sequences may contain additional residues not necessary for the binding of the antigen to the antibody, not allowing determination of the exact epitope position. In this case, further fine-mapping using short overlapping synthetic peptides might be useful.

Acknowledgements. We would like to thank Prof. George P. Smith (University of Missouri, U.S.A.) for fUSE5, bacteria, and a very detailed protocol and Dr. Barbara Hügle-Dörr (Heidelberg, Germany) for help in generating several libraries.

References

Blüthner M, Bautz EKF, Bautz FA (1996) Mapping of epitopes recognized by PM/Scl autoantibodies with gene-fragment phage display libraries. J Immunol Methods 198;187-198

Blüthner M, Schäfer C, Schneider C, Bautz FA (1999) Identification of major linear epitopes on the sp100 nuclear PBC autoantigen by the gene-fragment phage-display technology. Autoimmunity 29; 33-42

Klinger C, Huet J, Song D, Petersen G, Riva M, Bautz EKF, Sentenac A, Oudet P, Schultz P (1996) Localization of yeast RNA polymerase I core subunits by immunoelectron microscopy. EMBO J 15;4643-4653

Parmley SF, Smith,GP (1988) Antibody-selectable filamentous fd phages vectors: Affinity purification of target genes. Genes 73;305-318

Petersen G, Song D, Hügle-Dörr B, Oldenburg I, Bautz EK (1995) Mapping of linear epitopes recognized by monoclonal antibodies with gene-fragment phage display libraries. Mol. Gen. Genet., 249;425-431

Sambrook J, Fritsch EF, Maniatis T (1989) Molecular Cloning. Cold Spring Habor Laboratory Press

Smith GP (1992) Cloning in fUSE vectors. Division of Biological Sciences. University of Missouri. Laboratory manuscript

van Zonnefeld AJ, van den Berg BMM, van Mejer M, Pannekoek H (1995) Identification of functional interaction sites on proteins using bacteriophage-displayed random epitope libraries. Gene 167;49-52

Wang LF, Du Plessis DH, White JR, Hyatt AD, Eaton BT (1995) Use of a gene-targeted phage display random epitope library to map an antigenic determinant on the bluetongue virus outer capsid protein VP5. J Immunol Methods 178;1-12

Winter CH (1997) Klonierung und Charakterisierung von Fab- und scFv-Immunoglobulinfragmenten aus Mensch und Maus. PhD Thesis, University of Heidelberg

In vivo Experiments

Imaging Tumor Xenografts Using Radiolabeled Antibody Fragments

ANNA M. WU and ANDREW A. RAUBITSCHEK

Introduction

In vivo imaging remains an important counterpart to biodistribution studies, as a means for evaluating the targeting and distribution of radiolabeled engineered antibodies. Imaging allows evaluation of the entire, living animal simultaneously, and provides integrated information that cannot be obtained simply by well-counting of isolated organs and tissue. Images can also be obtained serially in the same animal to follow the distribution of a radiolabeled agent over time. This allows studies to be completed using smaller numbers of animals, and eliminates many issues of animal-to-animal (or xenograft-to-xenograft) variability. A major shortcoming of the approach is the challenge of obtaining quantitative information from scans, although newer imaging modes such as positron emission tomography (PET) can yield quantitative data. Furthermore, although imaging in small animals is hampered by the low resolution of clinical cameras and scanners, several dedicated small-animal imaging devices are under development as investigators realize the increasing value of studying animal models of disease.

A key consideration is choice of the imaging radionuclide. Many factors must be considered including half-life, emission spectrum, chemical properties, availability, and cost (for a review, see Hiltunen 1993). Standard hospital nuclear medicine cameras are optimized for detection of radioisotopes such as ^{99m}Tc (140 keV γ) and other nuclides with gamma emissions of similar energy.

✉ Anna M. Wu, Beckman Research Institute of the City of Hope, Department of Molecular Biology, 1450 East Duarte Road, Duarte, CA, 91010, USA
(*phone* +1-626-301-8287; *fax* +1-626-301-8280; *e-mail* awu@coh.org)
Andrew A. Raubitschek, City of Hope National Medical Center, Division of Radioimmunotherapy, 1500 East Duarte Road, Duarte, CA, 91010, USA

Radioiodinated antibodies for imaging

Radioiodination is popular due to ease of labeling and the small amounts of protein required (see Chapter 36). However, each isotope of radioiodine has advantages and disadvantages. For example, ^{125}I has a long half-life (60 d) and its low energy gamma emission is detected poorly by conventional gamma cameras. ^{131}I has significant beta decay and its gamma emission is of too high energy for efficient capture; nonetheless it is a popular choice. ^{123}I has favorable imaging and half life properties (159 keV; 13.2 hr) for gamma camera imaging using antibodies and fragments, but is costly and not as readily available. A detailed protocol for radioiodination of antibodies is presented in Chapter 36. The chemical linkage of the iodine to protein is labile, and dehalogenation can occur in tissues and tumors. Released iodine is scavenged by the thyroid, and to a lesser extent, the stomach. Uptake to these organs can be blocked by pretreatment with Lugol's solution and potassium perchlorate. The dehalogenation that occurs in liver and kidney can be advantageous, resulting in lower observed activities in these normal organs. Alternately, many laboratories have developed more stable iodination chemistries, generally involving radioiodination of a conjugate which is subsequently attached to the protein (see Hiltunen 1993 for review). However, in the absence of dehalogenation, the biodistribution properties will need to be re-evaluated for each labeled protein.

Procedure

Gamma camera imaging using radioiodinated antibodies

1. Implant athymic mice subcutaneously in the flank with antigen-positive tumor cells (usually 1-2 x 10^6 cells in a volume of 0.2 ml). Allow the xenograft to grow to 8-10 mm in diameter.

2. 24 hr prior to administration of radioiodinated antibody or fragment, pretreat mice by using 10 drops saturated KI/100 ml drinking water to block thyroid uptake.

3. 30 min prior to injection of radiolabeled antibody, administer 1.5 mg potassium perchlorate in 0.2 ml PBS by gastric lavage to block stomach uptake of radioiodine.

4. At time zero inject mice intravenously in the tail vein with 20-30 µCi of radiolabeled protein.

5. At the selected time, lightly anesthetize mice using a Bell jar charged withMetofane; a nose cone can be used to maintain anesthesia. Immobilize mice prone on a board and image using a nuclear medicine camera (e.g. Toshiba 901/HG, Toshiba America Medical Systems) equipped with a pinhole collimator. Image acquisition times of 5-15 min should be adequate.

Radiometal labeled antibodies for imaging

Radiometals provide a broad range of physical and chemical properties to choose from for labeling antibodies. Preferred isotopes for imaging include ^{99m}Tc (half-life, 6 hrs) for fragments and ^{111}In (half-life, 2.6 days) for intact antibodies. A protocol for labeling with ^{111}In is given in Chapter 36. ^{99m}Tc has been a favored radionuclide for labeling of antibody fragments due to the fact that it is generator-produced and readily available in hospital nuclear medicine departments. Several approaches have been used to attach ^{99m}Tc to antibodies, including direct labeling of proteins through endogenous cysteine residues, or indirect labeling by conjugation with preformed Tc-chelate complexes (e.g. BzMAG3, N3S, N2S2, or similar bifunctional chelates; see Reilly for a review).

Antibody engineering can be used to create specific binding sites for metals. Examples include the introduction of a minimal chelation site consisting of a Gly4Cys peptide to the C-terminus of an scFv for Tc-labeling, (George et al. 1995) and a general method for labeling His6-tagged scFv using a stable Tc(I)-carbonyl complex (Waibel et al. 1999).

Optimal imaging time

Selection of imaging time points will depend on the expected pharmacokinetics of the antibody being tested. If a biodistribution study has been conducted previously, time points can be selected based on high tumor uptake (%ID/g) or high tumor:blood uptake ratio. However, with smaller engineered fragments, highest tumor uptakes are achieved early, and highest tumor:blood ratios are reached later in the time course; clearly both are important. Throughout, one must also factor in the physical decay of the attached radionuclide. One approach is to calculate an Imaging Figure of Merit (IFOM) from the biodistribution, to predict the optimal imaging time. This indicator is proportional to the rapidity with which a tumor may be resolved from background statistically, and incorporates both the magnitude of the uptake and the tumor:blood ratios determined at each given time point (Williams et al. 1995). IFOM analysis can indicate the optimal time interval following injection for immunoscintigraphy, and also illustrates the importance of matching the physical half-life of the isotope with the biological half-life of the antibody agent. As an example, IFOMs have been calculated for a set of five antibodies and engineered fragments derived from the anti-CEA antibody T84.66 (intact, F(ab)'$_2$, minibody, diabody, scFv). Calculations suggest that if the preferred imaging radionuclide is iodine-123 ($t_{1/2}$ = 13.2 h), a minibody (scFv-C$_{H3}$) or diabody format would perform best, but for fluorine-18 (positron-emitter; $t_{1/2}$ = 110 min) a diabody would be preferred (Williams et al. 1995, Wu et al. 1999).

ImmunoPET

Positron emission tomography (PET) is gaining importance in clinical imaging, and the technique can be extended to antibodies and engineered fragments if they are labeled with positron-emitting radionuclides. As the isotope decays, a positron is ejected and after traveling a minimal distance, it will encounter an electron in ordinary matter and annihilate. This results in the simultaneous emission of two 511 KeV photons, in 180° opposite directions. Use of coincidence detection allows very precise localization and quantitation of the activity. To date, a number of laboratories have investigated labeling antibodies or fragments with positron-emitting radionuclides including: [18]F (110 min), [64]Cu (12.7 h), or [124]I (4.2 d half-life) (See Kairemo 1993 for review). Recent availability of a microPET instrument opens the possibility for high-resolution, sensitive, and quantitative imaging using radiolabeled antibody fragments in small animals such as mice (Chatziioannou et al. 1999, Wu et al. 2000).

References

Chatziioannou AF, Cherry SR, Shao Y, Silverman R, Meadors K, Farquhar TH, Pedarsani M, Phelps ME. (1999) Performance evaluation of microPET: A high-resolution lutetium oxyorthosilicate PET scanner for animal imaging. J. Nucl. Med. 40:1164-76.

George AJ, Jamar F, Tai MS, Heelan BT, Adams GP, McCartney JE, Houston LL, Weiner LM, Oppermann H, Peters AM. (1995) Radiometal labeling of recombinant proteins by a genetically engineered minimal chelation site: technetium-99m coordination by single-chain Fv antibody fusion proteins through a C-terminal cysteinyl peptide. Proc. Nat. Acad. Sci. 92:8358-62.

Hiltunen JV. (1993) Search for new and improved radiolabeling methods for monoclonal antibodies. A review of different methods. Acta Oncol. 32:831-839.

Kairemo KJA. (1993) Positron emission tomography of monoclonal antibodies. Acta Oncol. 32:825-830.

Reilly RM. Immunoscintigraphy of tumours using 99Tcm-labelled monoclonal antibodies: A review. (1993) Nucl. Med. Commun. 14:347-359.

Waibel R, Alberto R, Willuda J, Finnern R, Schibli R, Stichelberger A, Egli A, Abram U, Mach J-P, Pluckthun A, and Schubiger PA. (1999). Stable one-step technetium-99m labeling of His-tagged recombinant proteins with a novel Tc(I)-carbonyl complex.

Williams LE, Liu A, Wu AM, Odom-Maryon T, Chai A, Raubitschek AA, Wong JYC. (1995) Figures of merit (FOMs) for imaging and therapy using monoclonal antibodies. Med. Phys. 22:2025-2027.

Wu AM, Williams LE, Zieran L, Padma A, Sherman M, Bebb GG, Odom-Maryon T, Wong JYC, Shively JE, and Raubitschek AA. (1999) Anti-carcinoembryonic antigen (CEA) diabody for rapid tumor targeting and imaging. Tumor Targeting 4:47-58

Wu AM, Yazaki PJ, Tsai SW, Nguyen K, Anderson, A-L, McCarthy DW, Welch MJ, Shively JE, Williams LE, Raubitschek AA, Wong JYC, Toyokuni T, Phelps ME, and Gambhir SS. (2000) High-resolution microPET imaging of carcinoembryonic antigen- positive xenografts by using a copper-64-labeled engineered antibody fragment. Proc. Nat. Acad. Sci. USA 97:8495-8500

Xenograft Mouse Models for Tumour Targeting

GAIL ROWLINSON-BUSZA, JULIE COOK, and AGAMEMNON A. EPENETOS

Introduction

Human tumour xenografts in nude or SCID mice are a useful model for studying the tumour targeting ability of a new antibody, scFv or fusion protein. Nude mice are athymic and therefore immunologically incompetent, allowing the growth of implanted human tumours. These xenografts have been shown to retain the histology of the original tumour and the relevant human tumour markers. In order to compare different targeting molecules *in vivo*, the nude mouse provides a convenient model, since genetically identical tumours can be induced in several animals, allowing direct comparisons to be made.

There are clearly limitations of human tumour xenografts in immunologically incompetent mice as a model of the human disease. Since the xenografted tumour is the only human component in the model, there will be no cross-reactivity with other human antigens as may occur in patients. In addition, the lack of a functional immune system in the nude mouse prevents the formation of immune complexes which could affect the clearance of the antibody in an immunologically competent patient. This lack of immune system also precludes the use of the nude mouse model for strategies involving the recruitment of effector functions for tumour cell killing, such as antibody-directed cell-mediated cytotoxicity

✉ Gail Rowlinson-Busza, Antisoma, St. George's Hospital Medical School, Cranmer Terrace, London, SW17 0QS, UK (*phone* +44-20-8672-7200; *fax* +44-20-8767-1809; *e-mail* gail@antisoma.co.uk)

Julie Cook, Antisoma, St. George's Hospital Medical School, Cranmer Terrace, London, SW17 0QS, UK

Agamemnon A. Epenetos, Antisoma, St. George's Hospital Medical School, Cranmer Terrace, London, SW17 0QS, UK

(ADCC). In terms of imaging, tumours implanted subcutaneously in nude mice form discrete masses, which are more easily visualised by a gamma camera, compared with spontaneous tumours occurring in normal organs in patients. In addition, when designing therapeutic protocols in the nude mouse model, it is important to take into account the fact that the uptake of antibodies in xenografts is in excess of 10% of the injected dose/g, which is around a thousand times higher than found in tumours in patients.

In spite of these limitations, the xenograft model is invaluable for studying the antigen-binding capability of new antibodies and recombinant molecules *in vivo*, in a physiological situation. Different molecules can be compared within the system in terms of their stability, clearance, tumour uptake and efficacy and proof of principle established. For the foreseeable future this model will remain the most widely employed for tumour localisation studies prior to clinical trials.

Outline

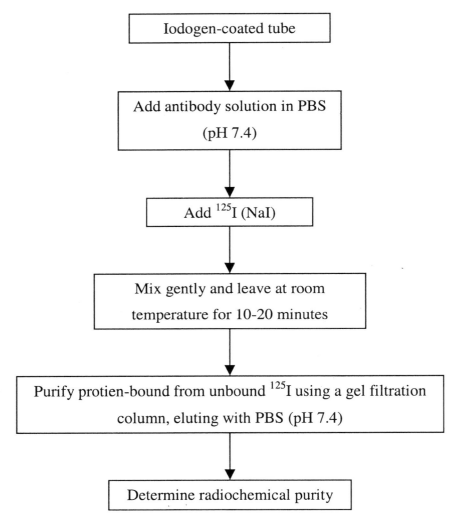

Fig. 1. Preparation of ^{125}I-labelled antibody

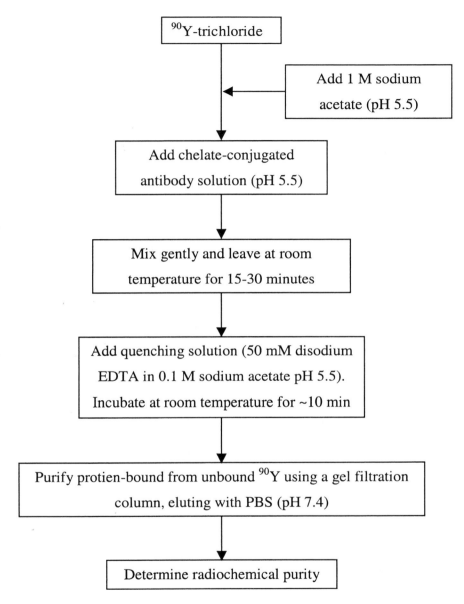

Fig. 2. Preparation of ^{90}Y-labelled antibody

Subprotocol 1
Radiolabelling of Proteins by Iodination

Iodination is one of the easiest methods for radiolabelling proteins and peptides. It involves the incorporation of radioactive iodine (usually ^{125}I) into a tyrosine residue by an oxidation reaction. This protocol uses Iodogen as the oxidising agent, as first described by Fraker and Speck (1978).

It is possible, by using two isotopes of iodine with different gamma energy spectra (e.g. ^{125}I and ^{131}I), to compare two different proteins in the same mice, such as a specific and a control antibody.

▪▪ Materials

- Iodogen (1,3,4,6-tetrachloro-3α, 6α-diphenylglycouril) (Pierce Europe B.V., Oud Beijerland, The Netherlands)

- Chromatography paper, such as grade 31ET Chr (Whatman Ltd., Maidstone, Kent, UK)

- [^{125}I] or [^{131}I] sodium iodide (Amersham, Pharmacia Biotech, Amersham, Bucks., UK)

▪▪ Procedure

For all procedures involving radioactivity, use adequate shielding (lead in the case of ^{125}I and ^{131}I) and adhere to the local rules in operation in your institute.

1. Dissolve Iodogen in chloroform at a concentration of 2 mg/ml. Transfer aliquots of 25 μl (50 μg) to round-bottomed plastic cryotubes (Nunc, Denmark). Allow the chloroform to evaporate to dryness, leaving a thin coating of Iodogen in the tube (leave in a fume hood overnight or use any other drying apparatus, such as a vacuum desiccator). Store Iodogen-coated tubes desiccated at -20°C until required for iodination.

2. Dilute the protein to be iodinated to 1-7 mg/ml in PBS or another suitable buffer (pH 7.4-8.0).

3. Place 100-400 μl of protein solution into an Iodogen-coated tube and add radioactive sodium iodide to a specific activity of 1-2 mCi/mg. Measure the activity in the tube (A1 mCi).

4. Agitate gently for several seconds and allow to react for 10-20 min at room temperature.

5. Purify the radiolabelled protein by size exclusion chromatography using a 20-ml Sephadex G-50 or G-25 (Pharmacia) column. This column can be made from a 20-ml syringe packed with pre-swelled Sephadex, placing a small circle of gauze in the bottom of the syringe before gently adding the gel. Wash the column with at least three bed volumes of PBS (pH 7.4). Remove the reaction mixture from the Iodogen tube and apply to the column. Rinse the Iodogen tube with 200 µl PBS and apply this to the column as well. Measure the activity in the empty tube (A2 mCi). Elute the column with PBS and collect 1-2 ml fractions.

6. Count the activity in each of the fractions and pool those containing the radiolabelled protein (the first peak of radioactivity). Measure the total recovered protein-bound activity (A3 mCi).

7. Calculate the radiolabelling efficiency (*RE*) of the radiolabelled protein as follows:

$$RE = \frac{A3}{A1 - A2} \times 100\,\%$$

8. Calculate the specific activity (*SA*) of the radiolabelled protein as follows.
   Assuming that the difference in activity between the full and empty Iodogen tube represents a proportionate loss of protein to be loaded onto the column, then the amount of protein loaded onto the column (*LP*) is:

$$LP = \frac{(A1 - A2) \times (protein\ added\ to\ Iodogen\ tube)}{A1}$$

9. Assuming all the loaded protein is recovered, then the specific activity (in mCi/mg) is:

$$SA = \frac{A3}{LP}$$

10. Sterilise the radiolabelled protein by filtration through a 0.22 µm filter before use *in vivo*.

Subprotocol 2
Radiolabelling of Proteins with Metals

Indium-111 (^{111}In) is a gamma-emitting radioisotope with photon energies (0.173 and 0.247 MeV) suitable for modern gamma-camera imaging. Yttrium-90 (^{90}Y) is a high energy (2.3 MeV) β-emitter which is frequently used in radioimmunotherapy. For radiolabelling with indium, yttrium and other radiometals, the protein must first be conjugated with a suitable chelating agent. The most commonly used for ^{111}In is diethylenetriaminepentaacetic acid (DTPA) as described by Hnatowich et al. (1983). More stable chelating agents are now available, such as backbone-substituted DTPA (Meares et al., 1984) and macrocycles (Deshpande et al., 1990). These are more suitable than DTPA for chelating ^{90}Y.

▨▨ Materials

– Instant thin layer chromatography (ITLC) silica gel impregnated sheets (Gelman Sciences, Michigan, USA) or thin layer chromatography (TLC) silica gel pre-coated plastic sheets (Sigma Chemical Co., Poole, Dorset, UK)

– Chelex-100 analytical grade resin (BioRad Laboratories, Hercules, CA, USA)

– Indium labelling: Reagent grade ^{111}In trichloride in 0.04 M HCl solution (Amersham Pharmacia Biotech, Amersham, Bucks., UK)

– Yttrium labelling: Carrier free ^{90}Y trichloride in 0.04 M HCl solution (AEA Technology, Oxford, UK)

All materials used for radiolabelling with metals must be free from metal contamination. Whenever feasible use polypropylene vessels for mixing and storing buffers. Wash all glassware used in buffer preparation in 6 M hydrochloric acid and rinse in ultra-pure water.

To remove metals from the buffers, pack a large (200 ml) column with Chelex-100. Wash the column with at least 3 bed volumes of ultra pure water. Pass all buffers through the column and then store at 4°C.

▪ ▪ Procedure

Indium labelling

For all procedures involving radioactivity, use adequate shielding (lead in the case of ^{111}In) and adhere to the local rules in operation in your institute.

1. Add 1-4 mCi ^{111}In to a reaction tube and adjust the pH to 6.0 with 1 M sodium acetate or 1 M sodium citrate.

2. Add up to 1 ml of the chelate-conjugated protein (e.g. DTPA-conjugated antibody) at a concentration of 300 µg/ml-5 mg/ml in an appropriate buffer (e.g. 0.1 M ammonium acetate, pH 5.5-6.0) to the ^{111}In chloride solution, and agitate gently to mix. Measure the activity in the reaction tube (A1 mCi).

3. Leave to stand at room temperature for up to 30 minutes.

Yttrium labelling

For all procedures involving radioactivity, use adequate shielding (1 cm perspex in the case of ^{90}Y) and adhere to the local rules in operation in your institute.

1. Add 12-25 µl of 1 M sodium acetate buffer (pH 5.5) to 50-100 µl of ^{90}Y such that the final acetate concentration is 0.2 M.

2. Add up to 1 ml of chelate-conjugated protein (300 µg-5 mg) dissolved in an appropriate buffer, such as 0.1 M ammonium acetate (pH 5.5), to the ^{90}Y solution and agitate gently to mix. Measure the activity in the reaction tube (A1 mCi).

3. Leave to stand at room temperature for up to 30 minutes.

4. Add an appropriate volume of 50 mM disodium EDTA in 0.1 M sodium acetate buffer (pH 5.5) to the reaction mixture such that the final EDTA concentration is 5 mM.

5. Leave to stand at room temperature for approximately 10 minutes.

Purification

1. Purify the radiolabelled protein by size exclusion chromatography using a 20-ml Sephadex G-50 or G-25 (Pharmacia) column. This column can be made from a 20-ml syringe packed with pre-swelled Sephadex, placing a small circle of gauze in the bottom of the syringe before gently adding the gel. Wash the column with at least three bed volumes of PBS (pH 7.4).

2. Apply the reaction mixture onto the top of the column. Add 1 ml PBS to the reaction tube and apply this to the column to ensure all of the antibody enters gel bed. Measure the activity in the empty tube (A2 mCi).

3. Collect and count the activity in 1-ml fractions. The first peak is the radiolabelled protein and the second peak is the free radioisotope.

4. Pool the 2-3 protein fractions with the highest activity. Measure the total recovered protein-bound activity in the vial (A3 mCi).

5. Calculate the radiolabelling efficiency (*RE*) and specific activity (*SA*) as described for iodination.

6. Sterilise the radiolabelled protein by filtration through a 0.22 μm filter before use *in vivo*.

Subprotocol 3
Determination of Radiochemical Purity

The radiochemical purity is determined by ascending paper chromatography. For iodinated proteins, chromatography paper is used as the solid phase and the solvent is 10% w/v trichloroacetic acid (TCA). For proteins labelled with ^{111}In or ^{90}Y, the solvent is 20-50 mM disodium EDTA in 0.1 M sodium acetate buffer (pH 5.5-6.0) and ITLC or TLC strips are used for the separation. In both cases, the protein-bound activity remains at the origin and the free radioisotope migrates up the strip with the solvent front.

▪▪ Procedure

1. Cut the chromatography paper into 1.5 x 10.0 cm strips. With a soft pencil gently mark the origin, (1 cm from the end of the strip) and the

solvent front (9 cm from the end of the strip). If any damage occurs to the coating of the ITLC or TLC strip, it should be discarded.

2. Add 1 ml of the appropriate solvent to a tube.

3. Place the chromatography strip on absorbent paper. Apply one drop of the labelled compound to the strip at the site of the origin pencil mark.

4. Place the strip in the vessel containing the solvent. Do not allow the origin to fall below the level of the solvent. Allow the solvent to migrate up the strip to the solvent front pencil mark (approximately 20 minutes in the case of chromatography paper or TLC strips and less than 5 min for ITLC strips).

5. Remove the strip from the tube and place on absorbent paper. Cut the strip into two equal halves (5 cm strips) and measure the radioactivity associated with each half. The radiochemical purity (*RP*) is:

$$RP = \frac{cpm \ at \ origin}{total \ cpm \ in \ the \ two \ halves} \times 100\%$$

Troubleshooting

Iodination using Iodogen may fail if there are insufficient tyrosine residues accessible in the protein. In that case, an alternative method should be employed, such as the Bolton-Hunter method, in which a ^{125}I-labelled acylating agent is conjugated to the protein via a lysine residue (Bolton and Hunter, 1973).

A common cause for the failure of labelling of proteins with radiometals is metal contamination of the buffers. Avoid any metal coming into contact with the buffer solutions; for example do not use metal spatulas or aluminium foil. Pay particular attention to the pH of the buffers, as this is another important factor in the successful labelling of proteins.

Another common problem is reduced immunoreactivity of the labelled compound. This is often due to the number of chelating agent molecules (e.g. DTPA) attached to the antibody. Usually optimal labelling can be achieved with an average of 1-2 chelating molecules per antibody.

Subprotocol 4
Biodistribution *in vivo*

For all experiments using animals, ensure that the relevant legislation is adhered to.

▪ ▪ Materials

- Inhalation anaesthetic, such as isoflurane (Abbott Laboratories Ltd., Queenborough, Kent, UK)

- Solvable or Soluene (Packard Biosciences, Groningen, The Netherlands) for solubilising tissues prior to counting ^{90}Y-labelled protein activity

▪ ▪ Procedure

1. Establish relevant human tumour xenografts in the flank of nude (or SCID) mice by subcutaneous injection of 10^6-10^7 cells in a volume of 100-200 μl sterile medium. Allow the tumour to grow to a diameter of 6-8 mm. The xenografts should not be allowed to grow too large or the centre of the tumour will become necrotic and may not bind the antibody.

2. Fill a 1-ml syringe fitted with 27-G needle with the radioactive injectate, which should be stored in a shielded container (lead or perspex, depending on the radioisotope).

3. Place the unanaesthetised mouse in an immobilisation jig with its tail protruding.

4. Heat the mouse tail by immersion in warm (42-44°C) water for a few seconds.

5. Apply pressure to the mouse's tail to restrict the blood flow along a lateral tail vein and insert the needle (with the aid of a magnifying light, if necessary).

6. Release the pressure on the vein and inject 10-20 μCi (5-20 μg) of radiolabelled protein in a volume of 100-200 μl.

7. Place a paper tissue over the puncture wound and apply pressure as the needle is removed. Release the pressure after a few seconds and check that the bleeding has stopped.

8. At times post-injection varying from 1 h to several days, depending on the clearance rate of the protein, kill groups of mice by exsanguination via cardiac puncture under terminal anaesthesia. This removes much of the blood pool from the tissues. Dissect 4 mice at each time-point to obtain tumour, blood and normal tissue samples (e.g. stomach, small intestine, colon, kidney, spleen, liver, lung, skin, heart, bone, muscle, brain).

9. Weigh the tissue samples, the stomach and intestine having been emptied, and measure their radioactivity content using a gamma-counter set to appropriate energy windows, along with serially diluted standards of the injectate. Results can then be expressed as a percentage of the injected dose/g of tissue and as tumour/normal tissue ratios. In the case of ^{90}Y-labelled tissues, the tissue samples should first be solubilised in a reagent such as Soluene or Solvable (1 ml of reagent solubilises 250 µg of tissue). In this case, Soluene or Solvable should also be added to the standards. The gamma-counter detects secondary X-rays produced by the interaction of the β-particles with the polypropylene tubes in which the samples are contained.

10. For dual-label experiments, count tissue samples containing both ^{125}I and ^{131}I in two channels of the gamma-counter, A and B, with energy windows set for ^{125}I and ^{131}I, respectively, and correct the counts for inter-channel crossover. Count standards of pure ^{125}I and ^{131}I and calculate correction coefficients (q) as follows:

$$q(A) = \frac{S_{125}(B)}{S_{125}(A)}$$

$$q(B) = \frac{S_{131}(A)}{S_{131}(B)}$$

where:
$S_n(A)$ = cpm of isotope n in channel A (similarly B),
then

$$N(125) = \frac{N(A) - q(B) \cdot N(B)}{1 - q(A) \cdot q(B)}$$

$$N(131) = \frac{N(B) - q(A) \cdot N(A)}{1 - q(A) \cdot q(B)}$$

where:
$N(125)$ = corrected cpm of ^{125}I (similarly ^{131}I)
$N(A)$ = measured cpm in channel A (similarly B).

In addition, the specificity index (*SI*) for a tissue is defined as:

$$SI_{tissue} = \frac{tumour/tissue\ (specific\ antibody)}{tumour/tissue\ (control\ antibody)}$$

The specificity index corrects for non-specific binding to a tissue. The higher the tissue *SI*, the more specific the antibody is for the tumour than for that tissue.

The localisation index (*LI*) for a tissue is defined as:

$$LI_{tissue} = \frac{tissue/blood\ (specific\ antibody)}{tissue/blood\ (control\ antibody)}$$

The localisation index corrects for the different blood pool in each tissue. The *LI* for a tissue is 1, unless there is increased binding of the specific antibody relative to the control antibody in that tissue, in which case it is greater than 1, which should be the case for the tumour, but not the normal tissues.

Subprotocol 5
Tumour Therapy

Radioimmunotherapy of xenografts can be achieved with ^{131}I- or ^{90}Y-labelled antibodies, or with other β-emitting radioisotopes if a suitable labelling method is available. The absorbed dose to the tumour can be calculated from the data obtained in a biodistribution experiment. The area under the tumour radioactivity curve (AUC) is proportional to the dose the tumour will receive if a therapeutic amount of radioactivity is injected.

▪▪ Procedure

1. Calculate the equilibrium absorbed dose to a tumour (D_{eq}) as follows:

$$D_{eq} = \phi \frac{A}{m} \sum_i \Delta_i$$

where:
ϕ = mean absorbed fraction
A = the time integral of activity (AUC)
Δ_i = the equilibrium dose constant
m = the tumour mass

Δ_i is a factor dependent on the energy emitted per disintegration in the form of i-type radiation (Loevinger and Berman, 1968); in the case of ^{131}I, this radiation is in the form of either β-particles or γ-rays, while ^{90}Y is a pure β-emitter. If the AUC is expressed in $\mu Ci \cdot h/g$, then $\Sigma \Delta_i$ is $0.4 \, g \cdot cGy/\mu Ci \cdot h$ for a tumour irradiated by ^{131}I uniformly distributed within the tumour volume, and $2.0 \, g \cdot cGy/\mu Ci \cdot h$ for ^{90}Y (Dillman and Von der Lage, 1975), taking into account only the contribution of the β-particles for ^{131}I, since most of the photon energy would not be deposited in tumour tissue. ϕ is a factor correcting for a non-infinite tumour volume and has values of 0.90 and 0.48 for a 7 mm tumour irradiated with ^{131}I and ^{90}Y, respectively.

2. Establish subcutaneous tumours as described for the biodistribution experiment.

3. Measure tumour diameters (d_1, d_2 and d_3) 2-3 times weekly in three orthogonal directions using a vernier calliper and calculate the tumour volume according to the formula for an ellipsoid:

$$v = \frac{\pi}{6}(d_1 \cdot d_2 \cdot d_3)$$

Commence tumour measurement at least one week before treatment.

4. When the tumours are around 7-8 mm in diameter (0.2 cm^3 in volume), divide mice into treatment groups of at least 6 mice each. The mean tumour volumes in each group at the time of treatment should not be significantly different. Leave one group untreated as a control.

5. Inject the calculated therapeutic doses of radiolabelled antibody into groups of mice as described for the biodistribution experiments. Pay particular attention to shielding, as the activity will be much higher than in the tracer experiments.

6. Divide each tumour volume by its respective volume on the day of treatment. Expressing the volume in this way as the relative tumour volume minimises any variation between the animals.

7. Measure tumour volumes until the tumours have at least tripled in volume (i.e. a relative tumour volume of 3). The tumour volume tripling time is taken as a measure of treatment efficacy.

8. Monitor the toxicity of the therapy by mouse weight and/or WBC counts.

This tumour therapy protocol can be used for non-radioactive therapeutic strategies. In that case, the administered dose is calculated from *in vitro* cell killing experiments.

References

Bolton AE, Hunter WM (1973) The labelling of proteins to high specific radioactivities by conjugation to a ^{125}I-containing acylating agent. Biochem J 133:529-539

Deshpande SV, DeNardo SJ, Kukis DL, Moi MK, McCall MJ, DeNardo GL, Meares CF (1990) Yttrium-90-labeled monoclonal antibody for therapy: labeling by a new macrocyclic bifunctional chelating agent. J Nucl Med 31:473-479

Dillman LT, Von der Lage FC (1975) Radionuclide decay schemes and nuclear parameters for use in radiation-dose estimation. Medical Internal Radiation Dose Pamphlet No10. Society of Nuclear Medicine, New York

Fraker PJ, Speck JC, Jr (1978) Protein and cell membrane iodinations with a sparingly soluble chloramide, 1,3,4,6-tetrachloro-3a,6a-diphenylglycoluril. Biochem Biophys Res Commun 80:849-857

Hnatowich DJ, Childs RL, Lanteigne D, Najafi A (1983) The preparation of DTPA-coupled antibodies radiolabeled with metallic radionuclides: an improved method. J Immunol Methods 65:147-157

Loevinger R, Berman M (1968) A schema for absorbed-dose calculations for biologically- distributed radionuclides. Medical Internal Radiation Dose Pamphlet No 1, J Nucl Med 9(Suppl 1):8-14

Meares CF, McCall MJ, Reardan DT, Goodwin DA, Diamanti CI, McTigue M (1984) Conjugation of antibodies with bifunctional chelating agents: isothiocyanate and bromoacetamide reagents, methods of analysis, and subsequent addition of metal ions. Anal Biochem 142:68-78

Affinity Maturation
of Antibody Fragments

Affinity Maturation by Chain Shuffling and Site Directed Mutagenesis

ULRIK B. NIELSEN and JAMES D. MARKS

Introduction

In the human immune system, antibodies with high affinities for antigen are created in two stages. A diverse primary repertoire of antibody structures is produced by the combinatorial rearrangement of germline V gene segments and antibodies are selected from this repertoire by binding to the antigen (Alt et al. 1987). Antibody affinities are then further improved by somatic hypermutation and further rounds of selection (Allen et al. 1987).

In vitro, phage display has been successfully used to increase antibody affinity more than 1000 fold (Schier et al. 1996c; Yang et al. 1995). The major decision for in vitro antibody affinity maturation is where and how to introduce mutations into the antibody V-genes. The easiest approach has been to introduce mutations randomly into the V-genes, thus mimicking the in vivo process of somatic hypermutation. Mutations have been introduced by chain shuffling (Marks et al. 1992), error prone PCR (Hawkins et al. 1992), by DNA shuffling (Crameri et al. 1996) or by propagation of phage in mutator strains of *E. coli* (Low et al. 1996). Whereas these approaches have yielded large increases in affinity for hapten antigens (up to 300 fold) (Low et al. 1996; Marks et al. 1992), results with protein antigens have been more modest (<10 fold) (Hawkins et al. 1992; Schier et al. 1996b). One limitation to random mutagenesis is that little useful information is generated with respect to the location of mutations which modulate affinity. Such information could be used to guide

Ulrik B. Nielsen, University of California, Departments of Anesthesia and Pharmaceutical Chemistry, San Francisco, USA

✉ James D. Marks, University of California, Departments of Anesthesia and Pharmaceutical Chemistry, San Francisco, USA (*phone* +01-415-206-3256; *fax* +01-415-206-3253; *e-mail* marksj@anesthesia.ucsf.edu)

subsequent mutagenesis efforts if the increase in affinity achieved is inadequate for the desired application. In addition, because of the random introduction of mutations into the framework regions, this approach may create problems of immunogenicity when the antibody is intended for therapeutic use. This problem could be minimized using DNA shuffling to remove deleterious or unnecessary mutations (Stemmer 1994).

Several groups have shown that targeting mutations to the CDRs can be an effective technique for increasing antibody affinity. During in vivo somatic hypermutation, mutations accumulate preferentially in the CDRs compared to framework residues (Loh et al. 1983). The location of these CDR mutations complements the locations where diversity is generated in the primary antibody repertoire (Tomlinson et al. 1996). However many CDR residues, especially in V_HCDR3 and V_LCDR3, are responsible for high energy interactions with antigen. Mutating residues in this region can in many cases abolish antigen binding. Fortunately, these residues will be recreated, albeit at low frequencies, given an adequate library size for the number of residues randomized. Mutant residues in the CDRs may increase affinity by introducing new contact residues or by replacing low affinity or "repulsive" contact residues with more favorable energetics (Novotny et al. 1989). However, it appears that many mutations introduced either by somatic hypermutation in vivo, or by random mutagenesis in vitro, instead exert their affect on affinity indirectly by repositioning the CDRs or the side-chains of contact residues for optimal interaction with the antigen (Foote and Winter 1992). Targeting mutations to the CDR regions of an antibody also may be less likely to generate immunogenic antibodies than mutations in the more conserved framework regions. Finally, with site directed mutagensis, sequence analysis of binding clones identifies both conserved structural and functional residues as well as those residues which modulate affinity (Schier et al. 1996a; Schier et al. 1996c). These results can help guide subsequent mutagenic efforts to further improve affinity.

Our lab has used both chain shuffling and site directed mutagenesis to increase antibody affinity and has compared these two techniques using an anti-ErbB2 sigle chain Fv (Marks et al. 1992; Schier et al. 1996a; Schier et al. 1996c). Overall, we find that site directed mutagenesis of the CDR3s of an antibody is the most efficient means to improve affinity by phage display. For example, we increased the affinity of the anti-ErbB2 antibody more than 1200-fold by sequentially targeting mutagenesis to the two CDR3's. Similarly, Yang et al. (Yang et al. 1995) increased the affinity of an anti-gp120 Fab 420-fold by mutating four CDRs (V_HCDR1, V_LCDR1, V_HCDR3 and V_LCDR3) in five libraries and combining independently selected mutations. However, they observed the largest affinity increases when optimizing the CDR3 regions.

Optimization of the two CDR3s is either done in parallel or sequentially. In the parallel strategy, the two CDR3 are randomized independently and beneficial mutations in single clones are combined. Although this strategy has been succesfully employed for optimization of several antibodies, not all mutations have been additive (Schier et al. 1996c; Yang et al. 1995). This is no surprise since many residues of the CDR3s pack against each other. Thus, we prefer sequential targeting of mutations to the CDR3 regions of the antibody.

Below we describe methods for phage displayed antibody affinity maturation using both chain shuffling and site directed mutagenesis. Our lab works exclusively with single chain Fv antibody fragments, and so the protocols use this format. The general principles can also be applied to maturation of Fabs.

Materials

- Vent DNA Polymerase (New England Biolabs (NEB))
- Geneclean kit (Bio 101, Inc.)
- Wizard PCR purification kit (Promega)
- Custom DNA primers (Genset)
- *Nco* I, *Not* I, *Sfi* I, *Bss*HII, *Xho*I restriction enzymes (NEB)
- T4 DNA ligase (NEB)
- *E. coli* strain TG1 (Stratagene)
- 37°C incubator (New Brunswick Scientific)
- Electroporator (Gene Pulser™ BioRad)
- PCR equipment
- 16°C water bath
- Helper phage (VCSM13, Stratagene)
- PEG 8000
- Biotinylation kit (Pierce, NHS-LC-Biotin)
- Non-fat milk powder (Safeway brand)

– Streptavidin-magnetic beads (Dynal)

– Magnetic rack (Dynal MPC-E)

– PD10 colums (Pharmacia)

– CM5 sensor chip (BIAcore)

– SOC (recipe: to 950 ml deionized H_2O add: 20 g bacto-tryptone, 5 g yeast extract, 0.5 g NaCl. Add 10 ml of a 250 mM KCl solution. Adjust the pH to 7.0 with 5 M NaOH and adjust the volume of the solution to 1 l with deionized H_2O. Sterilize by autoclaving for 20 min. Just before use add 1 ml of a sterile solution of 1 M $MgCl_2$ and 1 ml 1 M $MgSO_4$ each, and 20 mM glucose)

Subprotocol 1
Introducing Mutations by Chain Shuffling – Construction of Heavy Chain Shuffled Libraries

Chain shuffling relies on the natural mutation of V-genes that occurs in vivo in response to antigens that the host has been exposed to for the generation of sequence diversity. For chain shuffling of single chain Fv (scFv) from phage libraries, we first amplify V_H and V_L repertoires from either pre-existing naive scFv libraries or from total RNA prepared from volunteer human donors. Initially, we used splicing by overlap extension to splice either the wild type V_H or V_L to a gene repertoire of the complementary chain (Clackson et al. 1991; Marks et al. 1992). However this can artifactually generate shortened linker sequence (between VH and VL) leading to scFv dimers (diabodies) (Holliger et al. 1993) which may be preferentially selected on the basis of avidity (Schier et al. 1996b). Thus we now prefer to clone the wildtype V_H or V_L gene into a phage display vector containing a repertoire of the complementary chain (Schier et al. 1996b). Such repertoires have been described (Schier et al. 1996b) and it may be possible to obtain these from the authors, saving considerable time. Below we describe the generation of V_H and V_L repertoires in the vector pHEN-1, by amplifying from a pre-existing naive scFv library, and the subsequent cloning of wild type V_H or V_L into the appropriate vector for creation of chain shuffled libraries.

To facilitate heavy chain shuffling, libraries are constructed in pHEN-1 (Hoogenboom et al., 1991) containing human V_H gene segment repertoires (FR1 to FR3) and a cloning site at the end of V_H FR3 for inserting

the V$_H$ CDR3, V$_H$ FR4, linker DNA and light chain from a binding scFv as a
*Bss*HII-*Not*I fragment.

To create the libraries, three V$_H$ gene segment repertoires enriched for
human V$_H$1, V$_H$3, and V$_H$5 gene segments are amplified by PCR using as a
template single stranded DNA prepared from a 1.8 x 10^8 member scFv
phage antibody library in pHEN-1 (Marks et al. 1991) available from
Greg Winter's lab at the MRC Laboratory of Molecular Biology, Cambridge England. We used only VH1, 3 and 5 since we have not observed
any scFv phage antibodies derived from the VH2, 4 or 6 families. Primers
PVH1FOR1, PVH3FOR1, and PVH5FOR1 are designed to anneal to the
consensus V$_H$1, V$_H$3, or, V$_H$5 3' FR3 sequence respectively (Tomlinson
et al. 1992). A second set of primers add the *Bss*HII site for cloning.

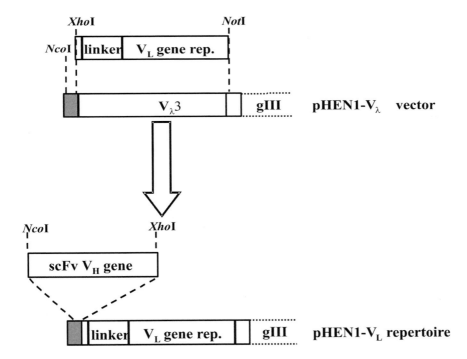

Fig. 1.

▨▨ Procedure

Construction of human VH gene segment repertoire libraries

First PCR   Primers for heavy chain shuffling

PVH1FOR1	5'-TCG CGC GCA GTA ATA CAC GGC CGT GTC-3'
PVH3FOR1	5'-TCG CGC GCA GTA ATA CAC AGC CGT GTC CTC-3'
PVH5FOR1	5'-TCG CGC GCA GTA ATA CAT GGC GGT GTC CGA-3'
PVH1FOR2	5'-GAG TCA TTC TCG ACT TGC GGC CGC TCG CGC GCA GTA ATA CAC GGC CGT GTC-3'
PVH3FOR2	5'-GAG TCA TTC TCG ACT TGC GGC CGC TCG CGC GCA GTA ATA CAC AGC CGT GTC CTC-3'
PVH5FOR2	5'-GAG TCA TTC TCG ACT TGC GGC CGC TCG CGC GCA GTA ATA CAT GGC GGT GTC CGA-3'
LMB3	5'- CAG GAA ACA GCT ATG AC -3'
scFvVL1BACK	5'-AGC GCC GTG TAT TTT TGC GCG CGA CAT GAC GTG GGA TAT TGC-3'

1. Make up three separate 50 μl PCR reaction mixes containing:

water	35.5 μl
20 x dNTPs (5 mM each)	2.5 μl
10 x Vent polymerase buffer (NEB)	5.0 μl
Lmb3 primer (10 pm/ μl)	2.5 μl
FORWARD primer* (10 pm/ μl)	2.5 μl
Single stranded scFv gene template (10 ng)	1.0 μl
Vent DNA polymerase (NEB, 2 units)	1.0 μl
*either PVH1FOR1, PVH3FOR1, or PVH5FOR1	

2. Heat to 94°C for 5 min in a PCR thermo-cycler.

3. Cycle 25 times to amplify the V_H genes at 94°C for 30 s, 42°C for 30 s min, 72°C for 1 min.

4. Gel-purify the V_H gene repertoires on a 1.5% agarose gel and extract the DNA using the Geneclean kit. Resuspend each product in 20 μl of water. Determine DNA concentration by analysis on a 1.5% agarose gel with markers of known size and concentration.

The DNA fragments from the first PCR are then used as templates for a second PCR to introduce a *Bss*HII site at the 3'-end of FR3 followed by a *Not*I site. The *Bss*HII site corresponds to amino acid residues 93 and 94 (Kabat numbering (Kabat et al., 1987) and does not change the amino acid sequence (alanine-arginine).

1. Make up 50 µl PCR reaction mixes containing: **Second PCR**

water	34.5 µl
20 x dNTPs (5 mM each)	2.5 µl
10 x Vent polymerase buffer (NEB)	5.0 µl
Lmb3 primer (10 pm/ µl)	2.5 µl
FORWARD primer* (10 pM/ µl)	2.5 µl
scFv gene template (100 ng)	2.0 µl
Vent DNA polymerase (NEB, 2 units)	1.0 µl
*either PVH1FOR2, PVH3FOR2, or PVH5FOR2	

2. Heat to 94°C for 5 min in a PCR thermo-cycler.

3. Cycle 25 times to amplify the V_H genes at 94°C for 30 s, 42°C for 30 s min, 72°C for 1 min.

4. Purify PCR product using the Wizard PCR purification kit and continue with restriction digest Subprotocol 5, digesting the fragments with *Nco*I and *Not*I. The digested fragments are gel purified and each V_H gene segment repertoire is ligated separately into gel purified pHEN-1 (Hoogenboom et al., 1991) digested with *Not*I and *Nco*I and ligated DNA used to transform *E coli* TG1 as described in Subprotocol 5.

Cloning the VL gene into VH gene segment repertoires to generate heavy chain shuffled libraries

To shuffle the heavy chain gene from a scFv, the light chain gene, linker DNA, and V_H CDR3 and FR4 are amplified using PCR from pHEN-1 containing the target scFv gene using the primers scFvVL1BACK and fdSEQ1. The protocol below uses primers that anneal to gene 3 in pHEN-1 (fdseq) and a VL1back primer designed to anneal to a V_H5 framework 3. For your specific scFv, it is likely that a different VL1BACK primer will need to be designed to anneal at the 5' end of the target scFv sequence, and possibly a different 3' primer will need to be designed as well, depending on the vector backbone that the target scFv is in.

1. Make up 50 μl PCR reaction mixes containing:

water	34.5 μl
20 x dNTPs (5 mM each)	2.5 μl
10 x Vent polymerase buffer (NEB)	5.0 μl
fdSEQ1 primer (10 pm/ μl)	2.5 μl
VL1BACK (10 pM/ μl)	2.5 μl
scFv light chain gene template (100 ng)	2.0 μl
Vent DNA polymerase (NEB, 2 units)	1.0 μl

2. Heat to 94°C for 5 min in a PCR thermo-cycler.

3. Cycle 25 times to amplify the V_H genes at 94°C for 30 s, 42°C for 30 s min, 72°C for 1 min.

4. Prepare vector DNA from V_H gene segment repertoires.

5. Purify PCR product using the Wizard PCR purification kit. The PCR product and the three repertoires are restriction digested with BssHII and NotI as described in Subprotocol 5. Libraries and insert are gel purified and ligated and transformed, also described in Subprotocol 5.

Subprotocol 2
Construction of Light Chain Shuffled Libraries

To facilitate light chain shuffling, a library is constructed in the vector pHEN1-V_λ3 (Hoogenboom and Winter, 1992) containing rearranged human V_κ and V_λ gene repertoires, linker DNA, and cloning sites for inserting a rearranged V_H gene as an NcoI-XhoI fragment. In this case, an XhoI site can be encoded at the end of FR4 without changing the amino acid sequence of residues 102 and 103 (serine-serine) (Kabat et al., 1987).

To create the libraries, V_κ and V_λ gene repertoires and linker DNA are amplified by PCR using as a template single stranded DNA prepared from a 1.8 x 10^8 member scFv phage antibody library in pHEN-1 (Marks et al., 1991) available from Greg Winter's lab at the MRC Laboratory of Molecular Biology, Cambridge England.

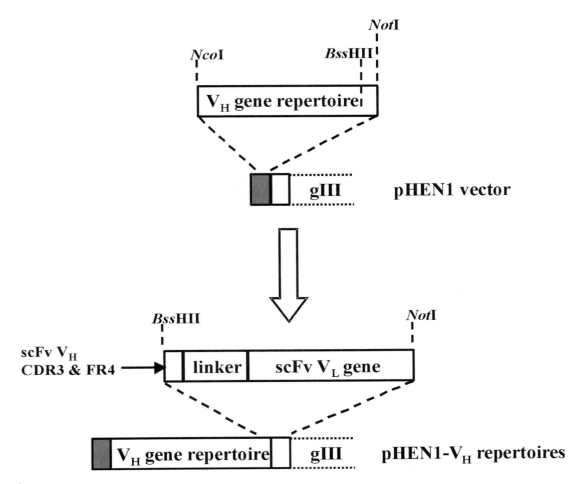

Fig. 2.

Procedure

Construction of human VL gene repertoire libraries

Primers for light chain shuffling

RJH1/2Xho	5'-GGC ACC CTG GTC ACC GTC TCG AGT GGT GGA-3'
RJH3Xho	5'-GGG ACA ATG GTC ACC GTC TCG AGT GGT GGA-3'
RJH4/5Xho	5'-GGA ACC CTG GTC ACC GTC TCG AGT GGT GGA-3'
RJH6Xho	5'-GGG ACC ACG GTC ACC GTC TCG AGT GGT GGA-3'
FdSEQ	5'- GAA TTT TCT GTA TGA GG -3'

1. Make up three separate 50 µl PCR reaction mixes containing:

water	35.5 µl
20 x dNTPs (5 mM each)	2.5 µl
10 x Vent polymerase buffer (NEB)	5.0 µl
fdSEQ1 primer (10 pm/ µl)	2.5 µl
BACK primer* (10 pm/ µl)	2.5 µl
Single stranded scFv gene template (10 ng)	1.0 µl
Vent DNA polymerase (NEB, 2 units)	1.0 µl

*either RJH1/2Xho, RJH3Xho, RJH4/5Xho, or RJH6Xho

The BACK primers were designed to anneal to the first 6 nucleotides of the $(G_4S)_3$ linker and either the J_H1, 2, J_H3, J_H 4,5 or J_H 6 segments respectively and contain the XhoI cloning site.

2. Heat to 94°C for 5 min in a PCR thermo-cycler.

3. Cycle 25 times to amplify the V_L genes at 94°C for 30 s, 42°C for 30 s min, 72°C for 1 min.

4. Purify PCR product using the Wizard PCR purification kit. The four PCR products and the pHEN1-$V_\lambda3$ vector are digested with XhoI and NotI. Vector and inserts are gel purified and ligated and transformed as described in Subprotocol 5.

Cloning the VH gene into VL gene repertoires to generate light chain shuffled libraries

To shuffle the light chain gene from a binding scFv, the rearranged heavy chain gene, is amplified using PCR from pHEN-1 containing the target scFv gene using the primers scFvJHXhoFOR and LMB3. The protocol below uses primers that anneal upstream of the pelB leader in pHEN-1 and a scFv specific primer that anneals to the J_H gene of the target scFv. One of the J_H primers below should be a perfect match for your target scFv V_H gene.

scFvJH1-2XhoFOR	5'-GAG TCA TTC TCG TCT CGA GAC GGT GAC CAG GGT GCC-3'
scFvJH3XhoFOR	5'-GAG TCA TTC TCG TCT CGA GAC GGT GAC CAT TGT CCC-3'
scFvJH4-5XhoFOR	5'-GAG TCA TTC TCG TCT CGA GAC GGT GAC CAG GGT TCC-3'
scFvJH6XhoFOR	5'-GAG TCA TTC TCG TCT CGA GAC GGT GAC CGT GGT CCC-3'

To create light chain shuffled libraries, follow Subprotocol 1 above, substitute in LMB3 for fdseq and one of the JH primers listed above for VL1BACK.

Subprotocol 3
Introducing Mutations by Site Directed Mutagenesis – Randomization of the scFvV$_L$CDR3

As described above, our preferred approach is to sequentially introduce mutations into V_L and V_H CDR3 using oligonucleotides. Since it is difficult to make libraries greater than 10^7 to 10^8 clones, decisions must be made as to which amino acids to diversify, and to what extent. Conventional oligo-directed mutagenesis uses the nucleotides NNS to randomize each amino acid residue. All parental contacts are discarded and the number of amino acids that can be scanned is limited to five, given typical transformation efficiencies. Instead, we prefer to minimize the number of non-viable structures by using nucleotide mixtures which bias for the wild-type residue, allowing more amino acids to be screened in one library. For V_L CDR3 mutagenesis, we also use molecular modeling using the most homologous V_L in the PDB database to distinguish between residues with solvent accessible side chains from those with buried side chains. Our previous results indicate that randomization of residues with buried side chains is a waste of sequence space, the wild type sequence returns in most cases (Schier et al. 1996a; Schier et al. 1996c). We also avoid mutating glycines and tryptophans as they invariably return as wild type sequence (Schier et al. 1996a; Schier et al. 1996c). Glycines are frequently critical residues in CDR turns and tryptophans frequently are essential structural or contact residues.

Procedure

We begin randomization with V_L CDR3 as it is typically shorter than V_H CDR3 and more importantly can be modeled based on homologous structures (see above). Randomization of the V_LCDR3 is also technically simpler than V_H CDR3 since it is located at the 3' end of the scFv gene. The randomization can therefore be carried out in just two PCRs. In the first PCR, a randomized primer (V_LFOR) and the primer LMB3 amplify most of the scFv gene and introduce mutations into the V_L CDR3. The second PCR amplifies the remainder of the scFv and appends a restriction site for cloning of the fragment, Figure 2.

Randomization Example of primers used for V_L-randomization:
In this example seven amino acids of the V_L are randomized (underlined region) with approximately 50% wild type amino acid at each position randomized:

```
GAT TAT TAC TGC CAG TTC TAT GAC AGC AGC CTG AGT GGT TGG GTG TTC GGC
CTA ATA ATG ACG GTC AAG ATA CTG TCG TCG GAC TCA CCA ACC CAC AAG CCG
 D   Y   Y   C   Q   F   Y   D   S   S   L   S   G   W   V   F   G
 VL-FR3                     randomized                  VL-FR4
```

V_LFOR random primer

```
5'- CCC TCC GCC GAA CAC CCA ACC 524 513 524 524 542 541
511 CTG GCA GTA ATA ATC AGC CTC -3'
```

V_L-NotI primer

```
5'- GAG TCA TTC TCG ACT TGC GGC CGC ACC TAG CAC GGT CAG
CTT GGT CCC TCC GCC GAA CAC CCA ACC -3'
```

Molar compositions:
1: A(70%), C(10%), G(10%), T(10%)
2: A(10%), C(70%), G(10%), T(10%)
3: A(10%), C(10%), G(70%), T(10%)
4: A(10%), C(10%), G(10%), T(70%)
5: G(50%), C(50%)

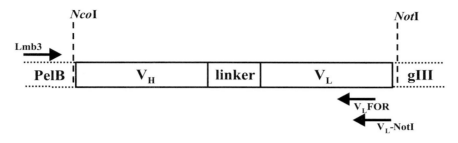

Fig. 3.

1. Make up 50 μl PCR reaction mixes containing:

water	35.5 μl
20 x dNTPs (5 mM each)	2.5 μl
10 x Vent polymerase buffer (NEB)	5.0 μl
Lmb3 primer (10 pm/ μl)	2.5 μl
V_LFOR primer (10 pm/ μl)	2.5 μl
scFv gene template (100 ng)	1.0 μl
Vent DNA polymerase (NEB, 2 units)	1.0 μl

2. Heat to 94°C for 5 min in a PCR thermo-cycler.

3. Cycle 30 times to amplify the V_L gene at 94°C for 1 min, 42°C for 1 min, 72°C for 2 min.

4. Gel-purify the V_L gene repertoires (app. 800 bp) on a 1% agarose gel and extract the DNA using the Geneclean kit. Resuspend each product in 20 μl of water. Determine DNA concentration by analysis on a 1% agarose gel with markers of known size and concentration.

The scFv V_L gene repertoire is now re-amplified with a primer which appends the restriction site *Not*I.

1. Make up 50 μl PCR reaction mixes containing: **Re-amplification**

water	35.5 μl
20 x dNTPs (5 mM each)	2.5 μl
10 x Vent polymerase buffer (NEB)	5.0 μl
Lmb3 primer (10 pm/ μl)	2.5 μl
V_L-NotI primer (10 pM/ μl)	2.5 μl
scFv gene template (100 ng)	1.0 μl
Vent DNA polymerase (NEB, 2 units)	1.0 μl

2. Heat to 94°C for 5 min in a PCR thermo-cycler.

3. Cycle 30 times to amplify the V_L gene at 94°C for 1 min, 42°C for 1 min, 72°C for 2 min.

4. Purify PCR product using the Wizard PCR purification kit and continue with restriction digest Subprotocol 5 using *Nco*I and *Not*I.

Subprotocol 4
Randomization of the scFv V$_H$CDR 3

The PCR for the repertoire is done in two steps. Initially, the V_H is amplified with the randomized primer (V_HFOR) and a primer based in the vector backbone, upstream of the scFv (LMB3). In a separate PCR, the V_L is amplified with a primer based in geneIII of the vector (Fdseq) plus a primer (V_HBACK) which is complementary to parts of the V_HFOR primer.

■ ■ Procedure

Randomization   Example of primers used for V_H-randomization:
In this example seven amino acids of the V_H are randomized (underlined region):

```
TAC TGT GCG AAA ATG ACA AGT AAC GCG TTC GCA TTT GAC TAC TGG GGC CAG GGA
ATG ACA CGC TTT TAC TGT TCA TTG CGC AAG CGT AAA CTG ATG ACC CCG GTC CCT
 Y   C   A   K   M   T   S   N   A   F   A   F   D   Y   W   G   Q   G
V_H-FR3             randomized                            V_H-FR4
```

The primer to introduce these changes had the following sequence:

```
V_HFOR primer 5'- CC CTG GCC CCA GTA GTC AAA 532 511 532
544 524 534 514 TTT CGC ACA GTA ATA AAC GGC -3'
```

Molar compositions:
1: A(70%), C(10%), G(10%), T(10%)
2: A(10%), C(70%), G(10%), T(10%)
3: A(10%), C(10%), G(70%), T(10%)
4: A(10%), C(10%), G(10%), T(70%)
5: G(50%), C(50%)
V$_H$BACK primer 5'- TTT GAC TAC TGG GGC CAG GG -3'

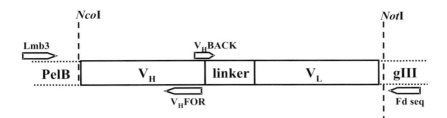

Fig. 4.

1. Make up two 50 μl PCR reaction mixes containing:

water	35.5 μl
20 x dNTPs (5 mM each)	2.5 μl
10 x Vent polymerase buffer (NEB)	5.0 μl
reverse primers* (10 pm/μl)	2.5 μl
forward primers# (10 pm/μl)	2.5 μl
scFv gene template (100 ng)	1.0 μl
Vent DNA polymerase (NEB, 2 units)	1.0 μl

*reverse primer for V_H is LMB3; for the V_L the reverse primer is V_HBACK.

#forward primer for V_H is V_HBACK; for the V_L the forward primer is FdSEQ.

2. Heat to 94°C for 5 min in a PCR thermo-cycler.

3. Cycle 30 times to amplify the V_H gene at 94°C for 1 min, 42°C for 1 min, 72°C for 2 min.

4. Gel-purify the V_H gene repertoires (app. 350 bp) and the V_L gene (app. 350 bp) on a 1.5% agarose gel and extract the DNA using the Geneclean kit. Resuspend each product in 30 μl of water. Determine DNA concentration by analysis on a 1.5% agarose gel with markers of known size and concentration.

The final V_HCDR3 full-length scFv repertoire is then constructed by PCR assembly of the amplified V_H and V_L fragments.

1. Make up 25 μl PCR reaction mixes containing: **PCR assembly**

water	6.5 μl
10 x Vent buffer	2.5 μl
20 x dNTP's (5 mM each)	1.0 μl
V_H repertoire (200 ng)	5.0 μl
V_L gene (200 ng)	5.0 μl
Vent DNA polymerase (NEB, 2 units)	1.0 μl

2. Cycle 7 times without amplification at 94°C for 1.5 min, 65°C for 1.5 min, and 72°C for 1.5 min to join the fragments.

3. After 7 cycles, hold at 94°C while adding the 25 μl mix containing the flanking primers:

water	15 μl
10 x Vent buffer	2.5 μl
20 x dNTP's (5 mM each)	1.3 μl
FdSEQ primer (10 pm/ μl)	2.5 μl
Lmb3 primer (10 pm/ μl)	2.5 μl
Vent DNA polymerase (NEB, 2 units)	1.0 μl

4. Cycle 30 times to amplify the assembled fragments at 94°C for 1 min, 42°C for 1 min and 72°C for 2 min.

5. Purify PCR product using the Wizard PCR purification kit (Promega).

6. Proceed to Subprotocol 5 for restriction digest with *Nco*I/*Not*I and ligation of fragment.

Subprotocol 5
Restriction Digest, Ligation, and Transformation of scFv Gene Repertoires

We recommend overdigestion of the PCR products due to the poor efficiency with which PCR fragments are digested. For cloning of scFv fragments into pHEN-1, use the enzyme indicated in the protocol.

■ ■ Procedure

1. Make up 100 μl reaction mix to digest scFv repertoires:

scFv DNA (1 to 4 μg)	50 μl
water	34 μl
10 x NEB Buffer	10 μl
restriction enzyme A (10 u/ μl)	3.0 μl
restriction enzyme B (10u/ μl)	3.0 μl

2. Incubate at 37°C overnight.

3. Gel-purify the gene repertoires on a 1% agarose gel and extract the DNA using the Geneclean kit. Resuspend each product in 30 μl of water. Determine DNA concentration by analysis on a 1% agarose gel with markers of known size and concentration.

4. Approximately 4 μg of cesium chloride purified pHEN 1 are digested with the appropiate enzyme. The digested vector DNA is purified on a 0.8% agarose gel, extracted from the gel , also using the geneclean kit. For optimal digestion, the restriction digest is carried out overnight and vector DNA is gel purified prior to ligation. Efficient digestion is important since a small amount of undigested vector leads to a very large background of non-recombinant clones. Use of vector DNA prepared by other techniques than cesium chloride will give lower transformation efficiencies!
   In ligation experiments, the molar ratio of insert to vector should be 2:1. Given that the ratio of sizes of assembled scFv (800 bp) to vector (4500 bp) is approximately 6:1, this translates into a ratio of insert to vector of 1:3 in weight terms.

5. Make up 50 μl ligation mixture:

10 x ligation buffer	5 μl
water	16 μl
digested pHEN 1 (100 ng/ μl)	20 μl
scFv gene repertoire (100 ng/ μl)	7 μl
T4 DNA ligase (400 u/ μl)	2 μl

6. Ligate overnight at 16°C.

7. Ethanol precipitate and wash pellet twice by thoroughly resuspending in 70% ethanol.

8. Resuspend DNA in 10 μl of water and use 2.5 μl/transformation (4 electroporations total) into 50 μl electrocompetent *E. coli* TG1. Set up a "no DNA" control for the electroporation to ensure the TG1 cells are not contaminated. Typical transformation efficiencies for *E. coli* TG1 cells are at least 5×10^9/μg of supercoiled pUC18 plasmid DNA.

9. Set the electroporator at 200 ohms (resistance), 25 μFD (capacitance) and 2.5 kilo volts. After electroporation the time constant should be 4.5 sec. If time constant is less than 4.3, repeat the DNA precipitation.

10. Grow bacteria from each electroporation in 1 ml SOC and incubate at 37°C for 1h shaking (250 rpm) and plate serial dilutions onto small TYE/amp/glu agar plates for determining the size of the library.

11. Centrifuge remaining bacteria solution at 1700 x g for 10 min at 4°C. Resuspend pellet in 250 μl and plate onto four or more large (16 cm) TYE/100 μg/ml ampicillin/2% glucose plates. Incubate overnight at 37°C.

12. Scrape bacteria from large plates by washing each plate with 3 ml 2xTY/ 100 μg/ml ampicillin/2% glu. Make glyerol stocks by adding 1.4 ml of bacteria and 0.6 ml 50% glycerol (sterilized by filtration through 0.2 μm filter). Save library stock at -70°C.

Note: We routinely obtain libraries with complexity ›10^7 cfu. If library is smaller than 10^6, the cloning should be repeated. It should be possible to routinely achieve a transformation efficiency for *E. coli* TG1 cells of greater than 5x10^9/μg of supercoiled plasmid DNA. This requires strict attention to the preparation of the electrocompetent cells.

Subprotocol 6
Rescuing Phagemid Libraries for Selection

■■ Procedure

1. Calculate the number of bacteria per ml from your library glycerol stock (OD$_{600}$ of 1.00 corresponds to approximately 1 x 10^8 cells). Usually the optical density of your bacterial stock is around 100, or 10^{10} cells/ ml. The input of cells for rescuing is dependent on the density of the original bacteria glycerol stock and should be in 10-fold excess of the number of different clones in the library, but should not exceed OD$_{600}$ of 0.05. This step ensures the diversity of the library for subsequent rounds of selections. For example, a library of 10^7 unique members requires at least 10^8 cells for starting the initial culture. A typical inoculum is 20 μl of glycerol stock into 50 ml of 2xTY/100 μg/ml ampicillin/2% glucose. Use of a starting OD greater than 0.05 will not result in an adequate number of doublings to ensure that the bacteria are healthy and can be easily infected by helper phage.

2. Grow with shaking (250 rpm) at 37°C to an $OD_{600} \sim 0.5$. If this inoculum has not grown to $OD_{600} \sim 0.5$ after 1.5-2 hrs increase inoculum size accordingly. It is also important that the culture is allowed to pass through several generations, thus if culture grows to $OD_{600} \sim 0.5$ in less than one hour, decrease the inoculum size.

3. Transfer 10 ml (about 5×10^8 bacteria total) to a 50 ml Falcon tube containing the appropriate number of helper phages. To ensure rescue of all clones in the library the ratio of helper phage : bacteria should be 10:1. Therefore add 5×10^9 plaque forming units (pfu) of helper phage to the bacterial solution. Incubate at 37°C without shaking for 30 min.

4. Plate 1 µl onto TYE/100 µg/ml kanamycin plate to check for infectivity. Incubate at 37°C overnight. The plate should be nearly confluent next day, indicating successful co-infection of the helper phage.

5. Centrifuge cells at 3000 x g to remove glucose and resuspend in 50 ml TYE/amp/kan.

6. Grow with shaking (250 rpm) overnight at 30°C.

7. Remove bacteria by centrifugation at 4000 x g 20 min, 4°C . Decant the clear supernatant containing phage particles (if cloudy, repeat centrifugation) into a 500 ml centrifuge bottle and add 10 ml 20% PEG8000/2.5 M NaCl per bottle. Mix and incubate on ice for 30 min.

8. Pellet phage by spinning for 15 min, 4000 x g at 4°C. Discard the supernatant.

9. Resuspend the white phage pellet in 10 ml PBS. To remove remaining bacteria debris, we recommend spinning down in the centrifuge for 20 min, 4000 x g at 4°C.

10. Transfer the supernatant to a 15 ml Falcon tube and repeat PEG precipitation with 1 ml 20% PEG 8000/2.5 M NaCl. Incubate on ice for 15 min.

11. Centrifuge 10 min, 4000 x g at 4°C and resuspend the white pellet in 1.5 ml PBS.

12. Filter supernatant through 0.45 µm syringe filter. The phage stock should be used at once for the next round of selection, but can be stored at 4°C for approximately 1 week without much loss in binding activity.

Subprotocol 7
Selection of Higher Affinity Clones from Diversified Libraries

Two approaches have been used to select rare higher affinity scFv from a background of lower affinity scFv or non-binding scFv: selections based on binding kinetics and selections based on equilibrium constant. In either case it is important to use labelled antigen in solution rather than antigen adsorbed to a solid matrix. This biases towards selections based on binding affinity or binding kinetics, rather than avidity. This is especially important when selecting scFv libraries, where it is known that the scFv can spontaneously dimerize in a sequence dependent manner (Schier et al. 1996b). Failure to use soluble antigen is likely to result in the selection of dimeric scFv whose monovalent binding constant is no higher than wild type. Even with the Fab format, soluble antigen should also be used to avoid selecting for phage displaying muliple copies of Fab which will have a higher functional affinity (avidity).

In selections based on binding kinetics, also termed off-rate selections, the phage population is allowed to saturate with the labelled antigen before a large molar excess of unlabelled antigen is added to the mix for a given amount of time. The duration of the competition with unlabelled antigen is chosen to allow the majority of the bound wild-type clones to dissociate while the improved mutants remain bound (Hawkins et al. 1992). This approach effectively selects for slower off-rates. Since a reduction in koff is typically the major kinetic mechanism resulting in higher affinity when V genes are mutated, both in vivo and in vitro (Marks et al. 1992), this approach should generally result in the selection of scFv's with improved Kd's.

We generally prefer equilibrium selections, in which phages are incubated with an antigen concentration below the equilibrium binding constant. This approach effectively selects for improved equilibrium constants. Reduction of the antigen concentration also helps ensure that the selection for higher affinity scFv occurs, rather than selection for scFv that express well on phage or are less toxic to E coli. The ability to use soluble antigen for selections affords the control over antigen concentration needed for the equilibrium screen. Phages are allowed to bind to biotinylated antigen and then recovered with streptavidin magnetic beads. Estimates of the optimal antigen concentration used for the selection can be estimated a priori (Boder and Wittrup 1998), but we prefer to use several different concentrations of antigen for the first round of selection. The antigen concentration for the following rounds of selection are guided by determining the fraction of binding phage present in the poly-clonal phage preparation. This is best done by surface plasmon resonance where the binding concentration is determined under mass transport limited conditions (Schier et al. 1996b). Alternatively, the percentage of binding clones can be determined by ELISA. From this data the optimal antigen concentration can be determined prior to the next round of selection. We typically lower the antigen concentration 10-100 fold after each round of selection. ScFv antibody clones with improved affinity are usually identified after 3-5 rounds of selection.

Procedure

Equilibrium selection

1. Prior to selection, block 50 μl streptavidin-magnetic beads with 1 ml 2% MPBS (2% milk powder in PBS) for 1h at RT in 1.5 ml microcentrifuge tube. Pull beads to one side with magnet. Discard buffer.

2. Block 1.5 ml microcentrifuge tube with 2% MPBS for 1h at RT and discard the blocking buffer. Incubate biotinylated antigen (for first round of selection concentrations should be: Kd/1, Kd/10, and Kd/100), prepared according to the manufacturer's instructions (Pierce), with 1 ml of phage preparation (approximately 10^{12} TU) in 2% MPBS (final conc.) by rocking at room temperature for one hour.

3. Add the streptavidin-magnetic beads to the phage-antigen mix and incubate on rotator at room temperature for 15 min. Place tube in magnetic rack for 30 seconds. Beads will migrate towards the magnet.

4. Aspirate tubes, leaving the beads on the side of the microcentrifuge tube. This is best done with a 200 μl pipette tip on a Pasteur pipette attached to a vacuum source. Wash beads (1 ml per wash) with PBS-Tween (0.1%) 7 times, followed by MPBS 2 times, then once with PBS. Transfer the beads after every second wash to a fresh eppendorf tube to facilitate efficient washing.

5. Elute phage with 100 μl 100 mM HCl for 10 min at room temperature. Place tube in magnetic rack for 30 sec. and beads migrate towards the magnet. Remove the supernatant containing eluted phage and neutralize with 1 ml of 1 M Tris, pH 7.4. Save on ice.

6. Add 0.75 ml of the phage stock to 10 ml of exponentially growing TG1 ($OD_{600} \sim 0.5$). (Store the remaining phage mix at 4°C).

7. Incubate at 37°C for 30 min without shaking.

8. Titer TG1 infection by plating 1 µl and 10 µl onto small TYE/100 µg/ml ampicillin/2% glucose plates (this is a 10^4 and 10^3 dilution, respectively).

9. Centrifuge the remaining bacteria solution at 1700 x g for 10 min. Resuspend pellet in 250 µl media and plate onto two large TYE/100 µg/ml ampicillin/2% glucose plates and incubate overnight at 37°C.

10. Add 3 ml of 2xTY/100 µg/ml ampicillin/2% glucose media to each plate, then scrape the bacteria from the plate with a bent glass rod. Make glycerol stocks by mixing 1.4 ml of bacteria and 0.6 ml 50% glycerol (filtered). Save stock at -70°C.

Note: For large, randomized libraries (complexity $> 10^7$), the selection titer from the first round of selection should be between 10^4 - 10^6 cfu. If the titers for all antigen concentrations are larger than 10^7 it is likely that the antigen concentration was to high or that the washing steps have been inadequate. If the titer is below $5x10^4$, you have either washed too many times, or too little antigen was used. In this case, reduce the number of washes.

Subprotocol 8
BIAcore Screen for scFv Clones with Improved Off-Rates

Even after stringent selections as described above, only a fraction of the ELISA positive clones will have a higher affinity than wild-type. The strength of the ELISA signal is a poor indicator of which clones have a lower K_d, and more typically correlates with expression level. Thus a technique is required to screen ELISA positive clones to identify those with a lower K_d. One such technique is a competition or inhibition ELISA (Friguet et al. 1985). Alternatively, one can take advantage of the fact that a reduction in k_{off} is typically the major kinetic mechanism resulting in higher affinity and thus antibodies with improved affinity can be identified by measuring the off-rate. Since k_{off} is concentration independent (unlike k_{on}), it is possible to measure k_{off}, for example using surface plasmon resonance (SPR) in instruments such as the BIAcore, without purifying the antibody fragment. We have found this a very useful technique for identifying higher affinity scFv and use it to rank affinity matured clones.

Procedure

First, we identify binding clones by ELISA as described elsewhere in this book. We then take approximately 20 to 50 randomly selected ELISA positive clones and rank them by k_{off} as described below. Those clones with the slowest k_{off} are then purified and the K_d is measured. Note that in the case of scFv, the shape of the dissociation curve can indicate whether there is a single or multiple k_{off}'s present. ScFv which have multiple k_{off}'s (a curve vs a straight line when plotting ln R1/R0 vs T) are a mixture of monomer and dimer and are best avoided for subsequent characterization.

1. Grow a 1 ml overnight culture in xTY/100 µg/ml ampicillin/2% glucose at 30°C (250 rpm).

2. Add 0.5 ml of the overnight culture to 50 ml 2xTY/amp/0.1% glu in an erlenmeyer flask and grow at 37°C shaking at 250 rpm for approximately 2 hours to O.D.$_{600}$ of 0.9.

3. Induce with 25 µl 1 M IPTG (final concentration of 500 µM) and grow at 30°C at 250 rpm for 4 hours. Collect cells by centrifuging in a 50 ml tube at 4000 x g for 15 minutes.

4. For the periplasmic preparation, resuspend the bacterial pellet in 1 ml PPB buffer (20% sucrose, 1 mM EDTA, 20 mM Tris-HCl, pH=8), transfer to 1.5 ml eppendorf tubes and keep on ice for 20 minutes. Spin down cells in microcentrifuge at 6000 rpm for 5 minutes. Discard the supernatant. Proceed with the bacterial pellet for the osmotic shock preparation.

5. For the osmotic shock preparation, resuspend the pellet in 1 ml 5 mM $MgSO_4$ (use 1/50 of total growth volume) and incubate on ice for 20 minutes. Spin down cells in microcentrifuge at full speed (14000 rpm) for 5 minutes.
   At this step the buffer of the osmotic shock fraction is exchanged to the BIAcore running buffer in order to avoid excessive refractive index change during the SPR analysis.

6. The buffer is most effectively exchanged using small columns such as PD10 (Pharmacia) or spin colums such as CentriSep colums (Princeton Separations).

7. In a BIAcore flow cell, immobilize antigen corresponding to 100-300 RU total scFv binding to a CM5 sensor chip using EDC/NHS chemistry

as described by the manufacturer. The immobilization of different antigens varies tremendously. However, 10 µg/ml in 10 mM NaAc, pH 4.5, is a good starting point for most protein antigens.

8. Dissociation of undiluted scFv in the periplasmic fraction is measured under a constant flow of 15 µl/min and HBS as running buffer. An apparent k_{off} can be determined from the dissociation part of the sensorgram for each scFv analyzed. The regeneration of the sensor chip between samples has to be determined for each antigen. However, 4 M MgCl$_2$ will regenerate most antigen surfaces without significant change in the sensorgram baseline after analysis of more than 100 samples.

References

Allen D, Cumano A, Dildrop R, Kocks C, Rajewsky K, Rajewsky N, Sablitzky F, Siekevitz M (1987) Timing, genetic requirements and functional consequences of somatic hypermutation. Immunol. Review 96: 5-22

Alt FW, Blackwell TK, Yancopoulos GD (1987) Development of the primary antibody repertoire. Science 238: 1079-87

Boder E, Wittrup K (1998) Optimal screening of surface-displayed polypeptide libraries. Biotechnology Progress 14: 55-62

Clackson T, Hoogenboom HR, Griffiths AD, Winter G (1991) Making antibody fragments using phage display libraries. Nature 352: 624-628

Crameri A, Cwirla S, Stemmer WP (1996) Construction and evolution of antibody-phage libraries by DNA shuffling. Nature Medicine 2: 100-102

Foote J, Winter G (1992) Antibody framework residues affecting the conformation of the hypervariable loops. J. Mol. Biol. 224: 487-499

Friguet B, Chaffotte AF, Djavadi-Ohaniance L, Goldberg ME (1985) Measurements of the true affinity constant in solution of antigen-antibody complexes by enzyme-linked immunosorbent assay. J. immunol. Meth. 77: 305-319

Hawkins RE, Russell SJ, Winter G (1992) Selection of phage antibodies by binding affinity. Mimicking affinity maturation. Journal Of Molecular Biology 226: 889-96

Holliger P, Prospero T, Winter G (1993) 'Diabodies': small bivalent and bispecific antibody fragments. Proc. Natl. Acad. Sci. 90: 6444-6448

Loh DY, Bothwell ALM, White-Scharf ME, Imanishi-Kari T, Baltimore D (1983) Molecular basis of a mouse strain specific anti-hapten response. Cell 33: 85-93

Low N, Holliger P, Winter G (1996) Mimicking somatic hypermutation: affinity maturation of antibodies displayed on bacteriophage using a bacterial mutator strain. J. Mol. Biol. 260: 359-368

Marks JD, Griffiths AD, Malmqvist M, Clackson TP, Bye JM, Winter G (1992) By-passing immunization: building high affinity human antibodies by chain shuffling. Biotechnology (N Y) 10: 779-83

Novotny J, Bruccoleri RE, Saul FA (1989) On the attribution of binding energy in antigen-antibody complexes McPC 603, D1.3, and HyHEL-5. Biochemistry 28: 4735-4749

Schier R, Balint RF, McCall A, Apell G, Larrick JW, Marks JD (1996a) Identification of functional and structural amino-acid residues by parsimonious mutagenesis. Gene 169: 147-55

Schier R, Bye J, Apell G, McCall A, Adams GP, Malmqvist M, Weiner LM, Marks JD (1996b) Isolation of high-affinity monomeric human anti-c-erbB-2 single chain Fv using affinity-driven selection. Journal Of Molecular Biology 255: 28-43

Schier R, McCall A, Adams GP, Marshall KW, Merritt H, Yim M, Crawford RS, Weiner LM, Marks C, Marks JD (1996c) Isolation of picomolar affinity anti-c-erbB-2 single-chain Fv by molecular evolution of the complementarity determining regions in the center of the antibody binding site. Journal Of Molecular Biology 263: 551-67

Stemmer WP (1994) Rapid evolution of a protein in vitro by DNA shuffling [see comments]. Nature 370: 389-91

Tomlinson IM, Walter G, Jones PT, Dear PH, Sonnhammer EL, Winter G (1996) The imprint of somatic hypermutation on the repertoire of human germline V genes. Journal Of Molecular Biology 256: 813-17

Yang WP, Green K, Pinz-Sweeney S, Briones AT, Burton DR, Barbas CFr (1995) CDR walking mutagenesis for the affinity maturation of a potent human anti-HIV-1 antibody into the picomolar range. Journal Of Molecular Biology 254: 392-403

Affinity and Specificity Maturation by CDR Walking

Kristiina Takkinen, Ari Hemminki, and Hans Söderlund

Introduction

Random mutagenesis and phage display selection provide excellent tools for rapid engineering of the binding properties of antibodies for different applications. One strategy to optimise the antibody combining site is to mutate the CDRs in a stepwise manner. The CDRs are mutated and selected sequentially, after every round the best mutant is used as the template for the subsequent round of mutagenesis and selection. Yang et al. (1995) developed a high affinity anti-HIV gp 120 Fab by the CDR walking strategy in four steps. They were able to further increase the affinity (altogether 420-fold, $K_d=1.5x10^{-11}$M) by combining independently selected mutations, however, in this case only one of the six combinations constructed resulted in higher affinity demonstrating the unpredictable behaviour of parallel CDR optimisation. By stepwise optimisations of the light and heavy CDR3 regions and by using low, decreasing amount of biotinylated antigen in the selection steps Schier et al (1996) isolated an anti-c-erbB-2 scFv with picomolar affinity ($K_d=1.3x10^{-11}$M). The CDR walking and parallel optimisation strategies have also been used to improve the affinity and specificity of a monoclonal anti-testosterone antibody in the development of a clinically useful recombinant antibody (Hemminki et al, 1998a, b). Improvement of the affinity and specificity was possible by the additivity effect of the optimised CDRs, by combining compatible mutant CDRs together the affinity was increased 12-fold with simultaneous decrease in the cross-reactivities with clinically relevant testosterone derivatives.

✉ Kristiina Takkinen, VTT Biotechnology, P.O.Box 1500, Espoo, -02044 VTT, Finland (*phone* +35-89-4565106; *fax* +35-89-4552103; *e-mail* kristiina.takkinen@vtt.fi)
Ari Hemminki, University of Helsinki, Novatreat Inc., Biocenter, P.O.Box 56, Helsinki, 00014, Finland
Hans Söderlund, VTT Biotechnology, P.O.Box 1500, Espoo, -02044 VTT, Finland

Outline

The steps of the affinity and specificity maturation by CDR walking are outlined in Figure 1. Protocols describing selection conditions used to improve the binding properties of the anti-testosterone antibody will be given here. The CDR regions of the anti-testosterone antibody were randomised using spiked oligonucleotides (Hermes et al, 1989, 1990, Thompson et al, 1996, Schier et al, 1996) and Fab display phage libraries were cloned by the overlapping PCR method (Barbas et al 1992). Mutants with improved binding properties were isolated by using two selection methods: (1) The specificity was improved by a competitive selection procedure (Hemminki et al 1998a) and (2) the affinity was improved using limiting, decreasing concentrations of biotinylated testosterone in the selection steps (Hemminki et al 1998b).

- clone the heavy and light chain genes (see Chapter 2)

⇩

- express soluble Fab fragments in *E.coli* (see Chapter 20)
- characterise the binding properties of the purified Fab and compare to the parental antibody (see Chapter 28)

⇩

- mutagenise the CDR region(s) by spiked oligonucleotides
- in the absence of 3-D structure, start with the CDR3 region(s)

⇩

- clone the randomised CDR(s) into a Fab display library by overlapping PCR
- verify the result of mutagenesis by sequencing 5-10 individual clones

⇩

- select binders by affinity and / or specificity panning

⇩

- produce soluble Fab fragments
- characterise the binding properties of isolated mutants by competitive ELISA and compare to the parental antibody (see Chapter 28)

⇩

- combine positive mutations into one Fab fragment
- determine possible additivity effects of the combined mutations by competitive ELISA

⇩

- for further improvement of the binding properties, target next CDR(s) to mutagenesis
- use the best mutant Fab isolated from the 1. round as the template

Fig. 1. The steps of the affinity and specificity maturation by CDR walking.

Procedure

Competitive selection

1. Using the competitive selection procedure heavy and light chain CDR3 mutations of the anti-testosterone antibody were isolated. By combining the isolated mutations in one Fab fragment the cross-reactivity with dehydroepiandrosterone sulfate (DHEAS) was decreased over 30-fold (from 1% to 0.03%) while the affinity (K_d=1.3x10^{-8}M) was preserved at the initial level.

2. Incubate 100 μl of the phage library stock (~ 10^{10} pfu) in PBS/1% BSA with 100 nM of the soluble cross-reacting antigen for 1 h at room temperature in a shaker.

3. Analyse the inhibition effect of the soluble cross-reacting antigen during each selection step by selecting the same phage stock with and without the cross-reacting antigen and titering the eluted binders on selective plates. Increase the concentration of the cross-reacting antigen to achieve at least 50% binding inhibition at subsequent selection steps.

4. Transfer the phage solutions to microtiterplate wells coated with 1 μg of the specific antigen-BSA conjugate and incubate 1 h at room temperature in a shaker.

5. Wash the wells 10 times with TBS/1% BSA/0.05% Tween 20.

6. Elute the binders with 100 mM HCl (pH adjusted to 2.2 with glycine)/ 0.1% BSA and incubate 15 min at room temperature in a shaker. Remove the eluate and neutralise with 2 M TRIS.

7. Amplify the eluted phage pool by infecting *E.coli* cells and prepare the phage stock for the next selection step.

8. After the last selection step produce soluble Fab fragments and characterise the cross-reactivities of individual clones by competitive ELISA (see Chapter 28) and compare to the parental antibody.

9. Sequence clones showing decreased cross-reactivity and evaluate additivity of individual mutations by combining those in one Fab template.

10. Use the best mutant Fab as the template when randomising next CDR(s).

Affinity selection

1. The best anti-testosterone Fab fragment (having mutations both in the heavy and light chain CDR3s) developed by the competitive selection was used as the template when the CDR1 and CDR2 regions of the heavy and light chain genes were randomised and cloned to individual phage libraries. Using the affinity selection protocol two different light chain CDR1 mutant clones were isolated with over 12-fold increase in the affinity while the cross-reactivities to related steroids were preserved at the same level as in the parental combined CDR3 mutant clone.

2. Incubate 100 μl of phage library stock ($\sim 10^{10}$ pfu) in PBS/1% BSA with 100-200 nM of biotinylated antigen for 1 h at room temperature in a shaker. Decrease the concentration of the biotinylated antigen 10-fold at each selection step. If the amount of the eluted phage titered after elution is decreasing significantly repeat the selection step with a higher antigen concentration.

3. Transfer the phage solution into streptavidin coated microtiter plate wells (Boehringer Mannheim GmbH, Germany) and incubate for 15 min.

4. Wash the wells 10 times with TBS/0.1% BSA/0.05% Tween 20.

5. Elute the binders with 50 mM NaOH, pH 12.6 and incubate for 15 min in a shaker. Remove the eluate and neutralise with 1 M TRIS pH 7.5.

6. Amplify the eluted phage pool by infecting *E.coli* cells and prepare the phage stock for the next selection step.

7. After the last selection step produce soluble Fab fragments and characterise the affinity of individual clones by competitive ELISA (see Chapter 28) and compare to the parental antibody.

8. Sequence clones showing increased affinity and evaluate additivity of individual mutations by combining those in one Fab template.

9. Use the best mutant Fab as the template when randomising next CDR(s).

Comments

When starting with a mAb, clone the genes encoding the Fab fragment and verify that the binding characteristics of the recombinant Fab fragment are comparable with the parental antibody. Fab fragment expression and display is recommended when accurate affinity and specificity ranking of isolated mutants is necessary.

Affinity selection using biotinylated antigen was first described by Hawkins et al (1992) and has been successfully applied e.g. by Schier et al (1996) and Thompson et al (1996). Competitive selection has been used to improve binding specificity e.g. by de Kruif et al (1995), Parsons et al (1996), Chames et al (1998) and Saviranta et al (1998).

References

Barbas, CFIII, Bain, JD, Hoekstra, DM and Lerner, RA (1992) Semisynthetic combinatorial antibody libraries: A chemical solution to the diversity problem. Proc Natl Acad Sci USA 89: 4457-4461

Chames, P, Coulon, S and Baty, D (1998) Improving the affinity and fine specificity of an anti-cortisol antibody by parsimonious mutagenesis and phage display. J Immunol 161: 5421-5429

De Kruif, J, Boel, E and Logtenberg,T (1995) Selection and application of human single chain Fv antibody fragments from a semi-synthetic phage antibody display library with designed CDR3 regions. J Mol Biol 248: 97-105

Hawkins, RE, Russell, SJ and Winter, G (1992) Selection of phage antibodies by binding affinity. Mimicking affinity maturation. J Mol Biol 226: 889-896

Hemminki, A, Niemi, S, Hoffrén, A-M, Hakalahti, L, Söderlund, H and Takkinen, K (1998a) Specificity improvement of a recombinant anti-testosterone Fab fragment by CDRIII mutagenesis and phage display selection. Protein Eng 11: 311-319

Hemminki, A, Niemi, S, Hautoniemi, L, Söderlund, H and Takkinen, K (1998b) Fine tuning of an anti-testosterone antibody binding site by stepwise optimisation of the CDRs. Immunotechnology 4: 59-69

Hermes, JD, Parekh, SM, Blacklow, SC, Köster, H and Knowles, JR (1989) A reliable method for random mutagenesis: the generation of mutant libraries using spiked oligodeoxyribonucleotide primers. Gene 84: 143-151

Hermes, JD, Blacklow, SC and Knowles, JR (1990) Searching sequence space by definably random mutagenesis: improving the catalytic potence of an enzyme. Proc Natl Acad Sci USA 87: 696-700

Parsons, HL, Earnshaw, JC, Wilton, J, Johnson, KS, Schueler, PA, Mahoney, W and McCafferty, J (1996) Directing phage selections towards specific epitopes. Protein Eng 9: 1043-1049

Saviranta, P, Pajunen, M, Jauria, P, Karp, M, Pettersson, K, Mäntsälä, P and Lövgren, T (1998) Engineering the steroid-specificity of an anti-17β-estradiol Fab by random mutagenesis and competitive phge panning. Protein Eng 11: 143-152

Schier, R, Mc Call, A, Adams, GP, Marshall, KW, Merritt, H, Yim, M, Crawford, RS, Weiner, LM, Marks, C and Marks, JD (1996) Isolation of picomolar afinity anti-c-erb-2 single-chain Fv by molecular evolution of the complementarity detremining regions in the center of the antibody binding site. J Mol Biol 263: 551-567

Thompson J, Pope, T, Tung, J-S, Chan, C, Hollis, G, Mark, G and Johnson, KS (1996) Affinity maturation of a high-affinity human monoclonal antibody against the third hypervariable loop of human immunodeficiency virus: Use of phage display to improve affinity and broaden strain specificity. J Mol Biol 256: 77-88

Yang, WP, Green, K, Pinz-Sweeney, S, Briones, AT, Burton, DR and Barbas, CFIII (1995) CDR walking mutagenesis for the affinity maturation of a potent human anti-HIV-1 antibody into the picomolar range. J Mol Biol 254: 392-403

Humanisation of Antibody Fragments

Humanisation by Guided Selection

Sigrid H.W. Beiboer and Hennie R. Hoogenboom

Introduction

The hybridoma technology has given rise to many rodent antibodies with exquisite specificity. Unfortunately, the use of murine antibodies for immunotherapy of human carcinomas has a disadvantage that during the repeated administration of the monoclonal antibodies to patients, a human anti-mouse antibody (HAMA) immune response can be induced. This HAMA response may cause a rapid blood clearance of the monoclonal antibodies, thereby reducing its efficacy (Khazaeli et al., 1994). Murine antibodies can be humanised by CDR-grafting (Jones et al., 1986; Chapter 40) where the hypervariable loops of the monoclonal antibody are transplanted to a human antibody, thereby retaining the epitope and specificity. However, such CDR grafted antibodies retain a large proportion of non-human sequences, which include the CDR's as well as framework residues, which are sometimes retained because they are involved (in)directly in antigen binding (Foote & Winter, 1992). Furthermore, humanisation by resurfacing (Roguska et al, 1996) yields the incorporation of even more non-human sequences. In all instances, some rodent antibody derived sequences are still included in the humanised antibody and thereby may enable the monoclonal antibody to provoke an immune response. For example, the transplantation of CDR-loops with non-human canonical forms, such as canonical structures 1 and 5 of murine light chain loop L1 for which no human homologue was found (Tomlinson et al., 1995), may both change the idiotype structure of the antibody combining surface as well as provide unique non-human T-cell epitopes. With the use of the

Sigrid H.W. Beiboer, Maastricht University, Department of Pathology, P.O. Box 5800, Maastricht, 6202 AZ, The Netherlands

✉ Hennie R. Hoogenboom, Maastricht University, Department of Pathology, P.O. Box 5800, Maastricht, 6202 AZ, The Netherlands (*phone* +31-43-3874630; *fax* +31-43-3876613; *e-mail* HHO@LPAT.AZM.NL)

phage display technique a new methodology for converting rodent antibodies into completely human antibodies with similar binding characteristics, was developed, based on chain shuffling of V-genes and selection on antigen (Jespers et al, 1994). The principle of the method, termed 'guided selection', is depicted in Figure 1. In a first step, one of the V-domains of the starting rodent antibody (heavy or light chain) is paired with a repertoire of naturally occurring human partner domains, derived, for example, from human donor peripheral blood lymphocyte (PBL) mRNA. This repertoire is selected on the original antigen; the rodent domain will guide the selection of a partner human chain(s) that will allow the binding of the antibody to the same epitope area on the antigen as the original rodent antibody. In a second step, the selected 'half-human' antibodies is used as a source of selected human V-genes, which are then allowed to pair with a repertoire of human partner domains and are again selected on antigen. The endpoint of this sequential chain shuffling procedure is a set of completely human antibodies, binding with high probability to the same epitope on the antigen as the original rodent antibody.

Even though most of the structural diversity of antibodies resides in the heavy chain, there is no rule of thumb as to whether the VH or the VL of the rodent antibody should be shuffled first with partner human chains. To be able to choose one route in preference to the other would require some prior knowledge of the relative contribution of heavy and light chains to the antigen-combining site, and of the property of the chain to form a successful pairing with domains of other species. In most instances one should assume equivalent importance of VH and VL in the rodent antibody, making both domains equally suitable for the first shuffle. Alternatively, both routes are followed at the same time, and two hybrid repertoires are made (murine VH-human VL repertoire and murine VL-human VH repertoire). As well as increasing the potential for diversity, this approach offers the opportunity of "parallel shuffling" later on, to combine the respective properties of selected human VH's and VL's into a single selected human repertoire. It may be that most of the clones selected from the parallel chain shuffled library are the same as those selected by sequential chain shuffling.

Several published examples demonstrate the power of the method. Figini et al. (Figini et al., 1994) shuffled the light chain of a murine anti-hapten phenyl-oxazolone antibody NQ10.12.5 with a repertoire of human heavy chains. After a second shuffle of the selected VH's with a repertoire of human VL's, entirely human antibodies were generated. A similar approach, starting with the murine heavy chain, was used to convert a murine anti-TNF antibody into a human version, yielding five human antibo-

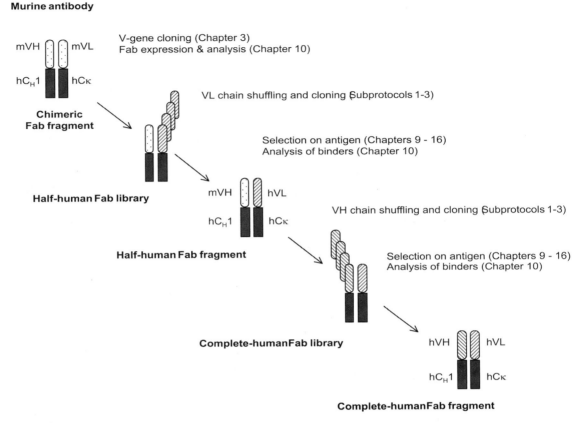

Murine antibody

mVH   mVL   V-gene cloning (Chapter 3)
             Fab expression & analysis (Chapter 10)

hC$_H$1   hCκ

**Chimeric
Fab fragment**

VL chain shuffling and cloning (Subprotocols 1-3)

Selection on antigen (Chapters 9 - 16)
Analysis of binders (Chapter 10)

Half-human Fab library

mVH   hVL

hC$_H$1   hCκ

VH chain shuffling and cloning (Subprotocols 1-3)

Half-human Fab fragment

Selection on antigen (Chapters 9 - 16)
Analysis of binders (Chapter 10)

Complete-human Fab library

hVH   hVL

hC$_H$1   hCκ

Complete-human Fab fragment

Fig. 1. The sequential chain shuffling strategy. Chapters and Subprotocol numbers are indicated for each step of the procedure of guided selection.

dies (Jespers et al, 1994); one of the full human antibodies has a similar affinity and competes with the murine antibody for binding to TNF. Also, it shows the same unique *in vivo* properties as the murine antibody. Since 1994, the method has been used more commonly. Three groups obtained human antibodies by sequential chain shuffling against folate-binding protein, a cell surface antigen overexpressed in many ovarium carcinomas (Figini et al., 1998), TNFα (Salfeld et al., WO9729131A1) and human interferon γ receptor 1 (Watzka et al., 1998). In the latter case, the second shuffle (of the VH) led to the selection of a humanised antibody that recognised an epitope present only on the immobilised (coated) antigen. In principle, the modulation of fine-specificity may be caused by the possible different chemistry interaction with a similar area on the antigen. However, in this case, it appears that a suboptimal selection procedure has interfered with the structure of the epitope, causing failure of the 'guided selection' procedure, and selection of antibodies to completely different epitopes on the antigen. Recently, we succeeded in humanising antibodies, which had the same epitope specificity as the original murine antibodies, by retaining the murine HCDR3 (Klimka et al., 2000; Beiboer et al., 2000). Rader et al. humanised the murine antibody (directed to human integrin $\alpha_v\beta_3$) by guided selection but succeeded only by retaining the murine CDR3's in both heavy and light chain (Rader et al., 1998). Thus, guided selection of rodent antibody V-genes provides a powerful tool for interspecies conversion of antibodies. The combination of shuffling of V-genes and selection on antigen provides a means to direct the isolation of a human V-gene pair with largely similar binding characteristics as the starting antibody.

Outline

As depicted in Figure 1, the first step in the guided selection procedure is to clone the variable domain genes of the rodent antibody. As one of the cornerstones of the technique is the selection of phage antibody fragments on antigen, it is advisable to clone the rodent antibody V-genes in the format, which will be used for the guided selection. Guided selection can be carried out in a number of different formats (Figure 2): either as single chain Fvs (scFv) or as Fab fragments, using a one or two replicon system. This chapter will concentrate on guided selection in Fab format, where the V-genes of a murine antibody are first cloned into a Fab format vector like pCES1 in which the human constant domains CL and C_H1 are present (de Haard et al. 1999). Some scFv's have a tendency to dimerise (Holliger et al., 1993), which leads to possible avidity-mediated artefacts during selection or screening. When using a Fab format for humanisation, we can screen the antibodies for affinity rather than avidity and also select with less avidity-mediated problems for high affinity binders. pUC119 based vectors, such as pCES1, mediate expression of soluble Fab's with a myc tag incorporated at the carboxyl terminus of the heavy chain, allowing immunodetection using the anti-c-myc monoclonal antibody 9E10 (Hoogenboom et al, 1991; Marks et al, 1991). Expression of the chimeric Fab (human constant domains and murine variable domains) should be confirmed by analysing a sufficient number of colonies from the cloning plate (20-50). These can be grown as individual cultures and induced from

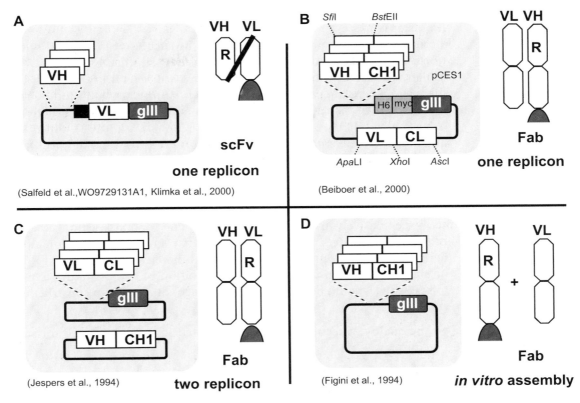

Fig. 2. Guided selection formats (heavy chain shuffling). Formats based on the scFv (**A**) and Fab (**B,C,D**) configuration and using one or two replicons have been developed. In each panel the DNA is depicted on the left and the antibody format as encoded proteins on the right. In **A** and **B** heavy and light chains are on the same replicon, in **C** heavy and light chains are on separate replicons, and in **D** the light chain is provided *in trans* as protein. Illustrated here is a heavy chain shuffle, the repertoire denoted by R.

the lac promoter as soluble Fab using IPTG. The initial analysis of binding of the chimeric Fab is an important step to ensure that the original antigen combining site has been maintained and that antigen can be detected in the assay of choice, be it ELISA, cell surface binding, immuno-precipitation etc. Issues to be aware of during hybridoma gene cloning have been extensively reviewed (Bradbury et al., 1995). It is also useful to check that the rodent antibody expressed as a phage antibody will work in the selection procedure of choice. Having isolated a number of clones scoring positive for antigen binding by ELISA, they should be sequenced and aligned to find a consensus sequence. The clone with the sequence closest to the consensus should be used as the building block for subsequent chain shuffling steps.

One of the critical steps in evaluating the success of the guided selection procedure is the construction of large chain shuffled repertoires with a high number of functional inserts. There are several examples in the literature where the ease with which a high affinity antibody can be isolated by phage display correlates with the size of the starting repertoire (Marks et al, 1991; Griffiths et al, 1994; Marks et al., 1992). Since the aim of guided selection is to confer all the properties of binding specificity and affinity from the rodent antibody onto a human equivalent, maximising library sizes during chain shuffling ensures that as many permutations as possible of heavy-light chain pairings are available for antigen selection. Protocols for the construction of naïve human antibody libraries are described extensively in Chapter 6. Human V-gene pools are easily accessible by PCR from a naïve human antibody library (Marks et al, 1991), or, alternatively, with appropriate oligonucleotides directly from mRNA of PBLs or spleen. The main criteria for the generation of chain shuffled repertoires, which should be optimised in all stages of repertoire construction, are:

- the quality of restriction enzyme digested PCR product used for sub-cloning

- the efficiency of ligation and recovery of ligated product

- the transformation efficiency of the *E.coli* host

Often the conditions used for cloning the initial cDNA library of the rodent hybridoma are sufficiently stringent for construction of subsequent half-human or fully human chain shuffled repertoires. The main distinction is the scale on which the procedure is carried out; relatively large quantities of V gene input material are essential in chain shuffling to ensure that the maximum library diversity is achieved. The rodent VH (or VL) gene will be obtained as a PCR fragment, which is cloned directly by restriction fragment cloning (Subprotocols 1-3). Here, suitable restriction sites need to be incorporated into the regions flanking the V-genes. Particular care has to be taken to choose sites that occur infrequently in V-genes. However, somatic mutation within V-genes may generate restriction sites, in which case it would be better to use PCR assembly as a means of chain shuffling. It may also be possible to build individual repertoires of heavy or light chains, such that rodent VH can be cloned into vectors containing repertoires of human VL-genes, or vice versa. We describe one system only, where the murine VH is combined with a human VL repertoire by cloning.

Murine V-genes are amplified by PCR as described in Chapters 2-4. A secondary PCR is performed to introduce restriction sites necessary for the cloning of the V-genes in the Fab format vector as described in Subprotocol 1. After the assurance that the antigen binding site of the chimeric Fab is maintained, the VH-gene is cloned into another vector absent in V-genes to avoid contamination of murine VL's. As described in Subprotocol 1, the human Vκ repertoire is amplified using HuVκBACK-APAmix and HuVCκFOR-ASC and the human Vλ repertoire is amplified using HuVλBACK-APAmix and HuVCλFOR-ASCmix from a highly diverse repertoire, preferably IgM. Subsequently, the light chain repertoires are digested with *Apa*LI/*Asc*I and ligated into phagemid vector pCES1 containing the murine VH. The obtained Fab repertoires are subjected to rounds of growth, rescue (for display) and selection on antigen, to enrich the population in antigen binders (Chapter 9-16). The concentration of antigen used in the selection procedure is an important factor. Schier et al. (Schier et al., 1995; Schier & Marks, 1996) obtained higher-affinity monomeric scFv only by selecting in solution using limiting concentrations of biotinylated antigen. At each round of selection, 24-48 clones are assayed for antigen binding by ELISA (Chapter 10). In certain cases, the ELISA signal obtained from soluble Fab's of half- or fully-human antibodies is too weak to analyse binding specificity and affinity in great detail. In these instances, the sensitivity of detection in ELISA can be increased by analysing the clones as phage: in general, we find assays in phage format give more consistent and reproducible results. Individual clones can be rapidly identified by *Bst*NI fingerprint (see Subprotocol 4) and the off-rate component of the measured affinity determined if required (see for example Chapters 28, 29 and de Haard et al., 1999). After achieving some 50 - 70% of positives for antigen binding, the murine VH of these antibodies is exchanged by a human VH repertoire and the same shuffling: A number of selected VL genes, which form binding pairs with the original murine VH domain, are combined individually or in batch with a repertoire of human VH genes. In order to avoid contamination of murine VH genes, the selected light chain(s) are cloned into a pCES1 vector absent in V-genes. As described in Subprotocol 1, the human VH repertoire is amplified using HuVHBACK-SFImix and HuVHFORmix. Subsequently, the VH repertoires are digested with *Sfi*I/*Bst*EII and cloned into the phagemid pCES1 containing the human VL (Subprotocols 2 and 3). Human antibodies binding to the antigen can be selected from the obtained repertoire and analysed.

All fully human clones still positive for antigen binding can be analysed further. Beside the useful fingerprint analysis with the restriction endonuclease *Bst*N1 (Subprotocol 4) and DNA sequencing (Chapter 30), is

the check to ensure that the human clones do not bind to a panel of irrelevant antigens, as unrequired cross-reactivity may have been introduced
during the guided selection. The choice of antigens to use for this panel is
up to the individual, but we generally use a panel of at least 10, to include
haptens, purified proteins from different species and antigens with structural similarities to the antigen of interest. Finally, a very indicative IC_{50}
value can be obtained by an ELISA based measurement. This method described in Chapter 28, allows the measurement of affinity using soluble
antigen to compete the binding of phage antibody to antigen coated
onto plastic.

Table 1. Oligonucleotide list

(a) Primers for secondary amplification of murine V genes
MVH1BACK-SFI (introduces a *Sfi*I site at the 5' end of the VH primary PCR products)
5'-CAT GCC ATG ACT CGC <u>GGC CCA GCC GGC</u> CAT GGC CSA GGT SMA RCT GCA GSA GTC WGG-3'
MVKFOR-XHO (introduces a *Xho*I site at the 3' end of the VL primary PCR products)
5'-CCG GAA TTC <u>CTC GAG</u> CCG TTT BAK YTC CAR CTT KGT SCC -3'
MVKBACK-APAmix, an equimolar mix of:

MVKABACK-APA	5'- ACC GCC TCC ACC A<u>GT GCA</u> CTT GAT GTT TTG ATG ACC CAA ACT CCA-3'
MVKCBACK-APA	5'- ACC GCC TCC ACC A<u>GT GCA</u> CTT GAC ATT GTG CTR ACC CAR TCT CCA-3'
MVKDBACK-APA	5'- ACC GCC TCC ACC A<u>GT GCA</u> CTT GAC ATC CAG ATG ACN CAG TCT CCA-3'
MVKEBACK-APA	5'- ACC GCC TCC ACC A<u>GT GCA</u> CTT CAA ATT GTT CTC ACC CAG TCT CCA-3'
MVKFBACK-APA	5'- ACC GCC TCC ACC A<u>GT GCA</u> CTT GAA AAT GTG CTC ACC CAG TCT CCA-3'

(b) Primers for secondary amplification of human V genes
HuVHBACK-SFImix, (introduces a *Sfi*I site at the 5' end of the VH primary PCR products), HuVκBACK-
APAmix (introduces a *Apa*LI site at the 5' end of the VκCκ primary PCR products), HuCκFOR-ASC
(introduces a *Asc*I site at the 3' end of the VκCκ primary PCR products), HuVλBACK-APAmix, (introduces a
*Apa*LI site at the 5' end of the VλCλ primary PCR products), HuCλFOR-ASCmix (introduces a *Asc*I site at the
3' end of the VλCλ primary PCR products), described by De Haard *et al.* (1999)

(c) Primers for amplification of cloned Fab's

FDTSEQ24	5'-TTT GTC GTC TTT CCA GAC GTT AGT-3'
PUCREVERSE	5'-AGC GGA TAA CAA TTT CAC ACA GG-3'

Ambiguity codes:
M=A or C; R=A or G; W=A or T; S= C or G; Y= C or T; K= G or T; B= C, G or T; N= A, C, G or T

Sequences encoding restriction sites are underlined

Materials

- Microcentrifuge
- Sterile H$_2$O
- Sterile microcentrifuge tubes
- Sterile pipette tips
- Agarose (normal and low melting point)
- 5 mg/ml ethidium bromide
- Agarose gel electrophoresis tank
- *Taq* DNA polymerase (Boehringer, Eurogentec)
- 10x *Taq* polymerase buffer as provided by the supplier of the *Taq* polymerase
- 5 mM dNTPs is an equimolar mixture of dATP, dCTP, dGTP and dTTP with a total concentration of 5 mM nucleotide (i.e. 1.25 mM of each dNTP)
- DNA thermal cycler for PCR
- Mineral oil (paraffin oil) (Sigma, cat. no. M-3516)
- Qiaex
- U.V. transilluminator
- Vortex mixer
- T4 DNA ligase + buffer (Promega)
- Phenol/Chloroform/Isoamylalcohol (P/C/I=25/24/1)
- Chloroform/Isoamylalcohol (C/I=24/1)
- Ethanol at -20°C
- 70% ethanol at -20°C
- 3 M NaAc, pH 5.2
- Sterile H$_2$O (ice-cold)
- Sterile 10% glycerol in H$_2$O (ice-cold)
- *E. coli* TG1

- 2 litre shake flask
- Shaking incubator at 37°C
- High-speed centrifuge (Sorvall or equivalent) and GSA rotor
- Benchtop centrifuge
- Sterile centrifuge bottles, prechilled
- Sterile 50 ml Falcon tubes
- Bio-Rad Gene Pulser Plus
- 0.2 cm cuvettes, prechilled
- pBR322 DNA at 10 ng
- Cell recovery medium: 2xTY + 2% glucose
- Sterile glycerol
- 15 ml polypropylene culture tubes
- Polypropylene freezing tubes
- *Bst*NI enzyme (New England Biolabs)
- 10x buffer for *Bst*NI (NEB2, New England Biolabs)
  - Acetylated BSA (New England Biolabs)

10x TBE, per litre

- 108 g Tris
- 55 g boric acid
- 40 ml 0.5 M EDTA, pH 8

2xTY broth, per litre

- 16 g tryptone
- 10 g yeast extract
- 5 g NaCl
- Sterilize by autoclaving

Minimal agar, per litre

- 12.8 g $Na_2HPO_4 \cdot 7H_2O$

- 3 g KH_2PO_4

- 0.2 g $MgCl_2$

- 4 g glucose

- 0.5 g NaCl

- 5 mg thiamine hydrochloride

- 1 g NH_4Cl

- 15 g agar

TYAG plates, per litre

- 2xTY

- 100 g/ml ampicillin

- 2% glucose

- 15 g agar

- Large plates (243 mm x 243 mm Nunc)

- Small plates (90 mm)

Subprotocol 1
Secondary PCR of Murine and Human V Genes, Introduction of Restriction Sites for Cloning

Procedure

1. Estimate the quantities of VH and VL DNA prepared by the primary PCR reactions (mouse: Chapter 5, human: Chapter 6) on an agarose gel.

2. Set up the following assembly PCR reaction:

	I	II
H$_2$O	32 µl	37 µl
10x *Taq* buffer	5 µl	5 µl
5 mM dNTP	2.5 µl	2.5 µl
10 µM BACK primer[a]	2.5 µl	2.5 µl
10 µM FOR primer[b]	2.5 µl	2.5 µl
50 ng DNA	5 µl	0 µl
Taq DNA polymerase (5 U/µl)	0.5 µl	2.5 µl

3. Overlay with 2 drops mineral oil. Amplify by PCR using 25 cycles of 94°C 1 min, 55°C 1 min, 72°C 2 min followed by incubation at 72°C for 10 min.

4. Check each PCR amplification with a 5 µl sample on a 2% gel.

5. Gel purify fragments on a 2% LMP (low melting point) agarose/TBE gel. Carefully excise the VH (400 bp) and VL (400 bp) bands (using a fresh sterile scalpel or razor blade for each) and transfer each band to a separate sterile microfuge tube.

6. The fragments are purified from the agarose using Qiaex. Recover the DNA in 50 µl H$_2$O.
a, b. Primer pairs (primers are described in Table 1):
Mouse VH: MVH1BACK-SFI; MVH1FOR-2(*Bst*EII)
Mouse VL: MVKBACK-APAmix; MVKFOR-XHOmix
Human VH: HuVHBACK-SFImix; HuVHFORmix
Human Vκ: HuVκBACK-APAmix; HuVCκFOR-ASC
Human Vλ: HuVλBACK-APAmix; HuVCλFOR-ASCmix

At this point the DNA is further treated as in Chapter 6 (digestion, purification), and cloned into pCES1 as described in the following Subprotocols, 2 and 3.

Subprotocol 2
Ligation

For the ligation, there is no way to guarantee high ligation efficiencies under all circumstances. In order to optimise conditions for a particular insert, it is best to keep the amount of vector constant and to use varying quantities of insert in the ligations. Molar ratios of vector to fragment of 1:2 appear to work best.

▨▨ Procedure

1. Set up the following ligation reactions:

	I	II
Vector DNA (50 ng/μl)	2 μl	2 μl
Digested fragment (20-50 ng/μl)	1 μl	0 μl
10 x ligase buffer	1 μl	1 μl
H$_2$O	5 μl	6 μl
Mix and spin briefly in a microfuge, then add:		
T4 DNA ligase (3 U/μl)	1 μl	1 μl

2. Incubate for one hour at RT or 4-6 h at 16°C or overnight at 4°C.

3. For optimum transformation efficiency, purify the ligation reaction as follows. Dilute the ligation mixture with 250 μl H$_2$O. P/C/I extract reaction once, C/I extract once, and ethanol precipitate (-70°C, 15 min) in the presence of NaAc and 20 μg glycogen (Boehringer); this greatly improves the recovery of precipitated ligation product.

4. Pellet ligated DNA in a microcentrifuge, 14000 rpm for 10 min. Aspirate ethanol, wash pellet once in 70% ethanol and dry pellet briefly.

5. Resuspend pellet in 10 μl H$_2$O and either store at -20°C or use directly for electroporation.

Subprotocol 3
Preparation and Electroporation of Competent TG1

Preparation of competent *E. coli* using this method routinely gives 1-3 x 10^{10} colonies /µg pBR322 DNA. Maximum efficiencies are obtained when the cells are prepared and electroporated on the same day; however, they may also be snap-frozen and stored at -70°C, resulting in approximately ten fold lower efficiencies.

Procedure

Preparation of electrocompetent cells

1. Inoculate an overnight culture of *E. coli* TG1 (from a single colony on minimal agar) in 2xTY broth. Grow at 37°C.
2. Next day, prewarm 500 ml 2xTY broth in a sterile 2 litre flask to 37°C, then add 5 ml of the fresh overnight culture.
3. Grow with shaking (minimum 250 rpm) for 105 min until the O.D. at 600 nm is approximately 0.5.
4. Chill flask on ice for at least 30 min. Transfer cultures to sterile pre-chilled centrifugation bottles, for use in the Sorvall GSA rotor.
5. Spin 4000 rpm, 4°C for 15 min. Decant supernatant and resuspend pellet (gently) in the starting volume ice-cold sterile H_2O.
6. Stand on ice for at least 15 min.
7. Spin 4000 rpm, 4°C for 15 min. Decant supernatant and resuspend in 0.5x starting volume ice-cold sterile H_2O.
8. Stand on ice for at least 15 min.
9. Spin 4000 rpm, 4°C for 15 min. Decant supernatant and resuspend in 10 ml ice-cold 10% glycerol. Transfer to 50 ml Falcon tube and stand on ice for at least 15 min.
10. Spin 3500 rpm, 4°C for 10 min in a benchtop centrifuge. Decant supernatant and resuspend final bacterial pellet in 1 ml 10% glycerol.
11. Divide into 160 µl aliquots in sterile microfuge tubes on ice; either use immediately or freeze as described above.

Electroporation of ligation reactions using the BIORAD Gene Pulser Plus

1. If frozen TG1 are to be used, thaw the appropriate number of aliquots on ice (allow 150 µl cells per 5 µl ligation reaction). If using fresh cells, add 150 µl directly to the ligation reaction on ice.

2. Transfer to a pre-chilled 0.2 cm cuvette (of a type compatible with the Biorad Gene Pulser Plus). Be sure to set up controls with 10 ng pBR322 DNA (to gauge efficiency) and without DNA (to guard against contaminating DNA in the libraries).

3. Set up the Gene Pulser Plus to give a 2.5 kV pulse, using the 25 µF capacitor and the pulse controller set to 200 Ohm.

4. Dry the cuvette with tissue and place in the electroporation chamber.

5. Pulse once. The registered time constant should be 4.5-5.0 msec.

6. Immediately add 1 ml fresh cell recovery medium to the cuvette, resuspend cells and transfer to a disposable culture tube.

7. Shake for 1 hour at 37°C to allow the cells to recover and to express antibiotic resistance.

8. Plate appropriate dilutions (in 2xTY) on 90 mm 2xTYAG plates. For plating out the entire library using pre-tested ligation mix and cells of known efficiency, plate aliquots containing at least 200000 clones on 243x243 mm (Nunc) 2xTYAG plates.

9. Grow overnight at 30°C.

10. Next day, to harvest the repertoire, flood plates with 2-10 ml 2xTY and detach cells by scraping using a sterile spreader. Transfer cells to a sterile polypropylene tube and disperse clumps by vortexing. Add sterile glycerol to a final concentration of 15%, mix, and freeze 1 ml aliquots at -70°C. At this stage, freshly-prepared aliquots of the library can be used to inoculate a culture to rescue phage and start selections.

From this repertoire, half murine/half human antibodies binding to the original antigen can be selected and analysed. For an extensive list of protocols for the rescue of phagemids, selection and screening see Chapters 9-16. After screening and identification of the best VL, the second shuffle can be performed.

Subprotocol 4
Fingerprinting of Clones with Restriction Endonuclease BstN1

This method is only intended as a rough guide to diversity in a panel of clones, in between sequential rounds of selection, or as a guide to diversity between individual clones prior to sequencing.

■ ■ Procedure

1. Make up 20 μl PCR mixes containing:

H₂O	15.0 μl
10x PCR buffer	2.0 μl
5 mM dNTPs	1.0 μl
FDTSEQ24 (10 μM)	1.0 μl
PUCREVERSE (10 μM)	1.0 μl
Taq polymerase (5 U/μl)	0.02 μl

2. With a toothpick touch a single colony and bring in contact with the PCR reaction, mix briefly. Throw away toothpick.

3. Repeat for all clones. Overlay reactions with mineral oil and heat to 94°C for 10 min. using the PCR-block.

4. Cycle 40 times to amplify the fragments: 94°C, 45 sec.; 55°C, 45 sec.; 72°C, 1.5 min. Optional: check PCR result, load 5 μl reaction mixture on 1% agarose.

5. After the PCR, add the following underneath the mineral oil:

Acetylated BSA (10 mg/ml)	0.2 μl
NEB2 buffer (10x)	4 μl
H₂O	15.5 μl
*Bst*NI (10 U/μl)	0.5 μl

6. Digest samples at 60°C for 2-3 hours.

7. Remove an aliquot of the digested sample from underneath the oil. Analyse on a 2% (w/v) agarose gel cast in TBE buffer containing 0.5 μg/ml ethidium bromide.

8. Compare the banding patterns of individual clones on a U.V. transilluminator.

Acknowledgements. This research was financially supported by the Netherlands Technology Foundation (STW), and was coordinated by the Earth and Life Sciences Foundation (ALW, project MGN55.3858, 805.17.753).

References

Beiboer SHW, Reurs A, Roovers RC, Arends JW, Whitelegg NR, Rees AR, Hoogenboom HR (2000) Guided selection of a pan carcinoma specific antibody reveals similar binding characteristics yer structural divergence between the original murine antibody and its human equivalent. J Mol Biol 296: 833-849

Bradbury A, Ruberti F, Werge T, Amati V, Di Luzio A, Gonfloni S, Hoogenboom HR, Piccioli P, Biocca S, Cattaneo (1995) The cloning of hybridoma V regions for their ectopic expression in intracellular and intercellular immunization. In: Borrebaeck CAK (Ed) Antibody Engineering. IRL Press, Oxford, UK, pp 295-361

Figini M, Marks JD, Winter G, Griffiths AD (1994) In vitro assembly of repertoires of antibody chains on the surface of phage by renaturation. J Mol Biol 239:68-78

Figini M, Obici L, Mezzanzanica D, Griffiths A, Colnaghi MI, Winter G, Canevari S (1998) Panning phage antibody libraries on cells: isolation of human Fab fragments against ovarium carcinoma using guided selection. Cancer Res 58: 991-996

Foote J, Winter G (1992) Antibody framework residues affecting the conformation of the hypervariable loops. J Mol Biol 224: 487-499

Griffiths AD, Williams SC, Hartley O, Tomlinson IM, Waterhouse P, Crosby WL, Kontermann R, Jones PT, Low NM, Allison TJ, Prospero TD, Hoogenboom HR, Nissim A, Cox JPL, Harrison JL, Zaccolo M, Gherardi E, Winter G (1994) Isolation of high affinity human antibodies directly from large synthetic repertoires. EMBO J 13: 3245-60

Haard HJW de, Neer N van, Reurs A, Hufton SE, Roovers RC, Henderikx P, Bruine AP de, Arends JW, Hoogenboom HR (1999) A large non-immunised human Fab fragment phage library that permits rapid isolation and kinetic analysis of high affinity antibodies. J Biol Chem, 274: 18218-18230

Holliger P, Prospero P, Winter G (1993) "Diabodies": small bivalent and bispecific antibody fragments. Proc Natl Acad Sci USA 90:6444-6448

Hoogenboom HR, Griffiths AD, Johnson KS, Chiswell DJ, Hudson P, Winter G (1991) Multi-subunit proteins on the surface of filamentous phage: methodologies for displaying antibody (Fab) heavy and light chains. Nucl Acids Res 19:4133-4137

Jespers LS, Roberts A, Mahler SM, Winter G, Hoogenboom HR (1994) Guiding the selection of human antibodies from phage display repertoires to a single epitope of an antigen. Bio/Technol 12:899-903

Jones PT, Dear PH, Foote J, Neuberger MS, Winter G (1986) Replacing the complementarity-determining regions in a human antibody with those from a mouse. Nature 321:522-525

Khazaeli MB, Conry RM, LoBuglio AF (1994) Human Immune response to monoclonal antibodies. J Immunother 15: 42-52

Klimka A, Matthey B, Roovers RC, Barth S, Arends JW, Engert A, Hoogenboom HR (2000) Human anti-CD30 recombinant antibodies by guided phage antibody selection using cell panning.Br J Cancer 83:252-260

Marks JD, Hoogenboom HR, Bonnert TP, McCafferty J, Griffiths AD, Winter G (1991) By-passing immunization. Human antibodies from V-gene libraries displayed on phage. J Mol Biol 222:581-597

Marks JD, Griffiths AD, Malmqvist M, Clackson TP, Bye JM, Winter G (1992) By-passing immunization: building high affinity human antibodies by chain shuffling.-Bio/Technol. 10:779-783

Rader C, Cheresh DA, Barbas CF (1998) A phage display approach for rapid antibody humanization: Designed combinatorial V gene libraries. Proc Natl Acad Sci USA 95: 8910-8915

Roguska MA, Pedersen JT, Henry AH, Searle SM, Roja CM, Avery B, Hoffee M, Cook S, Lambert JM, Blattler WA, Rees AR, Guild BC (1996) A comparison of two murine monoclonal antibodies humanised by CDR-grafting and variable domain resurfacing. Prot Eng 9: 895-904

Salfeld JG, Allen DJ, Kaymakcalan Z, Labkovski B, Mankovich JA, McGuiness BT, Roberts AJ, Sakorafas P, Hoogenboom HRJM, Schoenhaut D, Vaughan TJ, White M, Wilton AJ. Human antibodies that bind human TNF.agr. Patent WO9729131A1

Schier R, Bye J, Apell G, McCall, Adams GP, Malmqvist M, Weiner LM, Marks JD (1995) Isolation of high-affinity monomeric human anti-c-erB-2 single-chain Fv using affinity-driven selection. J Mol Biol 255:28-43

Schier R, Marks JD (1996) Efficient in vitro affinity maturation of phage antibodies using BIAcore guided selections. Hum Antibod 7:97-105

Tomlinson IM, Cox JPL, Gherardi E, Lesk AM, Chothia C (1995) The structural repertoire of the human Vκ domain. EMBO J 14: 4628-4638

Watzka H, Pfizenmaier K, Moosmayer D (1998) Guided selection of antibody fragments specific for human interferon γ receptor 1 from a human VH- and VL-gene repertoire. Immunotechnol 3: 279-291

Humanising Antibodies by CDR Grafting

SIOBHAN O'BRIEN and TARRAN JONES

Introduction

Following the original discovery of the process for making mouse mono-clonal antibodies by Kohler and Milstein in 1975 a tremendous amount of effort has been expended to isolate mouse antibodies against specific anti-genic targets, creating a huge reservoir of potential therapeutic agents. Unfortunately, the numerous limitations associated with the long term use of mouse monoclonal therapeutic antibodies have been so severe that until recently, very few monoclonal antibodies had been licensed for human use in the intervening time. The major stumbling blocks to their use as therapeutic agents include:

- Short serum half life.

- Poor utilisation of human effector functions via the mouse constant regions.

- Rapid onset of a human anti-globulin response to mouse specific residues - also known as the HAMA (Human Anti-Mouse Antibody) response.

Even chimerisation, the genetic linkage of murine variable regions to human constant regions, rarely solved the problems associated with the HAMA response, in particular. To overcome these limitations, Dr. Greg Winter and his colleagues (Jones, T.P. et al. 1986, Riechmann, L. et al. 1988) developed the concept of creating functional human-like antibodies by grafting the antigen binding complementarity determining re-

Siobhan O'Brien, AERES Biomedical Limited, 1-3 Burtonhole Lane, Mill Hill, London, NW7 1AD, UK
✉ Tarran Jones, AERES Biomedical Limited, 1-3 Burtonhole Lane, Mill Hill, London, NW7 1AD, UK (*phone* +44-20-8906-3811; *fax* +44-20-8906-1395; *e-mail* tarranjones@compuserve.com; *homepage* aeresbiomedical.com)

gions (CDRs) from the variable domains of rodent antibodies onto the framework regions (FRs) of human variable domains. Since then several different research groups, including the Antibody Engineering Research Group (Now AERES Biomedical Ltd.), have successfully developed and improved the methods for designing and constructing humanised antibodies via CDR grafting. Both the conservation of immunoglobulin structure and function across the species barrier as well as molecular modelling have contributed greatly to the success of the humanisation process. The wide base of knowledge acquired through each humanisation project has made it possible to accurately predict key residues important for the humanised antibody even before making the antibody.

To date, there are reports of well over 100 antibodies which have been humanised by CDR grafting and more than 40 of these have been, or are presently undergoing, clinical trials. In addition, at least 3 have already been approved for therapeutic use in the USA alone. These figures are a clear indication that humanisation via CDR grafting is now a clinically proven technology, which allows the utilisation of murine antigen binding domains within the framework of a human monoclonal antibody.

The relatively recent development of new technologies (e.g. combinatorial antibody phage display and transgenic mice expressing human antibodies) has now opened the door to the direct isolation of specific human monoclonal antibodies against novel targets. Even so, it is clear that where a well characterised, non-human antibody against an antigen already exists, then the humanisation of such an antibody to generate a therapeutic molecule may well offer one a greater expectation of success over these emerging technologies. This chapter describes the strategy used by AERES Biomedical Ltd. at the MRC Collaborative Centre to humanise murine monoclonal antibodies via CDR grafting.

Outline

The entire procedure followed by AERES for the humanisation of murine antibodies via CDR grafting is broadly outlined in Figure 1.

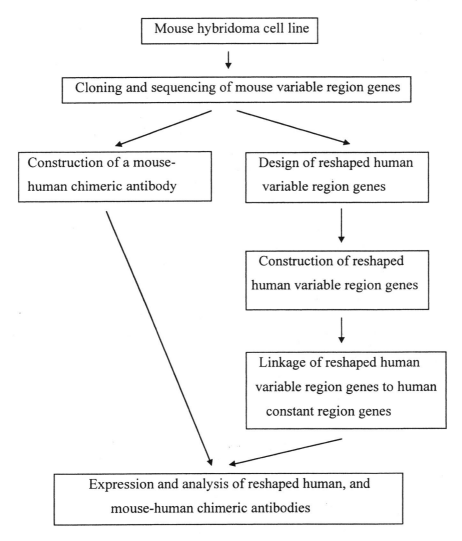

Fig. 1. MRC Collaborative Centre strategy for the humanisation of murine antibodies via CDR grafting.

Subprotocol 1
Isolation of Murine Variable Region Heavy and Light Chain Genes

The first step is to clone and sequence the cDNAs coding for the variable (V) domains of the mouse antibody to be humanised. The variable domain genes (V-genes) are cloned following a polymerase chain reaction (PCR) using specially designed primers that hybridise to the 5'-ends of the mouse constant regions and to the murine leader sequences upstream of the V regions. By using primers that hybridise to sequences external to the DNA sequences coding for the expressed variable domains, the full, accurate sequences of the mouse V-genes, are obtained. The cloned mouse leader and variable region sequences are used in the construction of chimeric mouse-human light and heavy chains and sometimes in the construction of the reshaped human light and heavy chains.

Materials

- RNA isolation kit (Stratagene, 2003345)

- First strand cDNA synthesis kit (Pharmacia, 27-9261-01)

- TA Cloning Kit (Invitrogen, K2000-01)

- PCR primers for cloning of variable region genes (Tables 1 and 2)

Table 1. PCR primers for cloning mouse kappa light chain variable regions

MKV1 (30mer)	ATGAAGATTGCCTGTTAGGCTGTTGGTGCTG
MKV2 (30mer)	ATGGAGWCAGACACACTCCTGYTATGGGTG
MKV3 (30mer)	ATGAGTGTGCTCACTCAGGTCCTGGSGTTG
MKV4 (33mer)	ATGAGGRCCCCTGCTCAGWTTYTTGGMWTCTTG
MKV5 (30mer)	ATGGATTTWCAGGTGCAGATTWTCAGCTTC
MKV6 (27mer)	ATGAGGTKCYYTGYTSAYCTYCTCTGRGG
MKV7 (31mer)	ATGGGCWTCAAAGATGGAGTCACAKWYYCWGG
MKV8 (25mer)	ATGTGGGGAYCTKTTTYCMMTTTTTCAATG
MKV9 (25mer)	ATGGTRTCCWCASCTCAGTTCCTTG
MKV10 (27mer)	ATGTATATATGTTTGTTGTCTATTTCT
MKV11 (28mer)	ATGGAAGCCCCAGCTCAGCTTCTCTTCC
MKC (20mer)	ACTGGATGGTGGGAAGATGG

MKV indicates primers that hybridise to leader sequences of mouse kappa chain V-genes. MKC indicates the primer that hybridises to the mouse kappa constant region gene.

Table 2. PCR primers for cloning mouse heavy chain variable regions

MHV1 (27mer)	ATGAAATGCAGCTGGGGCATSTTCTTC
MHV2 (26mer)	ATGGGATGGAGCTRTATCATSYTCTT
MHV3 (27mer)	ATGAAGWTGTGGTTAAACTGGGTTTTT
MHV4 (25mer)	ATGRACTTTGGGYTCAGCTTGRTTT
MHV5 (30mer)	ATGGGACTCCAGGCTTCAATTTAGTTTTCCTT
MHV6 (27mer)	ATGGCTTGTCYTTRGSGCTRCTCTTCTGC
MHV7 (26mer)	ATGGRATGGAGCKGGRGTCTTTMTCTT
MHV8 (23mer)	ATGAGAGTGCTGATTCTTTTGTG
MHV9 (30mer)	ATGGMTTGGGTGTGGAMCTTGCTTATTCCTG
MHV10 (27mer)	ATGGGCAGACTTACCATTCTCATTCCTG
MHV11 (28mer)	ATGGATTTTGGGCTGATTTTTTTTATTG
MHV12 (27mer)	ATGATGGTGTTAAGTCTTCTGTACCTG
MHCG1 (21mer)	CAGTGGATAGACAGATGGGGG
MHCG2a (21mer)	CAGTGGATAGACCGATGGGGG
MHCG2b (21mer)	CAGTGGATGAGCTGATGGGGG
MHCG3 (21mer)	CAAGGGATAGACAGATGGGGC

MHV indicates primers that hybridise to leader sequences of mouse heavy chain V-genes. MHCG indicates primers that hybridise to mouse gamma constant region genes.

Procedure

1. Grow the mouse hybridoma cell line in an appropriate culture medium to provide a total viable cell count of at least 10^8 cells.

2. Pellet the cells in a bench top centrifuge (250 g, 5 min). Gently resuspend the cells in 20 ml of PBS.

3. Add 100 µl of cells to 200 µl of PBS and 200 µl of trypan blue and mix gently. Pipette 10 µl of this mixture into a disposable cell-counting slide. Count the cells per square and determine the number of cells/ml according the manufacturer's instructions.

4. Pellet approximately 10^8 cells (250 g, 5 min).

5. Use the RNA isolation kit as described by the manufacturer to purify total RNA from the cells. The kit uses a guanidinium thiocyanate phenol-chloroform single-step extraction procedure (Chomczynski, P. et al. 1987).

6. Determine the quantity and quality of the total RNA by measuring the OD_{260} and OD_{280} and by testing 1 to 5 µg aliquots on a non-denaturing 1% (w/v) agarose gel in 1 x TBE buffer containing 0.5 µg/ml ethidium bromide. The concentration of RNA = OD_{260}x 40 µg/ml. The quality is satisfactory if OD_{260}: OD_{280}(1.9 and distinct bands are seen on the agarose gel representing 28S and 18S RNA - the 28S band being more intensely stained.

7. Following the manufacturer's instructions, use the first-strand cDNA synthesis kit to produce a single-stranded DNA copy of the hybridoma mRNA using the NotI-(dT)$_{18}$primer. Use 5 µg of total RNA in a 33 µl final reaction volume.

8. Following the reaction, heat at 90°C for 5 min to denature the RNA-cDNA duplex template and to inactivate the reverse transcriptase. Chill on ice.

9. Label eleven Gene Amp PCR reaction tubes MKV1 to MKV11. In each tube prepare a 100 µl PCR reaction mixture.

PCR reaction mixture
– 69.5 µl of sterile water
– 10.0 µl of 10 x PCR buffer II
– 6.0 µl of 25 mM $MgCl_2$
– 2.0 µl each of the 10 mM stock solutions of dNTPs
– 2.5 µl of 10 mM MKC primer
– 2.5 µl of one of the 10 mM MKV primers
– 1.0 µl of RNA-cDNA template mix
– Finally, add 0.5 µl of AmpliTaq DNA polymerase and overlay the reaction mix with 50 µl of mineral oil.

10. Prepare a similar series of reaction mixes to PCR clone the mouse heavy chain variable region gene using the twelve MHV primers and the appropriate MHC primer.

11. Place the reaction tubes into a DNA thermal cycler and cycle (after an initial melt at 94°C for 1.5 min) at 94°C for 1 min, 50°C for 1 min, and 72°C for 1 min over 25 cycles.

12. Follow the last cycle with a final extension step at 72°C for 10 min before cooling to 4°C. Use a ramp time of 2.5 min between the annealing (50°C) and extension (72°C) steps and a 30 sec ramp time between all other steps of the cycle.

13. Run a 10 µl aliquot from each PCR reaction on a 1% (w/v) agarose / 1 x TBE buffer gel, containing 0.5 µg/ml ethidium bromide, to determine which of the leader primers produces a PCR product. Positive PCR products will be about 420 to 500 bp in size.

14. For those PCR reactions that appear to produce full-length PCR products, repeat the procedure (steps 7 to 10) to obtain at least two independent PCR reactions that give full-length PCR products.

15. Directly clone a 1 µl aliquot of any potential PCR product into the pCRII™ vector provided by the TA Cloning kit as described in the manufacturer's instructions (This kit allows the direct cloning of PCR products without prior purification and takes advantage of the preference of AmpliTaq DNA polymerase to insert a 3'-overhanging thymidine (T) at each end of the PCR product.). Pipette out 10.0% (v/v), 1.0% (v/v) and 0.1% (v/v) aliquots of the transformed *E. coli* cells onto individual 90 mm diameter LB agar plates containing 50 µg/ml ampicillin and overlaid with 25 µl of the X-Gal stock solution. Incubate overnight at 37°C.

16. Identify positive colonies by PCR screening as described below.

Subprotocol 2
PCR Screening of Bacterial Colonies for Positive Clones

▨ ▨ Materials

- Techne PHC-3 DNA thermal cycler with Techne Hi-temp 96 multiwell plate (Techne, FPHC3MD)
- Techne Hi-Temp 96 microplate (Techne, FMW11)
- PCR screening primers that bracket the site of insertion (Table 3.)

Table 3. PCR screening primers

pCRII FOR (18mer)	CTAGATGCATGCTCGAGC
pCRII REV (21mer)	TACCGAGCTCGGATCCACTAG

■■ Procedure

1. Prepare a bulk solution of the PCR reaction mix (sufficient for 20 samples) as follows:

    PCR reaction mixture
    – 69.5 µl of sterile water
    – 10.0 µl of 10 x PCR buffer II
    – 2.0 µl each of 10 mM stock solutions of dNTPs
    – 2.5 µl of 10 mM MKC primer
    – 1.0 µl of one of the 10 mM MKV primers
    – 1.0 µl of mRNA-cDNA template mix
    – Finally add 0.5 µl of AmpliTaq DNA polymerase and overlay the reaction mix with 50 µl of mineral oil.

2. Dispense the above "mastermix" in 20 µl aliquots into the 96-well microplate.

3. Label twenty 30 ml universal containers and add to each 3 ml of LB media containing 50 µg/ml ampicillin.

4. Using an inoculating needle, gently "stab" an individual colony from a putative transformation mix grown overnight on selective agar plate. Then, stab the needle into one of the 20 µl aliquots of mastermix, making sure that the base of the microplate well is touched gently with the needle.

5. With the same needle, immediately inoculate a universal container (step 3) and incubate this culture overnight at 37°C and 300 rpm in the shaking incubator.

6. Prepare a negative control and a positive control where possible (As a positive control, add 1 µl of template DNA (5 ng/ µl) to 20 µl of mastermix.).

7. Overlay each of the inoculated PCR reactions with 2 drops of mineral oil per well.

8. Load the microplate into the Techne PHC-3 and cycle (after an initial melt at 94°C for 5 min) at 94°C for 1 min, 50°C for 1 min, and 72°C for 1 min over 25 cycles. Complete the PCR-reaction with a final extension step at 72°C for 10 min before cooling to 4°C. Use a ramp time of 30 sec between each step.

9. Run a 10 μl aliquot from each PCR reaction on a 1% (w/v) agarose / 1 x TBE buffer gel containing 0.5 μg/ml ethidium bromide, and estimate the size of any PCR products.

10. Sequence the DNA from at least two independently isolated positive clones of each variable region to identify possible errors introduced by PCR.

Comments

- In step 4, do not swirl the needle in the solution or rub it along the inside of the microplate well as excess template DNA can have a negative effect on the efficiency of a PCR reaction.

- Bands of 520 to 600 bp will be seen using the pCRII™ forward and reverse primers when the pCRII™ vector contains a variable region gene. A negative result (i.e. no insert in the pCRII™ vector) will produce an approximately 100 bp band. Treat any bands smaller than 500 bp with caution as they may be pseudogenes.

Subprotocol 3
Design of a reshaped Human Antibody

In most cases, a chimeric antibody is constructed and tested for its ability to bind to antigen prior to constructing a reshaped human antibody. There are two reasons for this:

- To confirm, in a functional assay, that the correct mouse variable regions have been cloned and sequenced.

- To create a valuable positive control for evaluating the reshaped human antibody.

In a chimeric antibody, no alterations are made to the protein domains that constitute the antigen binding site. In the vast majority of cases the chimeric antibody will bind to the antigen as well as the parent bivalent mouse antibody. In addition since the chimeric antibody is usually constructed with the same human constant regions that will be used in the reshaped human antibody, it is possible to compare directly the chimeric and reshaped human antibodies in antigen-binding assays that employ anti-human constant region antibody-enzyme conjugates for detection.

As a first step in the construction of the chimeric light and heavy chains, the cloned mouse leader-variable regions are modified at the 5'- and 3'-ends using PCR primers to create restriction enzyme sites for convenient insertion into the expression vectors, Kozak sequences for efficient eukaryotic translation (Kozak, M. 1987), and splice-donor sites for RNA splicing of the variable and constant regions. The adapted mouse light and heavy chain leader-variable regions are then inserted into vectors designed to express chimeric or reshaped human light and heavy chains in mammalian cells (Kettleborough, C.A. et al. 1991). These vectors contain the human cytomegalovirus (HCMV) enhancer and promoter for transcription, an appropriate human light or heavy chain constant region, a gene such as neo for selection of transformed cells, and the SV40 origin of replication for DNA replication in cos cells (Maeda, H. et al. 1991).

Analysis of mouse variable regions

Prior to actually beginning to design the reshaped human variable regions, it is important to carefully analyse the amino acid sequences of the mouse variable regions to identify the residues that are most critical in forming the antigen-binding site.

In addition to studying and comparing the primary amino acid sequences of the mouse variable regions, a structural model of the mouse variable regions is built based on homology to known protein structures, in particular, to the structures of other antibody variable regions. Molecular modelling is carried out using the AbM molecular modelling package supplied and utilised by Oxford Molecular Limited (OML). Antibody x-ray crystallographic structures available from the Brookhaven database along with some as yet unpublished immunoglobulin structures, were formatted to allow them to be used for modelling with AbM. As a first step in the modelling exercise, the framework regions (FRs) of the new variable regions are modelled on FRs from similar, structurally-solved immunoglobulin variable regions. Testing of AbM with known structures has shown that FR backbone homology is an important factor in the quality of any model, since the use of FR structures that poorly match a sequence being modelled can significantly and adversely affect the position and orientation of the CDR loops. Most of the CDRs of the new variable regions are modelled based on the canonical structures for CDRs (Chothia, C. et al. (1987), Chothia, C. et al. (1992), Chothia, C. et al. (1989), Tramontano, A. et al. (1990)). Testing of the performance of AbM predictions for known loop structures has shown that CDR loops which are created in this way are usually modelled very accurately, i.e. to within 1-1.5Å RMS deviation. Those CDRs which do not appear to belong to any known group of canonical structures, for example CDR3 of the heavy chain variable region, are modelled based on similar loop structures present in any structurally-solved protein. After adjusting the whole model for obvious steric clashes it is finally subjected to energy minimisation, as implemented in MACRO-MODEL, both to relieve unfavourable atomic contacts and to optimise van der Waals and electrostatic interactions.

Design of reshaped human antibody

The first step in the design process is to select the human light and heavy chain variable regions that will serve as templates for the design of the reshaped human variable regions. In most cases, the selected human light and heavy chains come from two different human antibodies. By not restricting the selection of human variable regions to variable regions that are paired in the same antibody, it is possible to obtain much better homologies between the mouse variable regions to be humanised and the human variable regions selected to serve as templates. In practice, the use of human variable regions from different antibodies as the basis of the design of a reshaped human antibody has not been a problem. This is probably because the packing of light and heavy chain variable regions is highly conserved.

The next step in the design process is to join the mouse CDRs to the FRs from the selected human variable regions. The preliminary amino acid sequences are then carefully analysed to judge whether or not they will recreate an antigen-binding site that mimics that present in the original mouse antibody. At this stage, the model of the mouse variable regions is particularly useful in evaluating the relative importance of each amino acid in the formation of the antigen-binding site. Within the FRs, each of amino acid differences between the mouse and the human sequences should be examined. In addition, any unusual amino acid sequences in the FRs of either the mouse or human sequences should be studied. Finally, any potential glycosylation sites in the FRs of either the mouse or human sequences should be identified and their possible influence on antigen binding considered. It is important to make the minimum number of changes in the human FRs. The goal is to achieve good binding to antigen while retaining human FRs that closely match the sequences from natural human antibodies.

Materials

- Genetics Computer Group (GCG) sequence analysis software package (Oxford Molecular Limited, UK)
- EMBL Data Library including the Kabat database (European Molecular Biology Laboratory, Heidelberg, Germany)
- Leeds database of protein sequences (Department of Biochemistry and Molecular Biology, University of Leeds, Leeds, UK)
- Molecular model of the mouse variable regions

Procedure

Analysis of the amino acid sequences of the mouse variable regions

1. Use the "SeqEd" program in the GCG package to create a series of files containing the consensus amino acid sequences of the subgroups of mouse and human light and heavy chain variable regions as defined by Kabat E.A. et al. (1991).

2. With the same program, create two files containing the amino acid sequences of the light chain and heavy chain variable regions of the mouse antibody.

3. Compare the amino acid sequences of the mouse variable regions to the mouse consensus sequences using the "Gap" program in the GCG package and identify the mouse subgroups to which the mouse variable regions belong.

4. Analyze the amino acid sequences of the mouse variable regions and locate the following features within them:
   - CDRs and FRs (Kabat, E.A. et al. 1991)
   - Residues that are part of the canonical sequences for loop structure (Table 4.)
   - Residues located at the V_L/V_H interface
   - Residues in the FRs that are unusual or unique for that position when compared to the consensus sequence for that mouse subgroup
   - Potential glycosylation sites

Table 4. Important residues for the maintenance of CDR loop conformation[a]

CDR Loop	Canonical Structure (Loop Size[b])	Residues Important for Loop Conformation[c,d] (Most Common Amino Acids)
L1	1 (10)[e]	**2**(I), 25(A), 30(V), 33(M, L) and **71**(Y)
	2 (11)	**2**(I), 25(A), 29(V, I), 33(L) and **71**(F, Y)
	2 (12)[f]	**2**(I, N), 25(A), 28(V, I), 33(L) and **71**(F, Y)
	3 (17)	**2**(I), 25(S), 27b(V, L), 33(L) and **71**(F)
	4 (15)[f]	**2**(I), 25(A), 27b(V), 33(M) and **71**(F)
	4 (16)	**2**(V, I), 25(S), 27b(I, L), 33(L) and **71**(F)
L2	1 (7)	**48**(I), 51(A, T,), 52(S, T) and **64**(G)
L3	1 (9)	90(Q, N, H) and 95(P)
	2 (9)[e]	90(Q) and 94(P)
	3 (8)	90(Q) and 95(P)
H1	1 (5)	**24**(A,V, G), **26**(G), **27**(F, Y), **29**(F), 34(M, W, I) and **94**(R, K)
	2 (6)	**24**(V, F), **26**(G), **27**(F, Y, G), **29**(I, L), 35(W, C) and **94**(R, H)
	3 (7)	**24**(G, F), **26**(G), **27**(G, F, D), **29**(L, I, V), 35a(W, V) and **94**(R, H)
H2	1 (16)	55(G, D) and **71**(V, K, R)
	2 (17)	52a(P, T, A), 55(G, S) and **71**(A, T, L)
	3 (17)	54(G, S, N) and **71**(R)
	4 (19)	54(S), 55(Y) and **71**(R)
	5 (18)	52a(Y), 54(K), 55(W) and **71**(P)

[a] This table summarises information presented in Chothia et al. 1987, Chothia et al. 1989, Tramontano et al. 1990, and Chothia et al. 1992.

[b] Loop size is the number of residues in the CDR loop as defined by Kabat et al. 1991.

[c] Numbering is according to Kabat et al. 1991. Note that in Chothia et al. 1989. L1 and H1 are numbered differently.

[d] The residue numbers printed in bold are located within the FRs of the variable region. Residues 26 to 30 of the heavy chain variable region are defined as FR residues by Kabat et al. 1991, however, structurally they are part of the H1 loop, Chothia et al. 1989.

[e] These canonical structures have been observed only in mouse antibodies and not in human antibodies.

[f] Approximately 25% of human and 20% of mouse sequences have 13 residues in canonical structure 2 or 14 residues in canonical structure 4. These minor variations in loop size result in changes at the tip of the L1 loop but do not significantly alter loop conformation Chothia et al 1989.

Selection of the human variable regions to serve as templates for reshaping

1. Compare the amino acid sequences of the mouse variable regions to the human consensus sequences and identify the most similar human subgroup for each mouse variable region.

2. Compare the amino acid sequences of the mouse variable regions to all human variable region sequences in the databases and, for each mouse variable region, identify the ten most similar human sequences. Use the "FastA" program in the GCG package.

3. Analyze the selected human variable regions for the following characteristics:
   - Percent similarity with the mouse variable region
   - Percent identity with the mouse variable region noting the location of regions of non-identity
   - Length of the CDRs in comparison to the mouse CDRs
   - Identity to the mouse sequence in the residues in the FRs that are part of the canonical sequences for loop structure (see Table 4)
   - Identity to the mouse sequence in the residues located at the V_L/V_H interface (see Table 5)
   - Residues in the FRs that are unusual or unique for a particular position when compared to the consensus sequence for that human subgroup (Kabat, E.A. et al., 1991)
   - Potential glycosylation sites

4. Make a subjective decision as to the most appropriate human sequences, one human light chain variable region sequence and one human heavy chain variable region sequence, to serve as templates for the design of a reshaped human antibody.

Design of the first versions of reshaped human variable regions

1. Write out the sequences of the proposed reshaped human variable regions with the CDRs from the mouse variable regions joined to the FRs from the selected human variable regions.

2. Highlight the amino acids in the human FRs that are different from those that were present in the mouse FRs. Use the structural model of the mouse variable regions to help evaluate the significance of

Table 5. Conserved residues found at the V_L/V_H interface[a]

Variable Region	Residue Position[b]	Number of sequences analysed	Number of different amino acids observed	Principal amino acids at this position (Number of occurrences[c])
V_L	34	1365	16	A(326), H(306), N(280)
	36	1324	7	Y(1057), F(143)
	38	1312	11	Q(1158)
	44[d]	1244	14	P(1060)
	46	1252	17	L(827)
	87	1222	8	Y(874), F319)
	89	1238	16	Q(654)
	91	1234	17	W(275), Y(216), G(209), S(169)
	96[d]	1034	20	L(220), Y(203), W(196), R(121)
	98[d]	1066	6	F(1058)
V_H	35	1459	19	H(378), N(356), S(287)
	37	1398	10	V(1212), I(151)
	39	1397	13	Q(1315)
	45[d]	1397	10	L(1362)
	47	1357	14	W(1252)
	91	1689	9	Y(1332), F(340)
	93	1683	16	A(1426)
	95	1451	20	D(285), G(212), S(187)
	100-100K[d,e]	1211	19	F(707), M(224)
	103[d]	1276	10	W(1251)

[a] The positions of interdomain residues were as defined by Chothia et al. 1985. The immunoglobulin sequences analysed were from the database of Kabat et al. 1991.

[b] Numbering is according to Kabat et al. 1991. The residue numbers printed in bold are located within the FRs of the variable region.

[c] Only those residues that displayed a frequency of occurrence of >10% are shown.

[d] One of six residues that constitute the core of the V_L/V_H interface as defined by Chothia et al. 1985.

[e] The residue that is immediately N-terminal to residue 101 in CDR3 is the amino acid that is part of the core of the V_L/V_H interface. The numbering of this residue varies.

the proposed amino acid changes. Consider conserving the following mouse residues:
- Residues that belong to canonical sequences for loop structure.
- Residues that the model suggests have a role in the supporting a CDR loop.
- Carefully examine buried residues and residues in the "Vernier" zone (Foote, J. et al. 1992).
- Carefully examine the H3 loop where there are no defined canonical structures to use for guidance.
- Residues that the model suggests are on the surface near the antigen-binding site.
- Residues are located at the V_L/V_H interface.

3. Examine the revised sequences and consider the following points:
- Removing any potential N-glycosylation sites within the human FRs and conserving, or removing, any potential N-glycosylation sites that were present in the mouse FRs. Use the model to predict whether potential glycosylation sites are located at positions that are on the surface and accessible and, therefore, likely to be used.
- Role of mouse residues that are atypical when compared to the consensus sequence for that subgroup of mouse variable regions. It is possible that atypical amino acid residues have been selected for, at certain positions, to improve binding to antigen.
- Location of human residues that are atypical when compared to the consensus sequence for that subgroup of human variable regions. It is possible that the potential immunogenicity of the reshaped human antibody will be increased if the human FRs contain unusual human sequences.

Design of the additional versions of reshaped human variable regions

1. When preliminary assays indicate that the first reshaped human antibody has a binding affinity equal or better than the mouse or chimeric antibody, additional versions may be made to further reduce the number of substitutions of mouse residues into the human FRs.

2. When preliminary assays indicate that the first reshaped human antibody has a poor binding affinity, determine whether one or both reshaped human variable regions is the cause. As already described, express the reshaped human light and heavy chains in all combinations

with the chimeric light and chains and determine the relative binding affinities of antibodies expressed.

3. Re-analyze the model and ask if any additional substitutions of mouse residues into the human FRs are required. Be particularly cautious about any amino acid differences between the mouse and human FRs that occur in buried residues.

4. Reconsider the removal or inclusion of any potential glycosylation sites.

Subprotocol 4
Construction of the Reshaped Human Variable Regions

Once the amino acid sequences of the reshaped human variable regions have been designed, it is necessary to decide how DNA sequences coding for these amino acid sequences will be constructed.

There are two fundamental approaches:

- Firstly, to take an existing DNA sequence coding for a variable region that is very similar to the newly designed reshaped human variable region and to modify the existing DNA sequence so that it will code for the newly designed reshaped human variable region. Modifications to existing sequences are usually carried out using PCR and specially designed synthetic oligonucleotide primers (Sato, K. et al. 1993).

- The second approach is to synthetically make a DNA sequence that will code for the newly designed reshaped human variable region, this is the strategy routinely used by AERES.

The method used is a variation on a gene assembly method first described by Stemmer, W.P.C. et al. (1995). Oligonucleotides of 40 bp in size are synthesised, these oligos collectively encode both strands of the desired V gene, arranged such that upon assembly complementary oligos will overlap by 20 bp.

The process involves four steps:

- Oligonucleotide synthesis

- Gene assembly

- Gene amplification

- Cloning
   An outline of the entire process can be seen in Figure 2.

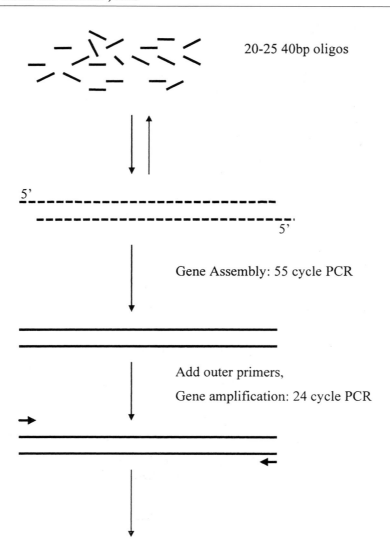

20-25 40bp oligos

Gene Assembly: 55 cycle PCR

Add outer primers,

Gene amplification: 24 cycle PCR

A-Tail and clone into pCRII™ vrctor

Fig. 2. Schematic representation of the gene synthesis method used to make humanised V-genes.

▪▪ Procedure

1. Using the "SeqEd" program in the GCG package, create a file of the DNA sequence of the human leader-variable region that has been selected as the template for reshaping.

2. Substitute the DNA sequences coding for the mouse CDRs for the DNA sequences coding for the human CDRs. Make minor modifications in the DNA sequences coding for the human FRs as required to make any amino acid changes that were specified in the design of the reshaped human variable region. Base the codon usage on the mouse variable region or refer to the database of immunoglobulin genes (Kabat, E.A. et al. 1991) and try to avoid rare codon usage.

3. In order to function in the expression vectors previously described (Maeda, H. et al. 1991), the reshaped human leader-variable region will require the following DNA sequences:
   – Kozak translation initiation sequence
   – Splice donor sequence
   – Restriction sites for cloning

4. Check the DNA sequence for the presence of unintentional splice donor sites that might interfere with RNA processing. Check that there are no internal restriction sites that will interfere with cloning into the expression vector. Remove undesirable DNA sequences by altering the codon usage.

5. The V-gene is divided into oligonucleotides 40 bp in size which overlap by 20 bp. Using the "Map" program in the GCG package, identify unique restriction sites already present in the DNA sequence and sites where unique restriction sites could be inserted without altering the amino acid sequence. Where possible, engineer a unique restriction site into each overlapping region.

6. Using the "Stemloop" program in the GCG package, identify potential stemloops within the DNA sequence. Remove any potential stemloops with a melting temperature of over $40°C$ by modifying the codon usage.

1. Prepare an Oligo Mix containing 1 µl of each assembly primer at 250 µM. 2. Prepare each gene assembly PCR reaction as follows:

PCR reaction mixture
   – 15.2 µl of sterile water
   – 2.0 µl of 10 x PCR buffer II

Design of the oligonucleotides and primers

PCR assembly of the oligonucleotides

- 0.6 µl of 100 mM $MgCl_2$
- 1.6 µl of the 10 mM stock solutions of dNTPs
- 0.25 µl of the Oligo Mix
- Overlay each reaction with 50 µl of mineral oil and place in a DNA thermal cycler.
- "Hot-start" the reactions by incubating for 5 min at 94°C, cooling to approximately 80°C, and adding 0.5 µl of Pfu DNA polymerase to each reaction.
- Cycle at 94°C for 30 sec and 52°C for 30 sec and 72°C for 1 min, for 55 cycles.

2. Immediately after the PCR reaction with Pfu polymerase add 1 µl of AmpliTaq polymerase, incubate at 72°C for 10 min.

3. Purify the resulting PCR products from using the Qiaquick PCR purification kit according to the manufacturer's instructions.

4. Evaluate the quality and quantity of the PCR products by measuring the OD_{260} and by testing aliquots on a 1.5% (w/v) agarose gel in 1 x TBE buffer containing 0.5 µg/ml ethidium bromide.

5. Clone the PCR product using the TA Cloning kit as already described.

6. PCR screen the transformants as previously described.

7. Sequence the DNA from several clones that have inserts of the correct size.

Comments

- If the DNA sequence of the selected human variable region is not available, use the DNA sequence from a closely related human variable region gene modifying the DNA sequence as necessary to obtain the required coding sequence.

- Pfu DNA polymerase has a proof reading facility which minimises insertion of mutations due to PCR error.

- The Pfu DNA polymerase is added last, i.e. after "hot start", because its 3'-5' exonuclease activity could potentially cause digestion of both primers and assembled gene products.

- The AmpliTaq DNA polymerase adds a poly A tail to the 3'-ends of the PCR product, thus facilitating cloning using the TA Cloning kit.

Subprotocol 5
Preliminary Expression and Analysis of Reshaped Human Antibodies

The reshaped human variable regions together with their leader sequences are cloned into mammalian cell vectors that already contain human constant regions. Each reshaped human variable region is linked via an intron to the desired human constant region. The expression vectors are identical or similar to the vectors that were described for the construction of chimeric light and heavy chains (Kettleborough, C.A. et al. (1991), Maeda, H. et al. (1991)).

The two mammalian cell expression vectors, one coding for the reshaped human light chain and one coding for the reshaped human heavy chain, are co-transfected into *cos* cells by electroporation using the method of Kettleborough et al. (1991), briefly, the DNA (10 µg each of the kappa light chain and heavy chain vectors) is added to a 0.8 ml aliquot of 1 x 10^7 cells/ml in PBS and pulsed at 1900 V, 25 µF capacitance using a Bio-Rad Gene Pulser apparatus. Following a 10 min recovery at RT the electroporated cells are added to 8 ml of DMEM containing 5% FCS and incubated for 72 hr in 5% CO_2 at 37°C.

The vectors will replicate in the cos cells and transiently express and secrete the reshaped human antibody. The medium is collected three days after transfection and analysed by ELISA to determine the approximate amount of antibody present (see following protocol). In most cases, cos cells will also have been transfected with the vectors that express the chimeric antibody. The chimeric and reshaped human antibodies as produced in the *cos* cells can be tested and compared for their relative abilities to bind to antigen. If purified antigen is available, the simplest approach is to use an ELISA format where the antigen is coated on the immunoplate and bound chimeric or reshaped human antibody is detected using a goat anti-human antibody-enzyme conjugate. However more accurate binding and pharmacokinetic measurements can be obtained using biosensors which can monitor antibody-antigen interaction in real time, one example of this technology being the Real-time BIA (BIAcore).

Materials

- Nunc-Immuno Plate MaxiSorp (Life Technologies, 43945A)

- Goat anti-human IgG antibody, Fc_γ fragment-specific (Jackson ImmunoResearch Laboratories Inc. via Stratech Scientific, 109-005-098)

- Human IgG1/kappa antibody (Sigma, I-3889)
- Goat anti-human kappa light chain peroxidase conjugate (Sigma, A-7164)
- K-BLUE substrate (Sky Bio, KB176)
- RED STOP solution (Sky Bio, RS20)
- Sample enzyme conjugate buffer (SEC buffer)

  SEC buffer recipe
  - 0.02% (v/v) TWEEN 20
  - 0.2% (w/v) BSA
  - In 1x PBS

■ ■ Procedure

1. Coat each well of a 96-well immunoplate with 100 µl aliquots of 0.4 µg/ml goat anti-human IgG antibody, diluted in SEC buffer, incubate overnight at 4°C.

2. Remove the excess coating solution and wash the plate three times with 200 µl/well of washing buffer (1 x PBS, 0.1% TWEEN).

3. Dispense 100 µl of SEC buffer into all wells except the wells in column 2, rows B to G.

4. Prepare a 1 µg/ml solution of the human IgG1/kappa antibody in SEC buffer to serve as a standard. Pipette 200 µl/well into the wells in column 2, rows B and C.

5. Centrifuge the medium from transfected cos cells (250g, 5 min) and save the supernatant.

6. Pipette 200 µl of the supernatant from the "no DNA" control (where cos cells were transfected in the absence of DNA) into the well in column 2, row D.

7. Pipette 200 µl/well of experimental supernatants into the wells in column 2, rows E, F, and G.

8. Mix the 200 µl aliquots in the wells of column 2, rows B to G, and then transfer 100 µl to the neighbouring wells in column 3. Continue to column 11 with a series of 2-fold dilutions of the standard, control, and experimental samples.

9. Incubate at 37°C for 1 hr. Rinse all the wells six times with 200 µl aliquots of washing buffer.

10. Dilute the goat anti-human kappa light chain peroxidase conjugate 5000-fold in SEC buffer and add 100 µl to each well. Repeat the incubation and washing steps (step 9).

11. Add 150 µl of K-BLUE substrate to each well, incubate in the dark at room temperature for 10 min.

12. Stop the reaction by adding 50 µl of RED STOP solution to each well. Read the optical density at 655 nm.

Comments

- The immunoplates prepared in step 1 maybe stored for up to 1 month at 4°C.

- To avoid possible aberrant results caused by "edge effects", the wells on the outside edges of the immunoplate are not used.

- The optimal dilution of any antibody or antibody-enzyme conjugate used should be determined for each lot.

Acknowledgements. The authors wish to acknowledge the valuable contributions of all previous workers in the Antibody Engineering Group at the MRC Collaborative Centre in developing and testing the methods outlined in this chapter, especially MRC Collaborative Centre scientists Dr. Katy Kettleborough, Dr. José Saldanha, Dr. Olivier Léger and Dr. Jon Chappel as well as visiting scientists Dr. Hiroshi Maeda of Kaketsuken, Dr. Frank Kolbinger of Novartis, and Dr. Masa Tsuchiya and Dr. Koh Sato of Chugai Pharmaceuticals.

References

Chomczynski P, and Sacchi N (1987) Anal Biochem 162: 156
Chothia C, and Lesk AM (1987) J Mol Bio196: 90
Chothia C, Lesk AM, Gherardi E, Tomlinson IM, Walter G, Marks JD, Llewelyn MB, and Winter G (1992) J Mol Bio 227: 799
Chothia C, Lesk AM, Tramontano A, Levitt M, Smith-Gill SJ, Air G, Sheriff S, Padlad EA, Davies A, Tulip WR, Colman PM, Spinelli S, Alzari PM, Poljak RJ (1989) Nature 34: 877
Chothia C, Novotny J, Bruccoleri R, and Karplus M (1985) J Mol Bio 186: 651
Emery SC, and Adair JR (1994) Exp Opin Invest Drugs 3: 241
Foote J, and Winter G (1992) J Mol Biol 224: 487

Gavel Y, von Heinje G (1990) Protein Eng 3: 43

Gooley AA, Classon BJ, Marschalek R, William KL (1991) Biochem Biophys Res Commun 178: 1194

Jones PT, Dear PH, Foote J, Neuberger M S, Winter G (1986) Nature 321: 522

Kabat E A, Wu TT, Perry HM, Gottesman KS, Foeller C (1991) Sequences of proteins of immunological interest, Fifth Edition, U.S. Department of Health and Human Services, U.S. Government Printing Office

Kettleborough CA, Saldanha J, Heath VJ, Morrison CJ, Bendig MM (1991) Protein Eng 4: 773

Kohler G, Milstein C 1975 Nature 256: 495

Kozak M (1987) J Mol Bio 196: 947

Maeda H, Matsushita S, Eda Y, Kimachi K, Tokiyoshi S, Bendig MM (1991) Hum Antibod Hybridomas 2: 124

Pisano A, Redmond JW, Williams KL, Gooley AA (1993) Glycobiology 3: 429

Riechmann L, Clark M, Waldmann H, Winter G (1988) Nature 332: 323

Sato K, Tsuchiya M, Saldanha J, Koishihara Y, Ohsugi Y, Kishimoto T, Bendig MM (1993) Cancer Research 53: 851

Stemmer WPC, (1995) Gene 146:49

Tramontano A, Chothia C, Lesk AM (1990) J Mol Biol 215:175

Antibody Engineering
to Improve Stability

Stabilization Strategies and Application of Recombinant Fvs and Fv Fusion Proteins

ULRICH BRINKMANN

Introduction

Antibodies are large (150 kDa) proteins with a small variable region (Fv) and a large constant region (Fc). The constant region harbors the effector function, e.g. for activation of complement, and confers stability to the antibody. The small amino-terminal Fv fragment (25 kDa), heterodimer of the variable region of the heavy and light chain of antibodies, is responsible for specific antigen binding. For many applications, particularly therapeutical applications which require tissue or tumor penetration of antigen binding proteins such as antibodies, size matters: often just the minimal region that is necessary for specific antigen binding is desired. For antibodies, this is the Fv region.

Unfortunately, most Fvs are inherently very instable without the accompanying Fc region and dissociate rapidly into their single domains, VH and VL. This results in a complete loss of the function of the Fv. Protein engineering and recombinant DNA cloning and expression techniques allow the production of small Fvs which have the domains of the heterodimer stabilized in various ways. Such recombinant Fvs may contain flexible inter- or intra-chain linkers, connectors peptides, or extra disulfide bonds, and they often fully retain the binding specificity and affinity of the corresponding antibody.

Here we describe and compare strategies that can be applied to generate functional stable recombinant Fvs and Fv fusion proteins with high binding affinity.

Ulrich Brinkmann, Epidauros Biotechnology, Pharmacogenetics Lab, Am Neuland 1, 82347 Bernried, Germany (*phone* +49-8158-998578; *fax* +49-8158-998548; *e-mail* uli@epidauros.com)

Strategies for stabilization of recombinant fv fragments

Various recombinant antibody fragments can be produced in *E. coli* using recombinant DNA expression techniques. Most similar to natural antibody fragments are recombinant Fab fragments, which contain the light chain and the heavy chain Fd fragment (VH and CH1), connected to each other via an interchain disulfide bond inbetween CL and CH1. This disulfide bond, combined with the remainder of the constant region confers great stability to the recombinant protein and therefore recombinant Fabs can be indistinguishable from Fabs that are derived by limited proteolysis (1). Fabs -stable as they are- still contain protein domains that are not necessary for specific antigen binding. For many applications it is desirable to have small molecules, e.g. the minimal region that is required for specific antigen binding. This smallest complete unit is the antibody Fv fragment, a heterodimer composed of the variable regions of the heavy and light chains, VH and VL (2-5). To generate Fvs, all of the constant region of an antibody is removed; this includes the CH and CL domains and the interchain disulfide connection which serve as the main stabilizers of the Fv region in the context of whole antibodies. Because of that, upon removal of CL and CH1, the remaining Fv is very instable and rapidly dissociates into VH and VL components (6,7). Stabilization strategies are necessary to prevent this VH-VL dissociation. Our lab has devised, applied and compared different strategies of Fv stabilization, not only on model antibodies, but also on a variety of recombinant Fvs and Fv fusion proteins that are produced clinical grade in large amounts and directly applied in experimental cancer therapy. The different methods for the generation of stable recombinant Fvs include peptide linkers, connectors, protein permutation, introduction of artificial interchain disulfide bonds and combinations of these methods.

Single-chain Fvs (scFv)

Fvs are heterodimers of the variable heavy chain (V_H) and the variable light chain domain (V_L). In contrast to Fabs in which the heterodimers are held together and stabilized by interchain disulfide bonds, in Fvs the VH and VL of Fvs are not covalently connected and because of that are unstable (6,7). Recombinant single-chain Fvs have their V_H and V_L domains covalently connected by a peptide linker (2-10), which fuses either the C-terminus of VH to the N-terminus of VL or the C-terminus of VL to the N-terminus of VH (Fig.1). Linker peptides that are commonly used are composed of 15 amino acids or longer and they are in most cases designed as a flexible chain. Frequently, a (gly4 ser)3 linker which is one of the first described "prototype" linkers, or variations of this linker type, are used to generate scFvs. Alternate linkers, e.g. with charged or hydrophobic residues have being tried. In many scFvs there seems to be little effect of these linker variations on affinity or stability of the Fv(8). However, the linker sequence can affect the yield of functional Fvs that are obtained from refolding of inclusion bodies (see below). Particularly hydrophobic residues in the linker may reduce the yield. Many scFvs retain the specificity and have similar affinity with the original antibody or the monovalent Fab fragment. In some cases, reduced affinity of scFvs when compared to the original antibody (Fab fragment) is observed (11). This may be caused by interference of the VH-VL joining peptide linker with antigen binding because the linker can come close to the binding CDR loops (Fig.1). In some instances reduced apparent affinity may also reflect a high degree of instability of the scFv (see disulfide-stabilized Fvs below).

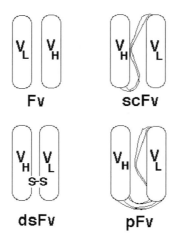

Fig. 1. Structure of stabilized Fvs. The portion of an antibody that is relevant for specific antigen binding is the Fv fragment. The Fv is composed of the variable region of heavy (VH) and light chain (VL) Fv fragments by themselves are instable because the VH/VL heterodimer has a strong tendency to dissociate. Recombinant single-chain Fvs have the VH and VL tethered by a peptide linker connecting either the C-terminus of VH to the N-terminus of VL or the C-terminus of VL to the N-terminus of VH. Disulfide-stabilized Fv fragments (dsFv) have no peptide linker but instead are stabilized by an interchain disulfide bond that connects VH and VL. DsFvs are more stable than scFvs. Base-loop connected permutated Fv. A peptide linker connects the C-terminus of VL to the N-terminus of VL and connecting peptides are set inbetween the base-loops at the bottom of VH and VL.

Disulfide-stabilized Fvs (dsFv)

ScFvs and scFv fusion proteins show in many cases a decent stability and production yields that are sufficient to obtain small amounts of protein for biochemical analyses and limited research applications. However, problems in the large scale production and clinical application of scFvs and scFv-immunotoxins, and in handling of scFvs, and stability analyses of scFvs under conditions that resemble clinical application (such as incubation in human serum at 37°C, Fig. 4) pointed towards the need of developing an alternative -and better- way to generate more stable recombinant Fv fragments. Another problem that is sometimes observed is that some scFvs lose part of their binding affinity once a linker connects the VH and VL domain (11). Disulfide-stabilized dsFvs differ from scFvs in that the VH/VL heterodimers are not stabilized by a linker peptide but instead by a disulfide bond that is placed inbetween structurally conserved framework residues of VL and VH (Fig.1). The VH and VL in those dsFvs are connected by a disulfide bond in a manner that does not interfere with the Fv structure and antigen binding. Because of that, the specificities and affinities of dsFvs are not reduced by the artificial stabilization method. The positions of cysteins to make recombinant disulfide-stabilized Fvs (dsFvs) were found by computer analyses. They are in the framework (FR) of VH and VL distant from the binding region. (12). Cysteins can be introduced either to connect FR 2 of VH at position 44 according to Kabat et al. (13) with FR 4 of VL at position 100, or FR 4 VH position 105 and FR 2 VL position 43. We have constructed many dsFvs using the VH44-VL100 positions which showed the same or better binding as the correspondent scFvs (11, 14-19). The alternative VH105-VL43 dsFv positions were also evaluated in one example and this dsFv form also retained the specificity and affinity of the original antibody (20). However, most dsFvs that were generated so far were constructed using the VH44-VL100 cystein combination. Subprotocol 1 details how these cystein positions can easily be identified (and mutated) in any Fv sequence. Because the cysteins are placed in structurally conserved regions, the disulfide-stabilization can be applied for Fvs of different species, such as for mouse or human antibodies. Because of the structural similarity of antibodies and T-cell receptors, the dsFv-design (using the VH44-VL100 positions) is also applicable for generation of stable recombinant TCR-Fvs (21). Fig.1 shows a comparison of the structures of scFv and dsFv of mAb B3, which is a murine antibody directed against a carcinoma related carbohydrate antigen (22, 23). The major advantage of dsFvs and dsFv fusion proteins over scFvs, in addition to sometimes improved affinity and their much higher production yields is their stability, which is greatly increased compared to scFv conterparts. The reason for increased stability is that dsFvs tend much less towards aggregation than scFvs and scFv fusion proteins because the conection between VH and VL is much tighter than in scFvs. In scFvs the domains are tethered by the linker and can upon separation either reassociate or aggregate with other molecules. In contrast, in dsFvs the domains are completely unable to separate. Direct comparisons of the stabilities of many different recombinant dsFvs/scFvs pairs which were generated from the same antibody consistantly demonstrated the superior stability of the disulfide-stabilization strategy. Most dsFvs and dsFv fusion proteins appear as "virtually indestructable", surviving weeks of incubation at 37°C without loss of activity, while the corresponding scFv proteins lose their activity within a couple of days, or in some cases within hours (Fig. 4).

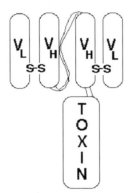

Fig. 2. Bivalent disulfide-stabilized Fvs (dsFv)2 and fusion proteins. Structure of a bivalent disulfide-stabilized Fv-toxin fusion protein. To generate this molecule, two VH domains, containing the VH44 cys mutation were linked to each other via a flexible peptide linker and then fused to a truncated derivative of Pseudomonas Exotoxin. The corresponding VL domains (with cys VL 100) were expressed in separate cultures and toxin fusion protein and the VL domains were combined in the refolding solution. Active bivalent dsFv-fusions can be purified with good yields. They are as stable as monovalent dsFvs but they display much higher apparent affinity than their monovalent conterparts. This translates into higher cytotoxic activity towards cultured human cancer cells.

Bivalent dsFv (dsFv2)

Fvs are the smallest complete units that confer specific antigen binding, and combined with disulfide-stabilization these protein modules can be obtained with the same extraordinary stability as whole antibodies. However there still is -in addition to size- one other significant difference between stabilized Fvs and antibodies: Fvs bind monovalent while antibodies bind multivalent (IgM with ten binding sites, IgGs with two). For some applications bi- or multivalent binding of recombinant Fvs is desirable. Bivalent binding to surface antigens can support antigen internalization (crosslinking effects), and bi- or multivalent antigen binding results in a greatly increased affinity (avidity effect). Because of the great advantages of disulfide-stabilization of Fvs over any other stabilization method, we developed a method to produce bivalent disulfide stabilized Fvs (bdsFv) and bdsFv fusion proteins. One example of the design and production of a stable bivalent disulfide-stabilized Fv and of a bdsFv -immunotoxin fusion proteins is the bdsFv of the anti erbB2 antibody e23. ErbB2 is a proteinous antigen (EGF-like receptor) that is overexpressed in many breast tumors, which is used as a target for immunotoxin therapy (9-11, see below). In a bdsFv, the VH and VL domains are linked, as in a classical dsFv monomer, by the VH44-VL100 interchain disulfide bond (Subprotocol 1). To obtain bivalency of the recombinant antibody, we have linked two VH (cys44) domains to each other via a flexible (gly4ser)x linker. Various lengths of the linker between the VH domains have been evaluated and we found that any length between (gly4ser)1 and (gly4ser)3 results in funktional bdsFvs. The connected VH domains can also be linked to fusion proteins, e.g. to truncated toxins to generate bivalent immunotoxins with high affinity (24, Fig.2). To produce bdsFvs, the connected VH domains and the corresponding VL (cys100) domains are produced in E. coli in separate fermentations and inclusion bodies are combined in the refolding reaction (see below). This results in high yields of stable bivalent disulfide stabilized Fv fragments and fusion proteins. The structure of a bivalent disulfide-stabilized Fv-toxin fusion protein is shown in Figure 2. Stability analyses showed that the bivalent dsFv has the same extraordinary stability as monovalent dsFvs. Binding studies that compared the monovalent and bivalent anti-erbB2 Fv showed that the bivalent molecule bound up to 20-fold stronger to antigen-positive human breast cancer cells than the monovalent disulfide-stabilized Fv (24). Consequently, immunotoxins containing the bdsFv with the higher affinity displayed a significantly improved cytotoxic activity towards cultured human cancer cells (24).

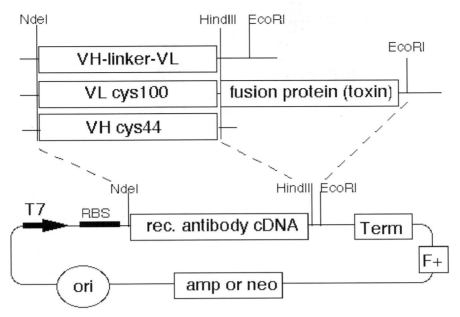

Fig. 3. Plasmids for expression of recombinant stabilized Fvs. Antibody coding cDNA fragments are obtained by reverse transcription of hybridoma mRNA or by phage display of scFv or dsFv libraries. PCR amplification of the VH and VL genes is performed with degenerate primers that are complementary to the beginning or end of VH and VL gene. Restriction sites for in-frame cloning into the immunotoxin vectors and the peptide linker sequence connecting the two variable regions are also introduced by that PCR for scFvs and the cysteins of dsFvs are generated by site directed mutagenesis. The positions and reading frames of the NdeI, HindIII and EcoRI sites into which the Fv-genes are cloned are identical in most of our expression vectors which facilitates switching of a given Fv in the vecoto and/or fusions to various fusion partners, e.g. toxin fragments. Single-chain scFvs, permutated pFvs and single-chain dsFvs and their fusion proteins are encoded by single plasmids. The components of dsFvs and bivalent dsFvs are in most cases generated with separate plasmids in separate fermentations.

Permutated Fvs (pFv)

One alternative, and rather experimental, method for Fv stabilization that we have devised is the generation of base-loop connected and permutated Fvs, which we term pFvs (for permutated Fv, 25). At the "bottom" of Fv fragments, i.e. distant from the binding region where the CL and CH1 domains have been removed, two loops of VH and VL come close to each other. To design base-loop connected pFvs, these two loops which in native antibodies connect the beta strand 3 with strand 3b in VH as well as

VL, were interconnected crosswise: The resulting pFv molecule contains the N-terminal half of VH - up to beta strand 3- connected at the base loop with the C-terminal half of VL (from beta strand 3b). The VL molecule is then circularly permutated by introducing a flexible peptide linker to connect the C-terminus with the N-terminus, and the N-terminal half of VL is then fused after beta strand 3 at the base loop with the strand 3b and the C-terminal remainder of the VH domain. This procedure generates an Fv molecule whose linear sequence appears to be mixed up by crosswise connection of half-domains. Nevertheless, this molecule can still fold into a functional Fv, although the yields in the production of such molecules are quite low. The reason for that is that the pathway by which such molecules fold is different from the pathway that nature has optimized for antibodies (pFvs have to fold in one reaction while Fvs are generated by separate domain folding and subsequent heterodimer association). However -as shown for the Fv fragment of the anti-Tac antibody (25, see also Table 1), pFvs can be recombinantly produced in decent yields, and they can retain full binding affinity, as shown by direct binding assays and Surface Plasmon Resonance (BIACORE) assays (26). A scheme of the structure of the pFv fragment is shown in Figure 1.

Single-chain disulfide-stabilized Fv (sdsFv)

DsFvs show many advantages over scFvs, and because of that for most applications it is desirable to replace the less stable scFv with dsFv molecules. However, the early steps in the production of the "classical" dsFv are somewhat more elaborate than the production of scFvs because the production of two protein components is required. To make these, we use two separate expression plasmids, two *E. coli* cultures and two separate fermentations. Although it is possible to construct dsFv "operons" for simultanous expression of VH and VL, we prefer the separate fermentation and separate purification of inclusion bodies of VH and VL. One important parameter that has to be carefully adjusted in our refolding procedure (see below) is that both protein components are present in the refolding assay in approximately equimolar concentrations. Separate preparations of inclusion bodies allow both components to be added in controllable amounts. This adjustment of the relative protein concentrations of VH and VL is not possible if both proteins are expressed in an operon simultanously and a mixture is purified from one fermentation. The construction of single-chain disulfide-stabilized (scdsFv) Fv molecules combines the advantages of both systems: high stability and higher yields associated with dsFv with the more economical (single-pot) fermentation techniques of scFvs (27, 28). ScdsFvs are single-chain molecules with a peptide linker joining VH and VL. This enables the production of both domains in one fermentation, with equimolar ratios of both domains. In addition, scdsFvs also contain the cystein mutations in VH and VL (VH44-VL100, see Subprotocol 1), which are capable to form a covalent disulfide between the domains. The stabilizing disulfide in scdsFvs can form effectively despite the fact that the peptide linker is still attached to the Fv (27). Obviously, the linker does not interfere with disulfide formation. One further conclusion that we are able to draw is that the peptide linker, at least with the (gly4-ser)3 sequence, does not destabilize Fvs by itself. The scdsFv displayed the same high stability as well as the same affinity as the dsFv without the linker (27, 28).

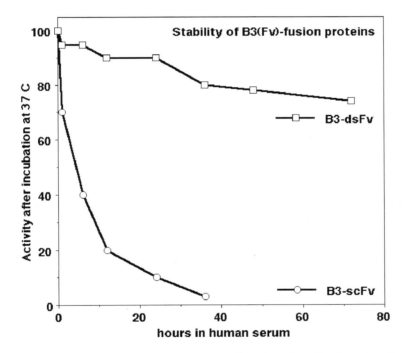

Fig. 4. Differences in Fv stability directly affect clinical applications. Equal amounts of immunotoxins containing a recombinant Fv region of the anti-LeY tumor associated carbohydrate antibody B3 were incubated at 37°C in human serum. The stability of such fusion proteins, which is determined by the stability of te Fv fragment, can be assayed by the anti-tumor activity of the immunotoxin. The comparison of the scFv and dsFv shows dramatic differences: The single-chainFv loses its activity almost completely after a couple of hours while the corresponding dsFv retains its activity for days. The stability of proteinous drugs in human serum at 37°C resembles the conditions in the human body, and this is relevant for predicting the behaviour of proteins in cancer therapy.

Subprotocol 1
Design, Construction and Expression of Disulfide-Stabilized Fv-Fragments

■■ Procedure

1. VH and VL coding cDNAs are obtained by PCR-based techniques, Determine the nucleotide sequences, the deduced protein sequences, and assign the framework and CDR-regions according to the Kabat scheme. These are standard techniques that are described elsewhere.

2. Design of the Cystein-mutation in VH:
   Align the protein sequence of the framework region 2 of your VH to one or more of the sequences below, allowing mismatches:

   ```
   W  V  R  Q  T  P  E  K  R  L  E  W  V  A
   W  V  R  Q  P  P  G  K  R  L  E  W  I  A
   W  V  K  Q  S  H  G  K  N  L  E  W  I  G
   F  V  K  Q  R  P  G  Q  G  L  E  W  I  G
   :  :  :  :  :  :  :  :  C  :  :  :  :  :
   ```

   The underlined amino acid represents position Kabat-VH44 and needs to be replaced by a cystein.

   Note: Note that the Kabat position refers to a consensus sequence. VH44 does not necessarily represent position 44 in your VH sequence.

3. Design of the Cystein-mutation in VL:
   Align the protein sequence of the framework region 4 of your VL to one or more of the sequences below, allowing some mismatches:

   ```
   P  L  T  F  G  A  G  T  K
   P  F  T  F  G  S  G  T  K
   P  P  T  F  G  G  G  S  K
   P  L  T  F  G  S  G  T  K
   :  :  :  :  :  C  :  :  :
   ```

The underlined amino acid represents position Kabat-VL100 and needs to be replaced by a cystein. Note that the Kabat position refers to a consensus sequence. VL100 does not necessarily represent position 100 in your VL sequence.

4. Use site-directed mutagenesis (standard techniques e.g. PCR or Kunkel mutagenesis) with primers that replace the VH44 and VL100 codons with cystein codons and clone the resulting mutated antibody chains into separate expression vectors.

   Note: pET vectors, or pET derived vectors are suitable for Fv expression because they combine the possibility of inducible very high expression (T7 system) with the possibility to directly mutate Fv sequences (single-stranded phage replication origin for Kunkel mutagenesis).

5. In the design of dsFv fusion proteins it may be of advantage to choose the VH domain as protein fusion partner and express the VL domain separately.

   Note: VH-domains, without VL have a strong tendency to aggregate. This prevents the co-purification of single-domain fusion proteins after refolding. VL-fusion proteins are frequently soluble after refolding and are often difficult to separate from the desired dsFv-fusion proteins.

Expression of stabilized Fvs and Fv fusion proteins

Recombinant stabilized Fv fragments are frequently produced in bacterial *E. coli* cultures that contain inducible expression plasmids (29, 30). Although in principle it is possible to produce Fvs in soluble form, e.g. by secretion into the periplasm or into the growth medium, not only in *E. coli* but also in other bacteria, in yeast, and even in cultured mammalian cells, we find it favourable to produce recombinant stabilized Fvs in bacterial inclusion bodies. Advantages of this strategy are well established fermentation techniques and very high yields, as well as the ease of purification of recombinant protein and removal of large amounts of bacterial lipopolysaccharide in a very early step of the procedure. Removal of bacterial lipopolysaccharide, (endotoxin) is especially important in the production of Fvs for clinical applications (31).

The composition of expression vectors to produce recombinant Fvs and Fv-fusion proteins is shown in Figure 3. The backbone vector for expression of single chain Fvs, permutated Fvs, or for the components of disulfide-stabilized Fvs and fusion proteins is a high copy number plasmid derived from vectors made by Studier (29). These plasmids contain a T7 promoter, translation initiation signals and a transcription terminator as well as an F+ replication origin to generate single stranded DNA upon infection with (M13) helper phage to allow site-directed mutagenesis of the inserted Fv sequences. The antibody coding region is obtained by PCR techniques, either from hybridoma cDNA or by phage display techniques, e.g. from mouse spleen libraries (32, 33). The VH and VL genes are inserted into the plasmid using a NdeI site which places it after a translation initiation signal and provides the translation start codon, and either a HindIII or EcoRI site. These restriction sites are introduced by PCR at the beginning or end of the VH and/or VL coding region. We do not find any influence of the orientation of the Fv fragment i.e. whether VH is positioned at the C- or N-terminus of the Fv. However, the orientation of the Fv may influence the yield of functional scFv- and dsFv-fusion proteins that can be obtained from refolding (8). The orientation, VH-VL or VL-VH, that results in the high expression is dependent on the particular antibody and has to be found empirically. Plasmids for production of Fv-fusion proteins have the Fv gene 3'-end connected at the HindIII site in frame (over a short connecting peptide, 34) to the fusion partner, e.g. a truncated toxin gene, growth factors, TNF, or other genes.

To produce dsFvs, two expression plasmids are required because we express the components of these molecules separately (11, 14-17). The cystein mutations that are required for intermolecular disulfide formation in dsFvs can be identified using Subprotocol 1, and are introduced either by PCR (e.g. pos. cys VL100) or by site directed mutagenesis (e.g. cys VH 44) using overlap extension PCR or Kunkel mutagenesis. The plasmids for expression of immunotoxins allow high expression levels of recombinant protein in IPTG induced cultures of *E. coli* BL21/DE3 (which harbors the T7 pol gene under control of the lacUV5 promoter for regulated T7 promoter expression (29). Generally, we obtain high cell densities in fermentations and the plasmids can be stably maintained in those cultures; not only in lab scale but also in large scale for the preparation of clinical grade protein.

Subprotocol 2
Refolding and Purification of Fv Fragments
from Bacterial Inclusion Bodies

Upon induction of recombinant protein expression, the Fvs (or their components) are produced in high yields. They accumulate like many other recombinant proteins in insoluble intracellular inclusion bodies (IB's) which are large aggregates of incorrectly folded inactive recombinant proteins. Inclusion bodies are microscopically visible, and can be separated from most other bacterial proteins by simple cell lysis and precipitation (e.g.centrifugation) techniques (30, 35-37). Subprotocol 2 provides a detailed method that we have used to obtain very clean IB preparations from many different Fvs. IBs contain almost pure recombinant immunotoxins. During the initial steps of purification from other bacterial proteins, also a great deal of the bacterial lipopolysaccharides becomes removed. This is an important advantage especially in the production of proteins for therapeutic applications in humans.

On the other hand, recombinant proteins from inclusion bodies are wrongly folded and inactive, and therefore need to be refolded to be active. Large amounts of highly active and pure Fvs, and particularly of Fv-fusion proteins can be obtained by an adaptation of a refolding protocol that was originally developed (by Boehringer Mannheim) for the industrial production of recombinant plasminogen activator and for antibody Fab fragments produced in *E. coli* (1, 30, 35-37). This protocol yields immunotoxins in sufficient quantity and excellent quality not only in lab scale for preclinical studies; but we have also successfully upscaled these refolding and purification principles to make large amounts of clinical grade material (30).

IB's are separated from soluble and membranous *E. coli* proteins by cell lysis with lysozyme and detergent (TritonX100), separation of the insoluble IBs from the soluble proteins by centrifugation and repeated washing of the IBs to remove remaining soluble proteins and detergent (see Subprotocol 2). IB's can be composed of nearly pure recombinant proteins, and IB preparations contain normally more than 70%, and if done carefully up to more than 90% recombinant protein. The detergent washing during preparation of IBs also effectively reduces the level of bacterial LPS (endotoxin) in the recombinant protein. Our refolding and purification protocol completely solubilizes and reduces IBs in Guanidine/HCl and DTE. This is followed by dilution in refolding buffer containing redox shuffling and aggregation preventing additives (GSH/GSSG and arginine, 35-37). Subprotocol 2 describes this procedure in detail. The yield of properly folded protein ranges between 10% to 15% of the recombinant Fv input protein.

Because of instability which many scFvs display, significant losses of scFvs or scFv fusion proteins are frequently observed in the subsequent purification procedure (30). The purification of properly folded antibody fragments consists of an initial removal of large aggregates, which precipitate once the arginine is taken out of the refolding solution, e.g. by dialysis. The refolding solution is then subjected to ion exchange (Q-sepharose and MonoQ or S-sepharose/MonoS, dependent on the IP of the Fv fragment or fusion protein) and size exclusion (TSK3000) chromatography (Subprotocol 2). This procedure not only separates monomeric active proteins from aggregates and wrongly folded and contaminanting protein, but it also removes effectively bacterial LPS that may still be present after the inclusion body preparation. The losses of scFvs during purification are mainly because scFvs and scFv fusion proteins, even in proper conformation, have a strong tendency to aggregate.

The production and refolding of scdsFvs and of base-loop connected pFvs follows the same scheme as described for scFvs. Differences are that scdsFvs have either the same or higher yields than scFvs, because they are more stable than scFvs. In contrast, the yield of properly folded pFv is much lower compared to the refolding yields of scFvs (25). The pathway by which a pFv folds is different from the natural pathway for antibodies (separate domain folding followed by heterodimer association), thus causing refolding problems that are reflected in the low yield of functional pFvs (25). The production and refolding of dsFvs and bivalent dsFvs and dsFv fusion proteins follows principally the same scheme as described for scFv-toxins, except that these molecules are composed of two or more components, each of which is produced in separate cultures and separately isolated in IBs. The solubilized and reduced IBs are combined in the refolding solution where the components effectively associate into heterodimers. DsFv-immunotoxins require a final oxidation step after completion of folding in the redox buffer, to oxidize the interchain disulfide bond, or they have to be refolded at alkaline pH (e.g. pH9.5, Subprotocol 2). The purification of dsFvs and fusion proteins is the same as for scFvs Because of their improved stability they show only marginal losses in the purification procedure that follows refolding (15-20). This results in higher yields. More than 20% of the total recombinant input protein can be recovered as active dsFv after refolding and purification.

▪▪ Procedure

Preparation of intracellular bacterial inclusion bodies containing Fvs and Fv-fusion proteins

1. Suspend the cell paste of 5000 "A600-units" (ca. 10 g wet weight) in 180 ml TES buffer (50 mMTris/Cl pH7.4, 20 mM EDTA, 100 mM NaCl) directly in a 250 ml centrifuge tube.
   We measure *E. coli* cell paste with "A600-units" (A600 x ml of culture) because the wet weight of harvested cultures varies significantly dependent on which centrifuge is used for how long. 5000 A600-units equals cells from 500 ml of a culture harvested at A600 of 10 (fermentor) or 1 liter of a culture harvested at A600 of 5 (Shaker flask).

2. Add 6 ml of a 10 mg/ml Lysozyme solution and incubate on a rocker for 60 min, then add 20 ml of a 25% TritonX100 solution (in water) and incubate on a rocker for 60-120 min.

   Note: Dissolving TritonX100 in water to a 25% solution takes time, so prepare ahead!

3. Shear DNA shortly (30 sec) with homogenizer (e.g. ultraturrax, large tip). Do not shear completely, leave it slightly viscous.

   Note: This prevents the pelleting of membrane fragments in the following centrifugation step.

4. Precipitate IBs by centrifugation for 45 min at 13.000 rpm, GSA rotor.

5. Resuspend pellet in 200 ml TE50/20 buffer (50 mMTris/Cl pH7.4, 20 mM EDTA) with homogenizer, add TritonX100 to 1% final, shake well and centrifuge again.

6. Repeat step 5.

7. Resuspend pellet in 200 ml TE50/20 buffer with homogenizer, centrifuge again.

8. Repeat step 7.

9. Transfer (scrape) pellet to a 50 ml centrifuge tube and save aliquot for SDS-PAGE analysis.

   Note: Protein assays with IB suspensions do not work!

10. IB pellets can be stored frozen at -70°C.

Refolding and Purification of Fvs and Fv-fusion proteins

1. Solubilize IB pellet in a 50 ml (SS34) centrifuge tube in solubilization buffer (6 M Guanidinium Chloride, 0.1 M Tris pH8, 2 mM EDTA, 5-7 ml, see below) with a homogenizer, small tip, and incubate overnight at room temperature.

2. Centrifuge at 15.000 rpm for 30 min.

   Note: The pellet is junk.

3. Determine protein concentration in supernatant (Bradford Assay works with GuaCl, reference proteins also in solubilization buffer!) and adjust with solubilization buffer to a final concentration of 10 mg/ml.
   The total volume needed to adjust to 10 mg/ml can be determined only after initial solubilization and protein assays and is dependent on the degree of expression of the recombinant protein. Usually, 5000 A600-units of cells that express protein with T7-vectors contain 100-300 mg solubilized inclusion body protein.

4. Add DTE to a final concentration of 10 mg/ml to 10 ml solubilized IBs, close tube.

5. Incubate for more than 3 hrs at room temperature. This completely reduces the disulfide bonds in the solubilized inclusion body protein.

6. Dilute the solubilized and reduced inclusion body protein solution 1:100 in refolding buffer:
   0.1 M Tris pH8, 2 mM EDTA, 0.5 M L-arginine/Cl (105 g/l), 0.9 mM GSSG (oxidized glutathione), pH8.) precooled to 10°C. Incubate for more than 20 hrs at 10°C (4°C is also ok for most proteins).

   Note: Arginine prevents aggregation, oxidized glutathione, combined with DTE from solub. solution results in redox-shuffling buffer, favours and stabilizes correct disulfides but reduces most incorrect disulfide bonds.

7. This is the basic protocol for single-chain Fvs- and Fv-fusion proteins. Modifications are necessary for disulfide-stabilized Fvs and fusion proteins that are refolded from two components: Inclusion body proteins of the two components are prepared, solubilized and reduced with DTE separately, and then combined in equimolar ratios to a final concentration of 10 mg/ml. We found two possibilities of "fixing" the interchain disulfide bond of the dsFv after refolding in the redox-shuffling solution. One is to add, after more than 20 hrs of refolding in

regular refolding buffer, a 10 fold excess of oxidized GSSG to the solution, followed by an additional incubation of more than 3 hrs. This is expensive (GSSG). Another possibility is to perform refolding in the arginine- redox-shuffling buffer not at pH 8 but at pH 9.5.

8. Change buffer by dialysis with standard membranes or cross flow (e.g. with amicon cartridges which works faster) against 20 mM Tris/Cl pH 7.4, 100 mM Urea at 4°C with several buffer changes until the conductivity is less than 3 mOhms.

    Note: Removal of Arg results in precipitation of many wrongly folded proteins, protein aggregates and contaminating proteins.This is often visible as white precipitate.

9. Centrifuge at 15.000 rpm for 30 min to remove the precipitates.

    Note: 45u filtration works also but the membranes become frequently clogged.

10. Concentrate supernatant to facilitate further purification (optional).

    Note: Whether protein concentration is useful depends on next steps to purify recombinant Fvs or Fv-fusion proteins. If affinity purifications available, concentration is often not necessary. For ion-exchange chromatography, concentration prior to coloumn loading reduces the loading times which is of advantage for the purification of instable Fvs.

11. Fvs and Fv-fusion proteins can be purified on ion exchange chromatography coloums, Q- or S- sepharose or MonoQ/S, dependent on the PI of the recombinant protein.

    Note: Do not overload coloums because the solution often still contains nucleic acids which bind to and compete for protein binding on ion exchange colums.

Applications

Importance of fv stabilization in clinical application of fvs and fv fusion proteins: recombinant immunotoxins

Recombinant immunotoxins are composed of stabilized Fvs of cancer specific antibodies which are recombinantly fused to cytotoxic proteins. The toxin moiety of the immunotoxin – we use a truncated form of Pseudomonas exotoxin – is modified in a manner which excludes its binding to normal human cells, but it retains all other functions of cytotoxicity (38-42). The recombinant stabilized Fv fragment, fused to the toxin, directs the immunotoxin specifically to cancer cells. Upon binding it becomes internalized and translocates through the membrane into the cytosol where it ADP-ribostranslates elongation factor EF2 at a modified histidine (diphthamide) residue. This inactivates EF2 and subsequently kills the cells by inhibition of protein synthesis and induction of apoptosis (43, 44). Because specific binding to cells is an absolutely required first step in this cytotoxic process, cells which do not carry the target antigen are not affected by tumor-specific immunotoxins. Our lab at the National Cancer Institute (NIH) has produced many different recombinant immunotoxins that are directed at various tumors. These show excellent activity and specificity towards cultured cancer cells and also in animal tumor models. Ongoing clinical trials indicate that in some cases the preclinical results can also be translated into promising applications of immunotoxins in cancer therapy (45-49). A list of recombinant Fv-toxin fusion proteins that are currently being evaluated in clinical trials for potential applications in cancer therapy is presented in Table 1.

Many factors influence the in vivo anti-tumor activity of recombinant immunotoxins. Among them are considerations of which types of cancer and at what stages may be the best targets for immunotoxin therapy. Important is also the tumor specificity of the antigen that is targeted by the recombinant antibody, the affinity, and the ability of the therapeutic proteins to enter and penetrate into tissues and tumors, which in turn is dependent on the size of the protein (hence the desire to generate recombinant antibody fragments as small as possible). One very important factor is the stability of immunotoxins. The toxin moiety of immunotoxins is very stable. Because of that, the in vivo stability of the recombinant Fv is the limiting factor that determines the stability, and thereby the usefulness of immunotoxins in therapeutical applications. Not surprisingly, the strategy that is applied for the generation of recombinant Fvs has a great and direct impact on the clinical efficiacy of immunotoxins. For therapeutical application of Fvs, one obvious and relevant assay to evaluate stability is the exposure to human serum at 37°C. In many examples, dramatic differences can be seen when the stability of scFvs and disulfide-stabilized Fvs is compared under such conditions. Fiure. 4 shows a typical example of a single-chain and disulfide-stabilized Fv fragment of an antibody directed at a cancer associated carbohydrate antigen (LeY). The scFv is significantly less stable than the dsFv counterpart. Consequently, in a clinical trial that was initiated with the scFv-immunotoxin, we had problems in reaching efficient concentrations of active immunotoxin in the tumor for a time period that is sufficient to kill most tumor cells. This prevented the observation of significant therapeutic anti-tumor activity, although some activity could still be observed in some patients. Another clinical trial with

Table 1. Specificity and application of some selected recombinant immunotoxins in cancer therapy.
Many recombinant Fv-containing immunotoxins have been designed and produced in our Lab, some with excellent specificity and activity towards human tumor cells. Only those immunotoxins with the best specific activity and lowest toxicity in preclinical development are further subjected to experimental cancer therapy in clinical trials which are performed at the National Cancer Institute. *LMB-1 is a first generation immunotoxin, a chemical toxin conjugate containing a whole IgG. This molecule is difficult to produce and shows low tumor penetration. The other immunotoxins of this table are recombinant Fv fusion proteins. All agents carry a truncated detoxified fragment of Pseudomonas Exotoxin A as cytotoxic moiety.

antibody	antigen	cancer	status	results	
B3 IgG*	LeY carbo-hydrate	breast, colon	phase I/II compl.	anti-tumor activity	too large for penetr.
B3 scFv	LeY carbo-hydrate	breast, colon	phase I/II compl.	minor responses	too instable
B3 dsFv	LeY carbo-hydrate	breast, colon	phase I/II pend.	pending	stable
anti-Tac scFv	IL-2 receptor p55	Leukemia, GVH	phase I/II pend.	anti tumor activity	promising
anti-Tac dsFv	IL-2 receptor p55	Leukemia, GVH	phase I/II pend.	pending	promising

the corresponding disulfide-stabilized Fv-immunotoxin, which shows a greatly increased stability, is currently pending (LMB9 in table 1). Other clinical trials that are pending at the National Cancer Institute include disulfide-stabilized immunotoxins which target the TAC antigen, a proteinous antigen on activated T cells which comprises the alpha subunit of IL2 receptor. In these trials, excellent activity of immunotoxins can be observed in the treatment of various leukemias (47-49).

Conclusions

We have designed, applied and evaluated various methods of Fv stabilization on many different Fvs derived from cancer specific antibodies, some of which are now applied in experimental cancer therapy. Our experience shows that among the different stabilization strategies, disulfide stabilization of Fvs generates the best results, not only in terms of stability but also in the ease and high yields of production of Fvs by refolding from bacterial inclusion bodies. And in applications where very high affinity is desired, bivalent dsFv molecules can be used which have the same stability as the monovalent dsFvs.

Acknowledgements. The described recombinant stabilized Fvs and Fv-fusion proteins were designed, produced and applied in the Laboratory of Molecular Biology at the National Cancer Institute, Division of Basic Sciences, National Institutes of Health. I thank Ira Pastan and all members of the Laboratory of Molecular Biology for their great support.

References

1. Buchner, J. and Rudolph, R. Renaturation, purification and characterization of re-combinant Fab fragments produced in *E. coli*. Biotechnology 9, 157-162, 1991
2. Bird RE., Hardman, KD., Jacobson, JW., Johnson S., Kaufman BM., Lee, SM., Pope, SH., Riordan, GS., and Whitlow, M. Single-chain antigen binding proteins. Science 242, 423-426, 1988
3. Johnson S. and Bird RE. Construction of single-chain Fv derivatives of monoclonal antibodies and their production in Escherichia coli. Meth. Enzymol. 203, 88-98, 1991.
4. Huston, JS., Levinson, D., Mudgett-Hunter, M., Tai, MS., Novotny, J., Margulies, MN., Ridge, RJ., Bruccoleri, RE., Haber, E. C, Crea, R. and Oppermann, H.. Protein engineering of antibody binding sites: recovery of specific activity in an anti-di-goxin single-chain Fv analogue produced in Escherichia coli. Proc. Natl. Acad. Sci USA 85, 5879-5883, 1988.
5. Chaudhary, VK., Queen, C., Junghans, RP., Waldman TA., Fitzgerald, D.J. and Pastan, I. A recombinant immunotoxin consisting of two antibody variable domains fused to pseudomonas exotoxin. Nature 339, 394-397, 1989.
6. Glockshuber R; Malia M; Pfitzinger I; Pluckthun A. A comparison of strategies to stabilize immunoglobulin Fv-fragments. Biochemistry 1990 Feb 13;29(6):1362-7
7. Webber K.O., Reiter,Y., Brinkmann, U., Kreitman, RJ. and Pastan,I. Preparation and characterization of a disulfide-stabilized Fv fragment of the anti-Tac antibody: Comparison with its single-chain analog. Mol. Immunol. 32, 249-258, 1995
8. Batra, JK., Fitzgerald, D., Gately, M., Chaudhary, VK and Pastan, I. Anti-Tac(Fv)-PE40, a single-chain antibody Pseudomonas exotoxin fusion protein directed at interleukin-2 receptor bearing cells. J. Biol. Chem. 265, 15198-15202, 1990.
9. Batra JK; Kasprzyk PG; Bird RE; Pastan I; King CR. Recombinant anti-erbB2 im-munotoxins containing Pseudomonas exotoxin. Proc Natl Acad Sci U S A 1992 89:5867-71
10. Wels W; Harwerth IM; Mueller M; Groner B; Hynes NE. Selective inhibition of tumor cell growth by a recombinant single-chain antibody-toxin specific for the erbB-2 receptor. Cancer Res 1992 52:6310-7

11. Reiter Y., Brinkmann U., Jung S-H., Lee, B., Kasprzyk P.G., King C.R. and Pastan I. Improved binding and anti tumor activity of a recombinant anti-erbB2 immunotoxin by disulfide-stabilization of the Fv fragment J. Biol. Chem. 269, 18327-18331, 1994

12. Jung SH; Pastan I; Lee B Design of interchain disulfide bonds in the framework region of the Fv fragment of the monoclonal antibody B3. Proteins 1994 19:35-47

13. Kabat, E.A., Wu, T.T., Perry, HM. , Gottesman, KS. and Foeller, C. Proteins of immunological Interest. US Dept. Health anh Human Services, Public Health Services, National Institutes of Health. NIH Publication. 5th ed., No. 91-3242, 1991.

14. Brinkmann, U., Reiter,Y., Jung, S.H, Lee B. and Pastan, I. A recombinant immunotoxin containing a disulfide-stabilized Fv fragment. Proc. Natl. Acad. Sci. USA 90, 7538-7542, 1993

15. Reiter,Y., Pai,L.H., Brinkmann, U. , Wang Q.C, and Pastan,I. Antitumor activity in mice of a recombinant immunotoxin containing a disulfide-stabilized Fv fragment. Cancer Research 54, 2714-2718, 1994

16. Benhar I; Reiter Y; Pai LH; Pastan I Administration of disulfide-stabilized Fv-immunotoxins B1(dsFv)-PE38 and B3(dsFv)-PE38 by continuous infusion increases their efficacy in curing large tumor xenografts in nude mice. Int J Cancer 1995 62:351-5

17. Reiter,Y., Brinkmann, U., Kreitman, R.J., Jung,S-H., Lee,B. and Pastan,I. Stabilization of the Fv fragments in recombinant immunotoxins by disulfide bonds engineered into conserved framework regions. Biochemistry 33, 5451-5459, 1994

18. Reiter,Y. and Pastan,I. Recombinant Fv immunotoxins and Fv fragments as novel agents for cancer therapy and diagnosis. Trends Biotechnol. 12, 513-520, 1998

19. Reiter,Y., Pai,L.H., Brinkmann, U. , Wang Q.C, and Pastan, I. Antitumor activity in mice of a recombinant immunotoxin containing a disulfide-stabilized Fv fragment. Cancer Research 54, 2714-2718, 1994

20. Reiter,Y., Brinkmann, U., Jung,S-H., Lee,B. and Pastan,I. Engineering disulfide bonds into conserved framework regions of Fv fragments: Recombinant Immunotoxins containing disulfide-stabilized Fv with improved biochemical characteristics. Protein Engineering 7, 697-704, 1994

21. Reiter, Y., Kurucz, I., Brinkmann, U., Lee, B., Segal, DM. and Pastan, I. Construction od a functional disulfide-stabilized TCR Fv indicates that antibody and TCR Fv frameworks are very similar in structure. Immunity 2, 281-287, 1995.

22. Pastan, I., Lovelace, ET., Gallo, M., Rutherford, AV., Magnani, JL., and Willingham, MC. Characterization of monoclonal antibodies B1 and B3 that react with mucinous adenocarcinomas. Cancer Res. 51, 3781-3787, 1991.

23. Brinkmann, U., Pai, L.H., FitzGerald, D.J., Willingham, M. and Pastan, I. B3(Fv)-PE38KDEL, a single chain immunotoxin that causes complete regression of a human carcinoma in mice. Proc. Natl. Acad. Sci. USA 88, 8616-8620, 1991

24. Bera TK, Onda, M., Brinkmann, U. and Pastan, I. A bivalent disulfide-stabilized Fv with improved antigen binding to erbB2. J. Mol. Biol. 1998 281, 475-483, 1998

25. Brinkmann, U., DiCarlo, A., Vasmatzis, G., Kurochkina, N., Beers, R., Lee, BK., and Pastan, I. Stabilization of a recombinant Fv fragment by base-loop interconnection and VH-VL permutation. J.Mol.Biol. 268, 107-117, 1997

26. Malmborg AC; Borrebaeck CA. BIAcore as a tool in antibody engineering. J Immunol Methods 1995 183:7-13

27. Rajagopal, V., Pastan, I. and Kreitman, RJ. A form of anti-Tac (Fv) which is both single-chain and disulfide-stabilized: comparison with its single-chain and disulfide-stabilized homologs. Protein Eng. 10, 1453-1459, 1997.

28. Kobayashi, H., Han, ES., Kim, IS., Le, N. Rajagopal, V., Kreitman, RJ., Pastan, I., Paik, CH., Carrasquillo, JA. Similarities in the biodistribution of iodene-labeled anti-tac single-chain disulfide-stabilized Fv fragment and anti-tac disulfide-stabilized Fv fragment. Nuc. Med. Biol. 25, 387-393, 1998.

29. Studier FW., Rosenberg, AH., Dunn, JJ., and Dubendorff JW. Use of T7 polymerase to direct expression of cloned genes. Meth. Enzymol. 185, 60-89, 1990

30. Brinkmann U. and Pastan I. Recombinant Immunotoxins: From Basic Research to Cancer Therapy. Methods 8, 143-156, 1995

31. Brinkmann U. Recombinant Toxins: Protein Engineering for Cancer Therapy Molecular Medicine Today 2, 439-446, 1996

32. McCafferty J; Griffiths AD; Winter G; Chiswell DJ Phage antibodies: filamentous phage displaying antibody variable domains. Nature 1990 348:552-4

33. Winter G; Griffiths AD; Hawkins RE; Hoogenboom HR. Making antibodies by phage display technology. Annu Rev Immunol 1994;12:433-5527a

34. Brinkmann, U., Buchner, J. and Pastan, I. Independent domain folding of Pseudomonas exotoxin and single chain immunotoxins: Influence of interdomain connections. Proc. Natl. Acad. Sci. USA 89, 3075-3079, 1992

35. Rudolph R and Lilie, H In vitro folding of inclusion body proteins FASEB J. 10, 49-56, 1996

36. Lilie, H, Schwarz, E. and Rudolph R Advances in refolding of proteins produced in *E. coli*. Curr. Opin. Biotechnol. 9, 497-501, 1998.

37. Buchner, J., Pastan, I. and Brinkmann, U. A method to increase the yield of properly folded recombinant fusion proteins: Single-chain immunotoxins from renaturation of bacterial inclusion bodies. Anal. Biochem. 205, 263-270, 1992

38. Reiter Y. Brinkmann, U., Lee, B. and Pastan I. Engineering Antibody Fv Fragments for Cancer Detection and Therapy: Disulfide-Stabilized Fv Fragments. Nature/Biotechnol. 14, 1239-1245, 1996

39. Vitetta ES From the basic science of B cells to biological missiles at the bedside. J Immunol 1994 153:1407-20

40. Pai LH; Pastan I . Immunotoxins and recombinant toxins for cancer treatment. Important Adv Oncol 1994:3-1912

41. Pastan IH; Archer GE; McLendon RE; Friedman HS; Fuchs HE Wang QC; Pai LH; Herndon J; Bigner DD. Intrathecal administration of single-chain immunotoxin, LMB-7 [B3(Fv)-PE38], produces cures of carcinomatous meningitis in a rat model. Proc Natl Acad Sci U S A 1995 92:2765-9

42. Kreitman R.J. and Pastan I. Recombinant single-chain immunotoxins against T and B cell leukemias. Leuk. Lymphoma 13, 1-10, 1994

43. Keppler-Hafkemeyer, A., Brinkmann U., and Pastan, I. Role of caspases in immunotoxin induced apoptosis of cancer cells. Biochemistry 37, 16934-16942, 1998

44. Hafkemeyer, P., Brinkmann, U., Gottesman, MM and Pastan, I. Apoptosis induced by Pseudomonas Exotoxin: a sensitive and rapid marker for gene delivery in vivo. Hum.Gene Ther. 10, 923-934, 1999

45. Pai LH., Wittes, R., Setser, A., Willingham, MC., and Pastan, I. Treatment of advanced solid tumors with immunotoxin LMB-1: an antibody linked to Pseudomonas Exotoxin. Nat. Med. 3, 350-353, 1996

46. Pai LH. and Pastan, I. Clinical trials with pseudomonas exotoxin immunotoxins Curr Topics Mikrobiol Immunol. 234, 83-96, 1998.

47. Kreitman RJ and Pastan I. Targeting Pseudomonas exotoxin to hematologic malignancies. 1995, Sem. Cancer Biol. 6, 297-306

48. Kreitman, RJ., Wilson, WH., White, JD., Stetler-Stevenson, M., Jaffe, E., Waldman, TA., and Pastan, I. Phase 1 trial of recombinant immunotoxin anti-tac(Fv)-PE38 (LMB2) in patients with hematologic malignancies. J. Clin. Oncol. 1999 in press

49. Kreitman, RJ., Wilson, WH., Robbins, D., Margulies, I., Stetler-Stevenson, M., Waldman, TA., and Pastan, I. Responses in refractory hairy cell leukemia to a recombinant immunotoxin. Blood 1999 in press

Bivalent and Bispecific
Antibody Fragments

Bivalent and Bispecific Diabodies and Single-chain Diabodies

TINA KORN, TINA VÖLKEL, and ROLAND E. KONTERMANN

Introduction

Diabodies are small dimeric bivalent or bispecific antibody fragments formed by cross-over pairing of two single-chain VH-VL fragments (Holliger et al., 1993, Whitlow et al., 1994). Dimer formation is favoured by reducing the linker length between the VH-VL domains from 15-20 amino acids, normally used to generate scFv fragments, to approximately 5 amino acids (Fig. 1A). Further reduction of the linker can result in the formation of trimeric or even tetrameric molecules (triabodies, tetrabodies) (Kortt et al., 1997; Le Gall et al., 1999). As shown by crystallographic studies, the two binding sites of a diabody molecule are facing away from each other (Perisic et al., 1994) (Fig. 1B). Bivalent diabodies are generated by dimeric assembly of two identical VH-VL chains (homodimers). Due to the presence of two antigen binding sites, bivalent diabodies exhibit an increased functional affinity (FitzGerald et al., 1997). The expression of two fragments of the format VHA-VLB and VHB-VLA in the same cell results in formation of heterodimers recognising two different antigens, but may also lead to the formation of non-functional homodimers (Fig. 1C). These homodimers can be easily separated by affinity chromatography. Bispecific diabodies have been successfully applied for diagnostic and therapeutic approaches through the recruitment of effector molecules and cells to specific targets (Holliger et al., 1996 and 1997; Kontermann et al., 1997a and b; FitzGerald et al., 1997; Zhu et al., 1996; Helfrich et al., 1998; Krebs et al., 1998). Various modifications have been added to the diabody

Tina Korn, Philipps-Universität, Institut für Molekularbiologie und Tumorforschung, Emil-Mannkopff-Straße 2, 35033 Marburg, Germany
Tina Völkel, Philipps-Universität, Institut für Molekularbiologie und Tumorforschung, Emil-Mannkopff-Straße 2, 35033 Marburg, Germany
✉ Roland E. Kontermann, Philipps-Universität, Institut für Molekularbiologie und Tumorforschung, Emil-Mannkopff-Straße 2, 35033 Marburg, Germany
(*phone* +49-6421-2866727; *fax* +49-6421-2868923; *e-mail* rek@imt.uni-marburg.de)

molecule in order to increase their stability. These include the introduction of interchain disulphide bonds (disulphide-stabilised diabodies; dsDb) (FitzGerald et al., 1997), the engineering of knobs-into-holes structures into the VH-VL interface (Zhu et al., 1997), and the fusion of the VHA-VLB and VHB-VLA chains by an additional linker generating single gene-encoded bispecific single-chain diabodies (scDb) (Brüsselbach et al., 1999) (Fig. 1D).

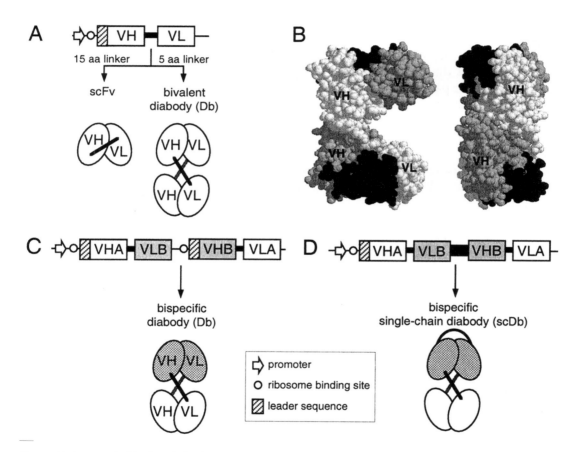

Fig. 1. A) Format of a bivalent diabody. A 15 amino acid linker generates mainly monomeric scFv fragments, while 5 amino acid linkers favour formation of diabody molecules. **B)** Space-filled structure of a diabody. The antigen-binding sites are shown in dark. **C)** Format of a bispecific diabody. **D)** Format of a bispecific single-chain diabody.

Outline

The general strategies to generate bivalent and bispecific diabodies as well as single-chain diabodies are depicted in Fig. 2, 3, and 4. Bivalent diabodies are generated by substituting the original linker between the VH and VL domain by a short peptide linker. The standard linker, which is also used in this protocol, is five amino acids long (with the sequence -GGGGS-). Variations in length and sequence are possible and are dependent mainly on your antibody. Two configurations are possible: the VH-VL configuration, fusing the VL domain C-terminal of the VH domain, and the VL-VH configuration, with the VH domain fused C-terminal of the VL domain. In some cases only one of these configurations will produce functional diabodies. In addition, the expression of soluble antibody fragments might be influenced by the order of the variable domains and the linker length (Desplancq et al., 1994; Alfthan et al., 1995). The strategies are based on the introduction of appropriate cloning sites into the VH and VL fragments. For the construction of bispecific molecules in the HL configuration, a BstEII site is introduced at the 3' region of the VH fragment (if not already present) and a SacI site into the 5' region of the VL fragment. For the construction of bispecific molecules in the LH configuration, a SacI site is introduced into the 3' region of the VL fragment and a BamHI site at the 5' end of the VH fragment. Bispecific molecules are then generated by combining the VH and VL fragments from two different antibodies. For this purpose an AscI site is used to combine the DNA fragments.

Materials

– 37°C incubator and shaker	**Equipment**
– ELISA plate reader	
– 1.5 ml reaction tubes	**Plastic ware**
– 96 well microtitre plates for ELISA (Nunc Maxisorp; Falcon Microtest III, etc.)	
– PBS (see Chapter 9)	**Buffers**
– TG1 (see Chapter 9)	**Bacteria**

Antibiotics	–	ampicillin-stock solution (1000 x): 100 mg/ml in H_2O
Bacterial media and plates	–	2xTY medium (see Chapter 9)
	–	9 cm round TYE plates containing 100 µg/ml ampicillin, 1% glucose (see Chapter 9)
Enzymes	–	T4 DNA ligase (3 u/µl; Promega)
	–	thermostable DNA polymerases (e.g. Taq, Vent (Biolabs), pfu (Stratagene))
	–	Calf intestine alkaline phosphatase (Gibco BRL)
Antibodies	–	anti-Myc antibody 9E10 (see Chapter 9)
	–	horseradish-peroxidase conjugated goat anti-mouse (see Chapter 9)
Other solutions	–	20 x dNTP mix (5 mM for each nucleotide) for PCR
	–	20% glucose in H_2O
	–	TMB substrate solution (see Chapter 9)
	–	30% H_2O_2
	–	1 M sulfuric acid
Oligonucleotides	–	LMB2 5'-GTA AAA CGA CGG CCA GT-3'
	–	LMB3 5'-CAG GAA ACA GCT ATG ACC-3'
	–	fdSeq1 5'-GAA TTT TCT GTA TGA GG-3'

Subprotocol 1
Restriction Site Analysis and Generation of Bivalent Diabodies

Check for restriction sites used for construction in the sequence of the VH and VL fragments of the antibody fragments you want to convert into a bivalent or bispecific diabody or single-chain diabody molecule. If you find additional sites you might have to use partial digests or multiple fragment ligation for the generation of antibody fragments. Alternatively, these sites can be deleted by site-directed mutagenesis. It is also possible to introduce other restriction sites suitable for cloning. We have found that most antibody fragments can be cloned as diabodies or single-chain diabodies using the above described restriction sites.

Bivalent diabodies are generated by linking a VH and a VL domain obtained from the same antibody with a short interdomain linker (Fig. 2). Use as starting material for the generation of bivalent diabody an antibody fragment (e.g. a scFv, or Fab fragment) cloned into a bacterial pUC19-derived expression vector such as pAB1 (Kontermann et al., 1997) containing a leader sequence (e.g. the pelB leader (Power et al., 1992)) or a phagemid vector such as pCANTAB6 (see Chapter 6 and Chapter 9).

Procedure

Generation of bivalent diabodies in the VH-VL configuration

1. Design the oligos for amplification of the VH and VL fragments (Fig. 2).   **strategy 1** Use approximately 20-30 nucleotides derived from your antibody sequence for annealing.

2. for the VH fragment you need primer LMB3 (annealing in the vector backbone) and a forward primer VH-BstEII-For (introducing a BstEII site in the 3' region of the VH fragment) (see Fig. 2).

3. for the VL fragment you need primer LMB2 (annealing in the vector backbone) or fdSeq1 (if the antibody fragment is fused to g3p, i.e. isolated from a phage library) and primer VL-Bst/Sac-Back annealing in the 5' region of the VL fragment and adding a BstEII site, the 5 amino acid linker encoding sequence, and a SacI site to the VL fragment.

4. Amplify the VH and the VL fragments with the respective primers by PCR. Use different polymerases including proof-reading ones. We routinely perform 25 cycles with an annealing temperature of 50 or 55°C.

5. Gel purify the PCR products. The two fragments should run at approximately 350 bp.

6. Digest the amplified VH fragment with SfiI and BstEII and the VL fragment with BstEII and NotI. Use the reaction conditions supplied by the manufacturer.

   Note: SfiI needs 50°C reaction temperature and BstEII 60°C.

7. Digest a bacterial expression vector, such as pAB1 (Kontermann et al.,

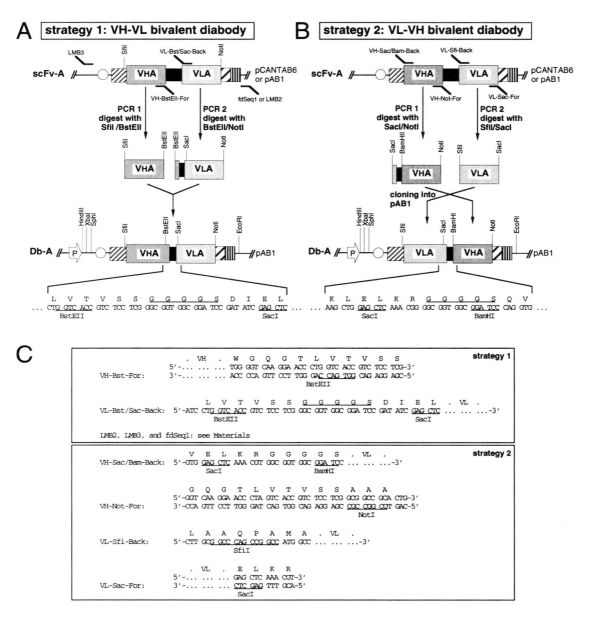

Fig. 2. Cloning strategies for the construction of bivalent diabodies. **A)** and **B)** Strategies 1 and 2. **C)** Sequences of the primers used for construction of bivalent diabodies by strategy 1 or 2. The sequences which have to be taken from your particular antibody are shown as dots. Add sufficient sequence to obtain specific annealing.

1997a) or similar vectors derived from pUC119mycHis (Low et al., 1996) which contain a pelB leader sequence and SfiI and NotI sites in the multiple cloning site. Dephosphorylate the digested vector with calf intestine alkaline phosphatase.

8. Purify digested fragments and vector by standard protocols (e.g. phenol/chloroform extraction and ethanol precipitation or by using commerically available spin-columns). Estimate amounts by running aliquots on a 1% agarose gel.

9. Ligate the VH and VL fragments together with the vector fragment at 15°C overnight using VH:VL:vector ratios of approximately 2:2:1 in a total volume of 20 µl.

10. Transform 10 µl of the ligation reaction into TG1 competent cells using standard protocols and plate cells onto TYE, 100 µg/ml amplicillin, 1% glucose plates. Incubate overnight at 37°C.

11. Screen for positive clones by PCR with primers LMB2 and LMB3. To perform the screen, pick 12-24 single colonies from the plate with sterile toothpicks, dip into 20 µl of PCR reaction mix (e.g. aliquoted into a 96 well PCR plate) and then streak it onto a TYE, 100 µg/ml amplicillin, 1% glucose plate (**master plate**). Run 30 cycles of PCR at an annealing temperature of 50°C and an extention time of 1 min. Analyse PCR products on a 1% agarose gel. Positive clones should give a product of approximately 900 bp (VH-VL insert plus flanking vector-derived sequences).

12. Analyse positive clones for expression of full-length antibody sequence by immunoblot experiments of bacterial pellets of induced overnight cultures with anti-Myc tag antibody 9E10. Alternatively, check directly the supernatant of an induced 2 ml culture in ELISA for antigen binding (Subprotocol 4, ELISA). Grow culture first in 2xTY, 100 µg/ml ampicillin, 0.1% glucose until an OD_{600} of 0.8-1.0 is reached, then add 1 mM IPTG and incubate shaking overnight at 30°C.

Generation of bivalent diabodies in the VL-VH configuration

1. Design the oligos for amplification of the VH and VL fragments (Fig. 2). Use approximately 20 nucleotides derived from your antibody sequence for annealing.   **strategy 2**

2. for the VH fragment you need backward primer VH-Sac/Bam-Back

which adds a SacI site, a five amino acid linker and a BamHI site at the 5' end, and forward primer VH-Not-For introducing a NotI site at the 3' end of the VH fragment (see Fig. 2).

3. for the VL fragment you need backward primer VL-Sfi-Back introducing a SfiI site at the 5' end of the VL fragment and forward primer VL-Sac-For annealing in the 5' region of the VL fragment and adding a SacI site to the VL fragment.

4. Amplify the VH domain and the VL domain with the respective primers by PCR and proceed as described in Subprotocol 1, strategy 1 using the restriction enzymes indicated in strategy 2.

Subprotocol 2
Generation of Bispecific Diabodies

Bispecific diabodies are generated by the expression of two chains of the format VHA-VLB and VHB-VLA (HL configuration) or of the format VLA-VHB and VLB-VHA (LH configuration) in the same cell. Each chain is preceded by a ribosome binding site and a leader sequence (strategies 3 and 4, Fig. 3). For construction, the VH and VL fragments of an antibody with a second specificity are amplified by PCR to introduce appropriate cloning sites, the second ribosome binding site as well as the five amino acid linkers. These fragments are then cloned into a plasmid containing a bivalent diabody in the same configuration (see Subprotocol 1).

▨▨ Procedure

Generation of bispecific diabodies in the VH-VL configuration

strategy 3
1. Design the oligos for amplification of the VH and VL fragments of the second antibody (Fig. 5). Use approximately 20-30 nucleotides derived from your antibody sequence for annealing.

2. For the VH fragment you need primers VH-Asc-Back (annealing in the leader sequence and adding an AscI site and a ribosome binding site upstream of the leader sequence) and VH-Sac-For (annealing at the 3' region of the VH fragment and adding a 5 amino acid linker and a SacI site) (see Fig. 3 and 5).

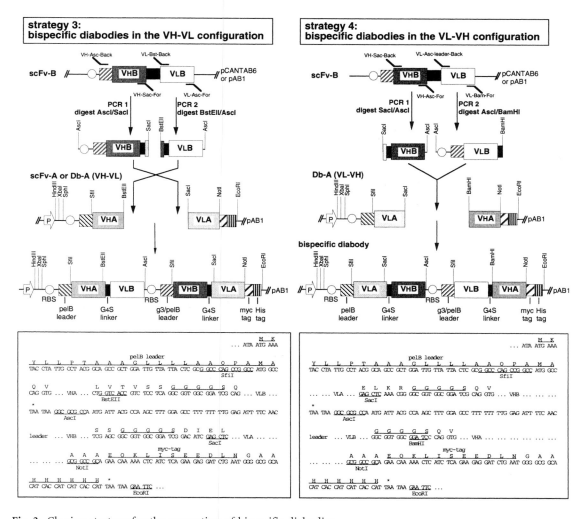

Fig. 3. Cloning strategy for the generation of bispecific diabodies

3. For the VL you need primers VL-Bst-Back (annealing in the 5' end of the VL fragment and adding a BstEII site and 5 amino acid linker) and VL-Asc-For (annealing in the 3' end of the VL fragment and adding an AscI site) (see Fig. 3 and 5).

4. Amplify the VH and VL fragments with the respective primers by PCR and purify fragments as described in Subprotocol 1.

5. Digest the VH fragment with AscI and SacI and the VL fragment with BstEII and AscI.

6. Digest the bivalent diabody construct in the HL configuration (generated as described in Subprotocol 1) with BstEII and SacI.

7. Proceed as described in Subprotocol 1 (from Step 8).

8. For the identification of positive clones, perform ELISA or other immunological test for both antigen specificities to ensure that both binding sites are assembled correctly (see Subprotocol 4).

Generation of bispecific diabodies in the VL-VH configuration

strategy 4

1. Design the oligos for amplification of the VH and VL fragments of the second antibody (Fig. 5). Use approximately 20-30 nucleotides derived from your antibody sequence for annealing.

2. For the VH fragment you need primers VH-Sac-Back (annealing in the 5' end of the VH fragment and adding a SacI site and a 5 amino acid linker sequence) and VH-Asc-For (annealing at the 3' region of the VH fragment and adding an AscI site) (see Fig. 3 and 5).

3. For the VL fragment you need primers VL-Asc-leader-Back (annealing in the 5' end of the VL fragment and adding an AscI site, a ribosome binding site and a leader sequence) and VL-Bam-For (annealing in the 3' end of the VL fragment and adding a 5 amino acid linker sequence and a BamHI site) (see Fig. 3 and 5).

4. Amplify the VH and VL fragments with the respective primers by PCR and purify fragments as described in Subprotocol 1.

5. Digest the VH fragment with SacI and AscI and the VL fragment with AscI and BamHI.

6. Digest the bivalent diabody construct in the LH configuration (generated as described in Subprotocol 1) with SacI and BamHI.

7. Proceed as described in Subprotocol 1, strategy 1 (from Step 8).

Subprotocol 3
Generation of Bispecific Single-Chain Diabodies

Bispecific single-chain diabodies are generated by the expression of a single fragment of the format VHA-VLB-VHB-VLA (HL configuration) or of the format VLA-VHB-VLB-VHA (LH configuration) (strategies 5 and 6, Fig. 4). For construction, the VH and VL fragments of an antibody with a second specificity are amplified by PCR to introduce appropriate cloning sites and the middle linker. These fragments are then cloned into a plasmid containing a bivalent diabody in the same configuration (see Subprotocol 1). Due to the presence of the middle linker, which is 20 amino acids in the HL configuration and 15 amino acids in the LH configuration, it is not necessary to introduce a second ribosome binding site and an additional leader sequence. These strategies can, of course, also be used to generate bivalent single-chain diabodies.

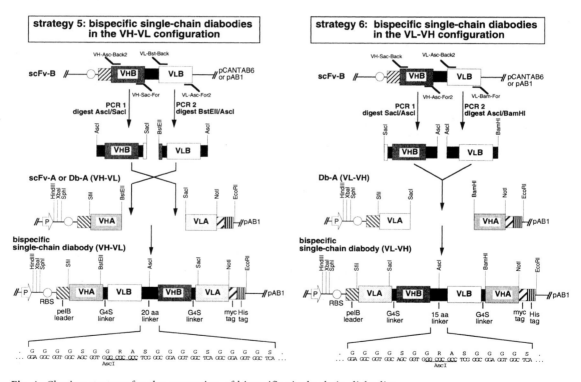

Fig. 4. Cloning strategy for the generation of bispecific single chain diabodies.

▪▪ Procedure

Generation of bispecific diabodies in the VH-VL configuration

strategy 5

1. Design the oligos for amplification of the VH and VL fragments of the second antibody (Fig. 5). Use approximately 20-30 nucleotides derived from your antibody sequence for annealing.

2. For the VH fragment you need primers VH-Asc-Back2 (annealing in the leader sequence and adding an AscI site and the second half of the middle linker sequence) and VH-Sac-For (annealing at the 3' region of the VH fragment and adding a 5 amino acid linker and a SacI site) (see Fig. 4 and 5).

3. For the VL fragment you need primers VL-Bst-Back (annealing in the 5' end of the VL fragment and adding a BstEII site and a 5 amino acid linker) and VL-Asc-For2 (annealing in the 3' end of the VL fragment and adding the first half of the middle linker sequence and an AscI site) (see Fig. 4).

4. Amplify the VH and VL fragments with the respective primers by PCR and purify fragments as described in Subprotocol 1.

5. Digest the VH fragment with AscI and SacI and the VL fragment with BstEII and AscI.

6. Digest the bivalent diabody construct in the HL configuration (generated as described in Subprotocol 1) with BstEII and SacI.

7. Proceed as described in Subprotocol 1, strategy 1 (from Step 8).

Generation of bispecific diabodies in the VL-VH configuration

strategy 6

1. Design the oligos for amplification of the VH and VL fragments of the second antibody (Fig. 5). Use approximately 20-30 nucleotides derived from your antibody sequence for annealing.

2. For the VH fragment you need primers VH-Sac-Back (annealing in the 5' end of the VH fragment and adding an SacI site and a 5 amino acid linker sequence) and VH-Asc-For2 (annealing at the 3' region of the VH fragment and adding the first half of the middle linker sequence and an AscI site) (see Fig. 4 and 5).

3. For the VL fragment you need primers VL-Asc-Back2 (annealing in the 5' end of the VL fragment and adding the second half of the middle linker sequence and an AscI site) and VL-Bam-For (annealing in the 3' end of the VL fragment and adding a 5 amino acid linker sequence and a BamHI site) (see Fig. 4 and 5).

4. Amplify the VH and VL fragments with the respective primers by PCR and purify fragments as described in Subprotocol 1.

5. Digest the VH fragment with SacI and AscI and the VL fragment with AscI and BamHI.

6. Digest the bivalent diabody construct in the LH configuration (generated as described in Subprotocol 1) with SacI and BamHI.

7. Proceed as described in Subprotocol 1, strategy 1 (from Step 8).

Subprotocol 4
Expression and Characterisation

▪▪ Procedure

Expression

1. Diabodies and single-chain diabodies can be purification from periplasmic preparations by IMAC as described in Chapter 17.

2. If the yield from the periplasmic preparation is low, check if you can purify more protein from the supernatant of an induced culture grown for 16-20 hrs at RT or 30°C (you can identify best temperature by growing a 2-5 ml culture at various temperatures and analysing the supernatant directly by ELISA (for ELISA procedure see below). For purification of diabodies or single-chain diabodies from bacterial supernatant you can concentrate proteins by precipitation with 50% saturated ammonium sulfate or by ultrafiltration.

3. Analyse purified proteins by 10-12% SDS-PAGE and by immunoblotting using a suitable anti-tag antibody (e.g anti-Myc-tag antibody 9E10 or anti-His-tag antibody) for detection of antibody fragments. Note that in bispecific diabodies only one chain contains tag sequences. Bivalent diabodies should run at a molecular mass of approximately

```
                                                                    strategy 3
VH-Asc-Back:    5'-TAA GGC GCG CCA ATG ATT ACG CCA AGC TTT GGA GCC TTT TTT TTG GAG-3'  (pCANTAB6 backbone)
                     AscI

                                                                        M  K
VH-Asc-Back:    5'-TAA GGC GCG CCA AAA TTC TAT TTC AAG GAG ACA GTC ATA ATG AAA-3'    (pAB1 backbone)
                     AscI

                 G   Q   G   T   L   V   T   V   S   S   G   G   G   G   S   D   I   E   L
                5'-GGC CAA GGT ACC CTA GTA ACC GTC TCG AGC GGC GGT GGC GGA TCG GAC ATC GAG CTC CGA-3'
VH-Sac-For:     3'-CCG GTT CCA TGG GAT CAT TGG CAG AGC TCG CCG CCA CCG CCT AGC CTG TAG CTC GAG GCT-5'
                                                                                SacI

                 L   V   T   V   S   S   G   G   G   G   S   .  VL   .
VL-Bst-Back:    5'-GAT CTG GTC ACC GTC TCC TCA GGC GGT GGC GGA TCG ... ... ...-3'
                     BstEII

                    .  VL   .       *
                5'-... ... ... TAA TAA GGC GCG CCA TGC-3'
VL-Asc-For:     3'-... ... ... ATT ATT CCG CGC GGT ACG-5'
                                   AscI
```
```
                                                                    strategy 4
                 E   L   K   R   G   G   G   G   S   .  VH   .
VH-Sac-Back:    5'-TAG GAG CTC AAA CGG GGC GGT GGC GGA TCG ... ... ...-3'
                     SacI

                    .  VH   .       *
                5'-... ... ... TAA TAA GGC GCG CCA ATG-3'
VH-Asc-For:     3'-... ... ... ATT ATT CCG CGC GGT TAC-5'
                                   AscI

VL-Asc-leader-Back:  5'-TAA GGC GCG CCA ATG ATT ACG CCA AGC TTT GGA GCC TTT TTT TTG GAG ATT TTC AAC
                          AscI

   V   K   K   L   L   F   A   I   P   L   V   V   P   F   V   A   A   Q   P   A   M   A   .  VL   .
  GTG AAA AAA TTA TTA TTC GCA ATT CCT TTA GTT GTT CCT TTC TAT GCG GCC CAG CCG GCC ATG GCC... ... ...-3'

                    .  VL   .   G   G   G   G   S
                5'-... ... ... GGC GGT GGC GGA TCC CAG-3'
VL-Bam-For:     3'-... ... ... CCG CCA CCG CCT AGG GAC-5'
                                   BamHI
```
```
                 G   R   A   S   G   G   G   G   S   G   G   G   G   S   .  VH   .    strategy 5
VH-Asc-Back2:   5'-GGT GGG CGC GCC TCG GGC GGA GGT GGC TCA GGC GGA GGT GGC TCA ... ... ...-3'
                       AscI

                    .  VL   .   G   G   G   G   S   G   G   R   A
                5'-... ... ... GGA GGC GGT GGC AGC GGT GGG CGC GCC TCG-3'
VL-Asc-For2:    3'-... ... ... CCT CCG CCA CCG TCG CCA CCC GCG CGG AGC-5'
                                                   AscI

VH-Sac-For and VL-Bst-Back:   see strategy 3
```
```
                 G   G   R   A   S   G   G   G   G   S   .  VL   .     strategy 6
VL-Asc-Back2:   5'-GGT GGG CGC GCC TCG GGC GGA GGT GGC TCA ... ... ...-3'
                       AscI

                    .  VL   .   G   G   G   G   S
                5'-... ... ... GGC GGT GGC GGA TCC ATC-3'
VL-Bam-For2:    3'  ... ... ... CCG CCA CCG CCT AGG TAG-5'
                                   BamHI

VH-Sac-Back and VL-Bam-For:   see strategy 4
```

Fig. 5. Primers for the generation of bispecific diabodies and single-chain diabodies.

30 kDa, bispecific diabodies should yield two bands with similar molecular weights, and single-chain diabodies should have a molecular mass of approximately 55 kDa.

4. Dimeric assembly can be further analysed by gel filtration, for example using a Superose 12 column. Diabodies and single-chain diabodies should migrate with a molecular mass of 50-55.000.

ELISA

1. Coat a microtitre plate with your antigen at 1-10 µg/ml overnight in a suitable buffer (PBS or carbonate buffer pH 9.6) at 4°C. Use one or more appropriate proteins as negative controls. Coat enough wells to analyse serial dilutions. Use two wells as blanks which are incubated without antibody fragment.

2. Next day, block remaining binding sites with 2% PBS containing 2% skimmed milk powder (MPBS) for 2 hrs.

3. Make serial dilutions (220-250 µl for each dilution) of your antibody fragment in 2% MPBS (e.g from 10 µg/ml - 0.01 µg/ml).

4. Pipette 100 µl of the dilution into the wells of the coated microtitre plate and incubate for 1 hr at RT.

5. Wash 6 times with PBS.

6. Add anti-Myc-tag antibody 9E10 diluted to 10 µg/ml in 2% MPBS and incubate for 1 hr at RT (you can also use anti-His-tag antibody or other anti-tag antibodies depending on your construct).

7. Wash 6 times with PBS.

8. Add HRP-conjugated goat-anti-mouse antibody diluted 1:5000 in 2% MPBS and incubate for 1 hr at RT.

9. Wash 6 times with PBS.

10. Add 100 µl of TMB/H_2O_2 per well and incubate until blue color has developed. Stop reaction by adding 50 µl of 1 M sulfuric acid. Read plate at 450 nm in a microtitre plate reader.

Sandwich-ELISA for the analysis of bispecific antibody fragments

1. Coat a microtitre plate with the first antigen as described above. Coat enough wells to analyse varying concentrations of your antibody fragment as well as of the second antigen. Proceed until step 5 of the ELISA procedure.

2. Add 100 µl of the second antigen diluted in 2% MPBS at varying concentrations and incubate for 1 hr at RT.

3. Wash 6 times with PBS.

4. Incubate plate with an antibody reacting specifically with your second antigen diluted to an appropriate concentration in 2% MPBS. Incubate for 1 hr at RT.

5. Wash 6 times with PBS.

6. Add a HRP-conjugated secondary antibody diluted in 2% MPBS and incubate for 1 hr at RT.

7. Wash 6 times with PBS and proceed as described above.

Immunoassays using bispecific diabodies or single-chain diabodies

If the second antigen is able to convert a chromogenic substrate (e.g. if it is an enzyme or conjugated with an enzyme you can use the antibody fragments as reagent in microtitre plate immunoassays (Kontermann et al., 1997a). This method can also be applied for immunocyto- and histochemical stainings using appropriate chromogenic substrates.

1. Coat a microtitre plate with the first antigen as described above. Coat enough wells to analyse varying concentrations of your antibody fragment as well as of the second antigen. Proceed until step 5 of the ELISA procedure.

2. Add 100 µl of the second antigen (i.e. an enzyme) diluted in 2% MPBS at varying concentrations and incubate for 1 hr at RT.

3. Wash 6 times with PBS.

4. Add a suitable chromogenic substrate in 100 µl of reaction buffer and incubate until colour reaction has developed. Read plate at an appropriate wave length.

Troubleshooting

- The diabody molecules are not active. Try to express the antibody molecules in the HL as well as the LH configuration to see whether this makes a difference. You can also try to increase (or decrease) the linker length or, in case of bispecific diabodies, swap the order of antibody specificity (i.e. from A-B to B-A).

- Low expression yields. Expression yields can be influenced by the configuration of the diabody molecules. In case of low yields, try different configurations. For single-chain diabodies we found that yields are reduced by approximately 50% compared to diabodies.

Applications

Various diagnostic and therapeutic applications of bivalent and bispecific diabodies and single-chain diabodies have been described. These applications include the use in immunoassays by recruitment of an enzyme (e.g. β-galactosidase) to disease-associated antigens, such as HIV gp120 or carcinoembryonic antigen (CEA), the recruitment of effector molecules of the immune system (e.g. immunoglobulins, complement C1q), and the retargeting of effector cells (e.g. cytotoxic T lymphocytes by binding to the T cell coreceptor CD3) to tumour cells (Zhu et al., 1996; FitzGerald et al., 1996; Holliger et al., 1996 and 1997; Kontermann et al., 1997a and b; Helfrich et al., 1998; Kipriyanov et al., 1998; Krebs et al., 1998; Brüsselbach et al., 1999). Furthermore, bispecific single-chain diabodies have been demonstrated to be active when expressed in the secretory pathway or when displayed in the plasma membrane of mammalian cells (Kontermann & Müller, 1999). In addition, IgG-like antibody molecules with increased functional affinity can be generated by fusing single-chain diabodies to the Ig γ1 Fc or CH3 region (Alt et al. 1999).

References

Alt M, Müller R, Kontermann RE (1999) Novel tetravalent and bispecific IgG-like antibody molecules combining single-chain diabodies with the immunoglobulin γ1 Fc or CH3 region. FEBS Lett 454;90-94

Alfthan K, Takkinen K, Sizmann D, Söderlund H, Teeri TT (1995) Properties of a single-chain antibody containing different linker peptides. Protein Eng 8;725-731

Brüsselbach S, Korn T, Völkel T, Müller R, Kontermann RE (1999) Enzyme recruitment and tumor cell killing in vitro by a secreted bispecific single-chain diabody. Tumor Targeting 4;115-123

Desplancq D, King DJ, Lawson ADG, Mountain A (1994) Multimerization behaviour of single chain Fv variants for the tumour-binding antibody B72.3. Protein Eng 8;1027-1033

FitzGerald K, Holliger P, Winter G (1997) Improved tumour targeting by disulphide stabilized diabodies expressed in Pichia pastoris. Protein Eng 10;1221-1225

Helfrich W, Kroesen BJ, Roovers RC, Westers L, Molema G, Hoogenboom HR, de Leij L (1998) Construction and characterization of a bispecific diabody for retargeting T cells to human carcinoma. Int J Cancer 76,232-239

Holliger P, Brissinck J, Williams RL, Thielemans K, Winter G (1996) Specific killing of lymphoma cells by cytotoxic T-cells mediated by a bispecific diabody. Protein Eng 9;299-305

Holliger P, Prospero T, Winter G (1993) "Diabodies": small bivalent and bispecific antibody fragments. Proc Natl Acad Sci USA 90;6444-6448

Holliger P, Wing M, Pound JD, Bohlen H, Winter G (1997) Retargeting serum immunoglobulin with bispecific diabodies. Nat Biotechn 15;632-636

Kipriyanov SM, Moldenhauer G, Strauss G, Little M (1998) Bispecific CD3xCD19 diabody for T cell-mediated lysis of malignant human B cells. Int J Cancer 77;763-772

Kontermann RE, Martineau P, Cummings CE, Karpas A, Allen D, Derbyshire E, Winter G (1997a) Enzyme immunoassays using bispecific diabodies. Immunotechnology 3;137-144

Kontermann RE, Müller R (1999) Intracellular and cell surface displayed single-chain diabodies. J Immunol Meth 226;179-188

Kontermann RE, Wing MG, Winter G (1997b) Complement recruitment using bispecific diabodies. Nat Biotechnol 15;629-631

Le Gall F, Kipriyanov SM, Moldenhauer G, Little M (1999) Di, tri and tetrameric single chain Fv antibody fragments against human CD19: effect of valency on cell binding. FEBS Lett 453;164-168

Low NM, Holliger P, Winter G (1996) Mimicking somatic hypermutation: Affinity maturation of antibodies displayed on bacteriophage using bacterial mutator strain. J Mol Biol 260;359-368

Krebs B, Griffin H, Winter G, Rose-John S (1998) Recombinant human single-chain Fv antibodies recoginizing human interleukin-6: specific targeting of cytokine-secreting cells. J Biol Chem 273;2858-2865

Kortt A, Lah M, Oddie GW, Gruen CL, Burns JE, Pearce LA, Atwell JL, McCoy AK, Howlett GJ, Metzger DW, Webster RG, Hudson PJ (1997) Single-chain Fv fragments of anti-neuraminidase antibody NC10 containing five- and ten-residue linkers form dimers and with zero-residue linker a trimer. Protein Eng 10; 423-433

Perisic O, Webb PA, Holliger P, Winter G, Williams RL (1994) Crystal structure of a diabody, a bivalent antibody fragment. Structure 2;1217-1226

Power BE, Ivancic N, Harley VR, Webster RG, Kortt AA, Irving RA, Hudson PJ (1992) High-level temperature-induced synthesis of an antibody VH domain in Escherichia coli using the PelB secretion signal. Gene 113;95-99

Whitlow M, Filpula D, Rollence ML, Feng SL, Wood JF (1994) Multivalent Fvs: characterization of single-chain Fv oligomers and preparation of a bispecific Fv. Protein Eng 7;1017-1026

Zhu Z, Presta LG, Zapata G, Carter P (1997) Remodeling domain interface to enhance heterodimer formation. Protein Sci 6;781-788

Zhu Z, Zapata G, Shalaby R, Snedecor B, Chen H, Carter P (1996) High level secretion of a humanized bispecific diabody from Escherichia coli. Bio/Technol 14;192-196

Miniantibodies

PETER LINDNER and ANDREAS PLÜCKTHUN

Introduction

The term "miniantibodies" describes artificial multivalent or multispecific recombinant antibody fragments. They resemble natural antibodies in carrying two or four binding sites with a flexible arrangement and a long "wing span" but they are much smaller, as they consist only of fusions of a scFv to an oligomerization module. Applications of multivalent and bispecific antibody fragments in a variety of formats have been reviewed (Plückthun and Pack 1997; Carter and Merchant 1997). Approaches to quantify the avidity gain achieved by the multivalency effect present in all the described miniantibody models have been reported (Crothers and Metzger 1972, Plückthun and Pack 1997, Müller et al. 1998b).

Briefly, there are three motivations to consider making miniantibodies:

- First, they will bind with significantly higher avidity than monomers to any surface carrying the antigen close enough that two (or more) binding sites can be engaged.

- Second, when these molecules are immobilized on plastic support, at least one binding site usually remains functional, while the other may denature upon binding to the plastic surface. In contrast, scFv fragments usually lose all binding under these conditions.

- Third, two different specificities can be combined, with numerous applications in biotechnology, diagnostics and, potentially, therapy.

Peter Lindner, Universität Zürich, Biochemisches Institut, Winterthurerstrasse 190, 8057 Zürich, Switzerland

✉ Andreas Plückthun, Universität Zürich, Biochemisches Institut, Winterthurerstrasse 190, 8057 Zürich, Switzerland (*phone* +41-1-6355570; *fax* +41-1-6355712; *e-mail* plueckthun@biocfebs.unizh.ch)

The basic element of all constructs is a fusion of a scFv fragment to an oligomerizing element, in the simplest case an amphipathic α-helix-forming stretch of amino acids (usually between 16 and 40 residues, but see Table 2 for details) via a flexible hinge region, giving the partners enough steric freedom to fold individually. As outlined schematically in Fig. 1, this leads to dimeric or tetrameric miniantibodies, depending on the oligomerization motif chosen. Most conveniently, the miniantibodies are expressed in the periplasm to allow the disulfide formation in the scFv part. This requires that the chosen oligomerization modules are compatible with periplasmic folding.

Dimeric miniantibody constructs

While all other principles for the formation of bivalent or bispecific antibody fragments (discussed in chapters 42 to 45) require a significant reconstruction of the format compared to the scFv, the generation of dimeric miniantibodies is simply achieved by adding a sequence to the C-terminus. Examples for such oligomerization modules are a naturally occurring dimerization helix from the yeast transcription factor GCN4 (O'Shea et al. 1991, Dürr et al. 1999), or C_H3/F_c-domains of antibodies (Chapter 44), or a synthetic 4-helix bundle element (Eisenberg et al. 1986).

The 4-helix bundle motif is obtained by the alignment of two double-helix pairs, each pair being fused to one scFv and "clasping" each other. As these helices are in antiparallel orientation, this results in the two scFvs emerging from opposite sides of the bundle (see Fig. 1, scdHLX). Parallel alignment of the helices is obtained if GCN4 "leucine zipper" or other coiled coils like e.g. JUN/FOS are used as fusion partners (Fig. 1).

Bispecific miniantibodies can be obtained if two different scFvs are chosen as fusion partners and fused to modules which form a specific heterodimer. Not all heterodimerizing modules work well in vivo, as problems of homodimerization and proteolytic susceptibility are often not considered. Recently, the question of specific heterodimerization was tackled using a helix-library vs. helix-library-selection approach in vivo (Arndt et al. 2000). The work resulted in a pair of coiled coil helices which showed excellent behavior with regard to stability, heterospecificity, and resistance to proteases.

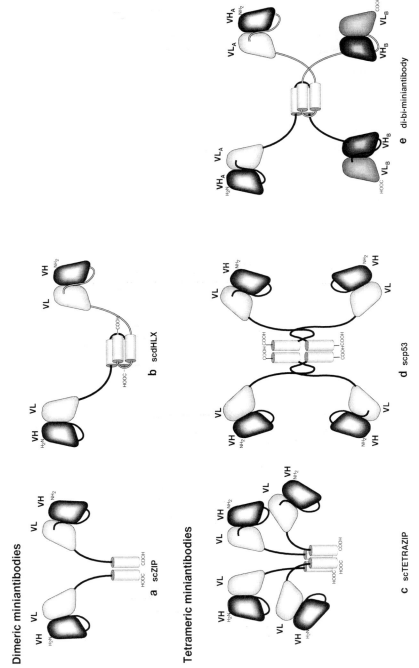

Fig. 1. Schematic representation of oligomeric miniantibody formats. VH and VL domains of the scFvs are represented in darker and lighter color, respectively. Linker and hinge regions are shown as black lines, and are either filled or not to indicate different polypeptide chains within one heterodimeric construct. Oligomerizing, helical elements are depicted as cylinders. The orientation is derived from the published crystal structures of the coiled coil (O'Shea et al. 1991), tetrazipper (Harbury et al. 1993), the NMR structure of the designed dHLX motif (Hill and DeGrado 1998) and the p53 tetramerization domain (Jeffrey et al. 1995). Dimeric miniantibodies: (*a*) GCN4 leucine zipper, scZIP; (*b*) four helix-bundle, scdHLX; Tetrameric miniantibodies: (*c*) modified GCN4 leucine zipper, scTETRAZIP; (*d*) oligomerizing domain of human p53, scp53; (*e*) (scFv)$_A$-hinge-dHLX-hinge-(scFv)$_B$ arrangement, di-bi-miniantibody.

Tetrameric miniantibody constructs

In principle, all fusion modules forming oligomers themselves appear to be suitable for designing oligomeric miniantibodies, provided their termini are accessible for attaching linker-scFv-cassettes and they are compatible with *in vivo* folding of the antibody.

Specific amino acid exchanges in the hydrophobic contact positions **a** and **d** of the GCN4 zipper results in the self-assembly of a stable tetrameric bundle (Harbury et al. 1993). A fusion of this zipper version to a scFv leads to tetrameric miniantibodies with improved affinity, compared to the corresponding dimeric construct (Pack et al. 1995, Fig. 1). The effect of amino acid exchanges in coiled coil helices on the oligomerization state have recently been reviewed (Müller et al. 2000).

Many fusions to enzymes and other proteins have been reported and have been summarized elsewhere (Plückthun and Pack, 1997). The lowest immunogenicity is to be expected with the fusion of a scFv to the oligomerization domain of human p53 (Jeffrey et al. 1995) via the human IgG3 hinge (Table 1; Burton 1985).

A combination of directed bivalency with bispecificity is obtained by using so-called "di-bi-miniantibodies" (Müller et al. 1998c). Here, a second scFv is fused after the dimerization motif, resulting in a $(scFv)_A$-hinge-dHLX-hinge-$(scFv)_B$ arrangement (Fig. 1).

In order to stabilize the oligomer formation by covalent linkage between either the zipper or the bundle-helix constructs, single cysteine residues may be added at their C-termini which lead to disulfide bridge formation (Pack and Plückthun 1992). It has to be noted, however, that also wrong disulfides may be formed with Fv-internal cysteine residues possibly leading to somewhat lower yields.

Materials

- French Pressure Cell (Aminco, Rochester, NY, USA)
- Molecular biology laboratory instrumentation

Procedure

Below, the bacterial expression of miniantibodies in shake flasks is described in detail. The procedure essentially follows the protocol as described in Pack and Plückthun (1992).

Table 1. Cross-references between oligomerizing elements and corresponding plasmid/literature.

construct	upper hinge	self-associating peptide	linker/tag	plasmid/s	reference
bivalent					
scHLX	murine IgG3	Helix		pACK01sc	C, D
scHLXc	murine IgG3	Helix	Cys tail	pACKHLXc	C, D, E
scdHLX	murine IgG3	Helix1-turn-Helix2		pACKdHLX, pHKK	B, I
scdHLX-His	murine IgG3	Helix1-turn-Helix2	spacer-His tail	pACKdHLXH, pAK500	F, J
scZIP	murine IgG3	GCN4 leucine zipper		pACKZIP	B, C, D, A, E
scZIPc	murine IgG3	GCN4 leucine zipper	Cys tail	pACKZIPc	C, D, E
bispecific					
scJUN	murine IgG3	JUN leucine zipper		pACKIHJUN	K, L
scFOS	murine IgG3	FOS leucine zipper		pACKFOS	K, L
CH1-CL	murine IgG3	CH1 and CL from IgG	His tail	pKM30245M1ChCl	G
tetravalent					
scTETRAZIP	murine IgG3	GCN4 leucine zipper, modified		pACKtZIP	A, E
scp53	human IgG3	oligomerization domain of human p53		pMS9-9 p53	K, L
scp53-His	human IgG3	oligomerization domain of human p53	spacer-His tail	pMStep53His	F
tetravalent/bispecific					
di-bi	murine IgG3	Helix1-turn-Helix2	His tail after 2nd scFv	pKM310M1dhlx425h	H

Important elements of various miniantibody formats are listed as overview. For exact amino acid sequences of the elements, see Table 2. Vectors carrying miniantibody genes in these formats and references are given. Letters in the reference column denote: (A) Pack et al. 1995; (B) Pack et al. 1993; (C) Pack and Plückthun 1992; (D) Pack et al. 1992; (E) Ge et al. 1995; (F) Rheinnecker et al. 1996; (G) Müller et al. 1998a; (H) Müller et al. 1998b; (I) Horn et al. 1996; (J) Krebber et al. 1997; (K) Plückthun and Pack 1997; (L) Pack 1994.

Table 2. Amino acid sequences of hinges and oligomerizing elements.

element	amino acid sequence	construct
murine IgG3 upper hinge	PKPSTPPGSS	
human IgG3 upper hinge	TPLGDTTHTSG	
helix	GELEELLKHLKELLKG-EF	scHLX
helix-Cys tail	GELEELLKHLKELLKG-PRKANSRNC	scHLXc
helix1-turn-helix2	GELEELLKHLKELLKG-PRK-GELEELLKHLKELLKG-EF	scdHLX or di-bi
helix1-turn-helix2-spacer-His tail	GELEELLKHLKELLKG-PRK-GELEELLKHLKELLKG-GSGGAP-HHHHH	scdHLX-His
GCN4 leucine zipper	RMKQLEDKVEELLSKNYHLENEVARLKKLVGER	scZIP
GCN4 leucine zipper-Cys tail	RMKQLEDKVEELLSKNYHLENEVARLKKLVGER-GGCGG	scZIPc
JUN leucine zipper	RIARLEEKVKTLKAQNSELASTANMLREQVAQLKQKVMNY	scJUN
FOS leucine zipper	LTDTLQAETDQLEDKKSALQTEIANLLKEKEKLEFILAAH	scFOS
GCN4 leucine zipper, modified	RLK**QIEDK**L**EEILSK**L**YHIENEL**A**RIKKLL**GER	scTETRAZIP
oligomerization domain human p53	KPLDGEYFTLQIRGRERFEMFRELNEALELKDAQAGKEP	scp53
oligomerization domain human p53-spacer-His tail	KPLDGEYFTLQIRGRERFEMFRELNEALELKDAQAGKEP-GGSGGAP-HHHHH	scp53-His

Amino acid sequences (one-letter-code) of various oligomerizing modules and hinges are given. In the modified GCN4 leucine zipper, which leads to tetramerization, the exchanged amino acids are in bold-face. The amino acids EF and the end of some constructs were introduced for an *Eco*RI-restriction site. Cross-references to the corresponding vectors and literature are listed in Table 1.

Detailed information on high-cell-density fermentation is given in Horn et al. (1996), Schroeckh et al. (1996), Plückthun et al. (1996), and Pack et al. (1993).

1. Select one of the presented formats (Table 1) for the chosen scFv and pick the appropriate vector (Table 2 and Fig. 2).

2. If no compatible restriction sites are present, design suitable PCR-primers.

   Note: A discussion on how to PCR amplify an antibody with an unknown sequence and convert it into a scFv format compatible with this vector series is given in Chapter 2

3. PCR amplify and clone the scFv in the selected vector. Confirm the correct DNA sequence.

   Note: Add 1% glucose in all growth media (steps 3 through 6) in order to reduce expression from potentially leaky CAP regulated promoters prior to induction.

4. Transform a suitable *E. coli* host (e.g. JM83 (Yanisch-Perron et al. 1985), RV308 (Maurer et al. 1980) or SB536 (Bass et al. 1996)).

5. From a single colony inoculate ca. 25 ml preculture in LB medium. For this volume use at least a 250 ml shaking flask. Shake at 25°C overnight.

6. From this overnight culture inoculate the main culture in rich medium with 1% glucose to a starting OD_{550} of 0.1. Use a baffled shake flask for higher final cell densities. Shake at 25°C and add 1 mM IPTG (final conc.) at an OD_{550} of 0.5. Continue growth for another 3 h or until the cells stop growing.

   Note: If the vector carries the *skp* or *fkpA* gene, much higher cell densities can be obtained, as the cells usually do not lyse or stop growth after induction. For details, see Chapter 23.

7. Harvest the cells by centrifugation (4500 x g, 10 min, 4°C).

   Note: From now on, all steps should be carried out at 4°C in order to minimize protease activity and to stabilize the protein of interest.

8. Resuspend the cell pellet carefully in 1/100 column volume of loading buffer. This, of course, depends on which purification method is chosen for the miniantibody. If the construct carries a his tag for IMAC purification (as for some of the vectors in Table 1 or pAK500, Fig. 2), it is recommended to use cold 50 mM Tris-Cl, 1 M NaCl, pH 8.0 (Lindner et al. 1992).

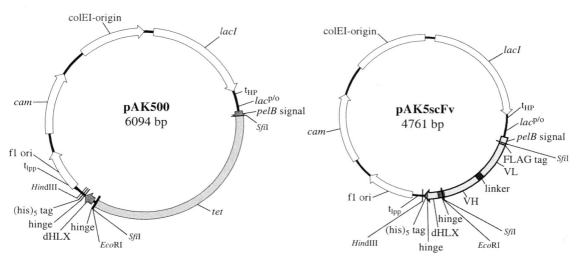

Fig. 2. pAK500 vector (Krebber et al. 1997, see also Chapter 2) containing the dHLX cassette and the tetracyclin-resistance cassette as stuffer. This stuffer, shown herre only schematically, contains the genes for *tetA* and *tetR*, and it does not make a fusion protein with upstream or downstream elements in the vector. For details, see Chapter 2. As an example for a potential miniantibody producing plasmid, the tet-cassette in pAK 500 was replaced by a scFv to build pAK5scFv. Size of the genetic elements is drawn to scale. *lacI*: lac repressor; t_{HP}: strong upstream terminator to prevent read-through from *lacI* expression; *lac* p/o: lac promoter/operator; *pelB* signal: leader sequence (pectate lyase gene of *Erwinia carotovora*), modified to contain a *SfiI* site; *tet*: tetracyclin resistance "stuffer" cassette (contains *tetA* and *tetR*-genes; 2101 bp), which is to be replaced by the antibody gene; FLAG tag: short (DYKD) version of FLAG tag for Western blot detection (Knappik and Plückthun, 1994); VH-linker-VL scFv gene; hinge: murine IgG3 hinge region; dHLX: double helix element, see Table 2; $(his)_5$ tag: stretch of 5 histidine residues for IMAC purification (Lindner et al. 1992) and detection with an anti-his tag antibody (e.g. 3D5-phosphatase fusion (Lindner et al. 1997)); t_{lpp}: downstream terminator; f1 ori: intergenic region of phage f1 (for production of single-stranded DNA; *cam*: chloramphenicol-acetyl-transferase gene; colEI-origin: plasmid replication origin (derived from pUC-plasmid series).

9. Break the cells by French Press (20000 psi, pass cells through twice) or sonification. In both methods, take care that the cell suspension is not exposed to heating.

10. Centrifuge the crude extract in order to separate insoluble cell debris from soluble protein (48000 x g, 30 min, 4°C). Carefully separate supernatant from pellet.

Note: The soluble/insoluble distribution of the miniantibody expressed may be checked by performing a Western blot (see also Chapter 23). If the gene of interest was inserted in a pAK-vector (see Table 1 and Fig. 2) via *SfiI* as described in Chapter 2, this will lead to periplasmically expressed protein carrying a short FLAG tag at its mature N-terminus (^+H_3N-DYKD...) which can be detected by the M1 anti-FLAG antibody (Kodak; Knappik and Plückthun 1994).

11. If the protein of interest is (mainly) in the soluble fraction, continue by applying the supernatant of step 10 to the appropriate chromatography column.

Note: Purification of scFvs via IMAC on the BioCAD-instrument with a rapid two-column procedure (PerkinElmer, Wellesley, MA, USA) is described in detail in Chapter 23. Conventional purification of antibody fragments via IMAC is described in detail in Lindner et al. (1992).

12. If the protein of interest is (mainly) insoluble, you may follow several lines:
  i. Co-express a molecular chaperone which may increase the soluble fraction (Chapter 23; Bothmann and Plückthun 1998, 2000).
  ii. Refold the protein from inclusion bodies. First, reclone it without signal sequence in a plasmid with the strong T7-expression system (Ge et al. 1995). Refolding, however, has to be optimized for each protein individually (for initial guidance, see Huston et al. 1991; Rudolph and Lilie 1996). However, there is a refolding kit available commercially, which may facilitate the screening for optimal conditions (Hampton Research, Laguna Niguel, CA, USA).
  iii. Probably most efficiently, introduce mutations in the scFv gene which may support proper folding or transplant the CDRs to a well folding framework, thus leading to reduced aggregation (see e.g. Knappik and Plückthun 1995; Nieba et al. 1997; Jung and Plückthun 1997; Willuda et al. 1999; Lindner, Blank and Honegger, unpublished experiments).

Acknowledgements. The authors are grateful to Kerstin Blank for helpful discussions and Barbara Klinger for excellent technical support and help in preparing Table 1.

References

Arndt KM, Pelletier JN, Müller KM, Alber T, Michnick SW, Plückthun A (2000) A heterodimeric coiled-coil peptide pair selected *in vivo* from a designed library-versus-library ensemble. J Mol Biol 295:627-639

Bass S, Gu Q, Christen A (1996) Multicopy suppressors of prc mutant Escherichia coli include two HtrA (DegP) protease homologs (HhoAB), DksA, and a truncated RlpA. J Bacteriol 178: 1154–1161

Bothmann H, Plückthun A (1998) Selection for a periplasmic factor improving phage display and functional periplasmic expression. Nature Biotechnol 16:376-380

Bothmann H, Plückthun A (2000) The periplasmic Escherichia coli peptidylprolyl cis, trans-isomerase FkpA. I. Increased functional expresssion of antibody fragments with and without cis-prolines. J Biol Chem 275:17100-17105

Burton DR (1985) Immunoglobulin G: functional sites. Mol Immunol 22:161-206

Carter P, Merchant AM (1997) Engineering antibodies for imaging and therapy. Curr Opinion Biotechnol 8:449-454

Crothers DM, Metzger H (1972) The influence of polyvalency on the binding properties of antibodies. Immunochemistry 9:341-357

Dürr E, Jelesarov I, Bosshard HR (1999) Extremely fast folding of a very stable leucine zipper with a strengthened hydrophobic core and lacking electrostatic interactions between helices. Biochemistry 38:870-880

Eisenberg D, Wilcox W, Eshita SM, Pryciak PM, Ho SP and DeGrado WF (1986) The design, synthesis, and crystallization of an alpha-helical peptide. Proteins 1:16-22

Ge L, Knappik A, Pack P, Freund C, Plückthun A (1995) Expressing antibodies in *Escherichia coli*. In: Antibody Engineering, 2nd edition, Borrebaeck CAK (ed) Oxford University Press, pp 229-266

Harbury PB, Zhang T, Kim PS, Alber T (1993) A switch between two-, three-, and four-stranded coiled coils in GCN4 leucine zipper mutants. Science 262:1401-1407

Hill RB, DeGrado WF (1998) Solution structure of α_2D, a nativelike de novo designed protein. J Am Chem Soc 120:1138-1145

Horn U, Strittmatter W, Krebber A, Knüpfer U, Kujau M, Wenderoth R, Müller K, Matzku S, Plückthun A, Riesenberg D (1996) High volumetric yields of functional dimeric miniantibodies in *Escherichia coli*, using an optimized expression vector and high-cell-density fermentation under non-limited growth conditions. (1996) Appl Microbiol Biotechnol 46:524-532

Huston JS, Mudgett-Hunter M, Tai M-S, McCartney J, Warren F, Haber E, Oppermann H (1991) Protein engineering of single-chain Fv analogs and fusion proteins. Methods Enzymol 203:46-88

Jeffrey PD, Gorina S, Pavletich NP (1995) Crystal structure of the tetramerization domain of the p53 tumor suppressor at 1.7 Ångstroms. Science 267:1498-1502

Jung S, Plückthun A (1997) Improving in vivo folding and stability of a single-chain Fv antibody fragment by loop grafting. Protein Engineering 10:959-966

Knappik A, Plückthun A (1994) An improved affinity tag based on the FLAG peptide for the detection and purification of recombinant antibody fragments. BioTechniques 17:754-761

Knappik A, Plückthun A (1995) Engineered turns of a recombinant antibody improve its *in vivo* folding. Protein Engineering 8:81-89

Krebber A, Bornhauser S, Burmester J, Honegger A, Willuda J, Bosshard HR, Plückthun A (1997) Reliable cloning of functional antibody variable domains from hybridomas and spleen cell repertoires employing a reengineered phage display system. J Immmunol Meth 201:35-55

Lindner P, Guth B, Wülfing C, Krebber C, Steipe B, Müller F, Plückthun A (1992) Purification of native proteins from the cytoplasm and periplasm of *Escherichia coli* using IMAC and histidine tails: a comparison of proteins and protocols. Methods 4:41-56

Lindner P, Bauer K, Krebber A, Nieba L, Kremmer E, Krebber C, Honegger A, Klinger B, Mocikat R, Plückthun A (1997) Specific detection of his-tagged proteins with recombinant anti-his tag scFv-phosphatase or scFv-phage fusions. BioTechniques 22:140-149

Maurer R, Meyer BJ, Ptashne M (1980) Gene regulation at the right operator (O_R) of bacteriophage λ. I. O_R3 and autogenous negative control by repressor. J Mol Biol 139:147-161

Müller KM, Arndt KM, Strittmatter W, Plückthun A (1998a) The first constant domain (C_H1 and C_L) of an antibody used as heterodimerization domain for bispecific miniantibodies. FEBS Lett 422:259-264

Müller KM, Arndt KM, Plückthun A (1998b) Model and simulation of multivalent binding to fixed ligands. Anal Biochem 261:149-158

Müller KM, Arndt KM, Plückthun A (1998c) A dimeric bispecific miniantibody combines two specificities with avidity. FEBS Lett 432:45-49

Müller KM, Arndt KM, Alber T (2000) Protein Fusions to coiled-coil domains. Methods Enzymol 328: 261–282

Nieba L, Honegger A, Krebber C, Plückthun A (1997) Disrupting the hydrophobic patches at the antibody variable/constant domain interface: improved in vivo folding and physical characterization of an engineered scFv fragment. Protein Engineering 10:435-444

O'Shea EK, Klemm JD, Kim PS, Alber T (1991) X-ray structure of the GCN4 leucine zipper, a two-stranded, parallel coiled coil. Science 254:539-544

Pack P (1994) PhD Thesis, Universität München, Germany

Pack P, Knappik A, Krebber C, Plückthun A (1992) Mono- and bivalent antibody fragments produced in E. coli: binding properties and folding in vivo. In: Harnessing Biotechnology for the 21st Century. American Chemical Society Conference Proceedings Series. Ladisch MR and Bose A (eds.) ACS

Pack P, Plückthun A (1992) Miniantibodies: Use of amphipathic helices to produce functional, flexibly linked dimeric Fv fragments with high avidity in Escherichia coli. Biochemistry 31:1579-1584

Pack P, Kujau M, Schroeckh V, Knüpfer U, Wenderoth R, Riesenberg D, Plückthun A (1993) Improved bivalent miniantibodies, with identical avidity as whole antibodies, produced by high cell density fermentation of Escherichia coli. Biotechnology 11:1271-1277

Pack P, Müller K, Zahn R, Plückthun A (1995) Tetravalent miniantibodies with a high avidity assembling in Escherichia coli. J Mol Biol 246:28-34

Plückthun A and Pack P (1997) New protein engineering approaches to multivalent and bispecific antibody fragments. Immunotechnology 3: 83-105

Plückthun A, Krebber A, Krebber C, Horn U, Knüpfer U, Wenderoth R, Nieba L, Proba K, Riesenberg D (1996) Producing antibodies in Escherichia coli: From PCR to fermentation. In: Antibody Engineering: A Practical approach. McCafferty J and Hoogenboom HR (eds), IRL press, Oxford, pp 203-252

Rheinnecker M, Hardt C, Ilag LL, Kufer P, Gruber R, Hoess A, Lupas A, Rottenberger C, Plückthun A, Pack P (1996) Multivalent antibody fragments with high functional affinity for a tumor-associated carbohydrate antigen. J Immunol 157:2989-2997

Rudolph R, Lilie H (1996) In vitro folding of inclusion body proteins. FASEB J 10:49-56

Schroeckh V, Kujau M, Knüpfer U, Wenderoth R, Mörbe J, Riesenberg D (1996) Formation of recombinant proteins in Escherichia coli under control of a nitrogen regulated promoter at low and high cell densities. J. Biotechnol. 49:45-58

Willuda J, Honegger A, Waibel R, Schubiger A, Stahel R, Zangemeister-Wittke U, Plückthun A (1999) High thermal stability is essential for tumor targeting of antibody fragments: Engineering of a humanized anti-epithelial glycoprotein-2 (epithelial cell adhesion molecule) single-chain Fv fragment. Cancer Research 59:5758-5767

Yanisch-Perron C, Vieira J, Messing J (1985) Improved M13 phage cloning vectors and host strains: nucleotide sequences of the M13mp18 and pUC19 vectors. Gene 33:103-119

ScFv-C$_{H3}$ and scFv-Fc Fusion Proteins

ANNA M. WU

Introduction

Earlier chapters (Chapters 2-16) have presented methods for producing small engineered antibodies (notably, scFv) from hybridomas or through selection from combinatorial libraries. It is now possible to generate fragments with a vast spectrum of specificities, including against targets previously inaccessible through hybridoma technology. For many final applications, however, it may be desirable to produce antibody-based molecules larger than scFv. A popular approach has been to fuse scFv to the hinge and Fc regions of immunoglobulins (scFv-Fc) or to just the third constant domain of IgG (scFv-C$_{H3}$), with or without a hinge region. Examples of antibody engineering issues that can be addressed include the following:

- Bivalent binding leading to improvement in avidity. Fusion of scFv to Fc regions or C$_{H3}$ domains themselves results in proteins that spontaneously dimerize and can exhibit bivalent binding (Shu et al., 1993; Hu et al., 1996).

- Bispecific binding. Use of a "knobs-into-holes" approach or similar strategies can generate bispecific molecules (Ridgeway et al., 1996).

- Chimerization. Incorporation of human constant domains can reduce immunogenicity.

- Control of serum half-life. Fc fusions generally retain the blood persistence properties of parental intact antibodies, whereas C$_{H3}$ fusion proteins have clearance properties intermediate between scFv and larger proteins. The serum half life is most likely modulated by interac-

Anna M. Wu, Beckman Research Institute of the City of Hope, Department of Molecular Biology, 1450 East Duarte Road, Duarte, CA, 91010, USA (*phone* +01-626-301-8287; *fax* +01-626-301-8280; *e-mail* awu@coh.org)

tions between the Fc domains with the FcRn/FcRp ("Brambell") receptor that controls serum persistence of immunoglobulins (Kim et al., 1999).

- Linkage of biological function. Antibody-dependent cellular cytolysis (ADCC) and complement-dependent cytolysis (CDC) can be linked to novel antigen specificities using scFv-Fc fusions.

- Ease of expression. Use of a single-chain format bypasses challenges associated with co-expression and assembly of separate antibody heavy and light chains.

Thus, production of scFv-Fc and scFv-C$_{H3}$ fusion proteins provides a simple means for producing antibody-like molecules that can retain desired characteristics of intact antibodies, while offering the ease of engineering and expresion of a single-gene, single-chain format. The final design, including selection of the Ig subclass and choice of constant domains to include in the fusion protein, will depend on the desired properties of the expressed fusion protein.

Overall design

Generation of scFv-Fc fusion proteins is fairly straightforward, with the most common constructs incorporating human IgG1 hinge and Fc (e.g., Shu et al., 1993). In addition, biological properties of chimeric antibodies of additional classes and subclasses have been evaluated (Bruggemann et al., 1987; Steplewski, 1988, Michaelsen et al., 1992) and provide a basis for producing additional scFv-Fc fusion proteins with a variety of biological properties. Aside from the usual considerations regarding the order of V$_H$ and V$_L$, and the length and sequence of the linker peptide in the scFv portion of the protein, the remaining design issue is treatment of the upper hinge region. In intact IgG1, the upper hinge contains a cys residue which is incorporated into a disulfide bond with the C-terminal cys of the kappa light chain. Since Cκ is absent from scFv-Fc fusion proteins, one can truncate the upper hinge, mutate the cys to another amino acid such as ser or pro, or maintain the "native" sequence, in which case it is possible that the extra cys residues could form an additional interchain disulfide bond.

Fusion of scFv to C$_{H3}$ domains (with human IgG1 C$_{H3}$ being the most frequent fusion partner) has been accomplished by two general approaches. One strategy is fusion of the C-terminus of the scFv to the

N-terminus of the C_{H3} domain using a very short peptide linker (2-4 aa; Hu et al., 1996; Li et al., 1997), which results in production of non-covalent dimers. Alternate strategies to further stabilize the molecule include incorporation of an IgG1 hinge region, allowing disulfide bond formation (Hu et al., 1996), or inclusion of cys residues at the end of the inter-variable domain linker (Li et al., 1997), to promote formation of covalent dimers. Further variations are possible, such as diabody-Fc fusions (Shan et al., 1999), diabody-C_{H3} fusions (Li et al., 1997), or larger, tetravalent single-chain diabody-C_{H3} or -Fc fusion proteins (Alt et al., 1999; Santos et al., 1999).

Gene assembly

Splice overlap PCR is the preferred method for generating antibody fusion proteins, allowing the experimenter full control over the final primary structure of the fusion protein. The construction methods for scFv outlined in Chapter 2 are easily adapted to the fusion of scFv gene segments to C_{H3} or Fc sequences. As noted above, additional amino acid residues (linkers) may be required, and these can be added analogously to the linkers in scFv molecules.

Expression of fusion proteins

Detailed methods for expression, purification, and characterization of recombinant antibodies are covered in other chapters (Chapters 17-19; 24-27) in this volume. In general, these larger recombinant fragments (scFv-Fc or scFv-C_{H3}) have been successfully expressed using mammalian systems, including myeloma cells, CHO, COS, and HEK293 cells. For glycosylation of the C_{H2} domain to occur, scFv-Fc fusion proteins must be expressed in eukaryotic cells. ScFv-C_{H3} proteins have also been expressed bacterially, but with low recoveries (Hu et al., 1996).

In particular, use of the NS0 myeloma/glutamine synthetase system of Bebbington et al. (1991, 1992; available from Lonza Biologics, Slough, Berkshire, UK) has consistently resulted in high expression of scFv-Fc and scFv-C_{H3} fusion proteins. The pEE12 expression vector contains a hCMV promotor to drive transcription of the recombinant antibody gene, and also carries the glutamine synthetase gene for selection. Host NS0 cells require glutamine for growth. Following introduction of the expression plasmid by standard electroporation methods, selection is accomplished by removing glutamine from the cell growth medium.

Purification and characterization

Purification of fusion proteins has been facilitated by incorporation of affinity tags (such as myc or 6His peptides, see Chapter 17). Non-tagged proteins have been purified by standard approaches including Protein A (for Fc fusion proteins), Protein L, anti-idiotype affinity chromatography, or conventional ion-exchange or other chromatography methods (see related Chapter 21 in this volume). Anti-Fc reagents are readily available for assay of scFv-Fc fusion proteins, and many commercial anti-Fc reagents retain sufficient reactivity for the C$_{H3}$ domain that they can also be useful for detection of scFv-C$_{H3}$ fusion proteins (e.g., anti-human Fc; Jackson ImmunoResearch, West Grove, PA, USA).

Characterization of purified proteins should include standard biochemical analysis, and determination of affinity and specificity. Depending on the final application, biological function in ADCC and CDC assays can be evaluated *in vitro*, and/or targeting and pharmacokinetic properties assessed *in vivo* (see related Chapter 35 and Chapter 36 elsewhere in this volume).

References

Alt M, Muller R, Konterman RE. (1999) Novel tetravalent and bispecific IgG-like antibody molecules combining single-chain diabodies with the immunoglobulin γ1 Fc or C$_{H3}$ region. FEBS Letters 454:90-91.

Bebbington CR. (1991) Expression of antibody genes in nonlymphoid mammalian cells. METHODS: A Companion to Methods in Enzymology 2:136-145.

Bebbington CR, Renner C, Thomson S., King D, Abrams D, Yarranton GT. (1992) High-level expression of a recombinant antibody from myeloma cells using a glutamine synthetase gene as an amplifiable selectable marker. Bio/Technology 10:169-175.

Bruggemann M, Williams GT, Bindon CI, Clark MR, Walker MR, Jefferis R, Waldmann H, Neuberger MS. (1987) Comparison of the effector functions of human immunoglobulins using a matched set of chimeric antibodies. J. Exp. Med. 166:1351-1361.

Hu SZ, Shively LE, Raubitschek A, Sherman M, Williams LE, Wong JYC, Shively JE, Wu AM. (1996) Minibody: a novel engineered anti-carcinoembryonic antigen antibody fragment (single-chain Fv-C$_{H3}$) which exhibits rapid, high-level targeting of xenografts. Cancer Res. 56:3055-3061.

Kim JK, Firan M, Radu CG, Kim CH, Ghetie V, Ward ES (1999) Mapping the site on human IgG for binding of the MHC Class I-related receptor, FcRn. Eur. J. Immunol. 29:2819-2825.

Li E, Pedraza A, Bestagno M, Mancardi S, Sanchez R, Burrone O. (1997) Mammalian cell expression of dimeric small immune proteins (SIP). Prot. Eng. 10:731-736.

Michaelsen GE, Aase A, Norderhaug L, Sandlie I. (1992) Antibody dependent cell-mediated cytotoxicity induced by chimeric mouse-human IgG subclasses and IgG3 antibodies with altered hinge region. Molec. Immunol. 29:319-326.

Ridgway JB, Presta LG, Carter P. (1996) "Knobs-into-holes" engineering of antibody C_{H3} domains for heavy chain heterodimerization. Prot. Eng. 9:617-21.

Santos AD, Kashmiri SVS, Hand PH, Schlom J, Padlan EA. (1999) Generation and characterization of a single gene-encoded single-chain-tetravalent antitumor antibody. Clin. Canc. Res. 5:3118s-3123s.

Shan D, Press OW, Tsu TT, Hayden MS Ledbetter JA. (1999) Characterization of scFv-Ig constructs generated from the anti-CD20 mAb 1F5 using linker peptides of varying lengths. J. Immunol. 162:6589-6595.

Shu L, Qi CF, Schlom J, Kashmiri SVS. (1993) Secretion of a single-gene-encoded immunoglobulin from myeloma cells. Proc. Nat. Acad. Sci. USA 90:7995-7999.

Steplewski Z, Sun LK, Shearman CW, Ghrayeb J, Daddona P, Koprowski H. (1988) Biological activity of human-mouse IgG1, IgG2, IgG3, and IgG4 chimeric monoclonal antibodies with antitumor specificity. Proc. Nat. Acad. Sci. USA 85:4852-4856.

CTL-Assays for Functional Testing of Bispecific Antibody Fragments

ANJA LÖFFLER, TORSTEN DREIER, and RALF C. BARGOU

Introduction

Bispecific antibody fragments, designed as therapeutic agents for the treatment of neoplastic diseases, usually combine the binding regions of antibodies against a specific antigen on the target cell with an antibody binding site against cell surface molecules on effector cells of the immune system such as NK cells or T lymphocytes. The establishment of appropriate in vitro assays for the determination of target cell lysis by the effector cells, activated by these molecules, are absolutely crucial for the functional testing. The choice of this assay mainly depends on the nature and availability of the target cells or the technical equipment available (e.g. the opportunity to work with radioactively labelled material).

In this chapter, three different assays for the determination of T cell cytotoxicity, mediated by bispecific antibody fragments, are described at the example of a bispecific single chain antibody CD19xCD3 (bscCD19xCD3).

- Subprotocol 1:
  The **chromium-51 release assay** is a technique developed in the 70s and applicable for different purposes and a variety of target cells and effector cells such as T cells, natural killer cells, neutrophils and macrophages. The target cells are intracellularly labelled by chromium-51.

✉ Anja Löffler, Medical School of the Humboldt University Berlin, Robert-Rössle-Clinic, Dept. of Hematology/Oncology/Tumorimmunology, Lindenberger Weg 80, 13122 Berlin, Germany (*phone* +49-30-94063519; *fax* +49-30-94063124; *e-mail* anja.loeffler@yahoo.com)
Torsten Dreier, Micromet GmbH, Am Klopferspitz 19, 82152 Martinsried, Germany
Ralf C. Bargou, Medical School of the Humboldt University Berlin, Robert-Rössle-Clinic, Dept. of Hematology/Oncology/Tumorimmunology, Lindenberger Weg 80, 13122 Berlin, Germany, Max Delbrück Center for Molecular Medicine, Robert-Rössle-Str. 10, 13092 Berlin, Germany

The chromium-51 release by the target cells after lysis is a determinant of T cell cytotoxicity. This technique is a reliable killing assay. It is easy to perform and results obtained are reproducible. A disadvantage is the need of facilities for work with radioactive material. Furthermore, it is our experience that primary cells are often not sufficiently labelled by chromium-51.

- Subprotocol 2 **PKH26 cytotoxicity assay:**
  This assay is based on labelling the target cell population with the lipophilic fluorescent dye PKH26 (orange) and differentiating between viable and non-viable cells via propidium iodine (PI) exclusion. In that way labelled target cells can easily be distinguished from non-labelled effector cells by their PKH26 fluorescence (FL-2). Dead cells with compromised membrane integrity incorporate propidium iodine in their DNA. This results in an intensive FL-3 signal compared to viable cells showing no signal for FL-3. In FACS analysis one can clearly differentiate the differentially stained cell populations in a FL-2 vs. FL-3 dot plot: Surviving target cells are FL-2 positive and FL-3 negative, killed target cells are both FL-2 and FL-3 positive, while viable effector cells are FL-2 and FL-3 negative and dead effector cells are FL-2 negative but FL-3 positive. Cytotoxicity is calculated depending on the number of surviving target cells. This method has a number of advantages:
  - It does not depend on radioactive labelling which is not feasable in all laboratories.
  - The use of a stably membrane-integrating lipophilic dye allows longer incubation times compared to intracellular labels with leaking tendency.
  - Cytotoxicity is calculated only upon the number of surviving target cells as killed target cells may desintegrate thereby generating more than a single FACS event.
  - In our hands primary cells tend to label better with PKH26 than with chromium-51. A disadvantage compared to a chromium release assay is that the assay is more labour intensive and that more cells are needed.

- Subprotocol 3 **Depletion assay:**
  This assay has been established for the measurement of T cell cytotoxicity mediated by bscCD19xCD3 in patient blood samples. In the blood of patients suffering from chronic lymphatic leukemia of the B cell type there are often very few T lymphocytes (<10%) found among the CD19 positive malignant B cells. Therefore the E:T ratio is very low (1:10 or even less) requiring very long assay incubation times for the detection

of T cell cytotoxicity against the autologous tumor cells. This makes the application of assays based on the labelling of target cells almost impossible. The principle of this method is very simple. Lymphocytes are isolated from peripheral blood of B-CLL patients, bscCD19xCD3 is added and after incubating for several days the portion of effector and remaining target cells is measured by FACS analysis. Although this method has been established for the special purpose of B-CLL patient samples, it should be applicable in all cases where only few effector cells are obtained or when it is impossible to separate effector and target cells or to specifically label the target cells. The technique is very easy to perform and requires only a flow cytometer. Disadvantages are the need for larger cell numbers and that the assay is very difficult to standardize.

Subprotocol 1
Chromium Release Assay

▪▪ Materials

- Gamma counter (e.g., Topcount, Canberra Packard)
- Chromium-51 as sodium chromate in saline (NEN Life Science Products Inc.)

RPMI 1640 medium (e.g., Biochrom)   Media

- 10% FCS
- 100 IU/ml penicillin
- 100 µg/ml streptomycin
- 50 µM β-mercaptoethanol

▪▪ Procedure

1. Isolate human peripheral blood mononuclear cells (PBMCs) as effector   Isolation of PBMCs
   cells from fresh buffy coats of random donors using Ficoll density gradient centrifugation with subsequent 100 x g centrifugation steps to remove thrombocytes.

2. Remove unwanted cell populations, in our example CD19-positive B cells, using immunomagnetic beads (Dynabeads M-450 CD19, Dynal).

3. Analyse the cell populations by flow cytometry (FACScan Becton Dickinson).

4. Incubate PBMCs over night at 37°C, 5% CO_2 to remove monocytes and macrophages which adhere to the culture flask.

5. Use these unstimulated PBLs directly as effector cells or stimulate T cells before performing the chromium release assay.

Target cell labelling

6. Centrifuge $1x10^6$ target cells, carefully remove the supernatant and label the cell pellet with 50µl chromium-51 in a round bottom tube 2 h at 37°C, 5% CO_2 (mix several times during this incubation period).

7. Wash cells twice with RPMI 1640 without supplements.

8. Add 10 ml RPMI 1640 without supplements, incubate cells 20 min at 37°C, 5% CO_2.

9. In the mean time, count effector cells and adjust cell number to $1x10^5$ cells/80µl in the case of an E:T ratio of 10:1 (Try different E:T ratios from 1:1 to 20:1.).

10. Centrifuge and resuspend the target cells in complete RPMI medium, count the target cells and adjust the cell number to $1x10^5$ cells/ml.

Chromium-51 release assay

11. Add 80 µl effector cells per well of a round-bottom 96 well microtiter-plate ($1x10^5$ cells for E:T ratio 10:1).

12. Add 20 µl antibody solution diluted in complete RPMI medium, or 20 µl 10% Triton-X100 or medium as controls.

13. Add 100 µl labelled target cells to each well.

14. For measuring the spontaneous release of chromium-51 by the target cells, incubate target cells alone. Maximal release is determined by incubation of target cells in 1% triton-X100. Background cytotoxicity of effector cells against the allogenic target cells is measured by incubation of effector cells and target cells without antibody.

15. Centrifuge plates for 3 min at 100 x g and incubate for 4 h (or 8 h) at 37°C, 5% CO_2.

16. Centifuge plates for 3 min at 100 x g and transfer 50 µl supernatant at a lumaplate.

17. Dry and seal plate and measure chromium-51 release in a gamma counter.

Results

The test should be carried out in triplicates, SD within the triplicates is usually below 6%.

Counts per minute (cpm) of maximal release should be at least tenfold of cpm of spontaneous release. The percentage specific lysis is calculated as specific lysis (%)= [(cpm, experimental release) - (cpm, spontaneous release)] / [(cpm, maximal release) - (cpm, spontaneous release)] x 100. An example of a result of such a chromium release assay using different target cell lines is given in Figure 1.

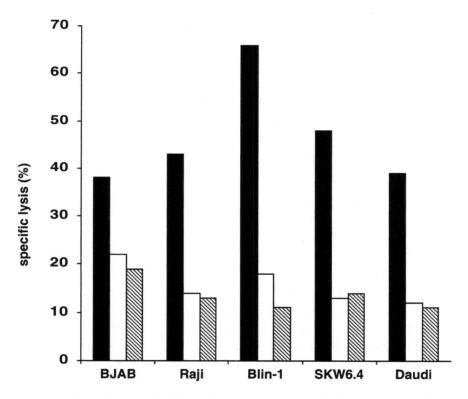

■bscCD19xCD3 (0.5 µg/ml) □control bscFv (0.5 µg/ml) ▨medium

Fig. 1. Cytotoxicity of bscCD19xCD3 in a chromium-51 release assay with primary human PBLs and different CD19-positive B cell lines (BJAB, Raji, Blin-1, SKW6.4, Daudi); E:T ratio 10:1; incubation time 8 h. Standard deviation in all triplicates was below 5%.

▨▨ Comments

- To establish this assay, it is recommended to examine different target cell lines to get the widest margin between allogenic cytotoxicity and experimental cytotoxicity.

- Effector T cells can be used either unstimulated or stimulated according to different protocols, e.g. stimulation with IL-2 (60 U/ml) alone, with PHA (2 µg/ml) + IL-2 (60 U/ml), with anti-CD3 antibodies (coated at 100 µg/ml) + anti-CD28 antibodies (1 µg/ml) or with the bispecific antibody itself.

- The assay can also be performed with isolated subpopulations of T cells (e.g., isolation of CD8-positive cells with immunomagnetic beads).

- The percentage of allogenic lysis varies between different donors within a range of 0 to 20%.

- A collection of possible specificity controls:
  - Incubation of target cells with antibody alone to exclude toxicity of the antibody solution itself.
  - Incubation of target and effector cells with the monospecific, parental antibodies and a combination of them and / or an irrelevant bsc antibody containing the same effector cell binding site.
  - Competition of bsc-mediated cytotoxicity by parental monoclonal antibodies (Pre-incubate target cells with the monoclonal for at least 30 min prior to the addition of effector cells and bsc.).
  - Use cells lacking the target antigen as target cells.
  - Exclusion of a bystander effect: label cells lacking the target antigen and use them in a mixture with unlabelled cells expressing the specific antigen as target cells.

Subprotocol 2
PKH26 Cytotoxicity Assay

▨▨ Materials

Equipment – Flow cytometer (e.g., FACSCalibur, Becton Dickinson)

Kits – PKH26-GL Fluorescent Cell Linker Kit (Sigma)

- Human T Cell Enrichment Columns (e.g., RD systems Cat-No. HTCC-500/525)

Erythrocyte Lysing Buffer   **Buffers**

- 8.29 g/l NH_4Cl
- 1.0 g/l $KHCO_3$
- 0.037 g/l EDTA

▒▒ Procedure

1. Isolate human peripheral blood mononuclear cells (PBMCs) as effector   **Isolation of PBMCs**
   cells from fresh buffy coats of random donors using Ficoll density gradient centrifugation with subsequent 100 x g centrifugation steps to
   remove thrombocytes.

2. Resuspend the pellet in 10 volumes erythrocyte lysing buffer. Incubate   **Erythrocyte lysis**
   at room temperature for 10 min.

3. Stop the lysing reaction by adding 10 volumes PBS.

4. Wash the pellet with 50 ml PBS.

5. Perform enrichment of T lymphocytes on Human T Cell Enrichment   **T cell enrichment**
   Columns (R&D Systems) with a maximum of $2.5x10^8$ cells according to
   the manufacturers protocol. Approximately 25% of the applied cells
   are eluted as T cells.

6. Wash the T cells, resuspend the T cells in RPMI medium. Adjust cell
   number to $1,25x10^7$ cells/ml.

7. Wash $5x10^6$ target cells twice with PBS. Gently resuspend the pellet in **Target cell labelling**
   0.5 ml diluent C (PKH26-GL Fluorescent Cell Linker Kit, Sigma).

8. Add PKH26 dye (2 µl of a $1x10^{-3}$ M stock solution in 0.5 ml diluent C) to
   a final concentration of $2x10^{-6}$ M.

9. Incubate at room temperature for 2 min. Periodically, invert the tube
   gently to assure mixing.

10. Stop the staining reaction by adding 0.5 ml FCS. Incubate 1 min at
    room temperature.

11. Add 10 ml RPMI medium. Centrifuge cells at 400 x g for 10 min. Wash
    three times with RPMI medium. Transfer cells to a new tube prior to
    the last washing step.

12. Adjust target cell number to $1x10^6$ cells/ml.

PKH26 assay 13. Add 100 µl target cell suspension to each well of a 96 well round bottom plate, add 80 µl T cell suspension and 20 µl bispecific antibody in different concentrations or 20 µl RPMI/10% FCS as a negative control.

14. Incubate plate for 4 h at 37°C, 5% CO_2.

15. Add 2 µl propidium iodine (stock solution 100 µg/ml, final concentration 1 µg/ml). Incubate 15 min at room temperature.

16. Analyse the samples by flow cytometry (e.g., FACSCalibur Becton Dickinson), count $1x10^5$ events (E:T ratio 1:10 or less events for lower E:T ratios).

Results

The samples are analysed by FACS in a FL-2 vs FL-3 diagram. This diagram shows the population of viable PKH26-labelled target cells in the lower right corner and the population of dead target cells in the upper right corner, respectively (Figure 2). Viable target cells are gated and the total cell number within this gate is determined. Specific toxicity is calculated as 100 x (1-(viable target cells of the test sample) / average number of viable target cells in the control samples (without antibody). All incubations are carried out in triplicates. Results and SD are comparable to standard chromium release assays.

In our experiments, dramatic target cell lysis could already be observed after 4 h. However, longer incubation periods (overnight) are easily applicable if lysis within 4 h is not sufficient.

Comments

- For the adjustment of the instrument settings use target cells + PBLs (for FSC and SSC), PKH26-labelled target cells (FL-2) and non-labelled target cells, completely lysed in 0.01% saponine solution (15 min at room temperature) and stained with PI (FL-3).

- Target cell labelling: after the final washing steps with PBS, the cell pellet should show a strong reddish color.

- Usage of frozen PBMCs as effector cells is possible, but results in an approximately 25% reduced cytotoxicity.

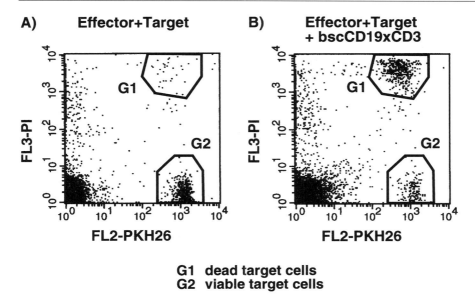

A) Effector+Target

B) Effector+Target + bscCD19xCD3

G1 dead target cells
G2 viable target cells

Fig. 2. Dot plots of a typical PKH26 cytotoxicity assay. Effector and target cells have been co-incubated for 4 h in the absence (A) or in the presence of 10 ng bscCD19xCD3 (B) prior to FACS analysis. Surviving target cells (region G2) can be clearly distinguished from the dead target cells (region G1) and effector cells (non-PKH26 stained cells on the left side of the dot plot) and quantified within region G2.

Subprotocol 3
Depletion Assay

▓▓ Materials

– Flow cytometer (e.g., FACScan, Becton Dickinson)   **Equipment Media**

– RPMI 1640 medium (e.g. Biochrom), supplemented with:
    – 10% FCS
    – 100 IU/ml Penicillin
    – 100 µg/ml Streptomycin
    – 50 µM β-mercaptoethanol

■ ■ Procedure

Isolation of patients PBMCs

1. Isolate primary lymphocytes of patients with chronic lymphatic leukemia of the B cell types (contain effector cells and target cells in our example) using Ficoll density gradient centrifugation.

Depletion assay

2. Block the surface of a 24 well plate by incubating with PBS/10% FCS (1.5 ml/well) for 1 h at 37°C, 5% CO_2, remove the PBS/ 10% FCS.

3. Adjust number of patients PBMCs to $3x10^6$ cells/ml in RPMI medium with supplements.

4. Add 1 ml cell suspension to each well.

5. Add bispecific antibody fragments, control antibodies, medium as a negative control and PHA + IL-2 as a proliferation control (final concentrations: PHA 2 µg/ml; IL-2 60 U/ml).

6. Incubate the assay for up to 10 days at 37°C, 5% CO_2.

   Note: It is recommended to measure the assay at two different time points depending on the T cell portion at the starting point, e.g. after 5 and 8 days in the case of T cell portion less than 20%. Generally, the medium color turning into yellow is a strong indication for finishing the assay.

7. Use 0.5 ml of the cell culture to perform a FACS analysis for measuring the T cell and target cell portion within the viable cell population (PI staining of dead cells).

   Note: Use an antibody against a different antigen to exclude that the target antigen is masked by bispecific antibody fragments at higher concentrations (in our example CD20 for B cells and CD4 + CD8 for T cells).

8. Perform a cell count of living and dead cells using tryptophane blue within the cell culture to determine the absolute cell numbers.

Results

The outcome and the incubation time of these experiments strongly depend on the patient, from which the cells are derived, and on the amount of T cells in the culture. In some cases, addition of low dose IL-2 (60 U/ml) is necessary, and the amount of antibody required for target cell depletion varies among different patients. Complete depletion of target cells can be expected after 5 or 6 days. If there is no cytotoxic effect after 10 days, the patients cells can not be activated by this experiment.

Comments

An important control is the determination of the T cell proliferation by stimulation with PHA + IL-2 in order to exclude that the effect is only due to T cell proliferation and not to specific target cell lysis.

References

Löffler A (2000): A recombinant bispecific single chain antibody CD19xCD3 induces rapid and high lymphoma-directed T cell cytotoxicity. Blood 95:2098-2103

Martin S (1980) A new neutrophil candida killing test: chromium -51 release from Candida guilliermondii. J. Clin. Pathol. 33(8): 757-61

Johnson TR (1972) Lymphocyte and antibody cytotoxicity to tumor cells measured by a micro- 51 chromium release assay. Immunol. Commun. 1 (3): 247-61

Capron A (1977) Interaction between IgE complexes and macrophages in the rat: a new mechanism of macrophage activation. Eur. J. Immunol. 7(5): 315-22

Targan S (1981) Activation of human NKCC by moderate exercise: increased frequency of NK cells with enhanced capability of effector–target lytic interactions. Clin. Exp. Immunol. 45(2): 352-60

Rowe M (1982) Monoclonal antibodies to Epstein-Barr virus-induced, transformation-associated cell surface antigens: Binding patterns and effect upon virus-specific T-cell cytotoxicity. Int. J. Cancer 29(4): 373-81

Slezak SE (1989) Cell- mediated cytotoxicity. A highly sensitive and informative flow cytometric assay. J. Immunol. Methods 117(2): 205-14

Recombinant Antibody-Fusion Proteins

Antibody-Enzyme Fusions

DIANNE L. NEWTON and SUSANNA M. RYBAK

Introduction

Toxins isolated from plants and bacteria have been coupled to monoclonal antibodies to produce potent selective cytotoxic agents (immunotoxins) for the treatment of cancer, AIDS, and immunological diseases (Vitetta 1993) (Brinkmann 1994). However a major obstacle to the successful use of these reagents has been toxic side effects and immunogenicity (Sawler 1985) (Schroff 1985) (Harkonen 1987) (Rybak 1991) (Soler-Rodriguez 1993) (Vitetta 1993). While much effort has been devoted to the humanization of the antibody portion of the reagent (See Chapters 6, 7, 14, 39, 40 and references therein), the toxin moiety still remains a problem. Human enzymes are now being used in lieu of plant and bacterial enzymes (reviewed in (Rybak 1999)). This chapter describes the preparation of RNase-based antibody fusion proteins. Non-cytotoxic human RNases such as eosinophil-derived RNase (EDN), pancreatic RNase, or angiogenin are equally or more cytotoxic than plant and bacterial toxins when injected into *Xenopus* oocytes (Saxena 1991). RNases linked chemically or fused genetically to cell surface binding ligands have potent anti-tumor effects both in vitro and in vivo. Furthermore therapeutic RNases do not cause non-specific toxic side-effects associated with the classical plant and bacterial toxins in mice (reviewed in (Rybak 1999)) or in humans (Mikulski 1993).

Dianne L. Newton, SAIC Frederick, National Cancer Institute-Frederick Cancer Research and Development Center, Frederick, MD, 21702, USA
✉ Susanna M. Rybak, National Cancer Institute - Frederick Cancer Research and Development Center, Developmental Therapeutics Program, Division of Cancer Treatment and Diagnosis, Bldg 567, Rm162, Frederick, MD, 21702, USA
(*phone* +01-301-846-5215; *fax* +01-301-846-7022; *e-mail* rybak@ncifcrf.gov)

Outline

As shown in Fig 1, this chapter describes the construction, expression, purification, and characterization of an RNase-sFv fusion protein. The RNase and sFv genes are each modified separately to contain complementary FB spacer DNA before they are combined and spliced together using the PCR technique of splicing by overlap extension (Horton 1990). Bacteria, BL21(DE3), specifically engineered for the expression of toxic proteins (Studier 1990), are transformed with the DNA encoding the fusion protein. Protein expression is induced by IPTG and the protein is expressed as insoluble aggregates in inclusion bodies. The inclusion bodies are purified, the fusion protein solubilized, denatured, and renatured before application to either a CM-Sephadex C-50 or heparin Sepharose column. Since RNases are basic proteins, a cationic exchanger resin such as CM-Sephadex C-50 is the recommended first column. The majority of contaminating proteins do not adhere to this resin and thus a relatively pure protein can be obtained. An affinity column such as Ni^{2+}-NTA agarose as the second column then results in a pure protein. The purified fusion protein is then characterized for binding and ribonuclease activity. The RNase assay described here uses tRNA as the substrate, is a relatively simple assay to perform which does not require any special equipment, and allows many samples to be processed simultaneously. Finally the RNase-sFv fusion protein is characterized for in vitro cytotoxic activity.

Materials

Equipment
- PCR thermocycler

- Bacterial shaker, temperature $34°C$

- Janke & Kunkel (Janke & Kunkel, GMBH, KG Staufen, Germany) polytron tissuemizer or similar model with a 100 mm long x 10 mm OD shaft

- 60 L vat equipped with a lower valve for easy emptying

- PHD cell harvester (Cambridge Technology Inc., Watertown, MA, USA)

- Scintillation counter for measuring [^{14}C]leucine incorporation or microtiter plate reader (MR4000; Dynatech Laboratories, Chantilly, VA, USA or equivalent) for WST-1 assay

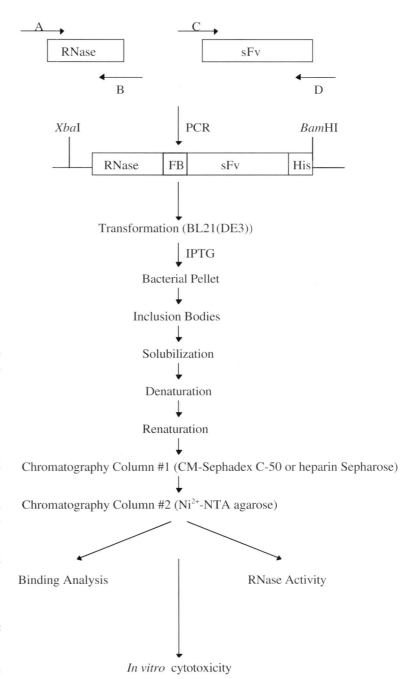

Fig. 1. Outline of construction, expression, purification, and characterization of an RNase-sFv fusion protein. The modified gene encoding the RNase protein is spliced onto the modified gene encoding the sFv (see Fig 2 and text). The resulting DNA is expressed in the bacteria, BL21(DE3). After induction by IPTG, the bacteria is pelleted, and inclusion bodies containing the protein are purified. The fusion protein is solubilized, denatured, and renatured before application to the first chromatography column. If the fusion protein contains a histidine tag, it is purified to homogeneity by Ni^{2+}-NTA agarose chromatography. The activities of both halves of the fusion protein are characterized (binding of the sFv and enzymatic activity of the RNase) before the activity of the total protein (cytotoxicity to cells) is determined.

- gel electrophoresis apparatus for running agarose gels

- gel electrophoresis apparatus for running protein gels

- FPLC system (optional) consisting of program controller, two pumps, a mixer, seven port M-7 valve, assorted sample loops, UV monitor, fraction collector, and appropriate columns such as cationic exchanger (Mono S HR 5/5) or anionic exchanger (Mono Q HR 5/5) (Pharmacia LKB Biotechnology Inc., Piscataway, NJ, USA)

- appropriate chromatography columns for purification of the RNase fusion protein (CM-Sephadex C-50, heparin Sepharose (both from Pharmacia LKB Biotechnology Inc., Piscataway, NJ, USA), affinity columns such as Ni^{2+}-NTA agarose (Qiagen, Chatsworth, CA, USA)

- H_2O bath or incubator at $37^{\circ}C$

- UV absorbance spectrophotometer

Reagents
- Genes encoding the sFv and RNase

- Reagents for performing PCR; GeneAmp PCR Reagent Kit (Perkin-Elmer, Foster City, CA, USA)

- Primers see Table 1 and **Procedure** for Construction of the RNase-sFv gene Step 1 for discussion of design

- Vector, pET-11d, or appropriate pET vector (Novagen, Madison, WI, USA)

  Note: The pET vector system was specifically designed for the expression of toxic genes (Studier 1990). A variety of pET vectors with different peptide tags for affinity purification are available.

- Appropriate restriction enzymes

- Agarose and NuSieve 3:1 Agarose (FMC BioProducts, Rockland, ME, USA)

- GeneClean II (Bio 101 Inc., La Jolla, CA, USA)

- Rapid DNA Ligation Kit (Boerhinger Mannheim, Indianapolis, IN, USA)

- DNA isolation kits: Wizard Plus Minipreps DNA Purification System (Promega, Madison, WI, USA), Qiagen Plasmid Maxi Kit (Qiagen, Chatsworth, CA, USA)

- Competent bacteria for generating plasmid DNA: XL1-Blue (Stratagene, La Jolla, CA, USA) or any host lacking the T7 RNA polymerase gene such as JM109 or HB 101.

- Competent bacteria for generating protein; BL21(DE3) (Novagen, Madison, WI, USA)

   Note: BL21(DE3) has been specifically engineered for the expression of toxic proteins and is not used for generating plasmid DNA. For expression of extremely toxic proteins, BL21(DE3)pLysS or pLysE should be tried.

- Luria broth (LB) and LB/amp plates (100 µg/ml) prepared as described in (Sambrook 1989)

- Terrific Broth (TB) prepared as described in (Sambrook 1989), add 1 ml of 100 mg/ml carbenicillin

- Antibiotics (Sigma, St Louis, MO, USA): ampicillin, 100 mg/ml in H_2O; carbenicillin, 100 mg/ml in H_2O

- Rifampicin (Sigma, St Louis, MO, USA), 20 mg/ml in methanol

- Isopropyl-β-D-thiogalactopyranoside (IPTG), 120 mg/ml (0.5 M) in H_2O (Gibco BRL, Grand Island, NY, USA)

- 4-20% Sodium dodecyl (SDS)-polyacrylamide gels (Novex, San Diego, CA, USA) or equivalent

- Sucrose buffer (ice cold): 30 mM Tris-HCl, pH 7.5, containing 20% sucrose and 1 mM EDTA Tris-EDTA buffer (TE): 50 mM Tris-HCl, pH 7.5, containing 20 mM EDTA

- Solubilization buffer: 0.1 M Tris-HCl, pH 8.0, containing 6 M guanidine-HCl (573.2 g/l), and 2 mM EDTA

- Renaturation buffer pre-chilled to 10°C: 0.1 M Tris-HCl, pH 8.0, containing 0.5 M L-arginine-HCl (105 g/l, Sigma, St Louis, MO, USA), 8 mM glutathione, oxidized (GSSG) (4.9 g/l, Boehringer Mannheim, Indianapolis, IN, USA), and 2 mM EDTA

- Dialysis buffer (10x): 0.2 M Tris-HCl, pH 7.5, containing 1 M urea (60 g/l)

   Note: Do not make more than 2-3 h in advance because cyanate ions form which can carbamylate amino groups. Dilute to 1x just before use.

- Lysozyme (Sigma, St Louis, MO, USA), 5 mg/ ml in H_2O, prepare just before use

- Triton X-100 (Sigma, St Louis, MO, USA), 25% solution in H_2O, use low heat to solubilize

- Dithioerythritol (DTE) (Sigma, St Louis, MO, USA)

- BCA protein assay reagent (Pierce Chemical Co., Rockford, IL, USA) or equivalent

- Chromatography buffer A: 20 mM Tris-HCl, pH 7.5, 10% glycerol

- Chromatography buffer B: Chromatography buffer A, containing 1 M NaCl

- Chromatography buffer with varying imidazole: Chromatography buffer A containing the following concentrations of imidazole; 40, 50, 60, 100, 200, 300, 400, 500 mM

- RNase-free tubes and pipette tips (disposable pre-sterilized tubes and tips may be used without autoclaving)

- RNase-free H_2O (deionized H_2O that has been autoclaved for 30 min is sufficient)

- tRNA (Sigma, St. Louis, MO, USA) in RNase-free H_2O

- Human serum albumin (HSA) (Sigma, St Louis, MO, USA)

- RNase assay buffer, for pH 7.5 or 8.0, 0.5 M Tris-HCl, pH 7.5 or 8.0, containing 5 mM EDTA and 0.5 mg/ml HSA; for pH 6.0, 0.4 M MES, pH 6.0, containing 0.5 mg/ml HSA

- Dilution buffer, 0.5 mg/ml HSA in RNase-free H_2O

- Ice-cold perchloric acid, 3.4% (2.4 ml 70% perchloric acid diluted to 50 ml with H_2O)

- 96-well microtiter plates (Corning Costar, Corning, N.Y., USA)

- Millipore Millex-HV filters (Millipore Products Division, Bedford, MA, USA)

- Serum- and leucine-free RPMI (Gibco BRL, Gaithersburg, MD, USA)

- [^{14}C]leucine, 0.1 mCi in 10 µl phosphate buffered saline (PBS) (NEN Life Science Products, Boston, MA, USA) or WST-1 (Boehringer Mannheim, Indianapolis, IN, USA)

- Glass fiber filters (Brandel Inc., Gaithersburg, MD, USA) (only if [^{14}C]leucine is used)

- Scintillation fluid (only if [^{14}C]leucine is used)

Procedure

Construction of the RNase-sFv gene

1. Design four oligonucleotide primers (see Fig 2 and Table 1) to incorporate a) an appropriate restriction enzyme site at the 5′ end of the RNase gene to facilitate cloning (sense direction, primer A), b) a spacer onto the 3′ end of the RNase gene to separate the RNase and the sFv genes (antisense direction, primer B), c) the same spacer described in b onto the 5′ of the sFv gene (sense direction, primer C), and d) the termination signal, 3-6 histidyl residues for affinity purification, and the appropriate restriction enzyme site to facilitate cloning at the 3′ end of the sFv gene (antisense direction, primer D).

   Primer design

   Note: Before beginning, the optimal orientation of the RNase with respect to the sFv should be determined. Fusion to a terminus of the RNase involved in the active site should be avoided. For example, the last three amino acid residues of the C-terminal region of the human RNase, angiogenin, are involved in an active center subsite (Russo 1996) while for the amphibian RNase, onconase, the N-terminal pyroglutamic acid folds back into the active site center (Mosimann 1994). Analyze the DNA encoding the RNase and sFv for naturally occurring restriction enzyme sites contained within the gene before designing primers A and D. The use of 2 different restriction enzyme sites will facilitate the directional insertion of the gene into the vector. Amino-acid residues 48-60 of fragment B of staphylococcal protein A have been used as a spacer sequence to separate the RNase and sFv (Tai 1990). RNase fusion proteins without a spacer have lower enzymatic, receptor binding and cytotoxic activities (Newton 1996). The histidyl residues in primer D may be eliminated if using a pET vector that already contains a tag for affinity purification.

2. Set up 2 separate PCR reactions to modify the RNase gene (PCR #1) and the sFv gene (PCR #2). Use primers A and B in PCR #1 and C and D in PCR #2 (see Fig 2). Each PCR reaction should contain the following per 100 μl reaction volume:

   PCR reaction

   Final Concentration

   - 1X reaction buffer (10 mM Tris-HCl, pH 8.3, 50 mM KCl, 1.5 mM MgCl$_2$)
   - 100 μM of each of the nucleotides

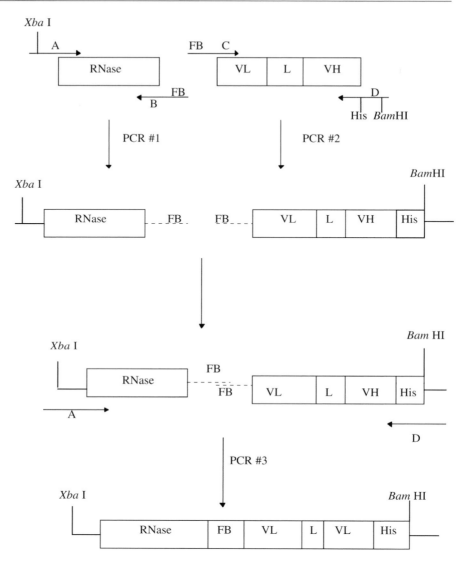

Fig. 2. Scheme of the strategy used for the synthesis of the RNase-sFv fusion protein. Primers A and D include the appropriate restriction sites for cloning into the pET-11d vector (see Table 1.). In addition primer D includes bp encoding 3 histidine residues, a tag for affinity purification of the fusion protein. Primers B and C contain complementary FB coding sequences used as a spacer to separate the RNase from the sFv. The RNase and sFv are each amplified separately by PCR (PCR #1 and #2). The PCR products are purified, mixed, and subjected to a second round of PCR (PCR #3) using Primers A and D. The final gene product is shown at the bottom of the figure. *Xba*I and *Bam*HI, restriction sites; RNase, ribonuclease; FB, 13 amino acid residue spacer peptide composed of residues 48-60 of Fragment B of staphylococcal protein A (AKKLNDAQAPKSD) (Tai 1990); VL, light chain variable region; L, flexible 15-amino acid residue linker, (GGGGS)$_3$ (Huston 1988); VH, heavy chain variable region; His, histidine tag for affinity purification of Ni^{2+}-NTA-agarose.

Table 1. Sequence of junction and primer oligonucleotides encoding EDNsFv (Zewe et al., 1997)

Primer	Sequence	Direction
A	5´-ATAT**ATCTAGA**-*AATAATTTTGTTTAACTTTAAGAAGGAGATATACAT-ATG*aaaccgccgcagttcacttgg-3´	Sense
D	5´-TATAT**GGATCC**-*CTATTA*- **ATGGTGATG** - tgaggagactgtgagagtggt-3´	Antisense
B	5´-**ATCACTCTTCGGCGCCCTGAGCGTCGTTCTTCTTGGC** -gatgatacggtccagatgaac-3´	Antisense
C	5´-**GCCAAGAAGAACGACGCTCAGGGCGCCGAAGAGTGAT** -gacatcaagatgacccagtct-3´	Sense

Lower case sequence, gene sequence (A and B, EDN; D and C, sFv); uppercase letters, clamp for later restriction enzyme digest; bold uppercase letters, restriction enzyme sites (*Xba*I (Primer A) and *Bam*HI (Primer D); italicized letters, modified cloning vector sequence; underlined uppercase letters, ATG start site; underlined italicized letters, two stop signals; underlined bold uppercase letters, spacer sequence (amino acid residues 48-60 of fragment B of staphylococcal protein A (Tai et al., 1990); italicized bold uppercase letters, histidine residues for affinity purification via Ni^{2+}-NTA agarose

- 0.5 µl AmpliTaq DNA polymerase to be added after the hot start
- 0.5 µM primer A or C
- 0.5 µM primer B or D
- 500 ng/ml DNA

Program the PCR machine for the following conditions: 94°C, 5 min hot start before beginning the program; 94°C, 1 min (denaturation); 55°C, 2 min (annealing); 72°C, 2 min (extension); 20 cycles.

Note: Keep the cycle number to a minimum (20 cycles) to minimize amplification of an error. High concentrations of template DNA help minimize PCR error.

3. Analyze 10 µl of PCR #1 by NuSieve 3:1 Agarose gel and 10 µl of PCR #2 by 1% agarose gel to determine the size of the PCR product.

   Note: DNA < 500 basepairs (bp) give faint bands on 1% agarose gels. NuSieve 3:1 Agarose is capable of resolving DNA fragments 30-1000 bp.

4. Purify the DNA using Gene Clean II following the manufacturer's instructions to remove the oligonucleotides and components of the PCR reaction. The final product should be in a volume of 15 µl sterile H_2O.

5. The two genes which now share a segment of identical sequence are ready to be spliced together using the PCR technique of splicing by overlap extension or SOEing (Horton 1990) (see Fig. 2, PCR #3). Use primers A (5′ primer of the RNase gene) and D (3′ primer of the sFv gene). The FB gene segment added to the 3′ end of the RNase gene and 5′ end of the sFv gene will hybridize together and allow the two genes to be spliced together. Perform the PCR reaction as described in step 2 using 1-2 µl (50-100 ng) of purified DNA from each of PCR reactions #1 and #2 as the DNA source and primers A and D.

   Note: The most common reason for PCR #3 not working well is a mistake in the primer sequence or in the design of the primers. If the primers are correct, the yield of the DNA after the purification procedure should be checked to determine if there was a loss in DNA at this step. Inclusion of 5-10 µl DMSO in the PCR reaction may help in the amplification of the DNA. If none of the above improve the amplification, a PCR optimizer kit (Invitrogen, San Diego, CA, USA) may be tried as well as varying the reaction times on the thermal cycler.

6. Repeat steps 3 and 4.

7. Restrict 7.5 μl (approximately 500 ng) of the spliced DNA from PCR #3 and 0.5 μg of the vector using 1 μl of each restriction enzyme in the appropriate restriction enzyme buffer and adjusting the final volume with sterile H_2O to 100 μl. Incubate at 37°C overnight.

8. Purify the restricted DNA and vector as described in step 4 (final volume 15 μl).

9. Ligate the restricted DNA and vector together using the Rapid DNA Ligation Kit according to manufacturer's instructions.

   Note: The optimal ratio of RNase-sFv DNA to vector DNA needs to be determined. If all was performed as described above, a good starting ratio is 3 μl RNase-sFv gene (approximately 100 ng) to 0.5 μl vector (15 ng).

10. Transform XL1-Blue competent bacteria with 1-2 μl (approximately 5-10 ng) of the ligated DNA according to the manufacturer's instructions (final volume 25 μl). Streak an LB/amp plate with ≤100 μl of the reaction and incubate overnight at 37°C.

    Note: If no colonies result or if using a single restriction enzyme site, the vector should be dephosphorylated using calf intestinal phosphatase as follows: add 2 μl of the phosphatase to the 100 μl restricted vector mixture before the vector is purified, incubate an additional 15 min at 37°C before inactivating the phosphatase by a 10 min incubation at 65°C. Purify the vector as described in step 4 above.

11. Inoculate 6 tubes each containing 5 ml LB/100 μg/ml ampicillin with 1 colony and incubate with shaking (225 rpm) overnight at 37°C. Place the plate containing the remaining colonies at 4°C for reuse if necessary.

12. Streak a fresh LB/amp plate with each of the overnight bacterial solutions and incubate the plate overnight at 37°C. Centrifuge the remaining bacterial solutions and isolate the plasmids containing the inserts using the Wizard Plus Minipreps DNA Purification System or other similar product following the manufacturer's instructions.

    Note: The second LB/amp plate contains a record of the miniprep DNA and can be used to generate more DNA. One plate can be divided into 6 sections, each section for one miniprep.

13. Restrict each miniprep with the appropriate restriction enzymes and identify those containing an insert of the correct size by running the digests on an agarose gel.

Restriction and ligation of DNA

Minipreps to identify clones with inserts

Note: Occasionally the DNA will ligate into the vector destroying the restriction enzyme site. When repeated ligation and transformation of a good PCR product fails to yield an insert, other restriction enzymes should be tried.

Protein expression 14. Transform BL21(DE3) competent bacteria with each of the plasmids that express a DNA insert of the correct size following the manufacturer's instructions.

Note: Not all plasmids containing an insert of the correct size will express a protein of the correct molecular weight for such varied reasons as an inappropriate termination signal or an insertion or deletion of a bp resulting in a truncated or nonsense protein.

15. Streak the entire reaction from step 14 onto LB/amp plates and incubate overnight at 37°C.

16. Place 5 ml LB containing 100 µg/ml ampicillin into 15 ml tubes; one tube for each transformation reaction. Add one colony from each plate into a tube and incubate 37°C with shaking at 225 rpm.

17. When the OD_{600}nm reaches 0.5-0.6, remove 2.5 ml from each tube and place into a fresh 15 ml tube so that each plasmid is now contained in two tubes. Induce one of the replicate tubes with 0.4 mM IPTG (final concentration), leaving the second tube alone. Continue the incubation of all tubes for 30 min at 37°C with shaking.

Note: No protein of the correct molecular weight should be expressed without IPTG induction.

18. Add 25 µl rifampicin (200 µg/ml final concentration) to each tube and continue incubation at 37°C for 90 min with shaking.

Note: Rifampicin inhibits bacterial protein expression by forming a very stable complex with RNA polymerases (Wehrli 1968). BL21 is sensitive to rifampicin (Studier 1990). Inhibition of bacterial protein expression allows foreign proteins that might not be expressed very well to be visualized better.

19. Remove 100 µl from each tube, centrifuge, aspirate the supernatant, add 25 µl protein denaturing dye, resuspend the pellet, boil 5 min, and analyze on a 4-20% SDS-polyacrylamide gel. Run the induced and non-induced bacteria containing the same plasmid side by side on the gel and compare the two lanes.

Note: Sequence the clones from bacteria expressing a protein of the appropriate molecular weight to ensure that no PCR errors occurred.

Purification of the RNase-sFv fusion protein

1. Transform BL21(DE3) competent bacteria with the plasmid containing the RNase-sFv as described in step 14 in the preceding section and incubate overnight at 37°C.

   Transformation and induction

   Note: Prepare 8 plates per 5 L TB.

2. Place 10 ml TB onto each LB/amp plate containing the colonies and with a tissue culture cell scraper, gently scrape the colonies from the surface of the agar. Distribute the TB solution containing the bacteria evenly into 5 L TB. Grow overnight at 34°C, shaking at 225 rpm.

   Note: The plates should be almost confluent in colony density. This step should be done late in the day.

3. The next morning, induce with 1 mM IPTG for approximately 5 hrs.

   Note: The OD_{600nm} should be ≥ 1.0 since TB is an enriched medium and bacteria can grow to a higher density than in the less enriched LB medium. The optimal length of time of induction should be determined for each construct.

4. Centrifuge the bacteria 10-20 min at 7000 rpm (8000 g) at 4°C.

   Note: The bacterial pellet may be frozen at -20°C if not ready to begin the extraction of protein. For optimal yield of protein, the pellet should be processed within one week. Frost-free freezers should be avoided.

5. Resuspend the pellet with ice-cold sucrose buffer (200 ml/l original bacterial culture) and leave on ice for 10 min. Centrifuge 20 min at 7000 rpm (8000g) at 4°C.

   Inclusion body purification

6. Pour off supernatant, resuspend the pellet with ice-cold H_2O (200 ml/l original bacterial culture), place 25 ml into each of 8 polypropylene tubes, and leave on ice for 20 min. Centrifuge at 12,000 rpm (17,000 g) for 20 min at 4°C. Pour off supernatant.

   Note: The pellet may be stored at -70°C at this step until ready to process further.

7. Resuspend the pellet in each tube with 9 ml TE buffer using a Janke & Kunkel polytron tissuemizer or similar model and combine 2 tubes together (1 l bacterial culture is now contained in 4 tubes). Incubate at room temperature for 30 min.

8. Add 0.9 ml of 5 mg/ml lysozyme per 18 ml tube (final concentration, 250 µg/ml) and incubate for 60 min at room temperature shaking occasionally.

9. Add 2.5 ml 5 M NaCl and 2.5 ml 25% Triton X-100 to each tube and incubate 30 min at room temperature shaking occasionally. Homogenize the sample just prior to centrifugation and centrifuge 12,000 rpm (17,000g) for 40 min at 4°C.

   Note: The solution becomes very viscous due to the presence of nucleic acids. Lack of homogenization with the tissuemizer just before centrifugation will result in a very loose pellet.

10. Carefully pour off supernatant and with the tissuemizer, resuspend the pellet in 15 ml TE buffer. Centrifuge for 30 min as in step 9. Repeat the washing procedure 3 more times decreasing the centrifugation time to 20 min for the last two washes.

    Note: The washed inclusion body may be stored at -70°C until ready for solubilization. The protein of interest should be the major band and in some cases, will be almost pure at this stage.

Solubilization, denaturation, and renaturation

11. Resuspend the inclusion bodies in 10-20 ml solubilization buffer using the tissuemizer and incubate for 2 h at room temperature.

    Note: The solution should be free flowing and not too viscous. Be careful not to dilute too much as the final protein concentration will be adjusted to 8 mg/ml as described below.

12. Centrifuge for 20 min at 12,000 rpm (17,000 g) at 4°C.

    Note: The supernatant can be stored at -70°C until ready for renaturation.

13. Determine the protein concentration of a 1:10 dilution of the supernatant using the BCA protein assay reagent or similar protein determination kit according to manufacturer's instructions. Adjust the protein concentration to 8 mg/ml with solubilization buffer. Add dry DTE to a final concentration of 0.3 M and incubate for ≥2 h at room temperature.

    Note: Use solubilization buffer to make the 1:10 dilution to avoid precipitation of the protein. Include the same amount of solubilization buffer in the standard curve as it does affect the protein determination reaction. The final concentration of protein in the renaturation buffer affects the yield of the protein and should be determined for each protein. The concentration described here (80 µg/ml) works well for many RNase immunofusion proteins.

14. Renature the protein by diluting it 100 fold with renaturation buffer which has been prechilled to 10°C. Incubate for 2-3 days at 10°C.

   Note: The optimal DTE to GSSG ratio has been carefully optimized to give a redox system of reduced and oxidized glutathione. To maintain this, the protein solution must be diluted exactly 100 fold (Buchner 1992). Temperature influences aggregation and thus yield of protein. Therefore, it is important that the buffer be prechilled to 10°C before use. The optimal length of time required for the protein to fold needs to be determined for each construct. Studies (Brinkmann 1992) indicate that the domains of the fusion protein fold independently of each other. Precipitation occurs at this step due to incorrect folding and aggregation of the protein.

15. Dialyze the refolded protein against 10 volumes of dialysis buffer at 4°C. Change the dialysis buffer a total of 3 times over a 24 h period.

   Note: A large loss of protein will occur at this step due to improper folding and aggregation. Urea (100 mM) helps limit the precipitation.

16. Centrifuge the dialyzed solution 20 min at 4°C at 7000 rpm (8000 g). Be careful when pouring off the supernatant as the precipitate does not tightly adhere to the walls of the centrifuge cup.

17. Apply the centrifuged protein to a CM-Sephadex C-50 or heparin-sepharose column. Use 4 ml resin per 160 mg of protein. Perform chromatography at 4°C.   **Chromatography**

   Note: Since RNases are very basic proteins, many (angiogenin, pancreatic RNase, frog RNases) will adhere to CM-Sephadex C-50 while the majority of contaminating proteins will not. Thus, whenever possible CM-Sephadex C-50 should be utilized as the first column. For the eosinophil-derived RNase (EDN), heparin-sepharose is the appropriate first column. Ni^{2+}-NTA agarose (for those proteins with a histidine tag) should not be the first column used in the purification procedure because of interference of contaminating proteins and should be reserved for the final purification step.

18. Wash the column with 2 column volumes of chromatography buffer A.

19. Elute the column with a 0-1.0 M NaCl gradient in chromatography buffer A.

Note: It is very important that the buffer contain 10% glycerol. The RNase-sFv will precipitate over time in the absence of glycerol. The buffer described here, 20 mM Tris-HCl, pH 7.5, containing 10% glycerol, has been found consistently to be the best buffer for RNase fusion proteins.

20. Analyze the eluted protein (15-20 μl) on an SDS-polyacrylamide gel and pool the fractions containing the protein of interest. If the protein contains a histidine tag, proceed to Step 21. If there is no tag, proceed to Step 24.

21. If the protein contains a histidine tag, add imidizole and Triton X-100 to final concentrations of 0.8 mM and 1%, respectively, and apply the sample to 0.6 ml Ni^{2+}-NTA agarose (for 320 mg of refolded protein as determined in Step 13).

    Note: Without the imidazole and Triton X-100, impure preparations of RNase fusion proteins result. Using the least amount of Ni^{2+}-NTA agarose possible minimizes the amount of nonspecific binding of other proteins to the column.

22. Rotate the slurry end over end for > 1 h at $4°C$ before collecting it as a column, washing with 10 ml chromatography buffer A containing 0.8 mM imidazole, and then step eluting with 2 column volumes of the same buffer made 40, 50, 60, 100, 200, 300, 400, and 500 mM imidazole.

    Note: For some RNase fusion proteins, a band of the appropriate molecular weight elutes with the 40-60 mM imidazole buffer. This material is discarded because it is usually too low in concentration and contains other impurities.

23. Analyze the fractions on an SDS-polyacrylamide gel and pool the appropriate tubes.

    Note: Samples may be stored at $4°C$ for up to 1 yr for some RNase fusion proteins. Freezing should be avoided and samples should be stored in the presence of 0.1 M NaCl or imidazole. Decreasing the concentration of either can result in precipitation of the fusion protein. It is important to achieve as high a protein concentration as possible on the final purification step as we have not found a successful means of increasing the concentration after this. Some RNase fusion proteins can be concentrated by using very small DEAE Sepharose or CM-Sephadex C-50 columns (0.2 ml) and strip eluting with 1 column volume aliquots. Before use in the cell cytotoxicity assays, however, dialysis or dilution will be required which usually results in some precipitation of the fusion protein. Concentration by Centricon cartridges, Diaflo ultrafiltration or surrounding a dialysis bag containing the fusion protein with dry G100 resin have been tried and do not concentrate without a great loss in protein.

24. If there is no histidine tag, dialyze the sample against chromatography buffer A overnight at 4°C.

25. Centrifuge the sample at 4°C for 10 min at 3000 rpm before applying it to either a Mono S HR 5/5 column or Mono Q HR 5/5 column equilibrated with buffer A. Elute the sample with a 30-min gradient between 0 and 80% chromatography buffer B with a final 10 min hold on 80% buffer B.

Note: Analyze the sample analytically first on each column to identify which column is the most appropriate for purification. Adjust the column conditions to best suit the sample. The conditions described here work for many different RNase fusion proteins. Chromatography conditions may have to be adjusted upon scale-up.

26. Analyze the fractions by SDS-polyacrylamide gel and pool the appropriate tubes.

Note: If contaminants remain, dialyze the sample as described in Step 24 and reapply to the same column adjusting the chromatography conditions so that the gradient is shallow in the area in which the protein elutes. The denaturation/renaturation procedure described here often produces different isomers of the same protein and these may be visualized by FPLC or HPLC. See Note to Step 23.

Characterization of the binding activity of the RNase-sFv

For the characterization of the binding activity of the sFv moiety of the RNase fusion protein, the reader is referred to the methodology described in Chapter 28 and Chapter 29 of this manual.

Characterization of the ribonuclease activity of the RNase immunofusion protein

1. Determine the pH at which the ribonuclease activity is optimal as follows:

pH optimum

Add 100 μl of each of the following solutions to numbered reaction tubes contained on ice:
- tRNA (stock concentration, 1.0 mg/ml)
- HSA (stock concentration, 0.5 mg/ml)
- Buffer (Stock concentration, 0.5 M Tris-HCl, pH 7.5, 5 mM EDTA, 0.5 mg/ml HSA or the same buffer at pH 8.0, or 0.4 M MES, pH 6.0, 0.5 mg/ml HSA)

Note: When performing this assay, gloves must be worn to avoid contamination of the assay with the RNases found on hands and in the environment.

2. Add 10 μl dilution buffer or RNase or RNase fusion protein.

Note: The dilution buffer (0.5 mg/ml HSA) is a good diluent for RNases and RNase immunofusion proteins. Very low blank values can be obtained and the RNases maintain full activity in its presence at very low concentrations. Keep the enzyme in the linear range of the assay. Do not limit the substrate (50-100 times the enzyme concentration). Include buffer components in the standards to control for buffer effects.

3. Initiate the reaction by placing the tubes at 37°C and incubating for the appropriate length of time.

Note: Incubation time varies with the RNase. Activity of fused RNases is generally lower than that of the free RNase, thus the incubation times may need to be adjusted. The following are the incubation times used for some non-fused RNases: EDN, 15 min; pancreatic RNase, 15 min; angiogenin, 18 hr; Onconase, 120 min.

4. Terminate the reaction by placing the tubes on ice and adding 700 μl 3.4% ice-cold perchloric acid. Incubate the tubes on ice for 10 min before centrifuging for 10 min in an eppendorf centrifuge at full speed.

5. Read the absorbance of the supernatant at OD_{260}nm. Average the replicate tubes and subtract the blank (those tubes containing only the dilution buffer) from those tubes containing the enzyme. Perform the studies described below at the appropriate pH, i.e., that pH yielding the highest RNase activity.

Determination of kinetic constants

6. Determine the Km (the substrate concentration at which the reaction velocity is half maximal) of the RNase as follows:

Add 100 μl of each solution to appropriately labelled reaction tubes:
- tRNA (stock concentration, vary from 0.25-10 mg/ml)
- HSA (stock concentration, 0.5 mg/ml)
- Buffer (optimal pH, see Step 1 above for description)

Note: Run a preliminary assay using a wide range of substrate concentrations to determine an approximate K_m. Choose 5 different substrate concentrations spanning 0.5 to 5 times the approximate K_m value. Be sure to prepare blanks (dilution buffer alone) for each substrate concentration as the blank will change with each substrate concentration.

7. Follow Steps 2-5 described above.

8. Prepare a Lineweaver-Burk plot by plotting the reciprocal of the substrate concentration (1/s) on the x-axis versus the reciprocal of the enzymatic activity (the $OD_{260}nm$ value minus the blank)($1/v_o$) obtained for each substrate concentration on the y-axis. Determine the V_{max} (maximal activity) from the intercept on the $1/v_o$ axis ($1/V_{max}$) and the K_m from the intercept on the 1/s axis ($-1/K_m$). Calculate the K_{cat} (catalytic constant or turnover number) and K_{cat}/K_m (an indicator of catalytic efficiency) as follows: convert K_m from mg/ml to a molar value using the M_r for tRNA of 28,100 (Rosenberg 1995) and V_{max} from absorbance to a molar value by dividing the $OD_{260}nm$ value by 7×10^5 (Newton 2000). Use the equation $K_{cat}(E_o) = V_{max}/t$ where E_o is the molar concentration of the enzyme used and t is the incubation time in seconds. The final units are s^{-1}. The efficiency of the enzyme is K_{cat}/K_m, the units are $M^{-1} s^{-1}$.

Characterization of the in vitro cytotoxic activity of the RNase immunofusion protein

1. The day before treatment, place 0.1 ml of 25,000 or 100,000 cells / ml for adherent or non-adherent cells, respectively, per well of a 96-well microtiter plate. The plating media is the media in which the cells are normally grown.

2. On the day of treatment prepare dilution curves of the non-fused RNase and RNase immunofusion protein in sterile dilution buffer. Be sure to include the buffer in which the RNase is stored as some cells are very sensitive to glycerol and imidazole. Apply 10 μl to the appropriate wells. Perform each sample in triplicate.

Note: Use Millipore Millex-HV filters to sterilize the RNase and RNase immunofusion protein. Other filters have been tried and substantial losses in protein have been observed.

3. Incubate the plate at 37°C in a humidified CO_2 incubator for 3 days. Shorter incubation times require higher concentration of protein.

4. Cell viability can be determined by the colorimetric assay, WST-1, following the manufacturer's instructions. Inhibition of protein synthesis is determined by replacing the media with 100 µl of serum- and leucine-free RPMI, adding 10 µl of [^{14}C] leucine (0.1 mCi), and continuing incubation for 2-4 hrs at 37°C. The cells are then harvested onto glass fiber filters using a PHD cell harvester, washing with H_2O, drying with ethanol, and counting in a scintillation counter.

5. Express the results as % buffer-treated cells calculated as follows:

$$\frac{\text{cpm (or absorbance for WST-1) of sample-treated cells}}{\text{cpm (or absorbance for WST-1) of buffer-treated cells}} \times 100$$

Plot protein synthesis (% of control) or cell viability (% of control) on the y-axis versus sample concentration on the x-axis. Determine the IC_{50}, the concentration of test sample which inhibits protein synthesis or cell viability by 50% from the plot.

Acknowledgements. The technical support of Dale Ruby is gratefully acknowledged. We very much appreciate our excellent administrative help and thank Ms. Beverly A. Bales, Robin L. Reese, and Jamie M. Tammariello. We are most grateful for the interest and support of Dr. Edward A. Sausville. This project has been funded in whole or in part with federal funds from the National Cancer Institute, National Institutes of Health, under Contract No. NO1-CO-56000. The content of this publication does not necessarily reflect the views or policies of the Department of Health and Human Services, nor does mention of trade names, commercial products, or organizations imply endorsement by the U.S. Government.
The publisher or recipient acknowledges right of the U.S. Government to retain a non-exclusive, royalty-free license in and to any copyright covering the article.

References

Brinkmann U, Buchner J, Pastan I (1992) Independent domain folding of *Pseudomonas* exotoxin and single-chain immunotoxins: Influence of interdomain connections. Proc. Natl. Acad. Sci. USA 89: 3075-3079
Brinkmann U, Pastan I (1994). Immunotoxins against cancer. Biochim. Biophys. Acta 1198: 27-45

Buchner J, Pastan I, Brinkmann U (1992) A method for increasing the yield of properly folded recombinant fusion proteins: Single-chain immunotoxins from renaturation of bacterial inclusion bodies. Anal. Biochem. 205: 263-270

Harkonen S, Stoudemire J, Mischak R, Spitler L, Lopez H, Scannon P (1987) Toxicity and immunogenicity of monoclonal antimelanoma antibody-ricin A chain immunotoxins in rats. Cancer Res. 47: 1377-1385

Horton R M, Cai Z L, Ho S N, Pease L R (1990) Gene Splicing by Overlap Extension: Tailor made genes using the polymerase chain reaction. BioTechniques 8: 528-535

Huston J S, Levinson D, Mudgett-Hunter M, Tai M S, Novotny J, Margolies M N, Ridge R J, Bruccoleri R E, Haber E, Crea R, Oppermann H (1988) Protein engineering of antibody binding sites: Recovery of specific activity in an anti-digoxin single-chain Fv analogue produced in Escherichia coli. Proc. Natl. Acad. Sci. USA 85: 5879-5883

Mikulski S M, Grossman A M, Carter P W, Shogen K, Costanzi J J (1993) Phase 1 human clinical trial of ONCONASE (P-30 protein) administered intravenously on a weekly schedule in cancer patients with solid tumors. Int. J. Oncol. 3:57-64

Mosimann S C, Ardelt W, James M N G (1994) Refined 1.7 A X-ray crystallographic structure of P-30 protein, an amphibian ribonuclease with anti-tumor activity. J. Mol. Biol. 236: 1141-1153

Newton D L, Xue Y, Olson K A, Fett J W, Rybak S M (1996) Angiogenin single-chain immunofusions: Influence of peptide linkers and spacers between fusion protein domains. Biochemistry 35: 545-553

Newton D L, Rybak, S M (2000) Preparation and preclinical characterization of RNase-based immunofusion proteins. Methods in Molecular Biology; Nuclease Methods and Protocols, in press

Rosenberg H F, Dyer K D (1995) Eosinophil cationic protein and eosinophil-derived neurotoxin. Evolution of novel function in a primate ribonuclease gene family. J. Biol. Chem. 270: 21539-21544

Russo N, Nobile V, DiDonato A, Riordan J F, Valee B L (1996) The C-terminal region of human angiogenin has a dual role in enzymatic activity. Proc. Natl. Acad. Sci. U.S.A. 93: 3243-3247

Rybak S M, Newton D L (1999) Immunoenzymes. In: Chamow SM, Ashkenazi A (eds) Antibody Fusion Proteins, John Wiley & Sons, New York, NY pp. 53-110

Rybak S M, Youle R J (1991) Clinical Use of Immunotoxins: Monoclonal Antibodies Conjugated to Protein Toxins. Immunol. and Allergy Clinics of North America 11: 359-380

Sambrook J, Fritsch E F, Maniatis T (1989) Molecular cloning. A laboratory manual. Cold Spring Harbor Laboratory Press

Sawler D L, Bartholomew R M, Smith L M, Dillman R (1985) Human immune response to multiple injections of murine monoclonal IgG. J. Immunol. 135: 1530-1535

Saxena S K, Rybak S M, Winkler G, Meade H M, McGray P, Youle R J, Ackerman E J (1991) Comparison of RNases and Toxins upon injection into Xenopus oocytes. J. Biol. Chem. 266: 21208-21214

Schroff R W, Foon K A, Beatty S M, Oldham R, Morgan A. (1985) Human anti-murine immunoglobulin response in patients receiving monoclonal antibody therapy. Cancer Res. 45: 879-885

Soler-Rodriguez A M, Ghetie M A, Oppenheimer-Marks N, Uhr J W, Vitetta E S (1993) Ricin A-chain and ricin A-chain immunotoxins rapidly damage human endothelial cells: Implications for vascular leak syndrome. Exp. Cell Res. 206: 227-234

Studier F W, Rosenberg A H, Dunn J J, Dubendorff J W (1990) Use of T7 RNA polymerase to direct expression of cloned genes. Methods Enzymol. 185: 60-89

Tai M S, Mudgett-Hunter M, Levinson D, Wu G M, Haber E, Oppermann H, Huston J S (1990) A bifunctional fusion protein containing Fc-binding fragment B of staphylococcal protein A amino terminal to antidigoxin single-chain Fv. Biochemistry 29: 8024-8030

Vitetta E S, Thorpe P E, Uhr J W (1993) Immunotoxins: magic bullets or misguided missiles? TiPS 14: 148-154

Wehrli W, Knusel F, Schmid K, Staehelin M (1968) Interaction of rifamycin with bacterial RNA polymerase. Proc. Natl. Acad. Sci. USA 61: 667-673

Zewe M, Rybak, S M, Dubel S, Coy J F, Welschof M, Newton D L, Little M (1997) Cloning and cytotoxicity of a human pancreatic RNase immunofusion. Immunotechnology 3: 127-136

Abbreviations

sFv	single chain antibody
RNase	ribonuclease
EDN	eosinophil-derived neurotoxin
HSA	human serum albumin
LB	Luria Broth
TB	terrific broth
IPTG	isopropyl-β-D-thiogalactopyranoside
DTE	dithioerythritol
tRNA	transfer ribonucleic acid
GSSG	oxidized glutathione
bp	basepairs
PCR	polymerase chain reaction
IC_{50}	concentration of test sample that inhibits protein synthesis by 50%
K_m	the substrate concentration at which the reaction velocity is half maximal
K_{cat}	catalytic constant
V_{max}	maximal velocity

Cytokine-scfv Fusion Proteins

CECILIA MELANI and DANIELA NICOSIA

Introduction

Systemic administration of interleukin-2 (IL-2) has been used for immunotherapy of a variety of human malignancies. However, its rapid blood clearance, that does not allow optimal concentration at the tumorsite, and severe toxicity at high dosage have limited its efficacy in vivo. Genetic manipulation of tumor cells or fibroblasts allowed IL-2 to be secreted in situ (Parmiani et al. 1997), but IL-2 secretion may vary according to the vector used, the strength of its promoter and the efficiency of protein synthesis in the targeted cell. The selective delivery of IL-2 at the tumor site would avoid toxicity associated with systemic administration, overcome the concerns due to ex vivo genetic manipulation of cells, and solve the difficulties of transducing tumor cells in vivo. Taking advantage of the targeting specificity of monoclonal antibodies (mAb) directed to tumor associated antigens (TAA), a fusion protein can be constructed that allows targeting IL-2 at the tumor microenvironments (Reisfeld et al. 1997). To this aim the use of the single chain Fv (scFv) of the Ab is advantageous because it maintains the binding specificity of the Ab, lacks fragments whose non-specific function could interfere with the desired effect, and is assembled as a single open reading frame, which is easily modified for subcloning strategies. The resulting fusion protein will have a small size that allows a better tissue penetration and distribution within the tumor (Yokota et al. 1992), reduces its immunogenicity, improves the local biologic effect of IL-2 and reduces its systemic toxicity.

✉ Cecilia Melani, Istituto Nazionale Tumori, Department of Experimental Oncology, Gene Therapy Unit, Via G. Venezian 1, Milano, 20133, Italy (*phone* +39-02-2390-212; *fax* +39-02-2390-630)
Daniela Nicosia, Istituto Nazionale Tumori, Department of Experimental Oncology, Gene Therapy Unit, Via G. Venezian 1, Milano, 20133, Italy

When choosing the cloning strategy to obtain a fusion protein between IL-2 and a scFv it should be considered that:

a. the expression vector should allow a high transcription rate

b. the producer cell should be easy to grow at high concentration in the least expensive medium

c. the expression system must provide the same post translational modifications that naturally occur to the two components of the fusion protein (i.e. glycosilation)

d. the fusion protein must be easy to purify

Bacterial expression has been obtained for some fusion proteins between IL-2 and antibodies or their fragments (Savage et al. 1993). Although very convenient in terms of expression level, that is easily inducible, and inexpensive culture conditions, it can result in low yields, when the protein is not released into the periplasmic space but accumulates in the inclusion bodies. Refolding of antibodies from inclusion bodies could be obtained, but it is a cumbersome procedure whose yield can vary greatly for distinct Fv regions (Boleti et al. 1995). Moreover, bacterial expression can generate endotoxins that may be difficult to remove, and protein glycosilation is not provided. The following protocol is designed to express a fusion protein between IL-2 and the scFv of the mAb MOV19 (Miotti et al. 1987), recognizing the alpha-folate receptor TAA. This system employs an eukaryotic expression vector carrying immunoglobulin specific promoter and enhancers for high expression rate in B cells and J558L murine myeloma cell line for production. Myeloma cells provide all the necessary post-translational modifications for eukaryotic proteins and can be grown at high concentration allowing for a continuous manufacturing of soluble protein with high yield (Dorai et al. 1994). Finally, by linking to the Ig constant kappa (Cκ) chain the IL-2/MOV19 fusion protein can be easily purified by affinity chromatography, without introducing exogenous, potentially immunogenic sequences. We found that the best configuration for a functional fusion protein between IL-2 and the MOV19 scFv, as far as antigen recognition and IL-2 bioactivity are concerned, contains the full length IL-2 cDNA with its leader sequence at the 5' end, an intervening oligonucleotide coding the 13 amino-acid spacer GSTSGSGKSSEGK to separate the IL-2 from the scFv, and the scFv sequence in the VH-linker-Vκ configuration. A splice donor site instead of the stop codon allows the junction to the Cκ coding sequence at 3' end of the expression cassette (Melani et al. 1998). The protocol, outlined in the Figure 1, consists of:

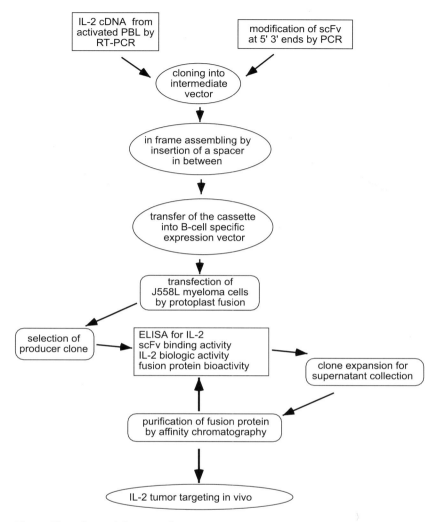

Fig. 1. Flow chart of the procedure

a. cloning of human IL-2 cDNA by RT-PCR from human activated peripheral blood lymphocytes (PBL) in an intermediate expression vector (pCDNA3)

b. insertion of appropriate restriction sites at 5' and 3' ends of the scFv by PCR and ligation at 3' end of IL-2 cDNA

c. insertion of the oligonucleotide spacer in between

d. transfer of the fusion protein cassette into the expression vector pH PCRγIII

e. transfection of J558L cells, screening of expressing clones and production of fusion protein

f. purification of the fusion protein from cell supernatant, assays for biologic function

The following protocol can be adapted to cytokines and scFv other than IL-2 and MOV19 taking into account that the restriction sites chosen for cloning (Hind III, BamHI, EcoRI and Xho I) must not cut into the cytokine nor in the scFv sequences; alternatively an intermediate cloning vector other than pCDNA3 can be used.

Materials

All the necessary instruments are usually available in a laboratory equipped for molecular biology and cell culture. Unless otherwise specified we used the following reagents:

- reverse transcription: MMLV-reverse transcriptase and buffer: GIBCO-BRL, Gaithersburg, MD, USA

- PCR: Taq polymerase and buffer, nucleotides: Promega Corp., Madison, WI, USA

- restriction enzyme digestion: enzymes, buffers and BSA: New England Biolabs, Beverly, MA, USA

- dephosphorilation: calf intestinal alkaline phosphatase and buffer: Boehringer Mannheim Italia, Monza, Italy or Promega

- ligation: T4 DNA ligase and buffer: GIBCO-BRL

- DNA blunting: T4 DNA polymerase and buffer: New England Biolabs

- DNA purification/desalting: Qiaquick Clean-up Spin Kit: Qiagen, Valencia, CA, USA

- DNA extraction from gel: Qiaex II gel extraction kit: Qiagen

Oligonucleotides and PCR primers were synthesized according to Table 1. Working dilution for PCR was 20 μM. Oligonucleotides for linker-spacer must be phosphorilated at 5' end. Dilute stock of oligonucleotides at 50 ng/ml just before annealing for ligation.

Table 1. PCR primers and oligonucleotides for spacer

name	sequence[a]	restriction site
IL-2 Direct	5'- CT<u>AAGCTT</u>CTGCCACAATGTACAGGATGC AAC-3'	Hind III
IL-2 Reverse	5'-GC<u>GGATCC</u>GTTAGTGTTGAGATGATGCTT TGAC-3'	Bam HI
MOV19 VH-ERI Direct	5'-CGG<u>GAATTC</u>CCCAGGTGCAGCTGCAGCAG TCTG -3'	Eco RI
MOV19 VκSD-XhoI Reverse	5'-CCA<u>CTCGAG</u>CACCTCATCACGCCCGTTTTA TTTCCAACTTTG -3'	Xho I
β2-microglobulin Direct	5'-CTCGCGCTACTCTCTCTTTCTGG-3'	
β2-microglobulin Reverse	5'-GCTTACATGTCTCGATCCCACTTAA-3'	
Spacer Direct	5'-<u>GATCC</u>TGG<u>GTCGAC</u>TTCCGGTAGCGGCAA ATCCTCTGAAGGCAAAG<u>G</u>-3'	Bam HI, Sal I, Eco RI
Spacer Reverse	5'-<u>AATT</u>CCTTTGCCTTCAGAGGATTTGCCGCT ACCGGAA<u>GTCGAC</u>CCA<u>G</u>-3'	Bam HI, Sal I, Eco RI

a) sequences that create the restriction site are underlined

Solutions, buffers and culture media were according to Current Protocols in Molecular Biology, Current Protocols in Immunology and (Sambrook et al., 1989). In some procedures we used modified standard solutions as follows:

Guanidine solution   **RNA preparation**

Guanidine isothiocyanate	75 g
Sarkosyl	0.75 g
Sodium citrate 1 M pH 7.0	3.75 ml
H_2O	up to 150 ml

Dissolve salts in H_2O by heating. Filter sterilize, keep at R.T in the dark. Just before use add beta-mercaptoethanol 7 μl/ml.

Cesium chloride solution (4.95 M)

Cesium Chloride	100 mg
Sodium acetate 4 M pH 5.0	3 ml
EDTA 0.5 M pH 8.0	1.2 ml
H_2O	up to 120 ml

Filter sterilize and autoclave. Keep at 4°C

Protoplast fusion

- Lysozyme solution: Dissolve lysozyme [1 mg/ml] in Tris.Cl 25 mM (pH 8.0) just before using

- PEG/DMSO solution: melt PEG-1500 (Sigma, St.Louis, MO, USA) in boiling water, mix 1 ml of PEG-1500 with 1 ml of DMEM 10% FCS by pipetting thoroughly then add 0.2 ml of DMSO and mix well.

Immuno-fluorescence

FACS buffer

BSA	2 g
PBS	100 ml
sodium azide	0.05%

Filter sterilize and keep at 4°C

Affinity chromatography

Elution buffer pH 3.0

H_3PO_4 (85%)	0.452 ml
Citric Acid-monohydrate	1.4 g
Boric Acid	0.7 g
H_2O	up to 1 l

Subprotocol 1
Cloning of Human IL-2 cDNA from Activated Peripheral Blood Lymphocytes (PBL)

■ ■ Procedure

1.1 Preparation of RNA from activated human PBL

1. Collect 10 ml of peripheral blood from healthy donor in heparinized tube.

2. Separate mononuclear cells by Ficoll-Hypaque gradient according to Current Protocols in Immunology.

3. Plate PBL at 0.5-1 x10^7/ml in 75 cm^2 flask in 15-20 ml of RPMI 1640 medium supplemented with 10% human AB serum and 10 μg/ml PHA and incubate at 37°C in 5 % CO_2.

4. After 48 h collect activated lymphocytes, wash twice with PBS and lyse the pellet in 4 ml of guanidine solution by vortexing and passing the cell lysate through a 22-G needle.

5. Overlay the cell lysate on 4 ml of CsCl solution in a 10 ml autoclaved polyallomer ultracentrifuge tube and spin in a swinging rotor (SW41 for Beckmann ultracentrifuge or similar) at 35000 rpm 16 h at 20°C.

6. Recover the total RNA and measure the concentration according to Current Protocol in Molecular Biology.

1.2 RT-PCR for IL-2

1. Dilute 1 μg of total RNA from PBL to a final volume of 12.5 μl with DEPC-treated H_2O and heat at 65°C for 10 min then put in ice.

2. Add 17.5 μl of the following mix:

DEPC-treated H_2O	2.5 μl
5x RT-buffer	6 μl
dNTP mix [10mM each]	1.5 μl
DTT [0.1 M]	0.3 μl
oligo dT (15-18 mer) [0.5 μg/ml]	0.2 μl
BSA [1 μg/ml]	3 μl
RNAse inhibitor [40 U/μl]	1 μl
MMLV-ReverseTranscriptase [200 U/ml]	3 μl

3. Incubate at 42°C for 2h, inactivate enzyme at 95°C for 10 min and store the cDNA at 4°C.

4. Prepare two PCR mixtures, one to amplify IL-2 and the other to amplify β-2 microglobulin as control of the reverse transcription (see Table 1 for primer sequences). In a PCR tube add:

cDNA	2 µl
10x PCR buffer	2.5 µl
MgCl$_2$ [25 µM]	1 µl
NTP mix [25 µM each]	0.2 µl
Direct primer [20µM]	1 µl
Reverse primer [20µM]	1 µl
sterile H$_2$O to final volume of 25 µl	14.7 µl
Taq Polimerase [5 U/ml]	0.125 µl

5. Cover with mineral oil if necessary (according to the PCR machine used) and perform 30 cycles of PCR by denaturing at 94°C for 1 min, annealing at 60°C for 1 min 30 sec, synthesizing at 72°C for 2 min; last step at 72°C for 7 min.

6. Analyze 5 µl of PCR on a 1% agarose gel (Current Protocol in Molecular Biology), check by U.V. the result of the amplification: β-2 microglobulin gives a band of 335 bp and IL-2 a band of 485 bp.

Troubleshooting   If no bands are visible check the quality of your RNA on a denaturing gel (Current Protocols in Molecular Biology) and if it is not degraded repeat cDNA synthesis. If there is no IL-2 amplification in the presence of β-2 microglobulin band prolong PBL stimulation for 72 h.

1.3 Purification and cloning of IL-2

1. Extract PCR product with 50 µl of chloroform and precipitate according to (Sambrook et al.1989) and resuspend in 38 µl of H$_2$O. Alternatively, use commercially available spun column to purify IL-2 DNA (Qiaquick, Qiagen).

2. Set up double digestion in a final volume of 50 µl:

DNA in H$_2$O	38 µl
10x BamHI buffer	5 µl
10x BSA solution	5 µl
Hind III [20U/µl]	1 µl
BamHI [20U/µl]	1 µl
incubate at 37°C for 1 h then inactivate 10 min at 65°C	

3. Run the digested DNA on a 1% agarose preparative gel in 1x TAE buffer, cut the band and elute the insert by Qiaex II gel extraction kit (Qiagen) according to manufacturer's instructions. Elute DNA in 20 µl of H$_2$O and store at -20°C.

4. Set up double digestion of 5 µg of vector pCDNA3 (Invitrogen) with Hind III and BamHI as stated before (Step 2 in section 1.3), check linearization of the plasmid by running 5 µl of digested DNA on a 0.8% agarose gel. In case digestion is partial add 1 µl of both enzymes and incubate at 37°C for additional 1-2 h.

5. Treatment of digested vector with alkaline phosphatase is optional, because the two ends are not compatible for ligation; however, it is highly recomended to avoid self-ligation of partially digested plasmid. Purify digested vector DNA by phenol extraction and proceed as in Step 1, Section 1.3; resuspend in 90 µl of H$_2$O and add:

Vector DNA in H$_2$O	90 µl
10x alkaline phosphatase buffer	10 µl
Calf intestinal alkaline phosphatase	1 µl

Incubate 30 min at 37°C; then add 1 µl of enzyme and incubate 50°C for additional 30 min. Inactivate the enzyme by adding 11 µl of EGTA 0.5 M (pH 8.0), for 30 min at 65°C

6. Extract DNA twice with 50 µl of Phenol plus 50 µl of chloroform, precipitate and wash DNA, resuspend in 50 µl of H$_2$O.

7. Estimate the amount of vector and insert DNA by running 1 and 3 µl of each on a 1% agarose gel: compare the intensity of the bands and calculate approximately the amount of DNA to be ligated at a vector:insert molar ratio of 1:3. Take into account that IL-2 insert length is roughly 1/10 of the pCDNA3 therefore when the two bands appear equally bright on the gel, the insert is 10 times more concentrated than the vector on a molar ratio. If 1 µl of vector appears as bright as 3 µl of insert, ligate an equal volume of each:

Vector (pCDNA3) DNA	1 µl (or as estimated)
Insert (IL-2) DNA	1 µl (or as estimated)
5x ligase buffer (Gibco-BRL)	4 µl
T4 Ligase (Gibco-BRL)	1 µl
H$_2$O	to a final volume of 20 µl

incubate over night at 4°C or 4 h at 25°C. Remember to include a vector self ligation as control

8. Transform competent DH5α E.Coli bacteria or equivalent strain with half (10 μl) of the ligation according to (Sambrook et al. 1989). Include appropriate controls: 10 pg of circular plasmid (positive control), linearized plasmid DNA (negative control), self ligated vector DNA.

9. Screen transformant by small scale plasmid DNA extraction and digestion with Hind III and BamHI (Sambrook et al. 1989). The resulting construct is pIL-2CDNA3 (Figure 2A). Choose a positive colony and prepare plasmid DNA on a larger scale with the Qiaprep kit (Qiagen) according to the protocol provided.

10. Sequencing of the IL-2 insert is necessary to confirm that the PCR did not introduce point mutations or other modifications. Radioactive sequencing can be performed following Sequenase kit instruction (Sequenase, USB), otherwise automated sequencing is provided as customer service by many companies (i.e. Invitrogen).

Subprotocol 2
Modification of scFv DNA and Cloning

▣▢ Procedure

Cloning of the scFv of interest is described elsewhere in this book. Usually the scFv DNA is cloned into a prokaryotic expression vector: in our case MOV19 is cloned Sfi I-Not I into the phagemid pCANTAB-E5 vector (Pharmacia).

1. Set up PCR as in Step 4 in Section 1.2. Use 0.1 μg of plasmid DNA containing the scFv and primers Direct VH-ERI and Reverse VκSD-XhoI (see Table 1 for primer sequences). Start 30 cycles of PCR as in Step 5 in Section 1.2. Check on an agarose gel for a band of 780 bp (as in Step 6, Section 1.2): the resulting MOV19 scFv DNA has now a splicing donor site (SD) at 3' end.

2. Purify the PCR product as in Step 1, Section 1.3.

3. Set up double digestion of MOV19SD with Eco RI and Xho I as in Step 2, Section 1.3.

4. Purify digested scFv DNA as in Step 6, Section 1.3

5. Prepare the pIL-2CDNA3 vector for ligation by digesting 5 µg with EcoRI and Xho I as in Step 2, Section 1.3 and proceed through Steps 4 to 6 of Section 1.3.

6. Set up ligation of the scFv MOV19SD into the pIL-2CDNA3 following Step 7, Section 1.3.

7. Proceed through Steps 8 to 10 of Section 1.3 to obtain the pIL-2/MOV19SD construct (Figure 2B).

Subprotocol 3
In Frame Assembling of the Fusion Protein

▓ ▓ Procedure

3.1 Insertion of the spacer

1. Digest 5 µg of pIL-2/MOV19SD DNA with Eco RI and Bam HI as in Step 2, Section 1.3.

2. Treat with alkaline Phosphatase as in Step 5, Section 1.3, purify and precipitate as in Step 6, Section 1.3, resuspend to a final concentration of 100 ng/µl.

3. Dilute the two phosphorilated oligonucleotides that give rise to the spacer (see Table 1) to 50 ng/µl in H_2O, mix 2 µl of each in an Eppendorf tube, heat at 37°C for 30 min to anneal then keep in ice.

4. Set up ligation by mixing in a 0.5 ml tube:

digested vector (from Step 2, Section 3.1)	1 µl
5x ligase buffer	4 µl
T4 Ligase	1 µl
H_2O	12 µl
annealed oligonucleotides (diluted 1:10)	2-4 µl

dilute 1:10 the annealed spacer oligos with H_2O immediately before adding to the ligation mix; incubate over night at 4°C. Include the vector self ligation as control

5. Transform bacteria as in Step 8, Section 1.3. Screen recombinants by digestion of plasmid DNA with Sal I, that cuts once into the spacer and once into the pCDNA3. The expected bands are 3.1, 2.1 and 1.3 Kb long.

6. Choose a positive colony and prepare "pIL-2-spacer-MOV19SD" (Figure 2C) plasmid DNA on a larger scale with the Qiaprep kit (Qiagen) according to the protocol provided.

3.2 Cloning into myeloma expression vector

The expression cassette is now transferred into the myeloma expression vector pH PCRγIII (Melani et al.1998), that contains the Cκ coding sequence and all the regulatory elements that improve fusion protein production (Figure 2D). Other expression vectors might also be suitable, such as Epi-Tag/V5 or Epi-Tag/His vectors (Invitrogen) that allows fusion with other "tags" than Cκ, but have not been tested in our case. With these vectors the 3' end of the scFv does not need a splicing donor site but must lack the stop codon: modify the reverse Vκprimer (Table 1) accordingly. The pH PCRγIII vector does not have restriction sites compatible with those of the expression cassette, since the cloning sites available are only Sac I and Hind III, at 5' and 3' ends respectively. Therefore, the cassette has to be cloned as a blunt ended fragment.

1. Digest 5 µg of "pIL-2-spacer-MOV19SD" DNA with Hind III and Xho I (see Step 2, Section 1.3). Check on a 1% agarose gel for the release of the 1.2 Kb insert from the 5.4 Kb vector.

2. Digest 5 µg of pH PCRγIII in a final volume of 50 µl with Sac I, as in Step 2, Section 1.3. Check on a 0.8% agarose gel: if linearization is completed proceed with the second restriction enzyme digestion by adding:

H₂O	40 µl
10x NEB buffer "2"	5 µl
Hind III [20U/ul]	1 µl

Incubate at 37°C for 1h then inactivate at 65°C 10 min. Check on 1% agarose gel for the release of two stuffer inserts of approximately 250 and 350 bp respectively.

3. Proceed to the blunting of vector and insert:

digested DNA	100 or 50 µl
10x T4 DNA polymerase buffer	10 µl
dNTP [25 mM each]	0.5 µl
T4 DNA polymerase [3 U/ul]	10 µl
H_2O	to a final volume of 200 µl

incubate at 12°C for 30 min, heat inactivate at 75°C for 10 min. Purify as in Step 6 of Section 1.3.

4. Proceed to dephosphorilation of the vector through Steps 5 and 6 of Section 1.3.

5. Recover vector (11.5 Kb) and insert (1.2 Kb) by a preparative gel purification as in Step 3, Section 1.3; use 0.7% agarose gel for vector and 1% for insert.

6. Estimate amount of vector and insert as in Step 7, Section 1.3 and calculate the amount of DNA to be ligated. Take into account that pH PCRγIII is 10 times the "IL-2-spacer-MOV19SD" insert on a molar ratio. Blunt ends ligation is more difficult than sticky ends ligation, and it is convenient to set up ligation at a vector:insert ratio of 1:1, 3:1 and 1:3.

7. Transform competent cells as in Step 8, Section 1.3 but use the K803 E.Coli strain. After transformation spin bacteria, resuspend in 200-300 µl and plate all bacteria (100 µl a time) onto ampicillin-LB plates. Plate positive and negative controls as usual.

8. Analyze colonies for the presence of the expression plasmid according to (Sambrook et al. 1989). Digest plasmid DNA with Eco RI which should give the following bands, according to the orientation of the insert respect to the promoter.

Direct orientation	Reverse orientation	No insert (self ligated vector)
5.2 Kb	4.9 Kb	7.8 Kb
4.2 Kb	4.4 Kb	
4.0 Kb	4.0 Kb	4.0 Kb

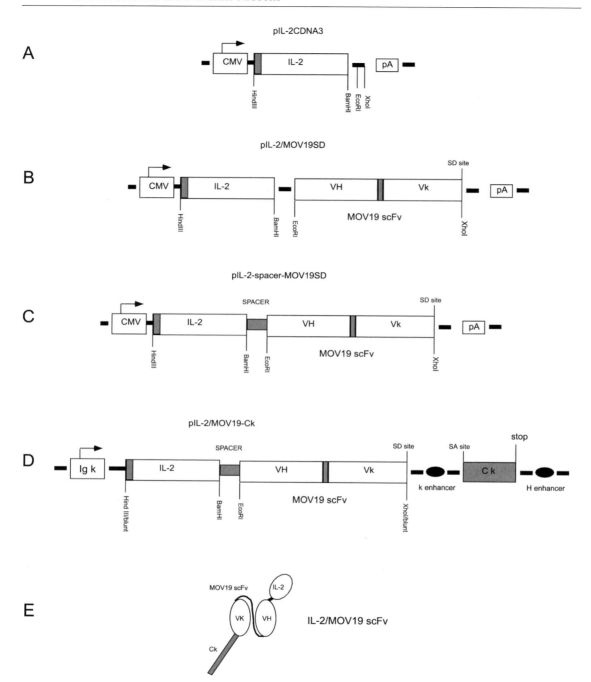

Fig. 2. Schematic representation of the intermediate constructs (A-C), the final expression vector (D) and the IL-2/ MOV19 scFv fusion protein (E).Arrow: promoter; pA: polyadenilation signal; SD/SA: splice donor/acceptor site; stop: stop codon.

Subprotocol 4
J558L Cell Transfection by Protoplast Fusion

▉ ▉ Procedure

4.1 Protoplast preparation

1. Inoculate a colony of transformed K803 in 5 ml of LB with ampicillin; grow over night at 37°C with shaking (200-220 rpm).

2. Transfer 100 µl of the over night culture in 100 ml di LB+0.2 % Maltose+ 0.1 mg/ml ampicillin in a 1 liter flask; grow until OD_{600} 0.5-0.7 then add 200 µl of a [60 mg/ml] chloramphenicol solution in EtOH and shake over night at 37°C.

3. Spin bacteria in a 50 ml tube 10 min at 3000 rpm 4°C, resuspend in 2.5 ml of 20% sucrose- 50mM Tris-Cl (pH 8.0) and add 0.5 ml of lysozyme solution: 10 min in ice.

4. Add 1 ml of 0.25 M EDTA pH 8.0; in ice 5 min then add 1 ml of 50 mM Tris-Cl pH 8.0.

5. While vortexing protoplasts at low speed, add slowly, dropwise 20 ml of prewarmed DMEM 10% FCS+ 10% sucrose+ 10 mM $MgCl_2$.

6. Add 0.2 ml of [1 mg/ml] DNAse solution and leave at R.T.

4.2 Protoplast fusion

1. Pellet protoplasts 30 min at 3000 rpm, discard the supernatant accurately.

2. Pellet 5 x 10^6 J558L cells, plated the day before in DMEM 10% FCS, resuspend in 5 ml of DMEM without serum and transfer onto the protoplasts pellet, without mixing.

3. Spin 10 min at 1000 rpm, suck the supernatant and resuspend the pellet.

4. While vortexing the tube at low speed, add slowly (in 90 sec), dropwise 2 ml of PEG 1500-DMSO, then, still vortexing, dilute slowly, dropwise with 10 ml of DMEM without serum.

5. Add 10 ml of DMEM 10% FCS and pellet 5 min at 1500 rpm.

6. Resuspend cells in 50 ml DMEM 10% FCS + 50 µg/ml Kanamycin and plate 50 and 100 µl in 96 well plates (flat bottom). Incubate at 37°C, 5% CO_2 over night

7. The day after start selection by adding 100 µl/well of DMEM 10% FCS with G418 (final concentration 2 mg/ml). Change selection medium every third day.

4.3 Screening of recombinant J558L

G418-resistant cells grow as clumps in a couple of weeks. A preliminary screening of fusion protein production can be made by ELISA for the Cκ tail or for human IL-2.

1. Set up ELISA using two antibodies that recognize the Cκ fragment of the murine Ig (Southern Biotechnology Associates, Birmingham, AL) or the human IL-2 (Duoset, Genzyme, Cambridge, MA) following the manufacturing instruction. Use dilutions of cell supernatant and test in duplicate.

2. Choose the colony that produces the highest amount of fusion protein and subclone at 0.5 cells/well. Select G418-resistant clones and repeat ELISA to identify the best producer.

3. Expand the clone, recover the supernatant and keep it at -20°C until purification.

Subprotocol 5
Functional Assays of the IL-2/MOV19-Ck Fusion Protein

The biological activity of the fusion protein can be tested in the supernatant of the transfected J558L cells and after purification, on different batches of the final product. The two functions of the fusion protein, the antigen binding capacity and the immuno stimulating activity of IL-2 can be tested either separately or all together.

■ ■ Procedure

5.1 Antigen binding

The antigen binding capacity of MOV19 scFv is tested inimmunofluorescence with antigen positive IGROV-1 (Bénard et al. 1985) and antigen negative SK-BR-3 (ATCC number HTB30) tumor cells.

1. Plate cells the day before testing in RPMI 1640 medium with 10% FCS.

2. Detach cells by treatment with trypsin, wash, resuspend FACS buffer and plate 1- 0.5 x 10^6 cells in a 96 well plate (U-bottom) in triplicate.

3. Spin the plate 3 min at 1500 rpm, remove medium by flicking the plate.

4. Dilute the fusion protein to 0.5 µg/ml (either purified or J558L cell supernatant) in FACS buffer, add 100 µl to one well and gently mix the cells with the pipette tip. Use the other wells for controls: use the MOV19 Ab, as positive control, and an unrelated Ab of the same isotype (IgG2a) as negative control, at [0.5 µg/ml]. Incubate 1 h in ice.

5. Wash 4 times by adding to each well 200 µl of FACS buffer and spin 3 min at 1500 rpm, remove medium by flicking the plate.

6. Dilute biotinylated anti Cκ (Southern Biotechnology Associates) to 2.5 µg/ml in FACS buffer, add 100 µl to each well, mix gently and incubate 1 h in ice.

7. Repeat Step 5 of this Section.

8. Dilute streptavidin-phycoeritrin conjugate (Pharmingen, San Diego, CA) to 0.5 µg/ml FACS buffer, add 100 µl/well, mix gently and incubate 1 h in ice in the dark.

9. Repeat Step 5 of this Section. Resuspend cells in 0.5 ml FACS buffer, transfer to appropriate tube and analyze at FACS scan.

5.2 Biological activity of IL-2 (CTLL-2 proliferation)

1. Prepare dilutions of fusion protein, either from supernatant of transformed J558L or purified, and of recombinant human IL-2 (Proleukin, Chiron Corp., Emeryville) in RPMI 1640 medium and plate 100 µl in a 96 well plate (U-bottom) in triplicate.

2. Proceed essentially as described for CTLL-2 proliferation assay, incubate 48 h at 37°C in 5% CO_2 and add 1 µCi/well of [^{3}H]thymidine (Amersham Life Science) during the last 18 h.

3. Harvest the cells and measure[^{3}H]thymidine incorporation by liquid scintillation.

4. Calculate the specific activity of the fusion protein by plotting the mean of cpm versus the concentration and comparing the curve obtained with standard IL-2 with IL-2/MOV19 scFv.

5.3 Targeting and bioactivity of IL-2/MOV19 scFv (Figure 3)

1. Plate antigen positive (IGROV-1) and antigen negative (SK-BR-3) tumor cells the day before testing.

2. Proceed through Steps 1 and 2 of Section 5.1 but resuspend in RPMI 1640 10% FCS and irradiate at 20000 rads; count and transfer 2×10^6 cells of each line into two 15 ml tubes.

3. Add 5 µg/ml of fusion protein to one tube and incubate 1 h in ice.

4. Wash the cells four times with 10 ml of medium, resuspend in 5 ml ad plate 1:2 dilutions (ranging from 20000 to 2000 cells) in 100 µl into a 96 well plate (U-bottom) in triplicate and leave 1 h at 37°C to adhere.

5. Wash CTLL-2 cells that have been starved of IL-2 (two days after the last feeding), count and plate 5000 cells/well in 100 µl onto the tumor cells. As control leave three wells of tumor without CTLL-2 and plate CTLL-2 with serial dilution of recombinant IL-2 in triplicate.

6. Proceed through Steps 2 and 3 of Section 5.2. Calculate activity as (mean cpm incorporated by CTLL-2) - (mean cpm incorporated by tumor cells). CTLL-2 cell proliferation, stimulated by antigen positive cells coated with the fusion protein, will be proportional to the number of stimulating tumor cells.

5.4 Purification of IL-2/MOV19 scFv fusion protein

The fusion protein is purified by affinity chromatography using the rat anti-mouse Ck moAb 187.1 (ATCC number HB58). The moAb is bound to CNBr-activated Sepharose 4B (Pharmacia Biotech, Uppsala, Sweden)

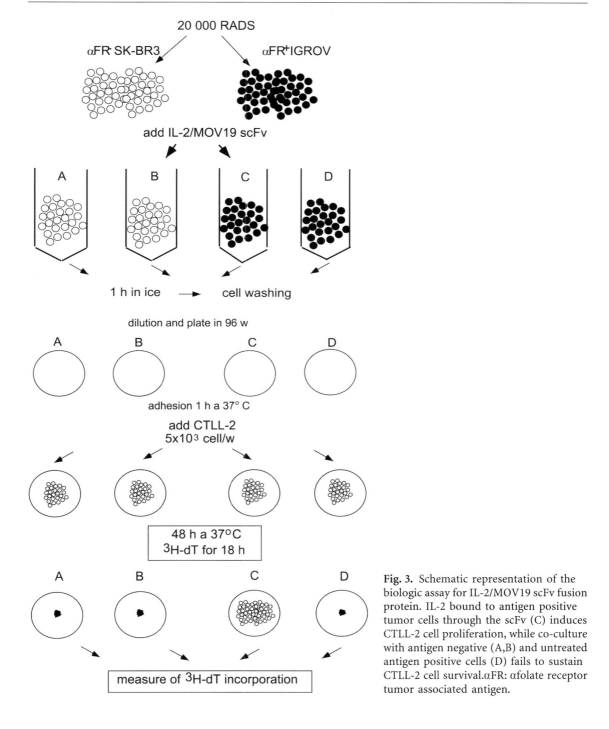

Fig. 3. Schematic representation of the biologic assay for IL-2/MOV19 scFv fusion protein. IL-2 bound to antigen positive tumor cells through the scFv (C) induces CTLL-2 cell proliferation, while co-culture with antigen negative (A,B) and untreated antigen positive cells (D) fails to sustain CTLL-2 cell survival.αFR: αfolate receptor tumor associated antigen.

according to manufacturing instruction. The purification is done according to Current Protocols in Immunology but IL-2/MOV19 is eluted in 5 column volumes of elution buffer at pH 3.0 and neutralized by Tris-Cl (pH 9.5). The eluted fusion protein is extensively dyalized in 3.4 mM EDTA/PBS and stored at -20°C.

Comments

The strategy outlined in this chapter can easily apply to the construction of other fusion proteins between a cytokine and the scFv of an antibody, taking into consideration a few critical steps. Choose accurately the restriction sites for cloning, which must not be present within any of the two sequences. The intermediate plasmid should provide not only the appropriate multiple cloning sites but also known flanking sequences for priming and sequencing of the inserted fragments. The in-frame cloning of the 3'sequence can be easily obtained by inserting the spacer in between, and therefore, the design of the two oligonucleotides is very critical because the spacer should encode neutral aminoacids to avoid unnatural folding of the fusion protein. The final expression vector should provide not only a strong and cell specific promoter but also a "tag" sequence that is very useful for protein purification, unless the fusion protein can be purified by binding to the soluble antigen.

Acknowledgements. The Associazione Italiana Ricerca sul Cancro and the Italian Ministry of Health supported this work.

References

Bénard J, De Silva J, De Blois MC, Boyer P, Duvilland P, Chiric E, Riou G (1985) Characterization of a human ovarian adenocarcinoma line, IGROV-1, in tissue culture and in nude mice. Cancer Res. 45: 4970-4979

Boleti E, Deonarain MP, Spooner RA, Smith AJ, Epenetos AA, George AJT (1995) Construction, expression and characterization of a single chain anti-tumour antibody (scFv)-IL-2 fusion protein. Ann. Oncol. 6: 945-947

Current Protocols in Immunology. (1994-1999) Cooligan JE, Kruisbeek AM, Margulies DH, Shevac EM, Strober W (eds). John Wiley and Sons, Inc.

Current Protocols in Molecular Biology. (1995-1997) Ausubel FM, Brent R, Kingston RE, Moore DD, Seidman JG, Smith JA, Struhl K (eds). John Wiley and Sons, Inc.

Dorai H, McCartney JE, Hudziak RM, Tai MS, Laminet AA, Houston LL, Huston JS, Oppermann H (1994) Mammalian cell expression of single-chain Fv (sFv) antibody proteins and their C-terminal fusion with interleukin-2 and other effector domains. BioTechnology,12: 890-897

Melani C, Figini M, Nicosia D, Luison E, Ramakrishna V, Casorati G, Parmiani G, Esh-har Z, Canevari S, Colombo MP (1998)Targeting of interleukin 2 to human ovarian carcinoma by fusion with a single-chain Fv of antifolate receptor antibody. Cancer Res. 58: 4146-4154

Miotti S, Canevari S, Mènard S, Mezzanzanica D, Porro G, Pupa SM, Regazzoni M, Tagliabue E, Colnaghi M.I (1987) Characterization of human ovarian carcinoma associated antigens defined by novel monoclonal antibodies with tumor restricted specificity. Int. J. Cancer 39: 297-303

Parmiani G, Colombo MP, Melani C, Arienti F (1997) Cytokine gene transduction in the immunotherapy of cancer. In: August JT (ed) Gene Therapy. Adv. Pharmacol., vol 40. Academic Press, London, pp 259-307

Reisfeld RA, Becker JC, Gillies SD (1997) Immunocytokines: a new approach to im-munotherapy of melanoma. Melanoma Res. 7 (suppl.2): S99-S106

Rosenberg SA, Lotze MT, Yang JC, Aebersold PM, Linehan WM, Seipp C, White DE (1989) Experience with the use of High-dose interleukin-2 in the treatment of 652 cancer patients. Ann Surg 210: 474-485

Sambrook J, Fritsch EF, Maniatis T.(1989) Molecular cloning. A laboratory manual. Cold Spring Harbor Laboratory Press, New York.

Savage P, So A, Spooner RA, Epenetos AA (1993) A recombinant single chain antibody interleukin-2 fusion protein. Br. J. Cancer, 67: 304-310

Yokota T, Milenic DE, Whitlow M, Schlom J (1992) Rapid tumor penetration of a single-chain Fv and comparison with other immunoglobulin forms. Cancer Res. 52: 3402-3408

Antibody Fusion Proteins for Targeted Gene Delivery

Christoph Uherek and Winfried Wels

Introduction

Molecular conjugates which employ antibodies or other ligands chemically coupled to polycations to deliver DNA into cells constitute an important group of non-viral vectors which are investigated as tools for therapeutic gene delivery (Wagner, 1998; Kircheis et al., 1999). The protein and DNA components of such artificial compounds assemble into stable complexes with defined target cell tropism. Thereby interaction with DNA is mediated by the binding of a polycationic reagent such as poly-L-lysine to negatively charged plasmid DNA resulting in a so-called polyplex (Felgner et al., 1997). The addition of a target cell-specific ligand facilitates cell recognition and uptake of the complex via receptor-mediated endocytosis. An important improvement of this concept was the incorporation of endosomolytic activities which upon internalization into cells greatly enhances the release of DNA from endocytic vesicles and the expression of the transferred gene (Wagner, 1998).

An alternative approach for the design of self-assembling systems for gene delivery is based on fusion proteins engineered to incorporate distinct activities required for DNA-binding and the different steps of cell recognition, intracellular delivery, and nuclear transport (Wels and Fominaya, 1998). Such fusion proteins in contrast to similar chemical conjugates can be produced in suitable expression systems in their final form and the resulting products are generally homogeneous in their composition. Recombinant antibody derivatives as integral components of DNA-binding fusion proteins have proven useful for targeting protein-DNA

Christoph Uherek, Chemotherapeutisches Forschungsinstitut Georg-Speyer-Haus, Paul-Ehrlich-Strasse 42-44, 60596 Frankfurt am Main, Germany
✉ Winfried Wels, Chemotherapeutisches Forschungsinstitut Georg-Speyer-Haus, Paul-Ehrlich-Strasse 42-44, 60596 Frankfurt am Main, Germany
(*phone* +49-69-63395-188; *fax* +49-69-63395-189; *e-mail* wels@em.uni-frankfurt.de)

complexes to a target cell population of choice (Fominaya and Wels, 1996; Uherek et al., 1998). To provide DNA carrier proteins with the ability to form a predictable and stable complex with plasmid DNA the natural DNA-binding domain of transcription factors can be employed. Fragments of the yeast Gal4 transcriptional activator as well as several well characterized DNA-binding activities from gene regulatory proteins of *E. coli* retain the ability to bind to their cognate DNA recognition sequence when expressed as part of a fusion protein. While the combination of an antibody domain and a DNA-binding activity might result in a protein capable of delivering plasmid DNA to cells (Chen et al., 1995), including an additional endosome release activity can greatly enhance successful DNA uptake and subsequent gene expression (Fominaya and Wels, 1996; Uherek et al., 1998).

Bacterial toxins such as *Pseudomonas* exotoxin A (ETA, PE) and diphtheria toxin (DT) bind to receptors on the surface of target cells and, after internalization via receptor-mediated endocytosis, utilize the acidification of intracellular vesicles as a signal to activate an endosome escape function. This results in the translocation of an enzymatically active toxin fragment from the endosome to the cytosol where it inactivates the cellular protein synthesis machinery. Distinct toxin domains provide the different functions required (Wilson and Collier, 1992). The acid activatable translocation domains of ETA and DT retain their endosome escape activity even when combined with completely unrelated cell recognition and intracellular effector functions (Wels and Fominaya, 1998). The resulting molecules are non-toxic but mimic the cell entry pathway of the parental toxins.

Below, a strategy is described for the construction, bacterial expression and functional characterization of target cell specific DNA carrier proteins which employ the yeast Gal4 DNA-binding domain for stable interaction with plasmid DNA, and a scFv domain for specific binding to receptors on the target cell surface. Upon receptor-mediated endocytosis of a protein-DNA complex containing such a fusion protein the translocation domain of diphtheria toxin included in the molecule enhances the release of the transferred DNA from the endosomal compartment thereby increasing transfection efficiency.

Materials

Equipment
- French press (e.g. SLM-Aminco French pressure cell press)
- Luminometer (e.g. Berthold AutoLumat LB 953)

Buffers

For protein expression and purification
- Buffer A: 8 M Urea, 50 mM Tris, pH 8.0, 150 mM NaCl, 10 µM $ZnCl_2$
- Buffer B: 8 M Urea, 50 mM Tris, pH 8.0, 150 mM NaCl, 10 µM $ZnCl_2$, 250 mM imidazole
- Refolding buffer: 100 mM Tris, pH 8.0, 8 mM oxidized glutathione, 0.5 M L-arginine, 150 mM NaCl
- Dialysis buffer: 50 mM Tris, pH 8.0, 50 mM KCl, 5 mM $MgCl_2$, 20 µM $ZnCl_2$, 20% glycerol

For electrophoretic mobility shift assay
- 5x EMSA-buffer: 50 mM Tris, pH 7.5, 100 mM NaCl, 5 mM EDTA, 25% glycerol, 5 mM dithiothreitol

For transfection
- 10x transfection buffer: 0.5 M Hepes, pH 7.5, 0.5 M KCl, 50 mM $MgCl_2$, 1 mM $ZnCl_2$

For luciferase assay
- Lysis buffer: 25 mM glycylglycine, pH 7.8, 1 mM dithiothreitol, 15% glycerol, 8 mM $MgSO_4$, 1% Triton X-100, 1 mM EDTA
- Dilution buffer: 25 mM glycylglycine, pH 7.8, 10 mM $MgSO_4$, 5 mM ATP
- Luciferin solution: 25 mM glycylglycine, 250 µM luciferin, 0.5 mM coenzyme A

Subprotocol 1
Construction and Functional Characterization of DNA-Binding Antibody Fusion Proteins

Construction of DNA-binding antibody fusion proteins

In wildtype DT the enzymatically active fragment is located N-terminal and the cell binding activity is located C-terminal of the translocation domain. Therefore a similar arrangement of domains is chosen for the construction of modular DNA carrier proteins. Fragments encoding the DNA binding domain of Gal4 and the translocation domain of DT are derived by PCR from appropriate templates (for details see (Uherek et al., 1998)). The cloning and selection of scFv antibody fragments is described in detail in several other chapters of this book. For a DNA carrier protein to function as a transfection reagent the chosen target receptor must be abundantly expressed on the cell surface to allow binding of the protein-DNA complex. Furthermore the target antigen must be internalized at a considerable rate to promote uptake of the complex into endosomes by receptor-mediated endocytosis.

The bacterial expression plasmid pSW55-GD5 schematically shown in Figure 1 contains coding sequences for Gal4 amino acids 2-147 fused to DT amino acids 195-383 and a scFv antibody domain (Uherek et al., 1998). A similar plasmid suitable for the insertion of a scFv fragment of choice can be obtained from the authors of this chapter. Cytoplasmic expression of the resulting fusion gene is controlled by an IPTG-inducible tac promoter. Synthetic sequences included at the 5' end of the fusion gene encode FLAG and polyhistidine tags for immunological detection and purification of recombinant protein by Ni^{2+} affinity chromatography, respectively.

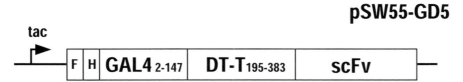

Fig. 1. Schematic representation of the *E. coli* expression plasmid pSW55-GD5 which encodes under the control of the IPTG-inducible tac promotor the modular fusion protein GD5. It consists of an N-terminal synthetic FLAG epitope (F), a polyhistidine tag (H) facilitating Ni^{2+} affinity chromatography, amino acids 2-147 of the yeast Gal4 protein (DNA-binding domain), amino acids 195-383 of the bacterial diphtheria toxin (translocation domain) and a C-terminal scFv antibody domain specific for the ErbB2 receptor protein (Uherek et al., 1998).

▣▣ Procedure

Bacterial expression and purification of DNA carrier proteins

Protocols for bacterial expression and purification of recombinant anti-body derivatives are described in detail in Chapters 17-23. Below a mod-ified procedure is outlined which works best for DNA carrier proteins containing Gal4 and scFv domains.

1. Transform an *E. coli* strain suitable for protein expression with the plasmid encoding the antibody fusion protein. For the DNA carrier protein GD5 (Uherek et al., 1998) best results were obtained using *E. coli* strain BL21(λDE3) trxB⁻ (Proba et al., 1995).

2. Inoculate 200 ml LB medium containing 100 µg/ml ampicillin with a single colony and incubate the culture overnight at 37°C and 220 rpm.

3. Dilute the culture to $A_{550} < 0.5$ in a final volume of 800 ml and grow to $A_{550} = 0.7\text{-}1$.

4. Induce protein expression by the addition of 0.25 mM IPTG (final concentration). For good aeration during protein expression it is recommended to use 2 l side-baffled Erlenmeyer flasks. If not available split the culture into two 2 l Erlenmeyer flasks. Incubate the culture for 90 min at room temperature and 150 rpm.

5. Harvest bacteria by centrifugation (10 min at 4,400 x g) and store pellets at -80°C until you proceed to step 6 (purification of the recombinant protein).

6. Resuspend pellet in 30 ml buffer A and lyse bacteria using a French press (3 cycles at 35,000 psi) or by sonication on ice (2 cycles for 3 min with 70% of maximal power). Clear the lysate by centrifugation for 30 min at 40,000 x g and 4°C and filtration of the supernatant though a prewetted folded filter.

7. Prepare a column for Ni²⁺-affinity chromatography (described in Chapter 17). Equilibrate the column with 5 column volumes of buffer A before you apply the cleared lysate at a slow flow rate (1 ml/min).

8. Remove unbound proteins by washing the column with 5 column volumes of buffer A followed by 3 column volumes of buffer A containing 40 mM imidazole.

9. Elute specifically bound proteins with 3 column volumes of buffer B and collect eluate in 1 ml aliquots.

10. Check protein content and purity of eluates by SDS-PAGE analysis followed by Coomassie staining or immunoblotting. Pool fractions containing the fusion protein.

11. To regain DNA binding activity of the Gal4 domain the fusion protein needs to be refolded in a redox buffer system. For complete reduction of disulfide bonds add dithioerythritol to a final concentration of 0.3 M and incubate at room temperature for 2 h.

12. Dilute 50-fold in ice-cold refolding buffer and incubate for at least 48 h at 10°C.

13. Concentrate refolded protein to the desired volume by using a concentrator. Alternatively the refolded protein can be filled into dialysis tubing and concentrated by incubation with polyethylene glycol 35,000.

14. Dialyze concentrated protein against dialysis buffer at 4°C for 12 h. Store dialyzed protein at -20°C.

Subprotocol 2
Functional Analysis

Binding to target receptors

Specific binding of the scFv domain to its target receptor can be determined by various methods including ELISA and FACS analysis. Basic protocols for these procedures are described elsewhere in this book. Alternatively, protocols for analysis of the binding specificity of DNA carrier proteins can also be found in (Fominaya and Wels, 1996; Fominaya et al., 1998; Uherek et al., 1998).

■ ■ Procedure

Binding to DNA

Sequence specific binding of the Gal4 domain to DNA can be analyzed in an electrophoretic mobility shift assay (EMSA) using a double-stranded DNA oligonucleotide which contains the Gal4 consensus binding sequence 5'-CGGAGGACAGTCCTCCG-3' (Carey et al., 1989). A typical example for the analysis of DNA binding is shown in Figure 2.

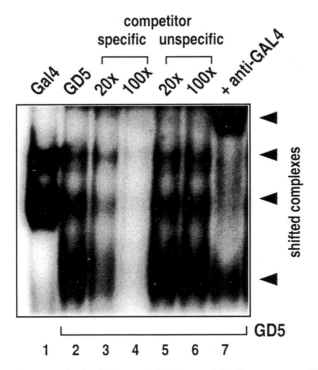

Fig. 2. Analysis of GD5 DNA-binding activity. Sequence-specific DNA-binding of GD5 was analyzed in electrophoretic mobility shift assays (EMSA). A double-stranded, ^{32}P-labelled Gal4-specific oligonucleotide was incubated with GD5 in the presence or absence of competitors. Complexes were analyzed in a 7.5% polyacrylamide gel and visualized by autoradiography. Lane 1, positive control: recombinant Gal4 DNA-binding domain (Santa Cruz Biotechnology); lanes 2-7, purified GD5 after refolding using a redox-buffer system; lanes 3 and 4, competition with a 20- or 100-fold molar excess of unlabelled Gal4-specific oligonucleotide; lanes 5 and 6, competition with a 20- or 100-fold molar excess of unlabelled control oligonucleotide; lane 7, super-shift of the complexes with an anti-Gal4 antibody. The positions of different GD5-oligonucleotide complexes are indicated by arrows. The area of the free oligonucleotide is not shown.

1. For radioactive labelling incubate 5 pmol of the double-stranded oligonucleotide with 1.5 MBq [γ-^{32}P]-ATP and 10 units T4 polynucleotide kinase in the buffer supplied by the manufacturer in a final volume of 20 μl for 30 min at 37°C.

2. Adjust the volume to 100 μl with water and separate oligonucleotide from unbound ATP using a MicroSpin S-200 HR column (Pharmacia Biotech). Measure incorporated radioactivity in a scintillation counter.

3. Incubate 100 ng of purified, refolded fusion protein in 1x EMSA-buffer, 400 µg/ml poly(dI-dC), 0.4 mg/ml bovine serum albumin and 5x 10^4 cpm of labelled oligonucleotide in a final volume of 20 µl for 30 min at room temperature.

4. Separate the samples in a non denaturing 7.5% polyacrylamide gel with 0.25x TBE as a running buffer.

5. Dry the gel and visualize radioactive bands by exposure to X-ray film.

Transfection of cells and luciferase assay

To achieve specific binding of the DNA carrier protein to a plasmid the Gal4 recognition sequence must be included in the DNA molecule. The construct pSV2G4LUC fulfills this requirement. It encodes under the control of the SV40 early promoter the firefly luciferase as a reporter gene and contains two Gal4 recognition motifs in the 3' untranslated region of the expression cassette (Fominaya and Wels, 1996). This construct and similar plasmids encoding CAT or β-galactosidase can be obtained from the authors of this chapter. Typical results obtained with the transfection procedure described below are illustrated in Figure 3.

1. Seed cells for transfection in 12-well tissue culture plates at a density of 30,000-50,000 cells per well and grow overnight at 37°C in 1 ml of complete growth medium. The chosen density should allow further cell growth for 2.5 days.

2. Exchange the growth medium with 1 ml/well fresh medium 0.5 to 2 h before the addition of transfection complexes.

3. Transfection complexes are prepared in a final volume of 110 µl/well. Duplicate or triplicate samples should be analyzed. Incubate 2 µg supercoiled DNA of a suitable luciferase reporter gene plasmid (e.g. pSV2G4LUC) with a variable amount of fusion protein in 1x transfection buffer and adjust the volume to 99 µl by the addition of dialysis buffer. Allow binding of the fusion protein to the DNA for 15 min at room temperature.

4. Add poly-L-lysine to facilitate condensation of the protein-DNA complex and to neutralize excess negative charge of the DNA. To obtain a charge ratio of 1 (electroneutral complex) add poly-L-lysine with an average chain length of 236 residues (poly-L-lysine HBr, Sigma, Mol. wt. 30,000-70,000; pL_{236}). For a 6 kb plasmid such as pSV2G4LUC

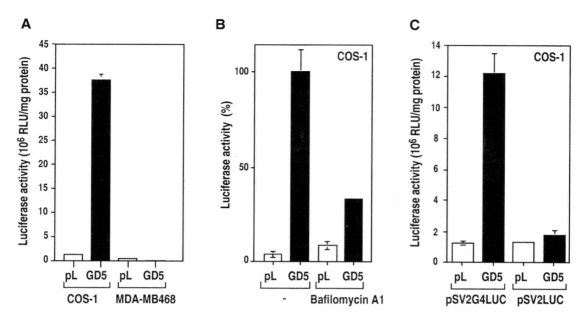

Fig. 3. GD5-mediated gene transfer. Preparation of transfection complexes, transfection of cells and determination of luciferase activity was performed as described in the protocol. Cells were seeded in 12-well tissue culture plates at a density of 5 x 10⁴ cells/well and grown overnight. Transfection complexes containing 4.5 µg GD5, 2 µg luciferase reporter plasmid and poly-L-lysine with an average degree of polymerization of 236 residues (250 nM) were added to the cells in normal growth medium. Control cells were treated with a reporter plasmid containing polyplex lacking the fusion protein (pL, open bars). After 4 h, the medium was exchanged, and the cells were grown for another 40 h before they were harvested for analysis. Luciferase activity is expressed in relative light units (RLU)/mg of total protein. A, Specificity of GD5-mediated gene transfer. COS-1 African green monkey kidney cells expressing ErbB2, and ErbB2-negative MDA-MB468 human breast carcinoma cells were transfected with GD5 polyplex. B, Contribution of the translocation domain to transfection efficiency. COS-1 cells were transfected in the presence or absence of 200 nM bafilomycin A$_1$, an inhibitor of vacuolar H$^+$-ATPases and endosomal acidification. Relative luciferase activities in comparison to cells treated with GD5-pSV2G4LUC polyplex in the absence of inhibitor are shown. C, dependence of GD5 mediated transfection on the presence of Gal4 consensus-binding sequences. COS-1 cells were transfected with polyplex containing either pSV2G4LUC or the similar pSV2LUC plasmid which lacks a Gal4-specific recognition sequence.

slowly add 11 µl of a 2.5 µM pL$_{236}$ solution under constant vortexing to achieve a molar ratio of 50:1, followed by incubation for 15 min at room temperature.

Since poly-L-lysine itself complexes DNA and serves as a transfection reagent, samples containing poly-L-lysine but lacking the DNA carrier protein should be included as a control.

5. Add 100 µl/well of the protein-DNA complex dropwise to the cells and mix with the growth medium by gently shaking the plate. Incubate for 4 to 16 h at 37°C.

6. Exchange the medium with 1 ml/well of fresh growth medium and incubate the cells at 37°C for 40 to 44 h after addition of the transfection complexes.

7. Remove the growth medium and wash the cells twice with 1 ml/well phosphate-buffered saline. Add 100 µl of lysis buffer to each well and incubate at room temperature for 15 min.

8. Harvest the cell lysates and transfer them to microspin tubes. Clear the lysates by centrifugation for 5 min at 13,000 rpm in a microcentrifuge.

9. Mix 50 µl of cleared lysate with an identical volume of dilution buffer in a 12 x 75 mm polystyrene round bottom tube suitable for luminometer measurement.

10. Measure luciferase activity for 30 s in a luminometer by automatic injection of 300 µl luciferin solution.

11. Determine the protein content of the cell lysates using the Bradford method (Bradford, 1976) and calculate luciferase activity as relative light units/mg of cellular protein.

Troubleshooting

Potential problems that may arise during the process of cloning a scFv antibody fragment or during protein expression and purification of antibody fusion proteins are addressed elsewhere in this book. Here we focus on topics that concern the functionality of a DNA carrier protein.

- **The fusion protein does not bind to target cells despite functionality of the scFv fragment when expressed as a single domain protein.**
  Possibly the DNA-binding and/or the translocation domain interfere with binding of the scFv to its target. The introduction of a flexible linker sequence separating the scFv from the other domains of the molecule might provide additional flexibility and improve binding. Alternatively, the scFv fragment might not be functional at all when expressed at the C-terminus of a fusion protein. In this case similar constructs containing the translocation domain of *Pseudomonas* exotoxin A might serve as an alternative since there the cell recognition domain

is located at the N-terminus of the fusion protein (Fominaya and Wels, 1996).

- **The fusion protein binds to cells irrespective of the presence of target antigen.**
  The DT translocation domain included in the fusion protein contains membrane-interacting hydrophobic helices. When these are exposed (typically upon a conformational change induced at acidic pH) they might facilitate non-specific interactions with cell membranes. Avoiding low pH environments and adjusting the refolding conditions during the preparation of the DNA carrier protein might help to prevent this problem.

- **The presence of fusion protein in the transfection complex does not enhance transfection efficiency in comparison to poly-L-lysine polyplex alone.**
  Analyze the DNA binding activity of the fusion protein. As shown in Figure 3C, sequence-specific binding of the fusion protein to vector DNA is a prerequisite for enhanced transfection efficiency. If the fusion protein does not bind to a Gal4-specific oligonucleotide in an EMSA, check the refolding conditions. In our hands complete reduction of the fusion protein prior to refolding is required to obtain functionality of the Gal4 DNA-binding domain. Binding of the fusion protein to the reporter gene construct can be analyzed by including the plasmid as a competitor in an EMSA. If the plasmid DNA does not compete binding of the radioactive oligonucleotide probe to the protein confirm the presence of Gal4 recognition sites in the reporter gene construct. If the Gal4 DNA-binding domain is functional but transfection efficiency remains low, check the endosome escape activity of the DT translocation domain. Intracellular activation of the translocation domain requires a drop in the endosomal pH below 5.5. Therefore its endosome escape activity is blocked by the presence of inhibitors of endosomal acidification during the transfection (Figure 3B). These inhibitors include weak acidotropic bases such as ammonium chloride or methylamine (both used at a final concentration of 50 mM), carboxylic ionophores like nigericin (5 μM), or inhibitors of vacuolar H^+-ATPases like bafilomycin A_1 (200 nM). If transfection is not affected by the inhibitors the translocation domain most probably is not functional. Check whether the protein was exposed to acidic pH after refolding or try to optimize refolding conditions for the particular protein used.

References

Bradford MM (1976) A rapid and sensitive method for the quantitation of microgram quantities of protein utilizing the principle of protein-dye binding. Anal Biochem 72:248-254

Carey M, Kakidani H, Leatherwood J, Mostashari F, Ptashne M (1989) An amino-terminal fragment of GAL4 binds DNA as a dimer. J Mol Biol 209:423-432

Chen SY, Zani C, Khouri Y, Marasco WA (1995) Design of a genetic immunotoxin to eliminate toxin immunogenicity. Gene Ther 2:116-123

Felgner PL, Barenholz Y, Behr JP, Cheng SH, Cullis P, Huang L, Jessee JA, Seymour L, Szoka F, Thierry AR, Wagner E, Wu G (1997) Nomenclature for synthetic gene delivery systems. Hum Gene Ther 8:511-512

Fominaya J, Uherek C, Wels W (1998) A chimeric fusion protein containing transforming growth factor-alpha mediates gene transfer via binding to the EGF receptor. Gene Ther 5:521-530

Fominaya J, Wels W (1996) Target cell-specific DNA transfer mediated by a chimeric multidomain protein. Novel non-viral gene delivery system. J Biol Chem 271:10560-10568

Kircheis R, Schüller S, Brunner S, Ogris M, Heider K-H, Zauner W, Wagner E (1999) Polycation-based DNA complexes for tumor-targeted gene delivery in vivo. J Gene Med 1:111-120

Proba K, Ge L, Plückthun A (1995) Functional antibody single-chain fragments from the cytoplasm of Escherichia coli: influence of thioredoxin reductase (TrxB). Gene 159:203-207

Uherek C, Fominaya J, Wels W (1998) A modular DNA carrier protein based on the structure of diphtheria toxin mediates target cell-specific gene delivery. J Biol Chem 273:8835-8841

Wagner E (1998) Polylysine-conjugate based DNA delivery. In: Kabanov AV, Felgner PL, Seymour LW (eds) Self-assembling complexes for gene delivery. From laboratory to clinical trial. John Wiley & Sons, Chichester, pp 309-322

Wels W, Fominaya J (1998) Peptides and fusion proteins as modular DNA carriers. In: Kabanov AV, Felgner PL, Seymour LW (eds) Self-assembling complexes for gene delivery. From laboratory to clinical trial. John Wiley & Sons, Chichester, pp 351-369

Wilson BA, Collier RJ (1992) Diphtheria toxin and Pseudomonas aeruginosa exotoxin A: active-site structure and enzymic mechanism. Curr Top Microbiol Immunol 175:27-41

Intracellular Targeting
of Antibody Fragments

Intracellular Single Chain Antibodies – Methods for Derivation and Employment

GUADALUPE BILBAO, JUAN LUIS CONTRERAS, and DAVID T. CURIEL

Introduction

The delineation of the molecular basis of cancer in general, allows for the possibility of specific intervention at the molecular level for therapeutic purposes. To this end, three main approaches have been developed: mutation compensation, molecular chemotherapy and genetic immunopotentiation. The strategy of mutation compensation aims at correcting the specific genetic defects in cancer cells. Such correction is accomplished by either ablation of oncogenic products, replacement of cellular tumor suppressor genes, or interference with dysregulated signal transduction pathways. A second strategy is molecular chemotherapy, which aims at increasing the specificity of drug delivery or to increase tolerance to standard chemotherapeutic regimens. A third strategy, genetic immunotherapy, aims at augmenting the specificity and/or the magnitude of the normal immune response to tumors. For each of these conceptual approaches, human clinical protocols have entered Phase I clinical trials to assess dose escalation, safety, and toxicity issues.

The genetic lesions etiologic of malignant transformation may be thought of as a critical compilation of two general types: aberrant expression of "dominant" oncogenes or loss of expression of "tumor suppressor" genes. Gene therapy strategies have been proposed to achieve correction of each of these lesions. For approaching the loss of function of a tumor suppressor gene, the logical intervention is replacement of the deficient function with a wild-type tumor suppressor gene counterpart. The disregula-

Guadalupe Bilbao, The University of Alabama at Birmingham, Gene Therapy Center, Birmingham, Alabama, USA

Juan Luis Contreras, The University of Alabama at Birmingham, Department of Surgery, Birmingham, Alabama, USA

✉ David T. Curiel, The University of Alabama at Birmingham, Gene Therapy Center, 1824 Sixth Ave. South, WTI 620, Birmingham, Alabama, 35394, USA

(*phone* +01-205-934-8627; *fax* +01-205-975-7476; *e-mail* david.curiel@ccc.uab.edu)

tion of oncogenes by mutation, gene amplification, gene rearrangement, or overexpression contributes the aberrant expression of the corresponding gene product that elicits the associated neoplastic transformation. One approach to gene therapy for cancer is to 'knock out' dominant oncogenes and thereby reduce the tumor's growth or invasive potential. Inhibition or ablation of oncogenic function can be attempted at three levels. First, the translation of the oncogene can be targeted. This strategy involves the use of the most universally employed methodology the "antisense" molecules to sequester oncogene mRNA (1-4). However, despite the fact that antisense inhibition can be demonstrated in many contexts, targeting the uptake of oligonucleotides into cells, the stability of the antisense molecules, and pharmokinetic considerations within animals have greatly limited translation of this approach into human clinical trials (5-6). Secondly, the function of the gene product can be targeted. This approach uses polypeptides containing dominant interfering mutations to downregulate signal transduction in tumor cells. Third, the nascent oncogenic protein can be prevented from reaching its proper intracellular location. This approach uses intracellular single chain Fv molecules (scFv) to preempt the cellular localization machinery and sequester proteins inside the cell. In this regard, single-chain immunoglobin (scFv) molecules retain the antigen-binding specificity of the immunoglobulin from which they were derived, however, they lack other functional domains characterizing the parent molecule. The basis of constructing scFvs has been established. Pastan et al. have developed methods to derive cDNAs which encode the variable regions of specific immunoglobins (7-8). Specifically, a single-chain antibody (scFv) gene is derived which contains the coding sequences for variable regions from the heavy chain (V_H) and the light chain (V_L) of the immunoglobulin separated by a short linker of hydrophilic amino acids. The resultant recombinant molecule, when expressed in prokaryotic systems, is a single-chain antibody (scFv) which retains the antigen recognition and binding profile of the parental antibody (9-11). The development of recombinant immunotoxins employing scFv moieties achieves cell-specific binding of the toxin to the exterior of the target cell, allowing receptor-mediated endocytosis to accomplish toxin internalization. A variety of strategies employing the recombinant scFv-directed immunotoxins have been developed by a number of investigators (7, 8, 12-15). In addition, it has recently been shown that scFv molecules may be expressed intracellularly in eukaryotic cells by gene transfer of scFv cDNAs. The encoded scFv may be expressed in the target cell and localized to specific, targeted subcellular compartments by appropriate signal molecules. Importantly, these intracellular scFvs may recognize and bind antigen within the target cell. Targets for the intracellular antibody knockout method have included viral antigens in the context of HIV infection (phase I clinical trial), transformed oncoprotein like erbB2 and c-Ras, and tumor associate antigens like CEA (16-18). In this regard, our group has recently exploited this technology to develop an anti-erbB-2 scFv that down-modulates the erbB-2 oncoprotein in erbB-2 overexpressing tumor cells (19-26). This method of genetic intervention achieved diverse anti-neoplastic effects specifically in tumor cells overexpressing the targeted oncoprotein. Importantly, this approach to human carcinoma of the ovary based on the anti-erbB-2 scFv strategy is now approved for a phase I clinical trial (http://www.nih.gov/od/ord/protocol.html).

This chapter will review in detail practical procedures to generate a single chain intracellular antibody. Most of the methods we employ in our lab use commercially available kits for convenience. We will emphasize in this review the different steps in our protocol that we have employed to develop scFvs to a variety of target proteins. The Recombinant Phage Antibody System (RPAS) from Pharmacia, is based on a phage-display technology where fragments of antibodies are expressed as fusion with gene 3 protein and displayed on the tips of M13 phage. Once antigen-positive clones have been identified by phage rescue, they are used to infect a nonsuppressor strain of *E. coli* HB2151 for the purpose of producing soluble antibodies. In our laboratory, we used a modified colony lift technology to identified antigen-positives clones directly into the nonsuppressor HB2151 *E. coli*. This modification in the expression module is a very simple way of screening your recombinant single chain antibodies, saving time-consuming steps.

Materials

All water used in these protocols should be sterile, deionized, and distilled. All reagents, plastic-and glassware should be sterile.

- mRNA Purification PolyATtract 1000 Kit (Promega cat # Z5420). Store **Kits**
at 4°C.

- First-Strand cDNA Synthesis Kit (Pharmacia Biotech cat # 27-9261-01). Store at -20°C.

- Mouse scFv Module Recombinant Phage Antibody System (Pharmacia Biotech cat # 27-9400-01). Store at -20°C.

- Mouse Ig-Prime Kit System (Novagen cat # 70082-3). Store at -20°C.

- QIAquick Gel Extraction Kit (QIAGEN cat # 28704). Store at room temperature.

- pShooter Mammalian Expression Vector collection (Invitrogen pCMV/*myc*/nuc cat. # V821-20, pCMV/*myc*/mito cat. # V822-20, pCMV/*myc*/cyto cat. # V820-20). Store at -20°C

Reagents

- β-Mercaptoethanol: is toxic, dispense in a fume hood and wear appropriate personal protective equipment. Store at 4°C.

- RNase Zap (Ambion cat. # 9780): Store at room temperature.

- Isopropylthio-β- D-galactopyranoside (IPTG): Store at -20°C.

- Ethanol: Store at room temperature.

- Mineral oil: Light Mineral Oil (Fisher cat # 0121-1). Store at room temperature.

- AmpliTaq® DNA Polymerase from (Perkin-Elmer cat # N808-0038). Store at -20°C.

- Ultrapure DNA grade Agarose (BioRad cat. # 162-0133). Store at room temperature.

- 0.5X and 1X TAE Buffer: 50X TAE 121g Tris, 50 ml 0.5M EDTA pH 8.00, 28.55 ml glacial acetic acid, QS to 500 ml with ddH$_2$O. Store at room temperature.

- RNase-Free Solutions: Add diethylpyrocarbonate [DEPC] (Sigma cat. # D-5758) to solution to a concentration of 0.05% (i.e. add 0.5 ml per liter of solution); shake well, incubate several hours to overnight at 37°C of 42°C; autoclave at least 45 minutes, or until DEPC scent is gone. Store at room temperature.

- 0.5M EDTA: 186.1g EDTA, 20g NaOH pellets, pH to 8.00 with NaOH, QS to 1L with ddH$_2$O. Store at room temperature.

- 100 Base-Pair ladder (Gibco cat. # 15-628-019). Store at -20°C.

- Ethidium bromide 100 mg/ ml (BioRad cat. # 161-0433). Store at room temperature.

- 6X Loading Dye: 30% glycerol and 0.25% bromophenol blue in 10 mM TE, pH 7.6. Store at room temperature.

- dNTP Mix (100 mM each of dATP, dCTP, dGTP, dTTP in sterile water). Store at -20°C.

- 25 mM MgCl$_2$ sterile. Store at room temperature.

- 10X PCR buffer (Perkin-Elmer cat # N808-0038). Store at -20°C.

- TE Buffer: 10 mM Tris-HCl (pH 8.0) and 0.1 mM EDTA. Store at room temperature.

- *Sfi* I (Boehringer Mannheim cat. # 1288-016). Store at -20°C.

- *Not* I (Boehringer Mannheim cat. # 1014-706). Store at -20°C.

- 3 M NaCl: 175.35 gr of NaCl, QS to 1L with ddH$_2$O. Store at room temperature.

- Phenol:chloroform:isoamyl alcohol (25:24:1), equilibrated against TE (Amresco cat. # 0883). Store at -20°C.

- Phosphate Buffered Saline (PBS): 0.2 g KCl, 0.24 g KH$_2$PO$_4$, 8 g NaCl, 1.44 g Na$_2$HPO$_4$, QS to 1 l with ddH$_2$O. Store at room temperature.

- T4 DNA Ligase (Boehringer Mannheim cat. # 481-220). Store at -20°C.

- 30% H$_2$O$_2$ (Any company). Store at 4°C.

- 2M glucose. Do not autoclave, filter sterilize. Store at room temperature.

- HRP/Anti-E Tag Conjugate (Pharmacia cat. # 27-9412-01). Store at -20°C.

- Chloro-1-Naphthol (4C1N) tablets (30 mg) (Sigma cat# C-6788). Store tablets at -20°C. Protect from heat, light and moisture. Allow to reach room temperature before use.

- Triethanolamine saline pH 7.5: Add 7.5 gr of NaCl, 2.8 ml of triethanolamine (Sigma cat. # T-1377), and 17 ml of 1M HCl to approximately 800 ml of dH$_2$O. Adjust the pH to 7.5 if necessary and QS to 1L with dH$_2$O. Store at room temperature.

- ABTS (2', 2'-Azino-Bis[3-Ethylbenzthiazoline-6-Sulphonic Acid] Diammonium) add the following to a 500 ml: 450 ml of 0.05M citric acid pH4.0, and 100 mg of ABTS. Filter sterilize the 1X ABTS stock solution and store at 4°C until needed.

- 40% Acrylamide/Bis Solution 37.5:1(BioRad cat. # 161-0148). Store at 4°C.

- 10% SDS (GIBCO cat. # 24730-012). Store at room temperature.

- Ammonium-Persulfate (APS) (Fisher cat. # BP179-25). Store at room temperature in the desicator.

- TEMED (BioRad cat. # 161-0800). Store at room temperature.

- 1X Western Blot Running Buffer: 25 mM Tris, and 192 mM Glycine. Store at room temperature.

- 1X Western Blot Transfer Buffer: 25 mM Tris, 192 mM Glycine, and 20% methanol. Store at room temperature.

- 4X Western Blot Sample Buffer (100 ml): Tris-HCl/SDS, pH 6.8 (20 ml ddH$_2$O, 3.03 g Tris Base, and 0.2 g SDS), 8 g SDS, 20 ml Glycerol, 0.1 g Bromophenol Blue. QS with ddH$_2$O to 100 ml. This is a non-reducing buffer. For reduccing conditions you need to add 6.2g of dithiothreotol (DTT) per 100 ml. Aliquots into 1 ml and store at -20°C.

Media   Store media at room temperature and all plates (and LBG medium) at 4°C. Use media and plates within 2 weeks if they contain antibiotics.

- Luria Broth Base Media "LB" (Sigma cat. # L-3522).

- LBG: LB Media + 20 mM glucose

- S.O.C. Media at room temperature (GIBCO cat # 15544-018)

- Minimal Medium Plates

- 2-YT Medium (GIBCO cat. # 22712-020)

- 2x YT-AG: 100 µg/ml ampicillin and 2% glucose.

- 2x YT-AI: 100 µg/ml ampicillin and 1 mM IPTG

- 2x YT-G: 2% glucose

Others   - Cluster tubes (96 tubes in a microtiter format, Costar cat # 4411)

- Plate sealer (Costar cat # 6524)

- RNase-free pipets (USA American Scientific Plastics cat. # 1010-8810)

- RNase-free user tubes (Ambion cat. # 12400)

- Colony lift butterfly S&S nitrocellulose membrane pore size 0.45 µm, diameter 82 mm (Schleicher & Schuell cat. # 401149)

- RNase Alert (Ambion cat. # 1960)

- Tranblot Tranfer Medium 0.2 μm Pure Nitrocellulose Membrane (BioRad cat. # 162-0146)

- Blot absorbent filter paper. (BioRad cat. # 170-3932)

Procedure

1. mRNA isolation from hybridoma cell line ($1x10^7$-$1x10^8$)

The success of antibody cloning depends on the purity of the mRNA. The source of the mRNA can be isolated from either mouse antibody produced hybridoma, established cell lines or spleen-derived B lymphocytes. Any kit that will provide high-quality mRNA is recommendable. In our laboratory we recommend the use of the Promega PolyATtract System 1000 because it isolates messenger RNA directly from crude cell or tissue lysates and eliminates the need for total RNA Isolation.

Before starting this protocol make sure that rotors and centrifuges are at room temperature to avoid precipitation of salts and detergents from solutions.

m-RNA isolation

1. Remove the GTC Extraction Buffer, Biotinylated Oligo (dT) Probe, Nuclease-free Water and SSC 0.5X Solution from the refrigerator and warm to room temperature. Preheat the Dilution Buffer to 70°C.

2. In a 50 ml sterile screw cap conical tube, add 41 μl of β-Mercaptoethanol (48.7%) per ml of Extraction Buffer and named "Extraction/BME Buffer". The final concentration of β-Mercaptoethanol is 2%. Use RNase-free pipettes and wear gloves to reduce the chance of contamination.

3. Collect $1x10^7$-$1x10^8$ cells in a sterile 50 ml conical tube by centrifugation at 300 x g for 5 minutes. Wash the cell pellet with 23 ml of ice cold, sterile 1X Phosphatase Buffer Saline (PBS) and centrifuge as above to collect the cells. Pour off the supernatant.

4. Add the "Extraction/BME Buffer" to the cells and mix by inversion in the 50 ml tube 4 times. Homogenizing the cells by high speed for 15-30 seconds using a small homogenizer, as Promega protocol recommends, is optional.

5. Aliquot the preheated Dilution Buffer to a sterile tube and add 20.5 µl of β-Mercaptoethanol (48.7%) per ml of Dilution Buffer. The final concentration of β-Mercaptoethanol is 1%. Add this to the homogenate and mix thoroughly by inversion. Add the Biotinylated Oligo (dT) Probe and mix well. Incubate this mixture at 70°C for 5 minutes.

6. Transfer the lysate to a clean, sterile 15 ml tube. Centrifuge at 12,000 x g for 10 minutes at room temperature to clean the homogenate of cell debris and precipitated proteins.

7. During the centrifugation, resuspend the SA-PMPs by gently rocking the bottle. The particles should appear as a homogeneous mixture and be fully suspended in the liquid. Transfer the 6 ml of SA-PMPs, to a sterile 50 ml conical screw cap tube away from the magnetic stand. Place the tube and the SA-PMPs on the magnetic stand. **Slowly** move the stand toward the horizontal position until the particles are collected at the tube side. Carefully pour off the storage buffer by tilting the tube so that the solution runs over the captured particles. Pouring in this manner decreases the chance of mixing the SA-PMPs into the solution again, which would decrease yields.

8. Resuspend the SA-PMPs to the original volume used (6 ml), in SSC 0.5X Solution. Capture using the magnetic stand. Pour off the SSC Solution as described in Step 7. Repeat the washing a total of three times. Resuspend in the original volume (6 ml) with SSC 0.5X Solution. Do not centrifuge these particles.

9. When the centrifugation of the homogenate is complete, carefully remove the supernatant with a sterile pipette, avoiding the pellet. The homogenate will be translucent.

10. Add the clear homogenate to the tube containing the washed SA-PMPs in SSC 0.5X Solution and mix by inversion. It is important to add the homogenate away from the magnetic separation stand to ensure proper mixing.

11. Incubate the homogenate/SA-PMPs mixture at room temperature for 2 minutes. Capture the SA-PMPs moving the magnetic stand toward the horizontal position until the homogenate clears and then carefully pour off the supernatant as in step 7. Save the supernatant in a sterile tube on ice until certain that satisfactory binding and elution of the mRNA has occurred.

12. Resuspend the particles with 2 ml of SSC 0.5X Solution by gently flicking the tube. Transfer the particle mixture to one of the 2 ml RNase-free user tubes. Capture the SA-PMPs by placing the tube in the magnetic stand. Carefully pipette off the SSC solution. Repeat this washing step twice. After the final wash, remove as much of the SSC solution as possible without disturbing the SA-PMPs cake.

13. To elute the mRNA, add 1 ml of RNase-free water and gently resuspend the particles by flicking the tube.

14. Magnetically capture the SA-PMPs by moving the magnetic stand toward the horizontal position, as before. Transfer the liquid containing the elute mRNA to a sterile RNase-free microcentrifuge tube.

Precipitation

15. To precipitate add 0.1 volume of 3M sodium acetate-DEPC treated and 1.0 volume of isopropanol to the eluate and incubate at -70°C overnight.

16. Centrifuge at 4°C at >12,000 x g for 30 minutes. Resuspend the RNA pellet in 1 ml of 70% ethanol fresh made with RNase-free water and centrifuge again. Be careful when decanting the supernatant to avoid losing the RNA pellet.

17. For short-term storage (<30 days) let the pellet dry at room temperature for about 20 minutes and resuspend in 50 μl of RNase-Free water and store at -70°C. For long-term storage (<30 days), resuspend the mRNA pellet in 70% ethanol at -70°C.

18. The concentration and the purity of the eluted mRNA can be determined by spectophotometry. Determine the absorbance readings at 230, 260 and 280 nm (A_{230}, A_{260} and A_{280}). Absorbance readings should be greater than 0.1 to ensure significance. To estimate the mRNA concentration, assume that a 40 μg/ml mRNA solution will have an absorbance of 1 at 260 nm. Also, determine the A_{260}/A_{230} ratio which will provide information on the purity of the sample. An A_{260}/A_{230} ratio less than 2 indicates that GTC or β-Mercaptoethanol from the Extraction Buffer is still present. If this is the case, precipitate the RNA again.

2. First-strand cDNA synthesis

This First-strand cDNA synthesis is catalyzed by Moloney Murine Leukemia Virus reverse transcriptase. The use of random hexamers eliminates the need for immunoglobulin-specific primers or oligo (dT) primers. Using these random haxamers, the resulting cDNAs are of sufficient length (7 kilobases or more) to clone the V regions from both the heavy and the light chain genes. We used the First-Strand cDNA Synthesis Kit from Pharmacia Biotech.

cDNA synthesis

1. Place the 20 μl mRNA sample.

2. Heat the mRNA solution to 65°C for 10 minutes, then chill on ice. Start the first-strand cDNA reaction **within** 2 minutes after placing on ice.

3. For each sample, label the tube as light chain or heavy chain in a 1.5 ml RNase-free microcentrifuge tube:

Reverse Transcriptase Reaction

mRNA	20 μl
Primed First-Strand Mix	11 μl
DTT Solution	1 μl
RNase-free Water	X μl
	33 μl

4. Incubate for 1 hour at 37°C. The completed First-strand cDNA reaction product is now ready for immediate PCR amplification.

3. Primary PCR amplification

Alignment of known gene sequence from the variable domains has demonstrated that there are regions of conservation within the variable domain, particularly at the 5' and 3' termini. This has led to the determination of species-specific consensus sequences, which have been used in the design of PCR primers. In 1991 Clackson et al. described primers for the PCR amplification of mouse variable regions. However, the light chains from some monoclonal antibodies are difficult to amplify and they have since been redesigned as part of the Recombinant Phage Antibody System (RPAS) from Pharmacia, Biotech, and Mouse Ig-Prime System from Novagen. In our laboratories, we use the RPAS, and when the light chain can not be amplified, we use the Mouse Ig-Prime Kit that contains primers for the μ, all γ, κ, and λ light chain. The First-strand antibody cDNA is used as a template for PCR amplification to generate suitable quantities of heavy (∼340bp) and light (∼325bp) chain DNA cloning.

1. Add to 500 µl microcentrifuge tubes labelled as "light chain" or "heavy chain" the following reagents:

Light chain PCR

First-strand Reaction	33 µl
Light Primer Mix	2 µl
Sterile distilled water	64 µl
total volume	99 µl

Heavy chain PCR

First-strand Reaction	33 µl
Heavy Primer 1	2 µl
Heavy Primer 2	2 µl
Sterile distilled water	64 µl
total volume	99 µl

2. Mix with a micropipettor and spin briefly.

3. Overlay each reaction with 2 drops of mineral oil. Place the tube in a thermocycler and heat at 95°C for 5 minutes.

4. Add 1 µl of AmpliTaq DNA polymerase to each reaction beneath the mineral oil, using separate pipette tips for each addition.

5. Run 30 cycles of 94°C for 1 minute, 55°C for 2 minutes and 72°C for 2 minutes.

4. Purification of primary PCR products

Before performing the assembly reaction, it is very important that the heavy and the light chain PCR products are isolated from the other reaction components. Purification of the DNA from an agarose gel can be done. In this context our laboratory used the QIAquick Gel Extraction Kit from QIAGEN. This QIAquick system combines the convenience of microspin technology with the selective binding properties of a specially adapted silica-gel membrane.

1. Prepare a 1% agarose gel in 1X TAE Buffer with wells sufficient to accommodate 100 µl samples.

2. Remove ~90 µl from the PCR amplification reaction tube and transfer to a new centrifuge tube.

3. Save as a backup 20 µl of the PCR product of each amplification at -20°C. Add 11 µl of 6X Loading Dye to the remaining 70 µl of PCR amplification reaction.

4. Load the PCR amplification reaction mix with the loading buffer in the well. Electrophoreses at 80V until the bromophenol blue dye has migrated ~1/3 the length of the gel.

QIAquick gel extraction

1. Excise the DNA fragment (heavy chain 340 bp and light chain ~325 bp) from the 1% agarose gel with a clean scalpel under a long wavelength ultraviolet light in the transilluminator. Try to minimize the size of the gel slice by removing extra agarose.

2. Weigh the gel slice in a tube. Add 3 volumes of Buffer QG to 1 volume of gel.

3. Incubate at 50°C for 10 minutes. To help to dissolve the gel you can mix by flicking or use a thermomixer. It is very important to solubilize the agarose gel completely. The maximum amount of agarose per column is 400 mg. At this time it is very important, to check the pH because the adsorption of the DNA to silica depends on the pH. Binding efficiency during the adsorption step is typically 95% if the pH is <7.5, and is reduced drastically at higher pH. However if the loading mixture pH is >7.5, it can be lowered by adding 10 µl of 3M sodium acetate pH 5.0.

4. Place a QIAquick spin column in a 2 ml collection tube and load the sample.

5. Centrifuge for 1 minute at 6,000 x g (~9,000 rpm) in an Eppendorf centrifuge. The maximum loading volume of the column is 800 µl. For larger sample volumes, multiple loading of the column are necessary.

6. Discard flow-through and place the QIAquick column back in the same collection tube.

7. Add 0.5 ml of Buffer QG to the QIAquick column and centrifuge for 1 minute.

8. Centrifuge for 1 minute at 6,000 x g (~9,000 rpm) in an Eppendorf centrifuge.

9. To wash, add 0.75 ml of Buffer PE to the QIAquick column and let the column stand 5 minutes.

10. Centrifuge for 1 minute at 6,000 x g (~9,000 rpm) in an Eppendorf centrifuge.

11. Discard the flow-through and centrifuge the QIAquick column for an additional 1 minute at 10,000 x g (~13,000 rpm).

12. Place the QIAquick column to air dry for 15-20 minutes.

13. To elute, add 50 µl of sterile water to the center of the column and let column stand for 5 minutes. DNA is an acid and will undergo auto-catalytic degradation in the absence of a buffering agent and must therefore be stored at -20°C when eluted with water.

14. Centrifuge the QIAquick column for 1 minute at 10,000 x g (~13,000 rpm).

5. Gel quantification of purified heavy and light chain from PCR amplification products

The success of assembly and fill-in reactions is dependent upon the molar concentration of both heavy and light chain PCR products. Agarose gel electrophoresis of aliquots of the purified heavy and light chain products alongside a known amount of V_H Marker (provided by the RPAS Kit) gives a visual estimate of the relative amounts of the fragments based on their band intensity in an ethidium-bromide-stained gel.

1. Prepare a small 1.5% TEA Agarose gel with 3 mm wells, and add ethidium bromide at the final concentration of 0.5µg/ ml.

Quantification

2. To prepare marker, add 2.5 µl or 5 µl (12.5 or 25 ng) of V_H marker (provided by the RAPS kit) with 1 µl of loading dye.

3. Load in separate wells 10 µl of each heavy and light chain with 2 µl of loading dye.

4. Load in separate well 1 µg of 100 bp ladder.

5. Electrophoreses TEA Agarose gel at 80 V for 30 minutes.

6. Photograph the gel under UV light. It may be necessary to over-expose the picture to visualize the marker and the heavy and light chain.

7. Compare the intensity of the heavy and light chain products with the V_H Marker.

8. Estimate the volume of purified heavy chain product that corresponds to 50 ng. Estimate the volume of purified light chain product that corresponds to 50 ng. This will be your 1:1 ratio.

6. Assembly and fill in reactions

In the assembly reaction, the heavy and light chain of the antibody are joined into a single chain antibody with a linker DNA ([Gly$_4$Ser]$_3$). When the linkers anneal to the heavy and light chain DNA, they prime a fill-in reaction in the presence of AmpliTaq DNA polymerase. For this reaction to proceed efficiently, approximately 1:1, 1:2, 1:3, 1:4 and 1:5 ratios of heavy to light chain DNA must be added. If you do not have enough PCR products you can reamplify heavy and light chain.

Assembly 1. Add the following components to a 500 µl microcentrifuge tube:

Heavy Chain Product (50 ng)	X µl
Light Chain Product (50 ng, 100 ng, etc.)	X µl
Linker-Primer Mix	4 µl
10X PCR Buffer I (without MgCl$_2$)	5 µl
dNTP Mix (10 mM of each)	5 µl
25 mM MgCl$_2$	5 µl
AmpliTaq DNA Polymerase (5U)	1 µl
Sterile distilled water	to 50 µl

2. Mix with a micropipettor and spin briefly.

3. Overlay each reaction with 2 drops of mineral oil. Place the tube in a thermocycler and heat at 95°C for 5 minutes.

4. Add 1 µl of AmpliTaq DNA polymerase to each reaction beneath the mineral oil, using separate pipette tips for each addition.

5. Run 7 cycles of 94°C for 1 minute, 63°C for 4 minutes. 1 cycle of 72°C for 10 minutes.

7. Second PCR amplification and purification

It is necessary to amplify the assembled scFv DNA for further cloning steps. In this second PCR amplification restriction sites are added. These restriction sites are used to clone into the phagemid vector. The restriction site primers (RS primers) contain the *Sfi I* annealing with the heavy chain in the 5' end, and the *Not I* site with the light chain in the 3' end.

1. Add to the Assembly and fill-in reaction the following:

AmpliTaq (5U)	1 µl
10X Buffer II (with MgCl₂)	5 µl
dNTP Mix (10 mM)	2 µl
RS Primers Mix	4 µl
Sterile distilled water	39 µl
total volume	50 µl

Second PCR amplification

2. Mix with a micropipettor and spin briefly.

3. Overlay each reaction with 2 drops of mineral oil. Place the tube in a thermocycler and heat at 95°C for 5 minutes.

4. Run 30 cycles of 94°C for 1 minute, 55°C for 2 minutes and 72°C for 2 minutes. 1 cycle of 72°C for 10 minutes.

5. Run a 1.5% TEA Agarose gel at 80 V for 30 minutes. A predominant band ∼750 bp in size should be present. Some heavy and light chain monomers may be visible.

6. Use the QIAquick gel extraction kit to extract the scFv DNA from the gel, and follow it by gel quantification.

8. Restriction digestion and purification

The assembly scFv contains the *Sfi I* and *Not I* sites introduced by the RS primer in the second PCR amplification. The scFv DNA have to be digested in order to be cloned into the phagemid plasmid pCANTAB 5E.

Digestion   1. *Sfi* I Digestion

Gel purified scFv product (0.25-1 µg)	up to 70.0 µl
10X *Sfi* I Restriction Buffer	8.5 µl
Sfi I (20U per reaction)	X µl
Sterile distilled water	to 85.0 µl

2. Mix gently and overlay with 2 drops of mineral oil. Incubate overnight at 50°C.

3. The next day equilibrate the *Sfi* I digested sample to room temperature and spin briefly.

1. Add to the scFv DNA *Sfi* I digested sample the following:

Not I Digestion

10X *Not* I Restriction Buffer	12.7 µl
Not I (20U per reaction)	X µl
Sterile distilled water	to 85.0 µl

2. Mix gently and overlay with 2 drops of mineral oil. Incubate during the day (8 hrs to overnight) at 37°C.

3. Use the QIAquick gel extraction kit to extract the scFv DNA from the gel, and follow it by gel quantification.

9. Ligation of the scFv cDNA into the phagemid plasmid pcantab5e

McCafferty et al. described in 1990 the fd-CAT1 original phage vector for antibody display (27). These vectors contain all the genetic information encoding the phage life cycle. In this context, an alternative system has been used up to now. This system involves cloning into phagemid vectors that contain a copy of the gene 3 and phage packaging signal sequence. Thus antibody fragments can be displayed as a fusion with the gene 3 protein and the genetic information is packaged thanks to the packaging signal. In our laboratory we have used the phagemid vector pCANTAB 5 E included in the Pharmacia Recombinant Phage Antibody System kit. This vector allows cloning of antibody genes into *Sfi* I and *Not* I sites. This vector incorporates an amber codon between the C-terminus of the cloned scFv and the start of gene 3 sequence allowing the recombinant antibody to be made as a soluble protein. This vector also includes a peptide tag, allowing the detection of the single chain antibody.

1. The assembled product should be gel-quantitated as described in the procedure "5 Gel quantification of purified heavy and light chain from PCR amplification products".

2. For ligation of the scFv gene to the pCANTAB 5 E vector (provided by the RPAS Kit from Pharmacia), add the following into a 1.5 ml micro-centrifuge tube:

scFv gene Fragment (100ng)	X µl
pCANTAB 5 E vector (250ng)	5.0 µl
10X Ligation Buffer	1.5 µl
T4 DNA Ligase (5U)	1.0 µl
Sterile distilled water	to 15.0 µl

3. Incubate all reactions at 14°C overnight.

4. Next day, add 2 µl of Sodium Acetate and 500 µl of 95% ethanol.

5. Incubate at -70°C 4 hrs.

6. Spin at 14,000 rpm for 30 minutes at 4°C.

7. Discard supernatant carefully so as not to lose the pellet and wash with 500 µl of 70% ethanol.

8. Spin at 14,000 rpm for 20 minutes at 4°C.

9. Discard supernatant carefully and let the pellet air dry.

10. Resuspend in 10 µl of sterile water (for electroporation you need a low ionic strength).

10. Transformation

Two *E. coli* strains are used for preparation of single chain antibody libraries. To produce phage-displayed recombinant antibodies, competent TG1 cells are used. This host strain produces a suppressor tRNA which allows readthrough (suppression) of the amber stop codon. The switch to soluble recombinant antibody production is accomplished by using the *E. coli* strain HB2151 cells. The HB2151 cells are the nonsuppressor strain, allowing the recognition of the amber stop codon; thus, only soluble single chain antibodies can be produced. The *E. coli* TG1 and HB2151 are supplied as lyophilized cultures in the Expression Module/RPAS Kit from Pharmacia for phage rescue. These E. coli cells will need to be prepared for an electroporation protocol. We recommend that 1 ng of uncut supercoiled vector (pcDNA3 or other irrelevant vector) be used to determine the efficiency of the competent cells.

Preparation of HB2151

1. Inoculate 1 liter of broth with a 1/100 volume of a fresh overnight culture.

2. Grow cells at 37°C in a shaking incubator until the OD_{600}= 0.5 - 1.0.

3. Chill the cells on ice for 15 to 30 minutes.

4. Transfer cell culture to sterile centrifuge tubes and centrifuge in a cold rotor at 4000 x g (max) for 15 minutes.

5. Remove supernatant and discard. Resuspend cell pellet in 1 liter of cold sterile 10% glycerol.

6. Centrifuge again as in step #4.

7. Remove supernatant and resuspend cell pellet in 500 ml of cold sterile 10% glycerol.

8. Centrifuge as in step #4.

9. Remove supernatant and resuspend cell pellet in 250 ml of cold sterile 10% glycerol.

10. Centrifuge as in step #4.

11. Resuspend cells in a final volume of 2 to 3 ml of cold 10% glycerol and dispense 40 µl per sterile Eppendorf tube.

12. Quick freeze cells in dry ice/ethanol bath before storing at -70°C. These cells should be good for six months.

Electroporation

13. Thaw the HB2151 cells (-70°C) on ice.

14. Add 1 to 2 µl of DNA (solution of DNA should be low in ionic strength) ligation mix and let sit on ice about 1 minute.

15. Set the electroporator at 1.8 kV.

16. Transfer the cell mixture to prechilled cuvettes (make sure the suspension is at the bottom of the cuvette).

17. Charge the pulser and then discharge (this should produce a pulse with a time constant of between 4 and 5 msec).

18. Add about 500 µl of S.O.C. media to cuvette, then transfer contents of cuvette to a snap capped tube containing 500 µl of media.

19. Incubate tube at 37°C (shaking) for one hour.

20. Plate onto 2xYT-AG selective media.

21. Incubate overnight at 30°C. These colonies will be used for the modified colony lift assay. Make sure that these colonies are well isolated: ~200-300 colonies per plate.

11. Modified colony lift assay

The following protocol describes a colony lift assay whereby scFv-expressing clones of *E. coli* can be rapidly identified. Due to the large number of colonies that can be simultaneously screened, it may be possible to recover not a single positive clone. Thus, we recommend that this step be repeated at least twice with each positive colon. After each panning, we strongly recommend a PCR amplification of each positive clone with the RS primer as indicated in the procedure 7 "Second PCR amplification and purification". Make sure you make a glycerol stock as soon as possible of each positive clone (28).

1. Dilute 50 µg of the target protein in 1 ml of PBS.   **Antigen-coated membrane**

2. Coat S&S nitrocellulose membrane adding the 1 ml of protein in the middle of the membrane and gently move it around until the entire membrane is wet. This side of the membrane is where your antigen is, make sure that you always have it face up.

3. Block filter with 5% milk in PBS for one hour.

4. Rinse the membrane with PBS and let it dry for 15 minutes.

5. Lay the antigen-coated membrane on top of the 2xYT-AI agar plate.

1. Put a new nitrocellulose membrane on top of the plate that has your   **Master membrane** colonies in the 2xYT-AG. Note: The colonies have to be well isolated. Invert "colony membrane" (colony-side up), and place on top of your "antigen-coated membrane" in the 2x-YT-AI.

2. Incubate at 30°C overnight.

3. Lift colony filter and place colony-side up onto a fresh 2x-YT-AG plate at 4°C until you have developed the antigen-coated membrane and be ready to select.

Development of antigen-coated membrane

1. Place antigen-coated membrane (face-up) in a petri dish and wash 3 times for 5 min with PBS-0.1% Tween20.

2. Block membrane in 5% nonfat milk in PBS for 1 hrs.

3. Wash membrane with PBS-0.1%Tween20 10min 3 times.

4. Add anti-E Tag-HRP antibody 1:1000 in 5% nonfat milk 1 hr. at room temperature.

5. Remove the antibody and wash 15 min 3 times with PBS-0.1%Tween20.

6. Develop with 4CN substrate (Sigma cat # C-6788).
   4CN substrate: Dissolve one tablet of 4-Choloro-1-Naphthol (4C1N) (30 mg) in 10 ml of methanol. This reagent is good for 2 weeks when kept in the dark at 4°C. Immediately prior to development add 2 ml of stock 4CN/MeOH to 10 ml of triethanolamine saline pH 7.5. Add 5 µl of 30% H_2O_2. Neutralize with water (rinse).

12. Production of soluble antibodies (periplasmic extract)

The single chain antibody is cloned into the pCANTAB 5 E phagemid vector and it can be expressed as a soluble protein from this vector. Although the localization and concentration of the scFv will vary, most in cases the bacterial periplasmic extract will yield the highest concentration of functional scFv antibodies (28).

Production of antibody

1. Grow up each of the positive clones in 2 ml of 2x YT-AG at 30°C to log phase (OD_{550} = 0.3-0.4).

2. At this point make a glycerol stock, and a PCR as indicated in the procedure 7 to check the presence of the scFv.

3. Spin down 1500 x g for 5min.

4. Resuspend cells in 2 ml of 2x YT-AI at 30°C overnight.

5. Spin 1500 g20 min in a clinical centrifuge and aspirate media.

6. Resuspend in 400 ml of cold 1 mM EDTA in PBS.

7. Incubate at 4°C for 30 min. with regular movement.

8. Spin at 16,000 x g for 10 min.

9. Transfer supernatant (soluble scFv) to a new tube.

13. Detection and binding specificity of E-tagged scFvs

Enzyme-linked immunosorbent assays (ELISA) can be used to characterize the positive clones obtained from the colony lift assay. In this ELISA
procedure, HRP/Anti-E Tag conjugate is used to detect E-tagged scFv from
the periplasmic extract bound to the antigen-coated microtiter well. Because the ELISA is quantitative in nature, the signal will vary with the expression level and affinity of the scFv. When the positive clones are identified these single chain antibodies can be cloned into eukaryotic expression vector for intracellular expression.

1. Coat wells of the microtiter plate for ELISA with 200 µl of target antigen   **Coating**
   in PBS (pH 8.0 to 8.5) yielding 100 ng to 10 µg per well.

2. Coat wells with appropriate controls (irrelevant protein).

3. Incubate overnight at 4°C, covered.

4. Shake out contents of plate and rinse once with PBS using a squirt
   bottle.

1. Add 200 µl of 3% BSA and 1% gelatin in PBS to each well.   **Blocking**

2. Incubate 1-2 hrs at room temperature.

3. Shake out contents of plate and rinse once with PBS using a squirt
   bottle.

1. Add 100 µl of periplasmic extract to each well.   **scFv Periplasmic**
                                                        Extract
2. Incubate at room temperature for 2 hrs or at 4°C overnight.

3. Wash 3 times for 15 min. with PBS.

1. Dilute HRP/Anti-E Tag Conjugated 1:1000 in 2% BSA, and add 100 µl   **Conjugated**
   to each well.   **antibody**

2. Incubate at room temperature for 1 hr.

3. Wash 3 times for 15 min. with PBS.

Developer

1. Develop with 1X ABTS substrate. Add 100 µl developer to all wells and allow color to develop 15 to 30 min or until color (green) reaction has occurred.

2. The reaction can be read in a microtiter plate reader set at 405-415 nm. If a microtiter plate reader is not available, a spectrophotometer set at an absorbency of 410 nm can be used to quantitate the results. The absorbency reading for your antigen should be at least two to three times higher than the absorbency reading for the negative control.

14. Western blot analysis of E-tagged scFvs

The principle behind SDS-PAGE is the electrophoretic separation of proteins based on mobility in an electrical field as well as their molecular size. Bigger proteins move through the gel slower than smaller ones. You can vary the concentration of the acrylamide to separate proteins of low, mid-range or high molecular size. For instance, if you are interested in a 29 kDa protein, use a 12-15% gel.

Electrophoresis

1. Set up the plates by placing the spacers on either side and clamping the white side-clamps by turning the screws tightly.

2. Install the plates into the bottom mold stand and turn the screws to clamp it down into the base.

3. Fill up with water to test seal and to insure that no leaks occur.

4. Pour the separating gel and let harden 30 min to 1hr. Place a layer of saturated isopropanol over the gel.

Resolving gel: 0.375 M Tris, pH 8.8 (~30 min)

	7%	8%	9%	10%	12%	15%	20%
1.5 M Tris (pH 8.8) ml	2.5	2.5	2.5	2.5	2.5	2.5	2.5
40% Acrylamide ml	1.75	2.0	2.25	2.5	3.0	3.75	5.0
ddH$_2$O ml	5.6	5.35	5.1	4.85	4.35	3.6	2.35
10% SDS	100 µl	100 µl	100 µl	100 µl	100 µl	100 µl	100 µl
10%APS (make fresh)	50 µl	50 µl	50 µl	50 µl	50 µl	50 µl	50 µl
TEMED	2.5 µl	2.5 µl	2.5 µl	2.5 µl	1.5 µl	1.5 µl	1 µl

5. Empty out the saturated isopropanol by inverting the plates and then pour in the stacker gel. You can either load the stacker gel into the plates using a 5 ml pipette with the combs in place or pour the stacker gel in and then place the comb between the glass plates. Be sure there are no bubbles in the bottom of the comb.

Stacker 4%: 0.125 M Tris, pH 6.8 (~25 min)

0.5 M Tris (pH 6.8)	1.0 ml
40% Acrylamide	0.4 ml
dd H$_2$O	2.54 ml
10% SDS	40 µl
10% APS	20 µl
TEMED	4 µl

6. Allow to set 30-60 minutes.

7. Gently pull out comb.

8. Place the top chamber over the gels and remove the bottom screws. Now put these screws into the top chamber and turn to tighten.

9. Fill up top chamber with 500 ml of 1X Western Blot Running Buffer.

10. Load samples using round sequencing pipette tips and put lid on. Prior to loading, samples should be boiled in 1X reducing Western Blot Sample Buffer.

11. Run the gels at 150 V for 1 hr.

Here the principle is to transfer proteins to a solid support for Western blotting. The entire assembly is done between the paddles of the transfer apparatus. Place the paddle with the protruding nubbin down first, then the backing, then all the things below followed by more backing and then the other paddle without the nubbin.

Western transfer

1. Carefully remove the gel from the plates.

2. Wet two pieces of blot absorbent filter paper, 1 nitrocellulose membrane, and the sponges in 1X Western Blot Transfer Buffer.

3. Lay the gel on top of the sponge and blot absorbent filter paper.

4. On top of gel, place a pre-wet in 1X Western Blot Transfer Buffer piece of nitrocellulose membrane and one blot absorbent filter paper.

5. Place one more piece of pre-wet blot absorbent filter paper on top of the nitrocellulose, and then 1 sponge.

6. Snap the paddles together.

7. Load into the transfer chamber which has been filled with 1 liter of 1X Western Blot Transfer Buffer.

8. Snap on power pack lid which has the back facing towards you (need to transfer from negative to positive). Make sure that the two protruding male electrodes in the bottom transfer tank are secured into the power-pack lid.

9. Run the transfer at 100V for 1 hr.

Developing Western

1. Block with PBS with 5% non-fat milk and 1% BSA for 30 min to overnight.

2. Add your primary antibody (HRP/Anti-E Tag) at 1:3,000 in PBS 5% non-fat milk and 1% BSA at 4°C shaking overnight.

3. Wash with PBS for 30 min to 1hr changing the PBS every 10 min (you will never over-wash, but you can under-wash).

4. Develop with 4CN substrate as indicated in procedure 11 "Modified colony lift assay".

15. Intracellular expression of the scFv

It has recently been shown that scFv molecules may be expressed intracellularly in eukaryotic cells by gene transfer of scFv cDNAs. The encoded scFv may be expressed in the target cell and localized to specific, targeted subcellular compartments by appropriate signal molecules. Invitrogen has mammalian expression vectors that allow targeting of your protein to different subcellular compartment. Each of the vectors in the pShooter collection incorporates a signal sequence that will direct the scFv to a specific subcellular location. These eukaryotic expression vectors contain a strong mammalian CMV promoter, and a C-terminal tag for rapid detection.

cDNA scFv Cloning in eukaryotic expression vector

Excise the scFv cDNA from the pCANTAB 5 E phage vector utilizing the *Sfi* I and *Not* I restriction sites as indicated in the procedure 8 "Restriction Digestion and Purification".

Set up ligation reactions as indicated in the procedure 9 "Ligation" of the scFv cDNA into the phagemid plasmid pCANTAB5E, follow up by transformation into *E. coli* competent cells of your choice (DH5α, SURE) as described in procedure 10 "Transformation".

Comments

We have identified 5 major sources of problems throughout the recombinant single chain antibody system:

Notes

1. Quality of isolated mRNA;

2. PCR amplification of V_H and V_L;

3. Assembly reaction of the scFv;

4. Ligation of the scFv to the phegamid plasmid pCANTAB5E;

5. Modified Colony Lift Assay.

In this section we will give guidelines on how to identify these problems and troubleshoot them.

- **Quality of isolated mRNA**
  The success of antibody cloning depends on the quality (purity) of the mRNA. Highly pure mRNA is required as starting material for generating a recombinant antibody. In our experience the PolyATtract System 1000 isolates this high quality mRNA. This system from Promega utilized Promega's MagneSphere technology for the purification of poly (A)+RNA, eliminating the need for oligo (dT) cellulose columns. For the successful isolation of intact mRNA four important steps must be performed: a) effective disruption of cells or tissue, b) denaturation of nucleoprotein complex, c) inactivation of endogenous ribonuclease (RNase) activity, and d) purification of RNA away from contaminating DNA and proteins. The most important of these is the inactivation of RNases. This system combines the disruptive and protective properties to inactivate ribonuclease present in the cell extract, as well as a high stringency method to obtain pure mRNA. The isolation of this high-quality mRNA with this kit is suitable for all molecular biology applications and the yields of poly (A)+ RNA is two-fold greater than with other methods.

Special precautions must be observed to avoid degradation of mRNA by RNases:

- Your hands are a major source of RNases so **always** wear gloves when working with RNA.
- The second major cause of RNase contamination is bacteria on molds that may be present on airborne dust particules, thus glassware, etc. that is to be used for RNA preparations should be purchased new and only used for working with RNA.
- Use RNase Zap (Ambion) to spray all work surfaces and equipment.
- All glassware should be washed thoroughly and soaked in a 0.1% solution of diethyl pyrocarbonate (DEPC) (inhibits RNases) for at least 15 min, autoclaved and then baked in an oven at 250°C for 3 hours (to destroy traces of DEPC–DEPC decomposes into CO_2 and ethanol when heated).
- Whenever possible use sterile single use plastic ware instead of glassware.
- Chemicals should be reserved for working with RNA and always handled with DEPC treated spatulas.
- All solutions should be treated with DEPC prior to use. The solutions should be adjusted to 0.1% in DEPC, shaken or stirred for at least 15 min and then autoclaved. Since DEPC decomposes in the presence of Tris, all solutions containing Tris should not be DEPC-treated but instead be made up with DEPC-treated and autoclaved ddH_2O. Likewise solutions that are not to be autoclaved should always be made up with DEPC-treated water or solutions. You may want to check your solution with RNase Alert from Ambion before using them.

- **PCR amplification of V_H and V_L**

  It can be possible that following the primary PCR amplification (procedure 3) step you can not see the light and heavy chain products. Since we have used the PolyATtract System 1000 we have not had this problem, however if you are using other RNA purification methods the reason to fail this fist PCR amplification can be:
  - The purity of the mRNA is not adequate: this problem may be resolved by using an RNA extraction kit followed by mRNA purification kit.
  - Ribonucleases are present in the mRNA preparation: follow RNase-free precautions.
  - The hybridoma was no longer producing antibody: check the hybridoma for the presence of soluble antibody.

- The primers are annealed nonspecifically to nucleic acids contaminating the mRNA: you will see a smearing in the gel electrophoresis, better purification of the mRNA preparation is required.
- If you can only see the heavy chain but not the light chain it is because the Pharmacia mouse scFv module kit contains only the kappa light chain primers since few lambda light chain sequences have been determined (~5%). The Novagen Mouse Ig-Primer Set (Novagen cat. # 69831-3) will allow you to PCR the light chain, however you may have to follow their protocols to assemble the scFv, and continue in the procedure 9.

Special precautions must be observed to avoid PCR contamination:
- Use the PCR hood when making reagents for PCR or putting together reactions. The pipetetmen in the hood have never been used to pipette DNA and should never be removed from the hood. Likewise, DNA–or anything that has been near DNA (dirty gloves, tube racks, etc.)-should never be brought into the PCR hood.
- Always make up your PCR reagents with stocks and solutions that have never been out on the lab bench. I always use disposable tubes and pipettes to make up stocks. It's often a good idea to make up a large batch of buffer or dNTPs, test them to make sure they are clean, and divide them into small aliquots.
- If you suspect that your PCR reagents (buffer, dNTPs, or water) are contaminated, try irradiating them on the short wave UV box (about 3-5 minutes). This will crosslink any contaminating DNA. You can also UV zap some primers without affecting their ability to prime, but this is primer specific. Do not UV zap the enzyme.
- As long as your reagents are clean, most contamination occurs when you add the template to your tubes. Take care in opening the tubes that contain your DNA template. A little bit of DNA aerosolized onto your glove can easily be transferred to your PCR tube. Use pipette tips with filters or positive displacement pipetess to add toyour template. Aerosolized DNA in pipetteman is the major source of contamination.
- Always include a negative control. Make sure to treat it like the other samples (e.g. add 1 µl of water at the lab bench with the pipetteman you're using to add DNA to the other tubes).
- For RT for PCR you should include 2 negative controls: One in which everything except RNA is added (H$_2$O control). (Be sure to actually add water to this tube at the same time you're adding

RNA to the other tubes); and a second one in which RNA is added, but no RT is included (this ensures that there is no contamination of your RNA).

- **Assembly reaction of the scFv**
  The assembly and fill in reaction of the scFv is without a doubt the trickiest step of all. We will give you some tips that in our experience have made this step easy.
  - Make sure a high quality of the cDNA of V_L and V_H is used as a start material. The QIAquick Gel Extraction Kit from QIAGEN will give you this high quality DNA. Two modifications we introduce to their protocol: i) Skip the isopropanol step after the solubilization of the agarose gel. It is very difficult to completely dry the isopropanol and it can interfere with downstream reactions; and ii) let the column air dry at room temperature for 20 min before elution of the DNA, thus no ethanol will be present.
  - The Pharmacia protocol suggest for the assembly reaction of the scFv an equivalent amount of the heavy and light chain products, however in our experience for this reaction to proceed efficiently we add an excess of light chain product. We proposed that approximately 1:1, 1:2, 1:3, 1:4 and 1:5 ratios of heavy to light chain DNA must be used for this assembly reaction to be successful.
  - The final concentration of the dNTPs in the assembly reaction is critical in this step. We have modified the dNTPs molarity to 10 mM using 1/10 of the total volume in the assembly reaction as indicated in the procedure 6.

- **Ligation of the scFv to the phegamid plasmid pCANTAB5E**
  After ligation of the scFv in the phegamid vector pCANTAB 5 E we recommended that Subprocedures 9 through 12 be done within 48 hrs. It is very important that you make a glycerol stock of the positive clones as soon as possible. PCR amplification of the scFv library after this time demonstrated different scFv cDNA sizes.

- **Modified Colony Lift Assay**
  Some antibodies produced by hybridoma cells cannot recognize their respective antigen if this is immobilized onto nitrocellulose. You need to try several times by colony lift assay if your hybridoma recognize the antigen-bound to the membrane. If this is the case, the colony lift assay cannot be used to detect the recombinant antibodies, and you may need to work with the Recombinant Phage Antibody System with the Phage Antibody Library by Pharmacia.

Acknowledgements. We wish to thank Patty Parker for her administrative assistance. This work is in support of the following grants: National Institutes of Health RO1 CA 72532; RO1 CA 68245, and the United States Department of Defense.

References

1. Krol, A.V., Stuitje, A.R. (1988) Modulation of eukaryotic gene expression by complement RNA or DNA sequences. Biotechniques. 6, 958-976.
2. Helene, C., Toulme, J-J. (1990) Specific regulation of gene expression by antisense, sense, and antigene nucleic acids. Biochemica and Biophysica, 1049, 99-125.
3. Gibson, I. (1996) Antisense approaches to the gene therapy of cancer. Cancer Metastasis Rev., 15, 287-299.
4. Milligan, J. F., R. J. Jones, B. C. Froehler, and M. D. Matteucci. 1994. Development of antisense therapeutics. Implications for cancer gene therapy. Ann. N. Y. Acad. Sci. 716:228-241.
5. Stein, C.A., Cheng, Y-C. (1993) Antisense oligonucleotides as therapeutic agents – Is the bullet really magic? Science, 261, 1004-1012.
6. Stein, C. A. (1995) Does antisense exist? Nat. Med., 1, 1119-1121.
7. Theuer, C.P., Pastan I. (1993) Immunotoxins and recombinant toxins in the treatment of solid carcinomas. Amer. J. Surg., 166, 284-288.
8. Brinkmann, U., Pai, L.H., FitzGerald, D.J. (1991) B3-(Fv)-PE38KDEL, a single chain immunotoxin that causes complete regression of a human carcinoma in mice. Proc. Natl. Acad. Sci. USA, 88, 8616-8620.
9. Hoogenboom, H. R., Marks, J. D., Griffiths, A. D., Winter, G. (1992) Building antibodies from their genes. Immunol. Rev., 130, 41-68.
10. Jost, C. R., Kurucz, I., Jacobus, C. M., Titus, J. A., George, A. J., Segal, D. M. (1994) Mammalian expression and secretion of functional single-chain Fv molecules. J. Biol. Chem., 269, 26267-26273.
11. Richardson, J. H., Marasco, W. A. (1995) Intracellular antibodies: development and therapeutic potential. Trends. Biotech., 13, 306-310.
12. Chen, S. Y., Bagley, J., Marasco, W. A. (1994) Intracellular antibodies as a new class of therapeutic molecules for gene therapy. Hum. Gene Ther., 5, 595-601.
13. Colcher, D., Bird, R., Roselli, M. (1990) In vivo tumor targeting of a recombinant single-chain antigen-binding protein. J. Natl. Can. Inst., 82, 1191-1197.
14. Wawrzynczak, E.J. (1992) Rational design of immunotoxins: current progress and future prospects. Anti-Cancer Drug Design, 7, 427-441.
15. Mykebust, A.T., Godal, A., Fodstad, O. (1994) Targeted therapy with immunotoxins in a nude rat model for leptomenineal growth of human small cell cancer. Can. Res. 54, 2146-2150.
16. Marasco, W.A., Haseltine, W.A., Chen, S-Y. (1993) Design, intracellular expression, and activity of human anti-human immunodeficiency virus type 1 gp120 single-chain antibody. Proc. Natl. Acad. Sci. USA, 90, 7889-7893.
17. Friedman, P.N., Chance, D.F., Trail, P.A. (1993) Antitumor activity of the single-chain immunotoxin BR96 sFv-PE40 against established breast and lung tumor xenografts. J. Immunol., 150, 3054-3061.
18. Werge, T.M., Biocca, S., Cattaneo, A. (1990) Cloning andintracellular expression of a monoclonal antibody to the p21ras protein. FEBS Lett., 274, 193-198.

19. Deshane, J., Loechel, F., Conry, R. M., Siegal, G. P., King, C. R., Curiel, D. T. (1994) Intracellular single-chain antibody directed against erbB2 down-regulates cell surface erbB2 and exhibits a selective anti-proliferative effect in erbB2 overexpressing cancer cell lines. Gen. Ther., 1, 332-337.

20. Deshane, J., G. P. Siegal, R. D. Alvarez, M. H. Wang, M. Feng, G. Cabrera, T. Liu, M. Kay, and D. T. Curiel. 1995. Targeted tumor killing via an intracellular antibody against erbB-2. Journal. of. Clinical. Investigation. 96:2980-2989.

21. Deshane, J., J. Grim, S. Loechel, G. P. Siegal, R. D. Alvarez, and D. T. Curiel. 1996. Intracellular antibody against erbB-2 mediates targeted tumor cell eradication by apoptosis. Cancer Gene Therapy. 3:89-98.

22. Grim, J., J. Deshane, M. Feng, A. Lieber, M. Kay, and D. T. Curiel. 1996. erbB-2 knockout employing an intracellular single-chain antibody (sFv) accomplishes specific toxicity in erbB-2-expressing lung cancer cells. American. Journal. of. Respiratory. Cell &. Molecular. Biology. 15:348-354.

23. Barnes, D. M., J. Deshane, G. P. Siegal, R. D. Alvarez, and D. T. Curiel. 1996. Novel gene therapy strategy to accomplish growth factor modualtion induces enhanced tumor cell chemosensitivity. Clinical Cancer Research 2:1089-1095.

24. Wright, M., J. Grim, M. Kim, T. V. Strong, G. P. Siegal, and D. T. Curiel. 1997. An intracellular anti-erbB-2 single-chain antibody is specifically cytotoxic to human breast carcinoma cells overexpressing erbB-2. Gene Therapy 4:317-322.

25. Kim, M., M. Wright, J. Deshane, M. A. Accavitti, A. Tilden, M. Saleh, W. P. Vaughan, M. H. Carabasi, M. D. Rogers, R. D. J. Hockett, W. E. Grizzle, and D. T. Curiel. 1997. A novel gene therapy strategy for elimination of prostate carcinoma cells from human bone marrow. Human Gene Therapy 8:157-170.

26. Curiel DT, Targeted tumor cytotoxicity Mediated by intracellular single-chain anti-oncogene antibodies. Gene Therapy in Advances in pharmacology Ed J.Thomas August. Academic Press Vol. 40:51-84.

27. McCafferty, J., Griffiths, A.D., Winter, G., and Chriswell, D.J. (1990). Phage antibodies: filamentous phage displaying antibody variable domains. Nature 348,552-554.

28. Rodenburg, C., Mernaugh, R., Bilbao, G., Khazaeli M.B. (1998) Production of a single cahin anti-CEA antibody from the hybridoma cell line T84.66 using a modified colony-lift selection procedure to detect antigen-positive scFv bacterial clones. Hybridoma 17, 1-8.

Expressing Intracellular Single-Chain Fv Fragments in Mammalian Cells

SILVIA BIOCCA, ALESSIO CARDINALE, and ANTONINO CATTANEO

Introduction

Intracellular antibodies are a new strategy for phenotypic knock-out in mammalian cells, based on the ectopic expression of recombinant forms of antibodies, to specifically interfere with the function of selected antigens. Antibody chains or domains, if equipped with suitable dominant and autonomous targeting signals, can be targeted towards new intracellular sites, where they have been shown to neutralize intracellular gene products in different animal or plant systems (Biocca et al. 1990; Biocca and Cattaneo, 1995; Richardson and Marasco 1995; Cattaneo and Biocca, 1997). This approach has been successfully applied to inhibit the function of several intracellular antigens in the cytoplasm, nucleus and in the secretory pathway of different biological systems (Biocca et al. 1993; 1994; Marasco et al. 1993; Tavladoraki et al. 1993; Duan et al. 1994; Mhashilkar et al. 1995).

In this chapter we shall describe all the steps required for expressing antibody domains in different intracellular compartments of mammalian cells, for their subsequent use in intracellular immunization experiments, starting from a cloned single-chain Fv (scFv) fragment. In particular, we shall provide examples and protocols for the expression in the cytoplasm of scFv fragments directed against the p21Ras protein.

✉ Silvia Biocca, University of Rome, "Tor Vergata", Department of Neuroscience, Via di Tor Vergata 135, Rome, 00133, Italy (*phone* +39-06-72596428; *fax* +39-06-72596407; *e-mail* biocca@uniroma2.it)

Alessio Cardinale, University of Rome, "Tor Vergata", Department of Neuroscience, Via di Tor Vergata 135, Rome, 00133, Italy

Antonino Cattaneo, International School for Advanced Studies (SISSA), Neuroscience Program, Via Beirut 2-4, Trieste, 34013, Italy

Vectors for the expression of scFvs in different intracellular compartments

The ability of dominant and autonomous targeting sequences to confer a new intracellular location to a reporter protein has been exploited to redirect antibodies or antibody domains to different intracellular compartments. In principle, different sorting signals can be used for antibody targeting. Some of these signals have been successfully used (reviewed in Biocca and Cattaneo, 1995; Cattaneo and Biocca, 1997), but more could be used, if needed.

A set of general vectors has been constructed to facilitate the expression of scFv fragments, linked to specific targeting signals, in different compartments including the endoplasmic reticulum (scFvex-ER), the cytoplasm (scFvex-cyt), the nucleus (scFvex-nuc), the mitochondria (scFvex-mit) and as secreted proteins (scFvex-sec). The integrated system of scFvexpress vectors, described in Persic et al. 1997a, is derived from the VHexpress vector, a vector used to produce secretory immunoglobulin heavy chains from cloned IgH regions (Persic et al.1997b).The scFvexpress-cyt is the prototype of these vectors and is shown in Figure 1.

This plasmid has no targeting signal (leader-less) and directs the expression of scFv in the cytoplasm, with an N-terminal methionine instead of the leader sequence for secretion. All other targeting vectors are derivatives of the scFvexpress-cyt and were obtained by the insertion of well characterized targeting signals (Biocca et al. 1995; Biocca and Cattaneo, 1995) either N- or C-terminal to the scFv, as appropriate. All the expression cassettes, which are shown in Figure 2, encode a C-terminal myc tag in frame with the scFv, in addition to any targeting signals, allowing their detection using the mAb 9E10 (Evans et al. 1985).

Given the different possible sources of scFv fragments, it is not possible to indicate a universal cloning procedure, since this depends on the source of the scFv and the strategy utilized for deriving it. Although phage antibodies are becoming the method of choice for deriving new antibodies, many hybridomas secreting antibodies with potentially useful specificities are available. A number of different methods have been described for cloning V regions from hybridomas using V region PCR (Orlandi et al.1989) or other PCR based techniques such as RACE (Frohman et al. 1988, Bradbury et al. 1994), oligoligation mediated PCR (Edwards et al. 1991) or phage display (McCafferty et al. 1990). Large libraries of scFv fragments displayed on phage have been made from both natural and synthetic V regions which are now available for deriving new antibody specificities (see Chapters 5-8 in this book).

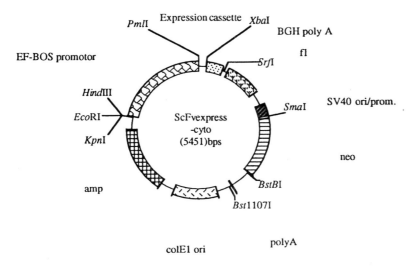

Fig. 1. Map of scFvexpress-cyt. This is the backbone plasmid from which the other scFv targeting plasmids are derived. Reprinted with permission from Persic et al. 1997a.

Each plasmid of the scFv express system has been designed to include at least two restriction sites at either end of the variable regions. For each set of cloning sites given, the outer pair does not modify the V regions sequence, while the inner pair involves the incorporation of amino acids which may not be present in the original V region sequence, but which in most cases will not alter the binding characteristics of the scFv.

Expressing scFv fragments in different intracellular compartments

These vectors can be used for transient or stable transfection of mammalian cells. We routinely use COS cells for transient transfection experiments, to initially validate a new intracellular scFv fragment. This system exploits the SV40 origin of replication which, in the presence of the large T antigen, allows the plasmid to be amplified to high levels. Other cell lines can be transiently transfected by using the two procedures described below with fair (C6 glioma, PC12, NIH fibroblasts, Jurkat etc) or comparable (3T3 K-Ras fibroblasts) efficiency.

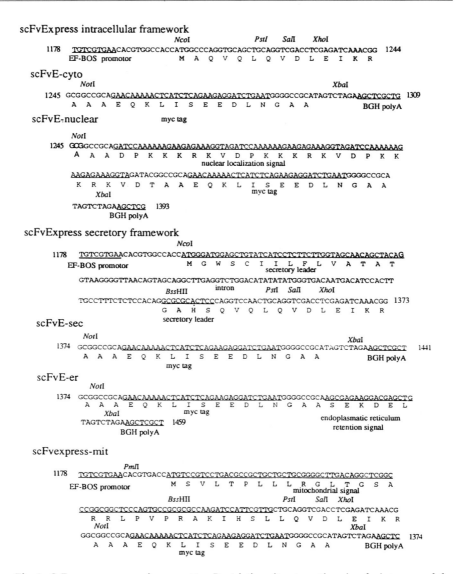

Fig. 2. ScFvexpress expression cassettes. Restriction sites, targeting signals, introns and the myc tag are indicated. Numbers refer to base pairs from the beginning of the scFvexpress expression cassette which includes the EF-BOS promoter. Reprinted with permission from Persic et al. 1997a.

Procedure

Procedure 1: Transient transfection of COS cells with PEI (polyethylenimine)

1. The day before transfection plate $5x10^5$ cells in a 100 mm tissue culture dish.

2. In eppendorf tubes prepare two solutions: solution A and solution B. Solution A containing 10 μg plasmid DNA in a final volume of 100 μl NaCl 150 mM. Solution B containing 2.8 μl pEi (polyethylenimine) 0.1 M (Aldrich, MW 25.000) in a final volume of 100 μl NaCl 150 mM.

3. Vortex briefly the contents of each eppendorf tube.

4. Add solution B to solution A (not viceversa) and vortex the final solution every 3 min. at room temperature for 12 min.

5. Add 3 ml DMEM to the pEi/DNA solution.

6. Wash the cells twice with PBS (Phosphate Buffered Saline).

7. Add the DMEM/pEi/DNA solution directly to the cells and incubate for 1 hr at 37°C.

8. Wash the cells twice with PBS.

9. Feed the cells with DMEM containing 10% FBS (Foetal Bovine Serum).

10. Harvest the cells 24 or 48 hrs later for further analysis.

Procedure 2: Transient transfection of NIH 3T3 fibroblasts with Superfect reagent

This protocol has been optimized for NIH 3T3 cells expressing K-Ras, but should be applicable for other cells as well.

Parameters that can be optimized are the ratio DNA/Superfect and the transfection time.

1. The day before transfection plate $5x10^5$ cells in a 100 mm tissue culture dish.

2. In polystyrene tubes add 10 μg plasmid DNA to 500 μl DMEM (antibiotics and serum free).

3. Add 100 μl Superfect reagent (Quiagen) to the DNA solution. Mix by pipetting up and down.

4. Incubate the Superfect/DNA/DMEM solution for 10 min. at room temperature.

5. Wash the cells once with PBS.

6. Add 3 ml of DMEM containing FBS and antibiotics to the reaction tube containing the transfection complexes. Mix by pipetting up and down and immediately add the total volume to the cells.

7. Incubate the cells with the complexes for 7 hrs at 37°C.

8. Wash the cells twice with PBS.

9. Feed the cells with cell growth medium (containing serum and antibiotics).

10. Harvest the cells 24 or 48 hrs later for further analysis.

Procedure 3: Analysis of intracellular scFv by indirect immunofluorescence

Indirect immunofluorescence of transfected cells allows to rapidly verify the intracellular location and the expression efficiency of the scFv fragment.

1. Place at the bottom of a 35 mm tissue culture dish 4 sterilized glass coverslips.

2. Add 2 ml of poli-L-lysine 0.1 mg/ml.

3. Irradiate with UV for 15 min.

4. Aspirate the poli-L-lysine and let the tissue culture dish dry under the hood.

5. Plate 2.0×10^5 COS cells on coated glass coverslips and transfect them.

6. Wash 3 times with PBS.

7. Fix the cells with 4% (w/v) paraformaldehyde in PBS for 10 min at room temperature.

8. Wash twice with PBS.

9. Permeabilize the cells with Tris-Cl 0.1 M pH 7.6/0.2% Triton X-100 for 5 min at room temperature.

10. Wash 3 times with PBS.

11. Incubate the cells for 1 h at room temperature with the appropriate dilution of affinity purified mouse anti-myc 9E10. Dilute the antibody in PBS containing BSA (bovine serum albumin) 0.2 mg/ml.

12. Wash the cells by dipping the glass coverslips serially into three beakers containing, respectively, PBS, PBST (PBS containing 0.05% Tween 20), PBS.

13. Incubate the cells for 30 min at room temperature with an appropriate dilution of FITC (fluorescein-isothiocyanate)-conjugated goat anti-mouse IgG (Sigma) in PBS-BSA 0.2 mg/ml.

14. Wash the cells as described in step 12, mount coverslips top down with mounting medium (Vectashield, Vector) and examine samples by confocal microscopy.

An example of a typical experiment is shown in Figure 3. The intracellular distribution of a secreted scFv fragment is illustrated in Fig. 3b. Removal of the hydrophobic leader sequence prevents its translocation to the endoplasmic reticulum, allowing its expression in the cytoplasm (Fig. 3d). The presence of a nuclear or a mitochondrial targeting signal allows their expression respectively in the nucleus (Fig. 3a) and in the mitochondria (Fig. 3 c). ScFv fragments can be targeted to the plasma membrane by incorporating the transmembrane and cytoplasmic domains of membrane receptors. The expression of a plasma membrane scFv fragment harbouring the transmembrane domain of the PDGF receptor (Chesnut et al. 1996) is shown in Fig. 3e and Fig. 3f.

Analysis with a confocal microscope make it possible to achieve a higher precision of intracellular location, particularly useful when performing intracellular co-localization studies of antibodies and their antigens (see below, Procedure 6).

Solubility analysis of intracellular scFv fragments

A systematic comparison of scFv fragments targeted to different subcellular compartments, as well as of different scFvs targeted to the same compartment, showed that the expression levels of the scFv vary according to the targeting signals and according to individual scFvs. The use of scFvexpress vectors optimized for the subcellular targeting of scFv antibody fragments in mammalian cells leads to higher expression levels, achieved through the optimization of many aspects of the vector, but this has not always resulted in higher levels of soluble protein (Persic et al. 1997a). In this respect, different scFv fragments have distinct properties when expressed in mammalian cells, in terms of solubility and propensity to aggregate (Cattaneo and Biocca, 1999). Although the propensity to ag-

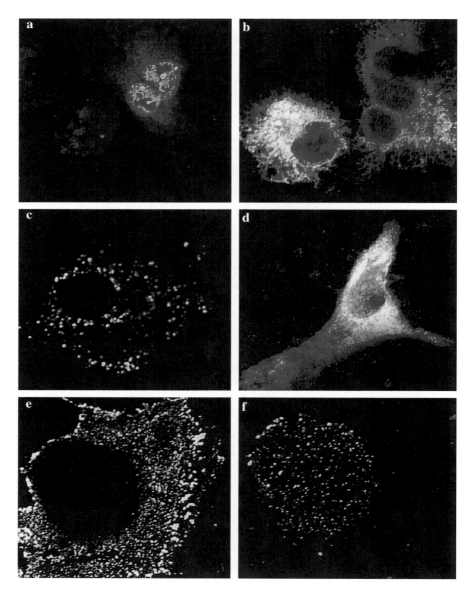

Fig. 3. ScFv fragments targeted to different intracellular compartments of COS cells, viewed by immunofluorescence and confocal microscopy, after labelling with antibody against the myc tag. ScFv fragments were equipped with the following targeting signals: nuclear (a), secretory (b), mitochondrial (c) and cytoplasmic (d). (e) and (f) are two focal planes of the same cell transfected with ScFv fragments targeted to the plasma membrane. Note the absence of fluorescence into the nucleus in section (e) and the presence of labelling in the plane of the plasma membrane in section (f).

gregate for scFvs is mostly confined to cytoplasmic expression, it was found that also secretory or nuclear scFv fragments can present solubility problems.

As will be shown below, also aggregating scFvs can be exploited for intracellular immunization, but it is important to verify their propensity to form intracellular aggregates with ad hoc assays. In order to discriminate between the soluble and the insoluble scFv fragments, we describe an immunoprecipitation protocol for soluble proteins and the analysis of the insoluble pool by Western blot.

Procedure 4: Analysis of soluble scFv fragments by immunoprecipitation

1. Incubate in Eppendorf tubes, for each sample, 20 µl of Protein A-Sepharose with 2 µg of affinity purified mAb 9E10 in Tris 25 mM pH 8.6, NaCl 150 mM (TBS) o/n at 4°C with gentle mixing.

   Preparation of 9E10-protein A-Sepharose

2. Wash 9E10-Protein A-Sepharose 3 times with TBS pH 8.0 by centrifugation for 1 min at 1500 rpm.

3. Wash 3 times with PBS a 100 mm tissue culture dish of COS transfected cells.

   Immunoprecipitation

4. Lyse the cells with 300 µl of ice cold extraction buffer containing Tris-Cl 20 mM, pH 8, $MgCl_2$ 20 mM, 0.5% NP40, 0.1 mg/ml leupeptin, chymostatin, aprotinin and 0.1 mM PMSF for 15 min.

5. Centrifuge cellular extracts for 15 min at 4°C (15.000 rpm) to separate soluble (supernatant) from the insoluble proteins (pellet).

6. Incubate the supernatant with 20 µl of 9E10-Protein A-Sepharose (see Steps 1 and 2) for 1.5 h at 4°C with gentle mixing.

7. Centrifuge the samples 1 min at 1500 rpm.

8. Wash the Sepharose beads 4 times with TBS containing 0.1% NP40 and once with Tris-Cl 5 mM pH 7.6.

9. Add 4 µl 4X SDS Sample buffer (Tris-Cl 125 mM pH 6.8, SDS 1%, glycerol 5%, DTT 10 mM, bromophenol blue 0.005%), boil 2 min, run a 10% SDS-polyacrylamide gel and further analyse by Western blot.

Procedure 5: Analysis of insoluble scFv fragments by Western blot

1. Perform Steps 3, 4 and 5 as described in Procedure 4.

2. Solubilize pellet fractions (insoluble pool) by adding SDS sample buffer, pipetting up and down, vortexing several times.

3. Boil the samples 2 min and run a 10% SDS polyacrylamide gel.

4. Transfer the samples from SDS-polyacrylamide gels to a nitrocellulose membrane or PVDF (Polyvynilidene difluoride) by electrophoretic transfer using Tris-glycine buffer 20% (v/v) methanol as transferring buffer for 2 h at 100 V.

5. Saturate non specific protein binding sites incubating the membrane in 5% non-fat dry milk in TBS containing Tween 0.1% (TBST) for 1 h at room temperature with gentle agitation.

6. Wash once briefly in TBST.

7. Incubate the membrane with an appropriate dilution of affinity purified mAb 9E10 in TBST for 1 h at room temperature with gentle agitation.

8. Wash the membrane 3 times in TBST for 10 min each to remove unbound antibody molecules.

9. Incubate the membrane with an appropriate dilution of goat anti-mouse peroxidase (Amersham) in TBST as secondary antibody for 30 min at room temperature with gentle agitation.

10. Wash the membrane 3 times in TBST for 10 min each to remove unbound secondary antibody molecules.

11. Visualize the blots by ECL (Emission Chemiluminescence, Amersham) following the manufacturer's instructions.

Figure 4 reports a typical analysis of cytosolic scFv fragments. In this experiment we have compared the intracellular distribution of three different cytoplasmic scFv fragments, expressed in COS cells. While two of them (Panel A (a) and (b)) concentrate in intracellular granules, the scFv shown in (c) shows a more diffuse distribution. The presence of intracellular granules is a property which is diagnostic of a high insolubility, as confirmed by western analysis of the soluble and insoluble pool of proteins. Only the scFv C12, which shows a diffuse pattern by immunofluorescence, is recovered in the soluble pool.

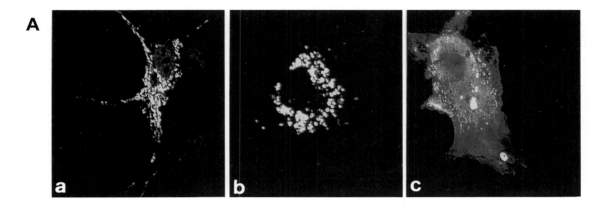

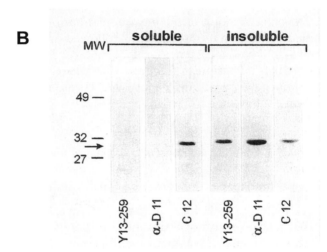

Fig. 4. Panel A: COS cells transfected with Y13-259 (a), α-D11 (b) and C12 (c) scFv fragments targeted to the cytoplasm, viewed by immunofluorescence with anti myc tag antibodies (monoclonal antibody 9E10) and confocal analysis. Panel B: Western blot analysis of the transfected cell populations shown in A), with anti myc tag antibodies (monoclonal antibody 9E10). Soluble and insoluble proteins were extracted as described in Protocol 4. Arrow points to the scFv fragment band. Reprinted with permission from Cardinale et al. 1998.

Expression of scFv fragments in the cytoplasm: the p21Ras case

We have recently correlated the solubility properties of different anti-Ras scFv fragments expressed in the cytoplasm of different cell lines with their capacity to interfere and inhibit Ras function upon intracellular expression. These experiments allowed us to describe a new mode of action for intracellular antibodies based on diverting the antigen from its normal location. Although these procedures have been systematically validated for the p21Ras system (Cardinale et al. 1998; Lener et al. 2000), we believe that traffic diversion represents a general mode of action for intracellular immunization with antibodies. In the secretory pathway, for example, the concept of traffic diversion has been demonstrated, by preventing the appearance of proteins to the plasma membrane through SEKDEL tagged antibodies (Graus-Porta et al. 1995; Deshane et al. 1994; Richardson et al. 1995; Cattaneo and Biocca, 1997).

The colocalization of the intracellular antibody and its antigen can be studied by co-immunoprecipitation followed by Western blot (Procedures 4 and 5) and by fluorescence confocal microscopy in transfected cells. The latter makes it possible to monitor in situ the efficacy of intracellular antigen-antibody interactions, before and regardless of the particular final application envisaged.

Procedure 6: Co-localization analysis by confocal microscopy

1. Perform Steps 1-10 as described in Procedure 3.

2. Incubate the cells for 1 h at room temperature with an appropriate dilution of a relevant specific antibody to reveal the antigen. For p21Ras protein use human recombinant anti-Ras Y13-259 (Werge et al. 1992).

3. Wash 3 times with PBS.

4. Incubate the cells for 30 min at room temperature with an appropriate dilution of a biotinylated conjugated secondary antibody (biotinylated-goat anti-human IgG γ specific).

5. Wash 3 times with PBS.

6. Incubate the cells for 30 min at room temperature with an appropriate dilution of Texas Red streptavidin.

7. Wash 3 times with PBS.

8. Incubate the cells for 1 h at room temperature with an appropriate dilution of affinity purified mouse mab 9E10 to reveal the scFv fragment.

9. Wash 3 times with PBS.

10. Incubate the cells for 30 min at room temperature with an appropriate dilution of FITC-conjugated goat anti-mouse as secondary antibody.

11. Wash 3 times with PBS.

12. Analyse the cells by confocal microscopy. Images should be recorded with simultaneus excitation and detection of both dyes to ensure the alignment to correct for possible crosstalk resulting from overlapping of the dyes. Recorded images are corrected, when necessary, with MultiColor analysis Package Software by Leica (TCS 4D system equipped with 40X 1.00-0.5 and 100X 1.3-0.6 oil immersion lens) and compared with images recorded with single dye excitation. The superimposition of the two chromophores in the same image results in a green/red color scale, leading to yellow color in case of co-localization.

The p21Ras protein is a prototype protein involved in signal transduction, located at the cytoplasmic face of the plasma membrane. This association is crucial for its function in stimulation of cellular proliferation, as demonstrated by mutagenesis of the C-terminal CAAX box, which inhibits membrane association and, consequently, Ras biological activity (Kato et al. 1992).

Figure 5 shows an example of a colocalization experiment, where we have studied whether an intracellularly aggregated anti-Ras scFv fragment is able to interact with the p21Ras in the cytoplasm. 3T3 transformed fibroblasts, expressing high levels of Ki-Ras, have been transfected with the anti-Ras antibody fragment and with a non relevant control scFv. The intracellular distribution of the scFv fragments and of the p21Ras protein was studied in the transfected cells by double immunofluorescence and analyzed by confocal microscopy (Figure 5A,B and C for the control scFv expression and D,E,F and G for the anti-Ras expression). Cells are viewed with anti p21Ras antibody (Fig. 5A and D) and with anti-myc antibody (Fig. 5B and E). The combination of the two fluorescence patterns is shown in Fig. 5C for the control scFv expression and F and G for the anti-Ras expression. In these images the signal viewed with the anti-Ras antibody (green) and the signal viewed by anti-myc (red) were combined and the two chromophores are superimposed in the same image with a green/red colour scale, leading to yellow colour in case of colocalization. As can

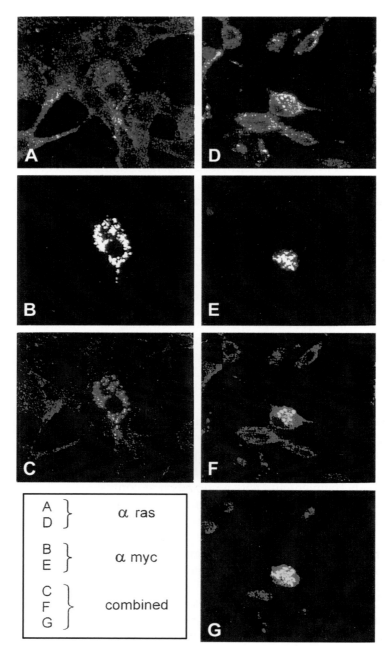

Fig. 5. Double immunofluorescence with anti-p21Ras antibody (A,D) and anti-myc (B,E) demonstrating the colocalization of cytoplasmic scFv fragment and p21Ras protein only in cells expressing anti-Ras scFv (F,G), but not in cells expressing the control scFv (C). A,B and C: 3T3 K-ras fibroblasts transfected with non relevant scFv fragment. D,E,F and G: 3T3 KRas fibroblasts transfected with anti-Ras scFv. F and G represent two images of a series of the same field. Reprinted with permission from Cardinale et al. 1998.

be seen, in cells transfected with a non relevant scFv, anti-Ras (green) and anti-myc (red) do not colocalize, whereas in anti-Ras transfected cells the two signals appear condensed in the same yellow aggregates.

Procedure 7: Bromodeoxyuridine (BrdU) incorporation in transfected 3T3 K-Ras cells

The activity of the endogenous p21Ras protein can be assayed by measuring the entry of cells in the S-phase of the cell cycle, by BrdU incorporation. This assay, an easy and fast double immunofluorescence analysis of transfected cells, allows to verify, in general, the inhibitory activity of all those scFvs which are directed against molecules which are crucially involved in cell proliferation.

1. Incubate transfected 3T3 K-Ras cells with 10 µM BrdU for 24 h at 37°C in CO_2 incubator.

2. Wash 3 times with PBS.

3. Fix and permeabilize the cells with 70% ethanol in 50 mM glycine buffer pH 2, for 30 min at -20°C.

4. Wash 3 times with PBS.

5. Incubate the cells for 30 min at 37°C with an appropriate dilution of mouse monoclonal antibody against BrdU (clone BMC 9318, IgG1 Boehringer-Mannheim) containing nucleases for DNA denaturation in incubation buffer (Tris-buffer 66 mM, $MgCl_2$ 0.66 mM, 2-mercaptoethanol 1 mM).

6. Wash 3 times with PBS.

7. Incubate the cells for 30 min at room temperature with an appropriate dilution of a biotinylated-anti-mouse antibody (Amersham).

8. Wash 3 times with PBS.

9. Incubate the cells for 30 min at room temperature with an appropriate dilution of streptavidin-Texas Red.

10. Wash 3 times with PBS.

11. Incubate the cells for 1 h at room temperature with an appropriate dilution of a rabbit anti-myc antibody (Clontech).

12. Wash 3 times with PBS.

13. Incubate the cells for 30 min at room temperature with an appropriate dilution of an anti-rabbit IgG FITC-conjugated (Sigma).

14. Wash 3 times with PBS.

15. Analyse the cells with a fluorescence or confocal microscopy.

16. Analyse at least 50-70 positively transfected cells and count, between them, the BrdU positive cells.

In the experiment presented in Figure 6 the expression of the anti-Ras scFv fragments resulted in the inhibition of DNA synthesis (less than 10%), as compared to non-transfected cells or to cells transfected with an irrelevant scFv fragment (more than 70%). The panel on the right shows an example of the double immunofluorescence experiment. Notice that all three cells transfected with a non relevant scFv, are also positive for the bromodeoxyuridine (positive signal in the nucleus), diagnostic of DNA synthesis, while the anti-Ras transfected cell shown in the photo below is negative for bromodeoxyuridine. At least 50-70 positively transfected cells were counted for each experiment.

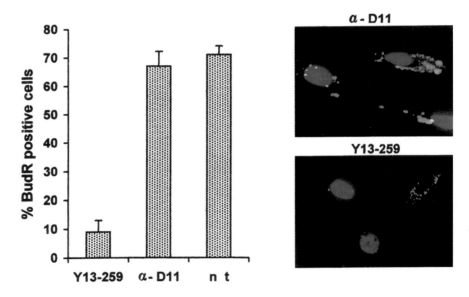

Fig. 6. Bromodeoxyuridine incorporation. BrdU positive cells were counted from non transfected cells (n t), pscFvexp-cyt-α-D11 (α-D11) and pscFvexp-cyt-Y13-259 (Y13-259) transfected cells. The figure shows the average from three different experiments. At least 50-70 positively transfected cells were counted in each experiment. Reprinted with permission from Cardinale et al. 1998.

Procedure 8: In situ analysis of apoptotic nuclei in cells expressing scFv fragments

In many cases of interest, intracellular antibodies are used in studies that have apoptosis as a final read-out (both inhibition or induction of apoptosis). The following is the protocol for the in situ identification of apoptotic cells. This method uses the in situ end labelling procedure for DNA fragmentation and is based on the specific binding of terminal deoxynucleotidyl transferase (TdT) to 3'-OH ends of DNA, ensuing the synthesis of a polydeoxynucleotide polymer. The protocol reported below (Procedure 8) for theTdT-mediated dUTP-biotin nick end labelling (TUNEL) (Gavrieli et al. 1992) is a double immunofluorescence analysis which allows to label apoptotic nuclei in cells expressing intracellular scFv fragments.

1. Perform Steps 1-6 as described in Procedure 3.

2. Fix the cells with 1% (w/v) paraformaldehyde in PBS pH 7.4 for 15 min on ice.

3. Wash twice in PBS.

4. Permeabilize the cells with chilled EtOH 70% for 10 min on ice.

5. Wash 4 times in PBS.

6. Add 2 µl of DNAase I (Boehringer) diluted in 38 µl of dH_2O for 45 min at 37°C in humid chamber only to cells to be used as positive controls.

7. Wash twice in PBS.

8. Pre-incubate the cells with 50 µl of Labelling Buffer (Trizmabase 30 mM pH 7.2, 140 mM sodium cacodylate, 1 mM cobalt chloride) for 10 min at room temperature.

9. Wash twice in PBS.

10. Incubate the cells with 50 µl of labelling mix containing labelling buffer, 1 µl of dNTP mix, containing biotynilated dUTP, 1 µl of Terminal deoxynucleotide Transferase (0.3 u/ µl) for 1 h at 37°C in a humid chamber.

11. Stop the reaction with 300 mM sodium chloride and 30 mM sodium citrate for 10 min at room temperature.

12. Wash 5 times with PBS and 5 times with PBST.

13. Incubate the cells for 20 min in humid chamber with an appropriate dilution in PBS of Streptavidin-FITC (Genzyme).

15. Saturate non specific binding sites incubating the cells with 3% BSA in PBS for 10 min RT.

16. Incubate the cells for 1 h at room temperature with an appropriate dilution of affinity purified mouse anti-myc 9E10.

17. Wash 3 times with PBS.

18. Incubate the cells for 30 min at room temperature with an appropriate dilution of an anti-mouse IgG FITC-conjugated (Sigma).

19. Wash 3 times with PBS.

20. Analyse the cells with a fluorescence or confocal microscopy.

21. Analyse at least 70-100 positively transfected cells and count, between them, the apoptotic positive nuclei.

Acknowledgements. The research described has been supported by two EU grants (Biotechnology BIO-CT-972285 and BIO4 980203), Ministero della Sanità (AIDS Project) and CNR (Progetto Finalizzato Biotecnologie). A.L. acknowledges a fellowship from EU Biotechnology BIO-CT- 972285.

References

Biocca, S., Neuberger, M.S. and Cattaneo, A. (1990) Expression and targeting of intracellular antibodies in mammalian cells. EMBO J. 9, 101-108.

Biocca, S., Pierandrei-Amaldi, P. and Cattaneo, A. (1993) Intracellular expression of anti-p21ras single chain Fv fragments inhibits meiotic maturation of Xenopus oocytes. Biochem Biophis. Res. Commun. 197, 422-427.

Biocca, S., Pierandrei-Amaldi, P., Campioni, N. and Cattaneo A. (1994) Intracellular immunization with cytosolic recombinant antibodies. Bio/Technology 12, 396-399.

Biocca, S. and Cattaneo, A. (1995) Intracellular immunization: antibody targeting to subcellular compartments. Trends Cell Biol. 5, 248-252.

Biocca, S, Ruberti, F, Tafani, M., Pierandrei-Amaldi, P. and Cattaneo, A. (1995) Redox state of single chain Fv fragments targeted to the endoplasmic reticulum, cytosol and mithocondria. Bio/Technology 13, 1110-1115.

Bradbury, A., Ruberti, F., Werge, T., Amati, V., Di Luzio, A., Gonfloni, S., Hoogenboom, H., Piccioli, P., Biocca, S. and Cattaneo, A. (1994) The cloning of hybridoma V Regions for their ectopic expression in intracellular and intercellular immunization. In: Borrebaeck, C.A.K. (Ed), Antibody Engineering. Oxford University Press, New York, Oxford pp 229-266.

Campbell, S.L., Khosravi-Far, R., Rossman, K.L., Clark, G.J. and Der, C.J. (1998) Increasing complexity of Ras signaling. Oncogene, 17, 1395-1413

Cardinale, A., Lener, M., Messina, S. Cattaneo, A. and S. Biocca (1998) The mode of action of Y13-259 scFv fragment intracellularly expressed in mammalian cells. FEBS Lett., 439: 197-202.

Cattaneo, A. and Biocca, S. (1997) Intracellular Antibodies: development and applications. Springer, Berlin Heidelberg.

Cattaneo, A. and Biocca, S. (1999) The selection of intracellular antibodies. Trends in Biotech., 17, 115-121

Chesnut, J.D., Baytan, A.R., Russel, M. et al. Selective isolation of transiently transfected cells from a mammalian cell population with vectors expressing a membrane anchored single-chain antibody. J. Immunol. Meth., 193, 17-27.

Deshane, J., Loechel, F., Conry, R.M., Siegal, G.P., King, C.R. and Curiel, D.T. (1994) Intracellular single-chain antibody directed against erbB2 down-regulates cell surface erbB2 and exhibits a selective anti-proliferative effect in erbB2 overexpressing cancer cell lines. Gene Therapy, 1, 332-337.

Duan, L., Bagasra, O., Laughlin, M.A., Oakes, J.W. and Pomerantz R.J. (1994) Potent inhibition of human immunodeficiency virus type 1 replication by an intracellular anti-Rev single-chain antibody. Proc. Natl. Acad. Sci. USA 91, 5075-5079.

Edwards, J.B., Delort, J. and Mallet, J. (1991) Oligodeoxyribonucleotide ligation to single stranded cDNAs: a new tool for cloning 5' ends of mRNAs and for constructing cDNA libraries by in vitro amplification. Nuc. Acids. Res. 19, 5227-5232.

Evan, G.I., Lewis, G.K., Ramsay, G. and Bishop, J. M. (1985) Isolation of monoclonal antibodies specific for human c-myc proto-oncogene product. Mol. Cell Biol. 5, 3610-3616.

Frohman, M.A., Dush, M.K. and Martin, G. (1988) Rapid production of full lenght cDNAs from rare transcripts: amplification using a single gene-specific oligonucleotide primer. Proc. Natl. Acad. Sci. U.S.A. 85, 8998-9002.

Furth, M. E., Davis, L. J., Fleurdelys, B., and Scolnick, E. M. (1982) Monoclonal antibodies to the p21 products of the tranforming gene of Harvey murine sarcoma virus and of the cellular ras gene family. J. Virol. 43, 294-304.

Gavrieli ,Y., Sherman, Y. And Ben-Sasson A. (1992) Identification of programmed cell death in situ via specific labelling of nuclear DNA fragmentation. J. Cell Biol. 119, 493-501.

Graus-Porta, D., Beerli, R.R. and Hynes, N.E. (1995) Single-chain antibody-mediated intracellular retention of ErbB-2 impairs Neu differentiation factor and epidermal growth factor signaling. Mol. Cell Biol. 15, 1182-1191.

Hoogenboom, H. R., Griffiths, A. D., Johnson, K. S., Chiswell, D. J., Hudson, P., and Winter, G. (1991) Multi-subunit proteins on the surface of filamentous phage: methodologies for displaying antibody (Fab) heavy and light chains. Nucleic Acids Res 19, 4133-4137

Kato, K., Cox, A.D., Hisaka, M.M., Grahm, S.M., Buss, J. E. and Der, C. J. (1992) Isoprenoid addition to Ras protein is the critical modification for its membrane association and transforming activity. Proc. Natl. Acad. Sci. USA 89, 6403-6407.

Lener M, Horn IR, Cardinale A, Messina S, Nielsen UB, Rybak SM, Hoogenboom HR, Cattaneo A and Biocca S (2000) Diverting a protein from its cellular location by intracellular antibodies: The case of p21Ras. Eur J Biochem 267:1196-1205

Marasco, W. A., Haseltine, W. A., Chen S. Y. (1993) Design, intracellular expression, and activity of a human anti-human immunodeficiency virus type 1 gp120 single chain antibody. Proc. Natl. Acad. Sci. USA 90, 7889-7893.

Marks, J. D., Hoogenboom, H. R., Bonnert, T. P., McCafferty, J., Griffiths, A. D., and Winter, G. (1991) By-passing immunization-human antibodies from V-gene libraries displayed on phage. J Mol Biol 222, 581-597

McCafferty, J., Griffith, A.D., Winter, G. and Chiswell, D.J. (1990) Phage antibodies: filamentous phage displaying antibody variable domains. Nature, 348, 552-554.

Mhashilkar, A.M., Bagley, J., Chen, S.J., Szilvay, A.M., Helland D.G. and Marasco W.A. (1995) Inhibition of HIV-1 Tat-mediated LTR transactivation and HIV-1 infection by anti-Tat single chain intrabodies. EMBO J. 14, 1542-1551.

Orlandi, R., Gussow, D.H., Jones, P.T. and Winter, G. (1989) Cloning immunoglobulin variable domains for expression by the polymerase chain reaction. Proc. Natl. Acad. Sci. U.S.A. 86, 3833-3837.

Persic, L, Righi, M, Roberts, A, Hoogenboom, H. R., Cattaneo, A. and Bradbury, A. (1997a) Targeting vectors for intracellular immunization. Gene 187, 1-8.

Persic, L, Righi, M, Roberts, A, Hoogenboom, H. R., Cattaneo, A. and Bradbury, A. (1997b) An integrated vector system for the eukaryotic expression of antibodies or their fragments after selection from phage display libraries. Gene 187, 9-18.

Richardson, J.H. and Marasco, W.A. (1995) Intracellular antibodies: development and therapeutic potential. Trends in Biotechnology 13, 306-310.

Richardson, J.H., Sodroski, J.G., Waldmann, T.A. and Marasco, W. (1995) Phenotypic knockout of the high-affinity human interleukin 2 receptor by intracellular single-chain antibodies against the alpha subunit of the receptor. Proc. Natl. Acad. Sci. USA 92, 3137-3143.

Tavladoraki P., Benvenuto, E., Trinca, S., De Martinis, D., Cattaneo, A. and Galeffi, P. (1993) Transgenic plants expressing a functional single chain Fv antibody are specifically protected from virus attack. Nature 366, 469-472.

Vaughan, T. J., Williams, A. J., Pritchard, K., Osbourn, J. K., Pope, A. R., Earnshaw, J. C., McCafferty, J., Hodits, R. A., Wilton, J., and Johnson, K. S. (1996) Human antibodies with sub-nanomolar affinities isolated from a large non-immunized phage display library. Nat. Biotechnol. 14, 309-314

Werge, T.M., Biocca, S. and Cattaneo, A. (1991) Intracellular immunization: cloning and intracellular expression of a monoclonal antibody to the p21 ras protein. FEBS Lett. 274, 193-198.

Werge, T.M., Bradbury,A.M., Di Luzio, A. and Cattaneo, A. (1992) A recombinant cell line expressing a form of the anti-p21 ras antibody. Oncogene 7, 1033-1035.

Appendices

Internet Resources

A recent review of internet resources related to antibodies can be found in Dübel 2000. All links featured in this article are accessible from the web site at http://www.mgen.uni-heidelberg.de/SD/IT/IT.html.

Databases

- **The Kabat Database of Sequences of Proteins of Immunological Interest**
  http://immuno.bme.nwu.edu/

- **V Base: The database of human antibody genes** at the MRC Centre for Protein Engineering (Cambridge, UK)
  http://www.mrc-cpe.cam.ac.uk/imt-doc/public/INTRO.html

- **ABG: Germline gene directories of mouse** at the National University of Mexico
  http://www.ibt.unam.mx/paginas/almagro/V_mice.html

- **ABG: Directory of 3D structures of antibodies** at the National University of Mexico
  http://www.ibt.unam.mx/paginas/almagro/structures.html

- **IMGT, the international ImMunoGeneTics database** at the Université Montpellier (France)
  http://imgt.cnusc.fr:8104/

- **Structural classification of proteins (SCOP)** entries of antibody variable domain-like structures at the MRC Centre for Protein Engineering (Cambridge, UK)
  http://sb10.llnl.gov/scop/data/scop.1.002.001.001.001.html

- **The National Center for Biotechnology Information** (NCBI) (including GenBank and PubMed)
  http://www.ncbi.nlm.nih.gov/

- **ExPASy** Molecular Biology Server at Geneva (including SwissProt)
  http://www.expasy.ch/

- **The Protein Data Bank** (PDB) of the Research Collaboratory for Structural Bioinformatics (RCSB)
  http://www.rcsb.org/pdb/

Information, Resources, and Tools

- **The Antibody Resource Page**
  http://www.antibodyresource.com/

- **Antibodies - Structure and Function,** maintained by Andrew Martin
  http://www.biochem.ucl.ac.uk/~martin/abs/index.html

- **IgBLAST** at the NCBI
  http://www.ncbi.nlm.nih.gov/igblast/

- **Fv modelling** at the University of Bath
  http://antibody.bath.ac.uk/wam.html

- **Canonical Structures and the confromations of CDR3 of the VH domain of immunoglobulins**
  http://www.predicct.sanger.ac.uk/irbm-course97/Biocomputing/Projects/h3.html

- **The hybridoma database** at the American Type Culture Collection
  http://www.atcc.org/hdb/hdb.html

- **ATCC American Type Culture Collection**
  http://www.atcc.org/

- **ECACC European Collection of Cell Cultures**
  http://www.biotech.ist.unige.it/cldb/descat5.html

- **Stefan Dübel's Recombinant Antibody Pages**
  http://www.recab.uni-hd.de

- **Roland Kontermann's Antibody Engineering Page**
  http://aximt1.imt.uni-marburg.de/~rek/introrapd.html

References

Dübel S (2000) The Antibody Web. Immunology Today 21: 355-357

Amino Acids: Nomenclature and Codons

Nomenclature and codons of amino acids

Name	Triple-letter code	Single-letter code	Codon(s)	Codon usage (human) (%/aa)	Codon usage (E. coli) (%/ aa)
Alanine	Ala	A	GCU	28.0	18.9
			GCC	41.7	24.4
			GCA	20.0	21.7
			GCG	10.3	35.0
Arginine	Arg	R	CGU	8.9	44.1
			CGC	21.1	37.5
			CGA	10.2	5.2
			CGG	19.7	7.6
			AGA	18.8	3.5
			AGG	21.0	2.1
Asparagine	Asn	N	AAU	42.4	39.3
			AAC	57.5	60.7
Aspartic acid	Asp	D	GAU	42.8	58.6
			GAC	57.2	41.4
Cysteine	Cys	C	UGU	40.9	43.5
			UGC	59.1	56.5
Glutamic acid	Glu	E	GAA	39.9	69.4
			GAG	60.7	30.6
Glutamine	Gln	Q	CAA	24.8	29.9
			CAG	75.2	70.1

Name	Triple-letter code	Single-letter code	Codon(s)	Codon usage (human) (%/aa)	Codon usage (E. coli) (%/ aa)
Glycine	Gly	G	GGU	15.8	38.1
			GGC	35.8	40.6
			GGA	24.1	8.8
			GGG	24.3	12.5
Histidine	His	H	CAU	39.6	51.1
			CAC	60.4	48.9
Isoleucine	Ile	I	AUU	33.1	46.2
			AUC	54.0	47.3
			AUA	12.9	6.5
Leucine	Leu	L	CUU	11.2	10.0
			CUC	20.8	9.7
			CUA	6.5	2.9
			CUG	44.5	55.6
			UUA	5.5	10.4
			UUG	11.5	11.4
Lysine	Lys	K	AAA	38.9	75.6
			AAG	61.1	24.4
Methionine	Met	M	AUG	100	100
Phenylalanine	Phe	F	UUU	41.1	50.4
			UUC	58.9	49.6
Proline	Pro	P	CCU	27.4	15.0
			CCC	35.3	9.4
			CCA	25.7	19.0
			CCG	11.6	56.7
Serine	Ser	S	AGU	13.0	12.8
			AGC	25.8	26.4
			UCU	18.2	18.6
			UCC	24.4	17.0
			UCA	12.8	11.4
			UCG	5.8	13.8
Threonine	Thr	T	ACU	22.5	19.9
			ACC	40.5	45.3
			ACA	25.3	12.0
			ACG	11.7	22.8
Tryptophan	Trp	W	UGG	100	100
Tyrosine	Tyr	Y	UAU	39.9	52.6
			UAC	60.1	47.4
Valine	Val	V	GUU	16.3	29.0
			GUC	25.7	19.5
			GUA	9.3	17.0
			GUG	48.7	34.5
Stop codons	-	-	UAA	29.2	66.7
			UAG	20.8	6.7
			UGA	50.0	26.6

Subject Index